Radar and Laser
Cross Section Engineering

Third Edition

Radar and Laser Cross Section Engineering

Third Edition

David C. Jenn
Naval Postgraduate School
Monterey, California

AIAA EDUCATION SERIES

Joseph A. Schetz, Editor-in-Chief
Virginia Polytechnic Institute and State University
Blacksburg, Virginia

AIAA
American Institute of Aeronautics and Astronautics, Inc.

American Institute of Aeronautics and Astronautics, Inc., Reston, Virginia

1 2 3 4 5

Library of Congress Cataloging-in-Publication Data

Names: Jenn, David C., author. | Schetz, Joseph A., editor.
Title: Radar and laser cross section engineering / David C. Jenn, Naval Postgraduate School, Monterey, California, Joseph A. Schetz, editor-in-chief Virginia Polytechnic Institute and State University, Blacksburg, Virginia
Description: Third edition. | Reston, Virginia : American Institute of Aeronautics and Astronautics, Inc., [2019] | Series: AIAA education series | Includes bibliographical references and index.
Identifiers: LCCN 2019005539 | ISBN 9781624105630 (hardcover)
Subjects: LCSH: Radar cross sections.
Classification: LCC TK6580 .J46 2019 | DDC 621.3848–dc22 LC record available at https://na01.safelinks.protection.outlook.com/?url=https%3A%2F%2Flccn.loc.gov%2F2019005539&data=02%7C01%7Ckatb%40aiaa.org%7Ca7e9ed6d341444fb71cf08d6902976d7%7C036e8604b674406a80987cb72daf432b%7C0%7C0%7C636854905102709901&sdata=tBe79qEiWra0ucq3aXMSryz19ZX5lPxq6qFTHk1J8sg%3D&reserved=0

FOREWORD

We are very happy to present the third edition of *Radar and Laser Cross Section Engineering* by David Jenn. This edition substantially updates the material in the earlier volume to reflect the rapid changes in this important field. In addition, more examples and homework problems have been added. The book now consists of nine chapters and six appendices, with a total of over 500 pages.

The AIAA Education Series aims to cover a very broad range of topics in the general aerospace field, including basic theory, applications, and design. A complete list of titles is available from AIAA and can be found through the link on the last page of this volume. The philosophy of the series is to develop textbooks that can be used in a university setting, instructional materials for continuing education and professional development courses, and also books that can serve as the basis for independent study. Suggestions for new topics or authors are always welcome.

Joseph A. Schetz
Editor-in-Chief
AIAA Education Series

FOREWORD TO THE FIRST EDITION

In designing stealthy weapon systems such as combat aircraft or missiles, the consideration of their radar and laser cross sections is of paramount importance. Strategic or tactical aircraft in combat operations are threatened by interceptor aircraft, surface-to-air missiles, and antiaircraft artillery. Even missiles are no longer immune from this threat. One of the countermeasures used to reduce or eliminate this threat is to minimize the radar and laser cross section. This text, *Radar and Laser Cross Section Engineering* by David C. Jenn, discusses both the physical and engineering aspects of this approach to stealthy design of weapon systems. It has been developed from the lecture notes used by the author in several courses on radar cross sections at the Naval Postgraduate School in Monterey, California. An important feature of this textbook is that it includes information on the reduction of both the radar cross section (RCS) and the laser cross section (LCS).

The first six chapters of this text discuss the basic theorems, concepts, and methods. Subsequent chapters focus on radar cross section reduction and measurement. Finally, the last chapter deals with the reduction and measurement of the laser cross section. Each chapter includes problems—a useful feature for teaching the course materials from this book. The primary feature of this text is the prediction, reduction, and measurement of electromagnetic scattering from complex three-dimensional targets.

The AIAA Education Series embraces a broad spectrum of theory and application of different disciplines in aerospace, including aerospace design practice. The Series has now been expanded to include defense science, engineering, and technology and is intended to serve as both teaching texts for students and reference materials for practicing engineers and scientists. *Radar and Laser Cross Section Engineering* is in this category. Additionally it complements a previously published text by AIAA entitled: *Radar Electronic Warfare*, by August Golden, Jr.

J. S. Przemieniecki
Editor-In-Chief (Retired)
AIAA Education Series

CONTENTS

PREFACE

Since the writing of the second edition, the importance of radar, and consequently radar cross section (RCS), has continued to grow. Many recent exciting developments have found applications in both the military and civilian sectors. The developments have been fueled by the growth in computer technology, signal processing techniques, and new materials having interesting and unusual electric and magnetic behavior. Many of the concepts conceived decades ago have finally become a reality because of the advances in technology.

In particular, the capabilities of modem radar to process and track a large number of targets in hostile battlefield conditions and severe clutter have continued to improve. In addition to radar, many other types of sensors are used to detect, track, and identify targets. Although the signatures presented to all sensors are an issue of concern, when it comes to long-range early warning, the radar signature remains the primary consideration. This has always been the case for aircraft, and to some extent ships, but even for other potential radar targets, like ground vehicles, tanks, and even personnel, RCS has become an important consideration in their design and operation.

Radars have found application in self-driving cars, robotics, and medical diagnosis. In all of these systems the target's RCS must be quantified and incorporated into the system design and operation. Commercial applications have helped to drive down the size, weight, and cost of radar sensor systems.

This edition has evolved from continuously updated course lecture notes on radar cross section at the Naval Postgraduate School. Since the first edition in 1995 there have been dramatic changes in many aspects of RCS. The first generation stealthy aircraft, the F-117, has been retired. New "fifth generation" aircraft and the U.S. Navy's 21st century *Zumwalt* destroyer are being deployed. The capabilities of commercial computer simulation tools, or more generally computational electromagnetics (CEM), continue to grow. Many sophisticated commercial software packages are available, and they have been integrated into aerodynamic and mechanical simulation modules. The increase in computer memory and speed has made RCS simulations practical and accessible for realistic complex targets.

The development of composite and artificial materials continues as well. Traditionally, composite material was formed by mixing multiple constituents in an effort to find just the right blend to satisfy the desired electrical and mechanical properties. Until recently the constituents themselves have largely been naturally occurring materials. Now, however, artificial metamaterials is an increasingly active area of research. Metamaterials have negative effective permittivity and permeability, leading to some very interesting reflection, refraction, and focusing behaviors. An electromagnetic (EM) cloaking device has been demonstrated using a metamaterial.

In addition to new discussions in the developing areas already mentioned, more homework problems and examples have been added to this new edition. As in the previous editions, the objective is to give the reader a relatively comprehensive exposure to the major aspects of RCS: EM theory, RCS design, modeling, simulation, reduction, and measurement. An effort was made to provide a sufficient connection to the EM theory without becoming too embroiled in the mathematics. The resulting formulas lie somewhere between those obtained by an intuitive "back of the envelope" argument presented by many RCS books, and what would be found in a graduate EM textbook. This approach has been applied uniformly to all the topics in the book, wherever possible. The presentation of the material assumes that the student has mastered electromagnetics at an undergraduate level, similar to what one would find in textbooks by Cheng, Ulaby, Rao, Sadiku, or Inan and Inan. Completion of a course in antennas also is a helpful prerequisite to this book, although students without an antenna background have successfully mastered the material presented here.

The topics and chapters essentially follow those of the earlier editions. Chapter 1 gives an introduction and overview of radar and RCS. More system issues have been introduced. In particular, the tradeoff between jamming and RCS reduction is discussed at some length. As in the first edition, Chapter 2 remains a review of the fundamental concepts and theorems that will be used time and again throughout the book. The physical optics approximation and surface impedance are both discussed in detail. Examples related to these topics also serve to demonstrate other techniques, such as the evaluation of radiation integrals and coordinate transformations.

Chapter 3 covers frequency domain numerical methods. The method of moments is discussed at length and is used as an opportunity to illustrate the general approach to the solution of integral equations. A brief discussion of the finite element method (FEM) has been added. Chapter 4 discusses time-domain solutions of the differential and integral forms of Maxwell's equations, primarily the finite difference-time-domain (FDTD) method and the finite integration technique (FIT). Chapter 5 introduces microwave optics with emphasis on geometrical optics (GO) and the geometrical theory of diffraction (GTD).

Chapter 6 deals with complex targets and the effects of errors and imperfections. Chapter 7 discusses the reduction methods of shaping, materials design and selection, and cancellation techniques. It includes some basic information on the electromagnetic behavior of metamaterials and how they are designed and constructed to provide a wide range of permittivity and permeability. Chapter 8 deals with the measurement of RCS, and Chapter 9 covers all aspects of laser cross section (LCS).

The third edition includes companion software with Matlab® software. The software collection includes a wide range of codes. Some have very general capability and are quite versatile, for example, *POFACETS* and *RAD JAM.* Others were written specifically to solve an example problem in the book or to aid in the solution of a problem at the end of a chapter.

The MKS system of units is used throughout the book. In the frequency domain discussions the $e^{j\omega t}$ time convention is used. Vectors are denoted by an overarrow ($\vec{A}$), unit vectors by a "hat" $\hat{n}$, and matrices by boldface (**B**). More details regarding the notation and conventions are given in the appendices.

The author would like to acknowledge the contribution of Jovan Lebaric (Chapter 4) and also the work of many students who contributed to examples, problems, and results that appear throughout the book. Students who made an especially significant contribution are

Filipos Chatzigeorgiadis
Christer Persson
Murray Regush
Qin ling
Mathew Yong
Kemal Yuzcelik
Georgeos Zafiropolus
Jeanette Olivia Tan

Finally, I am indebted to many coworkers at Hughes Aircraft (now Raytheon) and fellow faculty at the Naval Postgraduate School (NPS) and the University of Southern California (USC) for their encouragement and support. In particular I would like to acknowledge Prof. David Garren of NPS, Distinguished Professor Emeritus Allen Fuhs of NPS and the late Professor W. V. T. Rusch of USC.

David C. Jenn
July 2018

ABBREVIATIONS AND ACRONYMS

ABC	absorbing boundary condition
AF	array factor
BLL	backlobe level
BRDF	bidirectional reflectance distribution function
BSDF	bidirectional scatter distribution function
BW	bandwidth
CEM	computational electromagnetics
CP	circular polarization
CPR	co-planar ring
CW	continuous wave
DBIR	discrete boundary impulse response
dBm	decibels relative to a milliwatt
dBsm	decibels relative to a square
dBW	meter decibels relative to a watt
DD-X	designation for the U.S. Navy's next generation cruiser
DE	differential equation or difference equation
DNG	double negative
ECM	electronic countermeasures
EIRP	effective isotropic radiated power (same as ERP)
EHF	extremely high frequency
EM	electromagnetic
EMI	electromagnetic interference
ERP	effective radiated power (same as EIRP)
EW	electronic warfare
FAT	false alarm time
FD	frequency domain
FEBA	forward edge of battle area
FE-BI	finite element boundary integral
FEM	finite element method
FFT	fast Fourier transform

FIT	finite integration technique
FOD	f over D (f/D)
FOV	field of view
FSS	frequency selective surface
HFIE	H-field integral equation
HFSS	high frequency structures simulator
HPBW	half-power beamwidth
ILDC	incremental length diffraction coefficients
ISAR	inverse synthetic aperture radar
I	in-phase
JSF	joint strike fighter
LH	left handed
LO	low observable
LOS	line of sight
MDS	minimum detectable (discernable) signal
MEC	method of equivalent currents
MEW	method of edge waves
MKS	meter-kilogram-second
MM	method of moments
MTI	moving target indicator (indication)
NIM	negative index material
NRL	Naval Research Laboratory
PD	pulse Doppler
PEC	perfect electric conductor
PGF	path gain factor
PIM	positive index material
PLF	polarization loss factor
PMC	perfect magnetic conductor
PML	perfectly matched layer
PO	physical optics
PPI	plan position indicator (radar display)
PRF	pulse repetition frequency
PSD	power spectral density
PTD	physical theory of diffraction
PSS	phase switched screen
Q	quadrature
RAM	radar absorbing material
RCS	radar cross section

RF	radio frequency
RH	right handed
RRE	radar range equation
RWG	Rao-Wilton-Glisson basis function
SAR	synthetic aperture radar
SBR	shooting and bouncing rays
SF	subarray factor
SJR	signal-to-jam ratio
SLL	sidelobe level
SNR	signal-to-noise ratio
TBC	transparent boundary condition
TD	time domain or time difference
TDIE	time domain integral equation
TDPO	time domain physical optics
TE	transverse electric
TEM	transverse electromagnetic
TFM	thin film magnetodielectric
TGT	transparent grid termination
TM	transverse magnetic
TIS	total integrated scatter
UAV	unmanned air vehicle
UCAV	unmanned combat air vehicle
UHF	ultra-high frequency
ULO	ultra-low observable
UTD	uniform theory of diffraction
UWB	ultra-wideband
VHF	very high frequency
VSWR	voltage standing wave ratio
YAG	yttrium aluminum garnet (laser)

Chapter 1 Radar Cross Section

1.1 Introduction

The word *radar* was originally coined from the phrase *radio detection and ranging.* Since its invention during World War II, radar has played a crucial role in both military and civilian systems. In the civilian sector, radar is used for various aspects of navigation such as terrain avoidance, air traffic control, weather avoidance, and altimeters. In addition to these functions, radars on military *platforms,* such as planes, ships, and satellites, must perform reconnaissance, surveillance, and attack roles. Military missions that encounter an adversary's radar are most effectively performed when detection is avoided. Consequently, the reduction of radar cross section (RCS) has received high priority in the design of all new platforms.

Since the revelation of the stealth technology to the public in the early 1970s, the term *stealth* has been associated with *invisible to radar.* In fact, radar is only one of several sensors that is considered in the design of a *low-observable* (LO) platform. Others include infrared (IR), optical (visible), and acoustic (sound) sensors. It is also important that a low-observable target have low *emissions.* For example, a stealthy platform may be undetectable to an enemy radar, but, if a standard high-power search radar is operating on the platform, the search radar is likely to be detected by the enemy's electronic support measures (ESM).

Stealthy targets are not completely invisible to radar, as is often implied by the popular media. To be undetectable, it is only necessary that a target's RCS be low enough for its echo return to be below the detection threshold of the radar. Radar cross section reduction has evolved as a countermeasure against radars and, conversely, more sensitive radars have evolved to detect lower RCS targets. A point of diminishing return is quickly reached with regard to RCS reduction, however. After the strong scattering sources on a complex target are eliminated, the remaining RCS is due primarily to a large number of small scatterers. Treating these scatterers is much more difficult, and it eventually becomes a question of cost: Is an RCS reduction of a few percent worth the additional cost?

The financial aspect of low observability has caused a reexamination of the "stealth philosophy." In the early days of stealth, heavy emphasis was placed on reducing RCS, even if it came at the expense of other operational and performance parameters. The modern view of low observability is focused more on achieving an optimum balance between a whole host of performance measures, of which RCS is only one among equals. They include such things as IR and acoustic signatures, cost and maintainability, operational limitations, and the incorporation of electronic warfare (EW) techniques.

In this chapter the fundamentals of radar and EW systems are introduced, and the role of RCS in the performance of these systems is discussed. The general topics of electromagnetic scattering, RCS, and analysis methods are covered in Refs. [1–10].

1.2 Radar Systems and the Radar Range Equation

1.2.1 Derivation of the Radar Range Equation

The *radar equation* describes the performance of a radar for a given set of operational, environmental, and target parameters. In the most general case, the radar transmitter and receiver can be at different locations when viewed from the target, as shown in Fig. 1.1. This is referred to as *bistatic* radar. In most applications, the transmitter and receiver are located on the same platform and frequently share the same antenna. In this case, the radar is *monostatic*. When a radar uses two slightly separated antennas, it is called *quasi monostatic*. Even though the antennas are separated, they appear at essentially the same location in space when viewed from a target distant from the radar. Thus, for our purposes, quasi monostatic can be considered the same as monostatic.

The derivation of the radar range equation (RRE) requires some knowledge of wave propagation and scattering. The reader is referred to Appendix A for an overview of basic electromagnetics. To begin with, consider a monostatic radar located a distance R from a target, as illustrated in Fig. 1.2. The radar transmitter power is P_t and the antenna gain is G_t. The incident power density at the target W_i is given by

$$W_i = \frac{P_t G_t}{4\pi R^2} \tag{1.1}$$

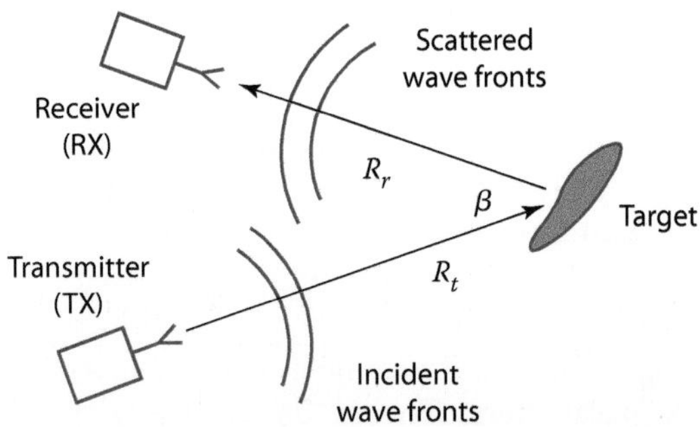

Fig. 1.1 Illustration of bistatic radar.

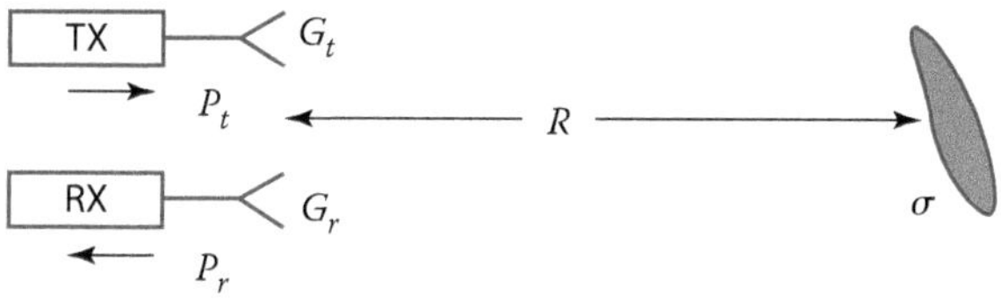

Fig. 1.2 Illustration of monostatic radar and geometry for derivation of the radar range equation.

The RCS σ relates the incident power density at the target to the power scattered back in the direction of the observer, assuming that the incident field is a plane wave

$$W_s = \left(\frac{P_t G_t}{4\pi R^2}\right)\left(\frac{\sigma}{4\pi R^2}\right) = \frac{P_t G_t \sigma}{(4\pi R^2)^2} \tag{1.2}$$

Note that the unit of σ is length squared. If the effective aperture of the receive antenna is A_{er}, the received target power is

$$P_r = W_s A_{er} = \frac{P_t G_t \sigma A_{er}}{(4\pi)^2 R^4} \tag{1.3}$$

For antennas with well-defined apertures, like reflectors and planar arrays, it is convenient to relate the effective aperture area to the physical aperture area A by an efficiency e such that

$$A_{er} = eA \tag{1.4}$$

Finally, using the relationship between gain and effective area for the receive antenna

$$G_r = \frac{4\pi A_{er}}{\lambda^2} \tag{1.5}$$

where $\lambda = c/f$ is the wavelength, the expression for the received power reduces to

$$P_r = \frac{P_t G_t G_r \sigma \lambda^2}{(4\pi)^3 R^4} \tag{1.6}$$

For monostatic systems a single antenna is generally used to transmit and receive so that $G_t = G_r \equiv G$.

Equation (1.6) is the classical form of the RRE, but it is too simplistic to accurately predict the performance of a complex modern radar. Many aspects of the radar system and its operating environment have been ignored in its derivation. Such things as losses and noise generated in the radar hardware, clutter and interference, and the motion of the target can degrade the radar's performance. On the other hand, signal processing techniques and pulse integration can improve the radar's capability to detect a target. A simple modification of Eq. (1.6) to include loss L and processing gain G_p gives

$$P_r = \frac{P_t G_t G_r \sigma \lambda^2 L G_p}{(4\pi)^3 R^4} \tag{1.7}$$

where it is assumed that $0 \leq L \leq 1$ and $G_p \geq 1$. The loss factor L is the product of the appropriate efficiency factors, which depend on the type of radar, its architecture, and its hardware. Equation (1.7) assumes an unobstructed line of sight (LOS) to the target.

Equation (1.7) does give some insight into the tradeoffs involved in radar design. The dominant feature of the monostatic RRE is the $1/R^4$ factor. Even for targets with relatively large RCS, high transmit power or a high gain antenna must be used to overcome the $1/R^4$ when the range becomes large.

1.2.2 Signal-to-Noise Ratio and Detection Range

In practice, the radar receiver will sense a nonzero signal even when there is no target present. This is due to other signal sources such as the following.

1. *Clutter*: Reflections from the ground, foliage, and other objects in the environment result in signals back to the radar receiver.
2. *Interference*: Signals from other electronic systems that radiate will be received. Their intended purpose might be to intentionally distract the radar (i.e., a jammer), or they may be unintentional interferers that occupy the same frequency band (e.g., radio stations, other radars, etc.).
3. *Noise*: The thermal motion of electrons gives rise to random voltages and currents. Surprisingly, for a well-designed radar operating at microwave frequencies, thermal noise generated in the radar's receive channel can be the limiting factor in detecting a low observable target.

A transmission line at room temperature can generate a significant random voltage at its input. In fact, any lossy circuit device generates thermal noise. The thermal noise power can be expressed as [1]

$$N_o = k_B T_e B \tag{1.8}$$

where $k_B = 1.38 \times 10^{-23}$ (J/K) is Boltzman's constant, B (Hz) is the bandwidth (BW) of the device, and T_e (K) is its effective temperature. When multiple devices are combined in the receiver, it is possible to come up with an effective temperature of the combination in terms of the effective temperatures of the individual devices [11].

There are several ways of defining bandwidth, and in the case of Eq. (1.8) the *noise bandwidth* should be used. For our purposes, the noise bandwidth can be considered to be the same as the *operating bandwidth* of the radar. Again, there are different ways of defining the operating bandwidth. A general definition is the range of frequencies over which the radar has acceptable performance. It is determined by many factors, such as antenna gain, amplifier gain, and matched filter loss, among others. When dealing with conventional radar architectures (i.e., not ultra wideband) it is sufficient to assume that the frequency characteristic of the radar *H(f)* is bandlimited and has constant amplitude as shown in Fig. 1.3. In this case the bandwidth of the radar is

$$B = f_2 - f_1 \tag{1.9}$$

and the center frequency, which normally corresponds to the carrier frequency of the waveform, is

$$f_c = \frac{f_2 + f_1}{2} \tag{1.10}$$

In addition to noise generated by the receiver itself, background noise collected by the antenna appears at the receiver. Most people are familiar with the fact that objects radiate heat in the IR spectrum. However, background objects such as the sun, cold sky, and earth also emit microwave radiation that is collected by the antenna. The antenna noise power can be described in terms of an *antenna noise temperature* T_A. The antenna temperature together with the effective temperature of the receiver defines *system noise temperature* that completely describes the noise for the purpose of radar systems analysis

$$T_s = T_e + T_A \tag{1.11}$$

Most radars use a type of *threshold detection*; that is, if P_r exceeds a given level above the noise [i.e., the signal-to-noise ratio (SNR)], then a target is declared present

$$\text{SNR}_{\min} = \frac{P_{r_{\min}}}{N_o} = \frac{P_t G_t G_r \sigma \lambda^2 L G_p}{(4\pi)^3 R^4 k_B T_e B} \tag{1.12}$$

The *minimum detectable signal* (MDS) $P_{r_{\min}}$ is the smallest received signal that is still above the detection threshold relative to the noise. As depicted

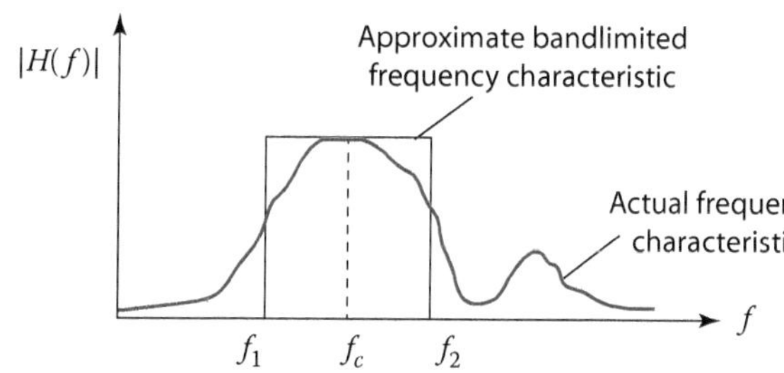

Fig. 1.3 A device frequency characteristic and its bandlimited approximation.

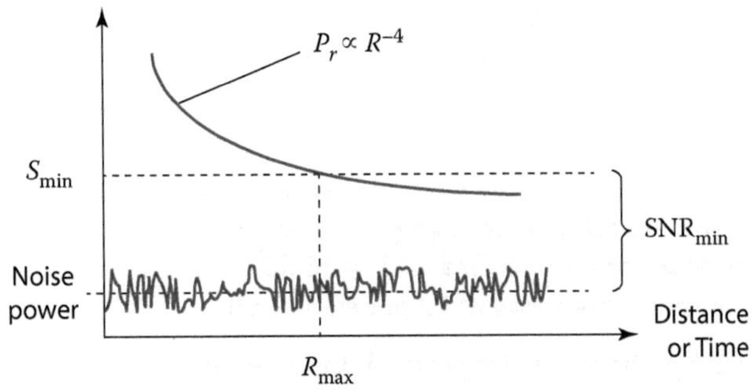

Fig. 1.4 Threshold detection in the presence of thermal noise.

in Fig. 1.4, there is a maximum range corresponding to the minimum SNR. Solving Eq. (1.12) for the range gives

$$R_{\text{max}} = \left[\frac{P_t G_t G_r \sigma \lambda^2 L G_p}{(4\pi)^3 k_B T_e B \text{SNR}_{\text{min}}}\right]^{1/4} \tag{1.13}$$

Typical minimum SNRs are in the range of 10 to 20 decibels (dB), where the decibel quantity is obtained from

$$\text{SNR}_{\text{min}} \text{ in dB} = 10 \log(\text{SNR}_{\text{min}}) \tag{1.14}$$

and log is the base 10 logarithm.

Example 1.1: Detection Range of a Search Radar

The AN/SPS-10 surface search radar has the following operating parameters:

peak transmitter power = 500 kW
antenna gain = 33.0 dB
frequency = 5.6 GHz
pulse width = 1.4 μs
PRF = 625 Hz
antenna scan rate = 16 rpm
azimuth half-power beamwidth = 1.5 deg

antenna noise temperature = 75 K
receiver noise bandwidth = 1 MHz

receiver effective temperature = 2900 K
system losses ahead of the receiver = 5 dB

(Continued)

Example 1.1: Detection Range of a Search Radar *(Continued)*

false alarm time (FAT) – average time between false alarms = 2 days
plan position indicator (PPI) display and operator
minimum SNR for the specified FAT = 16 dB
processing gain (integration of 10 pulses) = 9 dB

1. What is the thermal noise power in the receiver?

$$N_o = k_B T_s B = (1.38 \times 10^{-23})(75 + 2900)(10^6) = 4.1 \times 10^{-14}\ \text{W}$$
$$= -134\ \text{dBW}$$

Note: dBW is decibels relative to 1 W, P_r in dBW = 10 log (P_r in watts).

2. Calculate the MDS in dBW.

Convert the SNR from dB: $\text{SNR}_{\text{min}} = \dfrac{P_{r_{\text{min}}}}{N_o} = 10^{16/10} = 39.8$

$$P_{r_{\text{min}}} = (39.8)\left(4.1 \times 10^{-14}\right) = 1.6 \times 10^{-12}\ \text{W} = -118\ \text{dBW}$$

3. Calculate the peak effective radiated power (ERP) in dBW.

$$\text{ERP} = P_t G_t = \left(500 \times 10^3\right)\left(10^{33/10}\right) = 997.6\ \text{MW}$$

4. Calculate the effective area of the antenna in square meters.

$$A_{er} = \frac{G\lambda^2}{4\pi} = \frac{(1995.3)(0.054)^2}{4\pi} = 0.46\ \text{m}^2$$

5. Calculate the maximum free space detection range on a 0 dBsm target.

Convert the loss from decibels (note that "loss" implies a negative sign in the exponent): $L = 10^{-5/10} = 0.316$

Convert the processing gain from dB: $G_p = 10^{9/10} = 7.9$

The decibel unit of RCS is dBsm (decibels relative to a square meter), which is defined as

$$\sigma \text{ in dBsm} = 10 \log (\sigma \text{ in m}^2) \tag{1.15}$$

Therefore, the target RCS is $\sigma = 10^{0/10} = 1\ \text{m}^2$. Now using Eq. (1.13)

$$R_{\text{max}} = \left[\frac{P_t G^2 \sigma \lambda^2 L G_p}{(4\pi)^3 N_o \text{SNR}_{\text{min}}}\right]^{1/4}$$
$$= \left[\frac{500 \times 10^3 (1995.3)^2 (1)(0.054)^2 (0.316)(7.9)}{(4\pi)^3 (4.1 \times 10^{-14})(39.8)}\right]^{1/4}$$
$$= 4.6 \times 10^4 = 46\ \text{km}$$

1.2.3 Information Available from the Radar Echo

Sophisticated radar can provide a vast amount of information about a target and the environment. However, the fundamental measurement quantities are as follows: 1) *range,* usually by measuring the round-trip time delay of a pulse or series of pulses, 2) *velocity,* from the Doppler frequency shift or range rate (i.e., measuring the range at two times and estimating the velocity by the change in range divided by time interval), and 3) *direction,* by antenna pointing.

Other information might include the following:

1. The size of the target based on the magnitude of its return.
2. Target shape based on the echo strength as a function of viewing angle, also called *aspect angle.* Furthermore, shape information can also be obtained by illuminating the target with different polarizations and observing the change in echo.
3. Target moving parts, such as propellers and jet turbines, inferred from their Doppler frequency components.

In practice, radars transmit continuously to cover search patterns, track moving targets, and improve detection by integrating (summing) several target returns. Figure 1.5 shows a simple pulse train that might be used in a pulse-Doppler (PD) or moving target indication (MTI) radar. The frequency of the sinusoid is the radar's carrier frequency f_c. The pulse amplitude (voltage) is V_0, the width τ, the period T_p and the pulse repetition frequency (PRF)

$$f_p = \frac{1}{T_p} \tag{1.16}$$

Target range is the fundamental quantity measured by most radars. It is obtained by recording the round-trip travel time of a pulse T and computing range from

$$R_t + R_r = cT \quad \text{(bistatic case)} \tag{1.17}$$

$$R = cT/2 \quad (\text{monostatic case } R_r = R_t \equiv R) \tag{1.18}$$

where $c \approx 3 \times 10^8$ m/s is the velocity of light in free space.

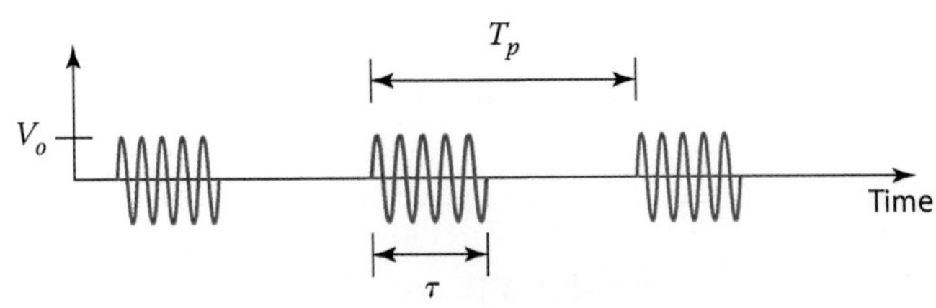

Fig. 1.5 Pulse train waveform.

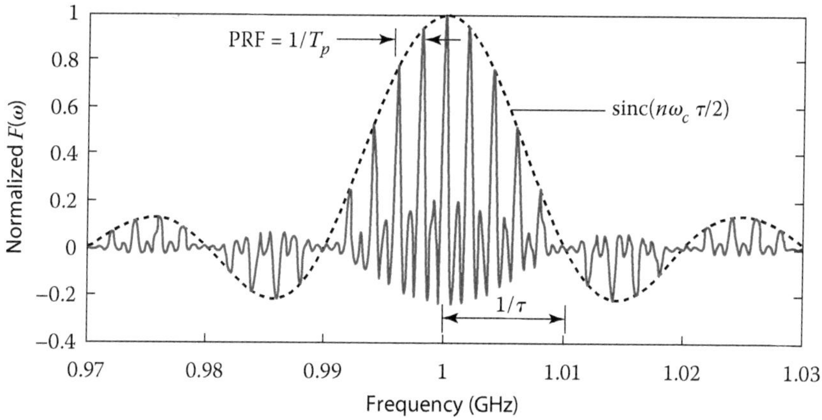

Fig. 1.6 Positive frequency spectrum of a pulse train with $N_p = 5$ pulses, carrier frequency $f_c = \omega_c/2\pi = 1$ GHz, pulse width $\tau = 0.1 \times 10^{-6}$ s, and period $T_p = 5\tau$ s.

The frequency spectrum of a pulse train with a carrier frequency of 1 GHz is shown in Fig. 1.6. The envelope of the spectrum is the Fourier transform of a pulse of width τ. From Fourier analysis [12] this is a "sinc" function, where

$$\text{sinc}(X) \equiv \sin(X)/X \tag{1.19}$$

The spread in frequency varies as $1/\tau$. For the radar to transmit a relatively undistorted waveform, the hardware must have a bandwidth that essentially matches that of the waveform. Thus, the required bandwidth of the radar is set primarily by the pulse width τ. A good rule of thumb for a matched receiver is that the bandwidth should be approximately the inverse of the pulse width, or

$$B \approx 1/\tau \tag{1.20}$$

A radar estimates the range to a target by measuring the round-trip time delay of a pulse and then applying Eq. (1.18). Generally, the time interval between two pulses is quantized in "bins" as shown in Fig. 1.7. Because the range and time axes differ only by a scale factor, they are referred to as

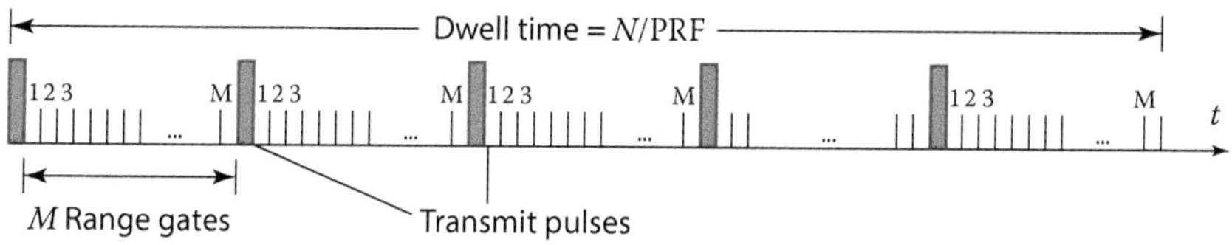

Fig. 1.7 Pulse train and range gates.

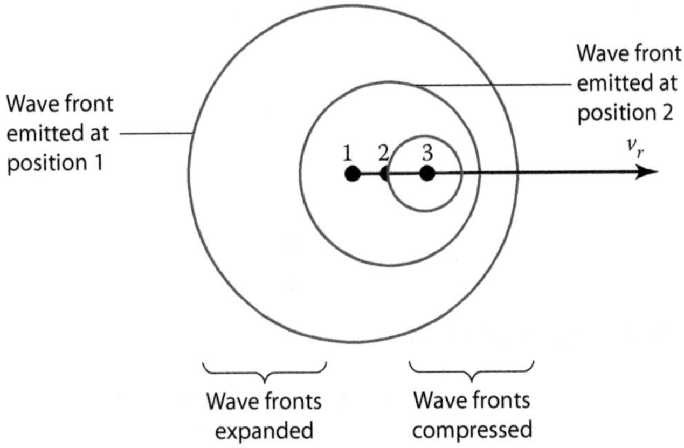

Fig. 1.8 Illustration of the Doppler effect for a moving target.

range bins or *range gates.* The range of a target is estimated by measuring which range bin its echo pulse falls into on return. The radar may have difficulty separating out returns from multiple targets that fall into the same range bin. For example, if a B-747 and a Cessna 172 are at the same range, the returns will overlap, and only one large target will be seen. Of course, the two aircraft will most likely have different velocities and directions, so that they can be separated by examining the Doppler frequencies as described in the following sections.

A target's velocity can be estimated by measuring its Doppler shift. Targets in motion relative to the radar cause the return signal frequency to be shifted as shown Fig. 1.8. The wavefronts are compressed for targets approaching the radar (increased frequency relative to the carrier) and expanded for receding targets (decreased frequency relative to the carrier). The time-harmonic transmitted electric field has the form

$$E_t \propto \cos(\omega_c t) \tag{1.21}$$

and, therefore, the received signal has the form

$$E_s \propto \cos(\omega_c t - 2kR) = \cos[\omega_c t + \Phi(t)] \tag{1.22}$$

where

$R = R_o + v_r t$
$k = 2\pi/\lambda =$ the wave number or propagation constant
$v_r =$ the radial component of the relative velocity vector
$R_o =$ the range at $t = 0$
$\omega_c = 2\pi f_c$

The factor 2 in the cosine of Eq. (1.22) arises from the round-trip path delay for monostatic radar.

The Doppler frequency shift is given by the time rate-of-change of the phase

$$\omega_d = \frac{d\Phi(t)}{dt} = \frac{d}{dt}(-2kR) = -2k\frac{dR}{dt} = -2kv_r \tag{1.23}$$

or in Hertz,

$$f_d = -\frac{2v_r}{\lambda} \tag{1.24}$$

Now rewrite the signal phase as

$$\Phi(t) = -2k\left[R_o + \left(-\frac{\omega_d}{2k}\right)t\right] \tag{1.25}$$

so that

$$E_s \propto \cos[(\omega_c + \omega_d)t - 2kR_o] \tag{1.26}$$

Equation (1.26) verifies that the radar measures the Doppler-shifted frequency

$$\omega_c + \omega_d = 2\pi(f_c + f_d) \tag{1.27}$$

A Doppler shift occurs only when the *relative velocity* vector has a *radial* component. In general there will be both radial and tangential components to the velocity, as shown in Fig. 1.9. However, only the radial component will be measured by the radar.

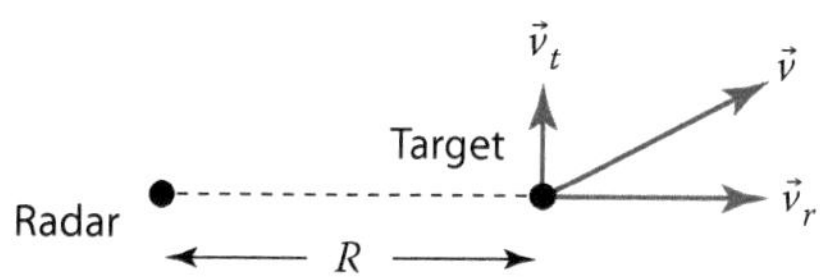

Fig. 1.9 Radial and tangential components of the relative velocity vector.

Example 1.2: Calculating the Doppler Shift

A stationary radar at the origin of a cartesian coordinate system detects moving targets with the velocities $\vec{v}_1 = 60(\hat{x} + \hat{y})$ m/s and $\vec{v}_2 = -100\hat{x}$ m/s, as shown in Fig. 1.10. What is the difference between Doppler frequencies if the radar operates at 1 GHz?

Target 1's velocity is completely radial (receding):

$$\vec{v}_{r_1} = |60(\hat{x} + \hat{y}) \cdot \hat{\rho}| = |60(\hat{x} + \hat{y}) \cdot (\hat{x}\cos 45\text{ deg} + \hat{y}\sin 45\text{ deg})|$$
$$= \sqrt{60^2 + 60^2} = 84.9\text{ m/s}$$
$$f_d = -\frac{2v_r}{\lambda} = -\frac{2(84.9)}{0.3} = -566\text{ Hz}$$

(Continued)

Example 1.2: Calculating the Doppler Shift *(Continued)*

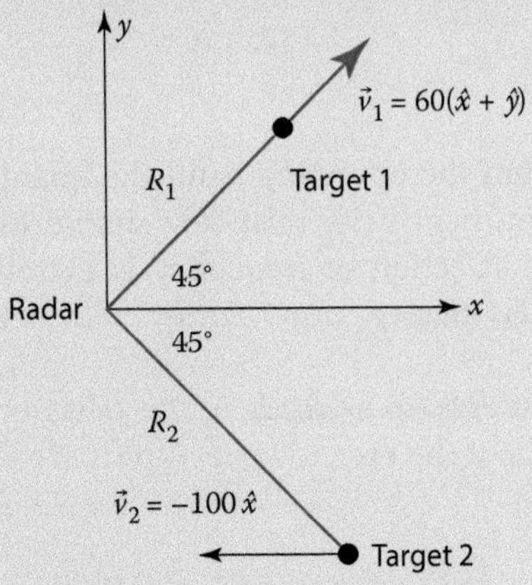

Fig. 1.10 Velocity vectors for Example 1.2.

The radial unit vector in the cylindrical coordinate system is $\hat{\rho}$, and the transformation between coordinate systems is discussed in Appendix B. The radial component of velocity for the second target is

$$\vec{v}_{r_2} = -100\hat{x} \cdot \hat{\rho} = -100\hat{x} \cdot (0.707\hat{x}) = -70.7 \text{ m/s}$$

and its Doppler shift is

$$f_d = -\frac{2v_r}{\lambda} = -\frac{2(-70.7)}{0.3} = 471.3 \text{ Hz}$$

Therefore, the difference in Doppler frequencies is 1037.3 Hz. These two targets can be discriminated based on their Doppler frequencies, even if they are in the same range bin.

Even though a pulsed waveform is transmitted by a radar, it is still possible to approach the scattering analysis as a time-harmonic problem. As long as the entire target is completely illuminated by each pulse in the train, the scattering from a target is essentially the same as that for a continuous sinusoid. Therefore, in the frequency domain, phasor notation can be used.

1.2.4 Issues in the Selection of Radar Frequency

Traditionally radar waveform bandwidths have been relatively small; that is, they have been *narrowband* radars. Although there is no standard definition of narrowband, it is generally applied to operating bandwidths less than about 10 percent:

$$\frac{B}{f_c} \times 100 \lesssim 10\% \tag{1.28}$$

At all frequencies within the operating band the quantities in the RRE can be considered constant, hence the relatively simple form of Eq. (1.7). For narrowband radar the selection of frequency is extremely important. The frequency spectrum is divided into *bands,* with letter designations as shown in Table 1.1.

There are many tradeoffs to be made in the selection of a radar's operating frequency. The RCS of the target is one; others are the *antenna gain and beamwidth, transmitter power, ambient noise, Doppler shift, size constraints,* and *atmospheric attenuation.* Often there are conflicting requirements on the radar parameters. For example, suppose it is necessary to maximize the detection range of the radar for search. A large antenna gain increases the detection range, and to maximize the antenna gain for a fixed area, a high frequency (small wavelength) is desired. However, there are disadvantages to high frequencies. One is that the atmospheric attenuation increases, and because the path is round trip, the factor $\exp(-4\alpha R)$ must be added to the RRE, where α (Np/m) is the one-way electric field attenuation coefficient defined in Appendix A.

Maximizing the antenna's gain is not the only consideration in its design. A high-gain antenna has a narrow beamwidth, which is undesirable for search. A small beamwidth requires more time to search a fixed region of

Table 1.1 IEEE Radar Frequency Bands (After Ref. [13])

Band Designation	Frequency Range
HF	3–30 MHz
VHF	30–300 MHz
UHF	300–1000 MHz
L	1–2 GHz
S	2–4 GHz
C	4–8 GHz
X	8–12 GHz
Ku	12–18 GHz
K	18–27 GHz
Ka	27–40 GHz
V	40–75 GHz
W	75–110 GHz
Millimeter (MM)	110–300 GHz

space compared to a large beamwidth. In this application it may be more efficient to use a wider beamwidth (lower gain) and make up the difference in gain by increasing the transmitter power.

A few long-range ballistic missile defense radars operate in the 300-MHz region [ultra high frequency (UHF)], but most others use frequencies greater than 1 GHz (L band and above). Devices at low frequencies are larger than similar devices at a higher frequency, and, therefore, the system's components require more volume. Likewise, the antenna must be larger than it is at a higher frequency to achieve the same beamwidth and gain. Low frequencies are capable of handling more power because the applied voltages can be higher without causing breakdown. Finally, ambient noise is lowest in the 1–10-GHz range, and low-altitude atmospheric attenuation favors frequencies below 18 GHz.

The function of the radar also influences the choice of frequency. For instance, an airborne imaging radar must be compact and lightweight but must also have a narrow antenna beam for resolving scatters. These all favor a high frequency. On the other hand, a ground-based search radar must radiate high power to achieve a large detection range. In this case, the larger and heavier equipment that is necessary for high-power operation is not a problem.

1.3 Polarization Definitions

It will be shown that a target's RCS is a function of many parameters. Among them are the 1) radar's frequency and bandwidth; 2) radar's transmit antenna polarization, which sets the polarization of the wave incident on the target; 3) radar's receive antenna polarization, which determines the components of the scattered field that are received by the radar; 4) shape of the target; 5) materials of which the target is composed; and 6) radar/target geometry.

The polarization of a plane wave is defined in Appendix A. Two orthogonal polarization components are required to completely describe any arbitrary wave polarization. There are primarily three polarization references applied to RCS. The Cartesian and spherical components are defined in Fig. 1.11.

1. *Horizontal (H) and vertical (V)*: Horizontal and vertical are convenient when dealing with targets and radars that are near the ground ($x-y$ plane) or at altitude flying parallel to the ground. From the figure it is apparent that the observer is on the ground $\theta \rightarrow 90$ deg; the correspondence between H–V and $\theta - \phi$ becomes $E_V \approx E_\theta$ and $E_H \approx E_\phi$.
2. *Transverse electric (TE) and transverse magnetic (TM)*: The reference axis for defining the transverse plane must be specified. Most often the z-axis is used. For TE_z the electric field vector lies in a plane transverse

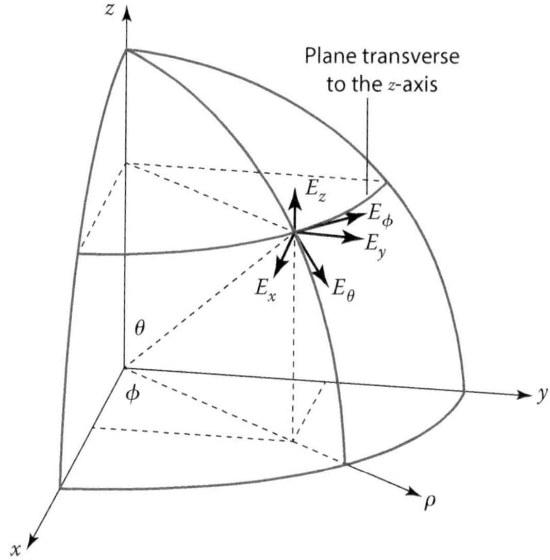

Fig. 1.11 Spherical coordinate system.

to the z-axis (i.e., in a plane parallel to the x–y plane). Similarly, for TM_z the magnetic field vector lies in a plane transverse to the z-axis. From Fig. 1.12 it is evident that when the electric field is TE_z polarized, $E_\phi \neq 0$, $E_\theta = 0$; when the electric field is TM_z polarized, $E_\theta \neq 0$, $E_\phi = 0$. TM and TE to the x or y coordinate axis is sometimes used, but the relationships to E_ϕ and E_θ are not as simple as for the z axis. TE and TM are convenient when dealing with isolated targets and are generally not used in radar system analysis.

3. *Any two orthogonal coordinate systems variables*: Examples are x and y, or θ and ϕ. The spherical variables θ and ϕ are perhaps the most useful. When computing RCS, the origin of the spherical system is placed on or near the target. When analyzing radar performance, the origin of the spherical system is placed at the radar antenna.

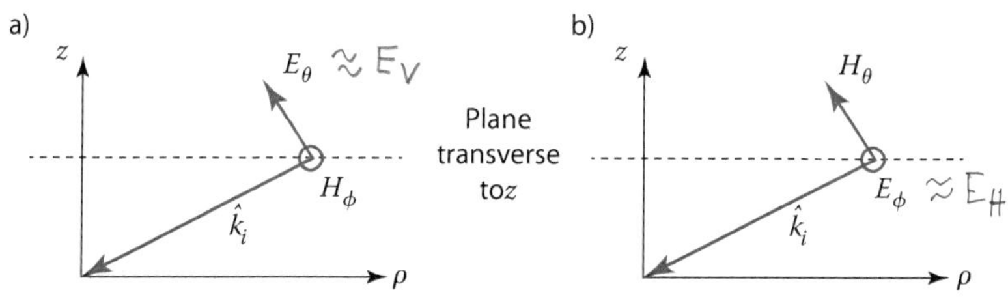

Fig. 1.12 Cut through the ρ–z plane (side view) illustrating a) TM and b) TE polarizations.

1.4 Multipath and Other Environmental Effects

A potential problem for the radar is *clutter* return that falls into the same range bin as a target. The clutter (e.g., ground, buildings, etc.) can be illuminated by the main beam of the antenna or one of its sidelobes, as shown in Fig. 1.13. Even though the clutter is in a sidelobe, and therefore the antenna gain is much lower than the main-beam gain, the clutter area can be extremely large, and hence its echo much larger than that of a low-observable target in the main beam. Again, it is possible to discriminate the target from the clutter based on Doppler frequency.

As shown in Fig. 1.14, when a radar is tracking a target near the ground, *multipath* (i.e., reflections from the ground and other reflecting surfaces) presents a problem. It is most pronounced when the ground is flat and the surface material is a good reflector (e.g., water or wet soil). The reflection from the ground can combine constructively (add) or destructively (subtract) with the direct signal. In some cases complete cancellation can occur, and the radar cannot detect the target, which is a *blind* condition.

To account for multipath and other environmental effects, a path-gain factor F can be applied to the RRE. This factor gives the one-way electric field between the target and radar relative to the direct path field. Because power is proportional to electric field squared and the round trip requires another squaring, Eq. (1.7) becomes

$$P_r = \frac{P_t G_t G_r \lambda^2 \sigma L G_p}{(4\pi)^3 R^4} |F|^4 \tag{1.29}$$

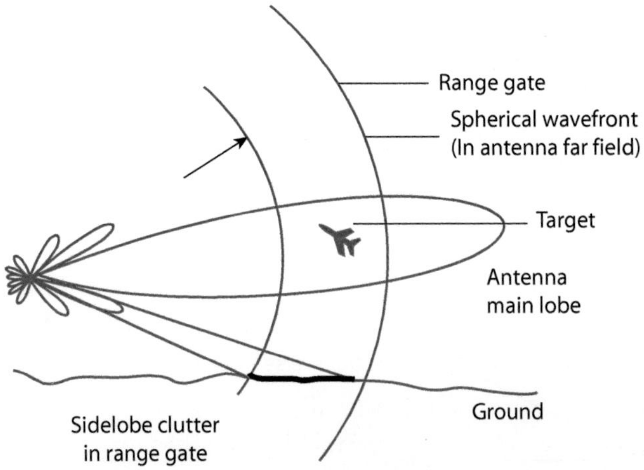

Fig. 1.13 Example of sidelobe clutter in the same range gate as a target.

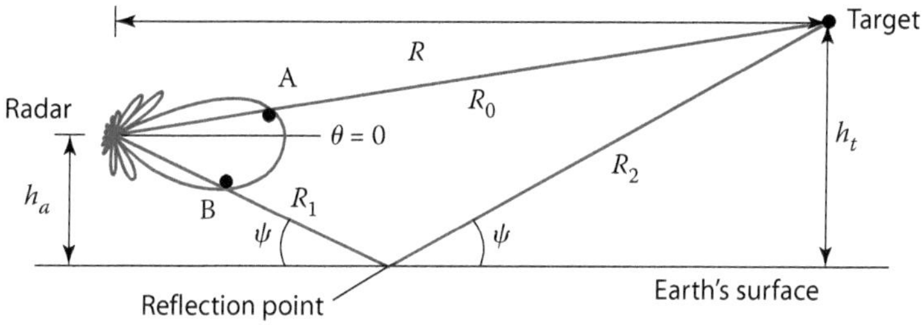

Fig. 1.14 Ground bounce for a flat Earth.

When both the radar and target are at low altitudes, the following assumptions can be made:

1. The reflection coefficient $\Gamma \approx -1$ for both polarizations (horizontal and vertical) when the *grazing angle* is small, $\psi <$ several degrees.
2. The direct and reflected path gains are equal $G(\theta_A) \approx G(\theta_B)$.
3. The target RCS versus aspect is constant.

The difference between the direct and reflected path lengths is

$$\Delta R = \underbrace{(R_1 + R_2)}_{\text{REFLECTED}} - \underbrace{R_o}_{\text{DIRECT}} \tag{1.30}$$

The total signal at the target is

$$\begin{aligned} E_{\text{total}} &= E_{\text{reflected}} + E_{\text{direct}} \approx E(\theta_A) + \Gamma E(\theta_B) e^{-jk\Delta R} \\ &\approx E_{\text{dir}} \Big| \underbrace{\left[1 + \Gamma e^{-jk\Delta R}\right]}_{\equiv F,\text{PGF}} \Big| \end{aligned} \tag{1.31}$$

The path gain factor takes on the values $0 \le |F| \le 2$. If $|F| = 0$, the direct and reflected rays cancel (destructive interference); if $|F| = 2$, the two waves add (constructive interference).

Reference [1] shows that if $h_a \ll R$ and $h_t \ll R$, then an approximate expression for the path difference is

$$\Delta R \approx \frac{2 h_a h_t}{R} \tag{1.32}$$

and

$$|F|^4 = 16 \sin^4\left(\frac{k h_t h_a}{R}\right) \tag{1.33}$$

Multipath can be extremely important when operating near the surface, such as surface search on the ocean. Using the free space form of the RRE without consideration of reflections from the ocean surface will give erroneous detection ranges. A more complete four-path model has been developed that includes the mixing of the direct and reflected paths [14].

1.5 Radar Countermeasures

1.5.1 Reducing the Radar's Maximum Detection Range

An example of a radar network used to protect a high-value target is depicted in Fig. 1.15. The circles represent the maximum detection ranges of the individual radars, which are configured to provide uninterrupted coverage of the target. Conventional aircraft without electronic countermeasures cannot penetrate the radar network without being detected.

There are basically two methods of defeating the radar, that is, effectively reducing its maximum detection range:

1. *Low observability or "stealth."* Here the approach is to reduce the target's RCS, and thus R_{max} in Eq. (1.13). However, because of the fourth root, a significant reduction in the maximum detection range requires an even larger reduction in RCS. For example, to halve the detection range the RCS must be reduced by 12 dB.
2. *Electronic countermeasures (ECM)** These include passive measures like deploying *chaff* or active techniques such as *jamming*. Chaff is essentially clutter whose return is so large that it hides the target echo. Similarly, a jammer injects interference into the radar, which masks the target's return.

Note that it is not necessary to use one method at the exclusion of the other. Various combinations of the two can be used. However, when ECM is employed, the radar will be aware of the attack and may possibly

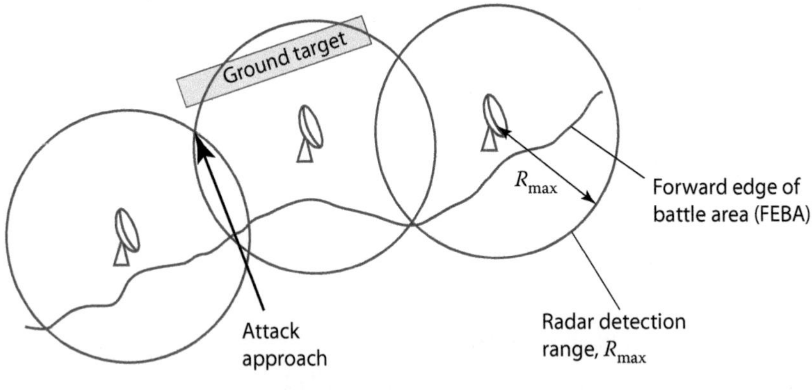

Fig. 1.15 A network of radars arranged to provide continuous coverage of a ground target.

*There are many other techniques that fall into the category of ECM, for example *deception*, but here only jamming and chaff are considered.

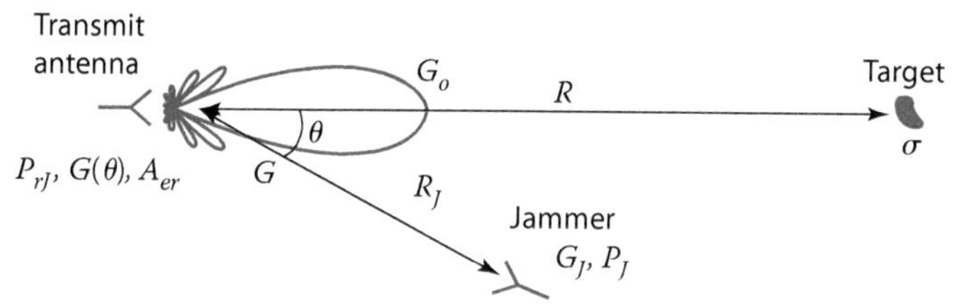

Fig. 1.16 Geometry for computing the jammer burnthrough range.

defend against it. On the other hand, a stealthy platform does not reveal its presence.

1.5.2 Jammer Burnthrough Range

To evaluate how a jammer's signal affects a radar, a transmission equation must be derived. Consider the standoff jammer shown in Fig. 1.16. The radar antenna main beam, with gain G_0, is pointed directly at the target with RCS σ. The power transmitted by the jammer is P_J and its antenna gain G_J. The radar antenna's gain in the direction of the jammer is $G(\theta)$. The jammer power received by the radar is

$$P_{rJ} = W_i A_{er} = \left(\frac{P_J G_J}{4\pi R_J^2}\right)\left(\frac{\lambda^2 G(\theta)}{4\pi}\right) = \frac{P_J G_J \lambda^2 G(\theta)}{(4\pi R_J)^2} \tag{1.34}$$

The target return, neglecting loss and processing gain, is given by Eq. (1.6)

$$P_r = \frac{P_t G_0^2 \lambda^2 \sigma}{(4\pi)^3 R^4} \tag{1.35}$$

If the jammer is effective, its received signal will be much larger than the thermal noise, and therefore noise can be neglected. The ratio of the target signal power to jammer power [i.e., the *signal-to-jam ratio* (SJR)] is

$$\mathrm{SJR} = \frac{S}{J} = \frac{P_r}{P_{rJ}} = \left(\frac{P_t G_0}{P_J G_J}\right)\left(\frac{R_J^2}{R^4}\right)\left(\frac{\sigma}{4\pi}\right)\left(\frac{G_0}{G(\theta)}\right) \tag{1.36}$$

The *burnthrough range* occurs when SJR = 1.

Several important points can be made with regard to Eq. (1.36):

1. The difference in range, R_J^2 versus R^4, is a big advantage for the jammer.
2. The difference in gains, G versus $G(\theta)$, is usually a big disadvantage for the jammer, unless the jammer can position itself in the main beam of the radar antenna. Low sidelobe radar antennas reduce a jammer's effectiveness.
3. Given that the geometry of the engagement is fixed, the only parameter that the jammer has control of is the ERP (P_J G_J).

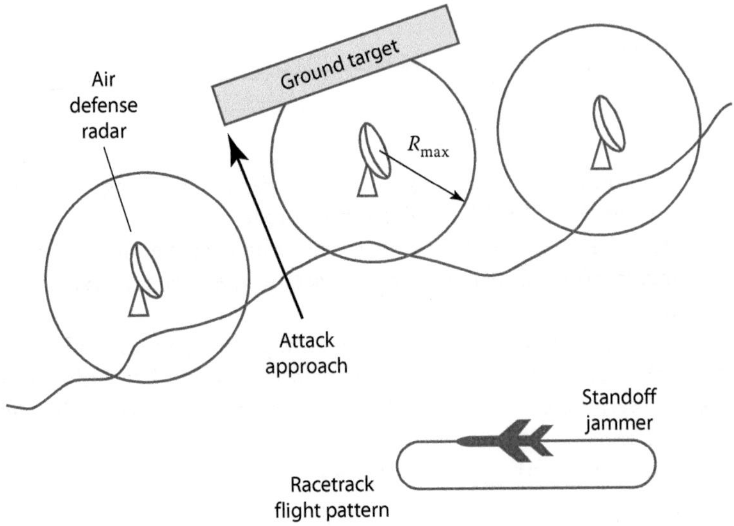

Fig. 1.17 Standoff jammer operating against a network of radars.

4. As pointed out already, the radar knows it is being jammed. The jammer can be countered using waveform selection and signal processing techniques.

Thus, when jamming is effective, the situation depicted in Fig. 1.15 becomes similar to the one shown in Fig. 1.17. Corridors of undetectability have been opened.

In obtaining Eq. (1.36), the radar processing gain has been ignored. Signal processing of the received target echo can give the radar a big advantage over a jammer that is transmitting simple broadband noise. Another important factor is the ratio of the radar bandwidth to jammer bandwidth. If the jammer is spreading its power over a bandwidth B_J that is wider than the radar's bandwidth B, then only a fraction of the jammer power, $P_J B/B_J$, enters the radar's receiver.

1.5.3 Tradeoff Between Low Observability and Electronic Warfare

Equation (1.35) clearly shows that low observability and jamming complement each other. Of course, a radar can be defeated entirely by stealth techniques; however, the cost may be prohibitively high in terms of 1) design and engineering, 2) vehicle performance and operational limitations, and 3) manufacturing and maintenance.

The demands on RCS can be relaxed if jamming or some other electronic warfare technique is employed, as demonstrated by the following example.

Example 1.3: Comparison of Jamming for Conventional and LO Targets Using RADJAM

As an example, the MATLAB® software RADJAM is used to examine the detection contours of a radar under several different operating conditions. The data for the AN/SPS-10 are given in Example 1.1. For a noise jammer, the thermal noise generated in the receiver can be neglected in comparison to the jammer noise. The noise floor is set by the jammer and SNR_{min} becomes SJR.

The antenna sidelobe level is set at 15 dB, which is typical for a standard reflector antenna. The specified antenna gain can be used to estimate the dimensions. Because the AN/SPS-10 tracks in azimuth, the azimuth beamwidth is much narrower than the elevation beamwidth. (Azimuth and elevation coordinates are discussed in Appendix B.) Assuming a length of $a = 1.6$ m in azimuth and an efficiency of $e = 0.55$ (typical for a reflector),

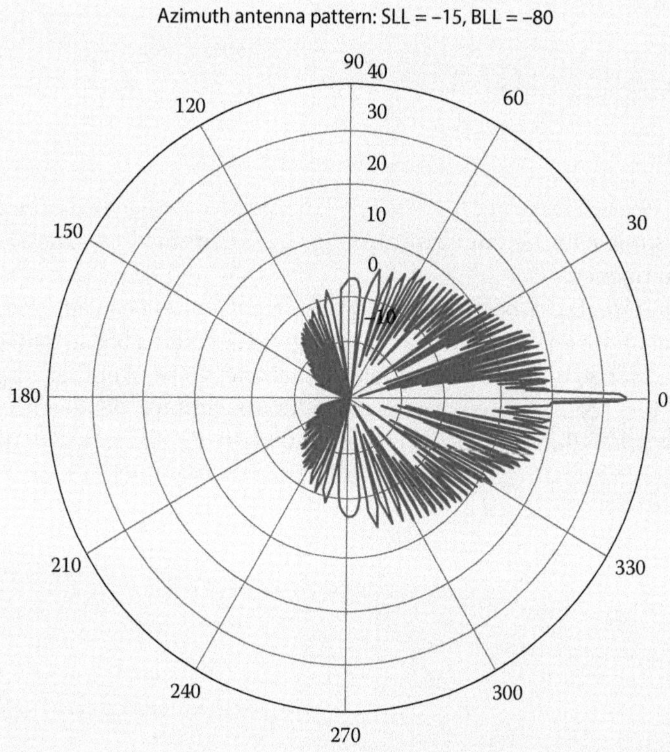

Fig. 1.18 Polar plot of the AN/SPS-10 antenna pattern generated by RADJAM.

(Continued)

Example 1.3: Comparison of Jamming for Conventional and LO Targets Using RADJAM *(Continued)*

the elevation dimension b can be computed from Eq. (1.4):

$$A = ab = \frac{A_{er}}{e} = \frac{G_r \lambda^2}{4\pi e} = \frac{1995.3(0.054)^2}{4\pi(0.55)}$$
$$\approx 0.84 \text{ m}^2 \rightarrow b = 0.84/1.6 = 0.526 \text{ m}$$

The radar and jammer parameters can be entered into RADJAM and the detection contour plotted. The radar is fixed at the origin and represented by the "R" on the detection contour plot. The jammer location is also fixed and is represented by a "J" on the plot. The target moves in a circle around the radar (in azimuth, at a constant elevation). The radar main beam is fixed on the target, and for each azimuth angle R_{max} is computed. After the calculation is completed for all azimuth angles, the contour of all R_{max} values is displayed on a polar grid.

Using the efficiency and computed antenna dimensions, RADJAM plots the pattern shown in Fig. 1.18. The detection contour without jamming ($P_J = 0$) is simply a circle of radius 46 km, as computed in Example 1.1 and plotted by RADJAM in Fig. 1.19. Now adding a 10 W jammer 10 km from the radar

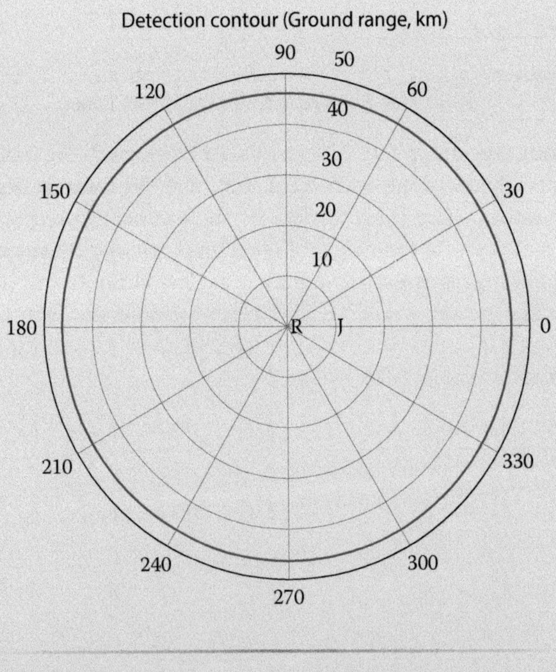

Fig. 1.19 Detection contour for the AN/SPS-10 without jamming.

(Continued)

Example 1.3: Comparison of Jamming for Conventional and LO Targets Using RADJAM *(Continued)*

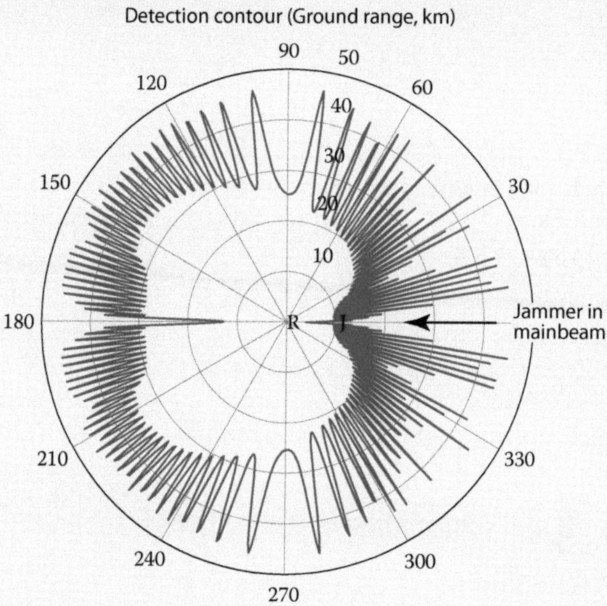

Fig. 1.20 Detection contour for the AN/SPS-10 with a jammer at 10 km ($P_J = 10$ W, $G_J = 3$ dB, $B_J = 10$ MHz, and a 0 dBsm target).

at 0° azimuth, the plot in Fig. 1.20 results. At 0° azimuth the jammer is in the antenna main beam along with the target, and the jammer signal swamps the target return. The sharp spikes in the pattern correspond to angles where the jammer is in the nulls of the antenna pattern. At these null angles, the detection range is essentially the same as that without jamming.

What radar power would be required to restore the detection range in the main beam to its original value of 46 km? Referring to Eq. (1.36), for a main-beam jammer $G_0/G(\theta) = 1$, and

$$P_t = \left[\frac{4\pi R^4 P_J G_J}{G_0}\left(\frac{\text{SJR}}{R_J^2}\right)\right] = \left[\frac{4\pi(10)(2)(39.8)(46\times10^3)^4}{(1995.3)(10\times10^3)^2}\right]$$
$$= 2.2446\times10^{11} = 224.5\ \text{GW} = 113\ \text{dBW}$$

Alternately, what increase in RCS is required to restore the detection range to its original value?

$$\sigma = \left[\frac{4\pi R^4 P_J G_J}{P_t G_0}\left(\frac{\text{SJR}}{R_J^2}\right)\right] = \left[\frac{4\pi(46\times10^3)^4(10)(2)(39.8)}{(200\times10^3)(1995.3)(10\times10^3)^2}\right]$$
$$= 1.1\times10^6\ \text{m}^2 = 60.5\ \text{dBsm}$$

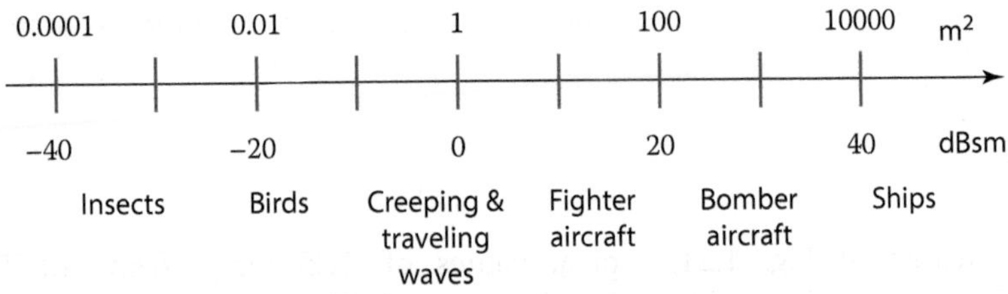

Fig. 1.21 Typical values of RCS for some natural and man-made objects.

Some typical values of RCS are shown in Fig. 1.21 for a variety of natural and man-made targets. In the last example, the target's RCS can be increased six orders of magnitude with jamming before it is detected at the same range as the original 0 dBsm target without jamming. This is equivalent to exchanging the RCS of a sea-skimming missile with that of a ship! However, as pointed out previously, the radar is aware of the jammer's presence and it can take steps to counter the effects of jamming [15].

1.6 General Characteristics of Radar Cross Sections

1.6.1 Definition

The definition of RCS can be stated as

$$\frac{\text{Power reflected to receiver per unit solid angle}}{\text{Incident power density}/4\pi}$$

A more convenient form for RCS calculations is

$$\sigma = \lim_{R\to\infty} 4\pi R^2 \frac{|\vec{W}_s|}{|\vec{W}_i|} \tag{1.37}$$

or, in terms of the incident and scattered electric field intensities $\vec{E}_i$ and $\vec{E}_s$,

$$\sigma = \lim_{R\to\infty} 4\pi R^2 \frac{|\vec{E}_s|^2}{|\vec{E}_i|^2} \tag{1.38}$$

Calculation of RCS is essentially a matter of finding the scattered electric field from a target. If the current induced on the target by the incident plane wave can be determined, the same radiation integrals used in antenna analysis can be applied to compute the scattered field. In the far zone of the target, the scattered field dependence on range will approach $1/R$ and, therefore, σ will be range independent. However, determining the induced current is a difficult problem because Maxwell's equations must be solved for complicated boundary conditions. In most cases, only a numerical solution is possible.

The unit of RCS most commonly used is decibels relative to a square meter (dBsm), which was defined in Eq. (1.15) and repeated here for convenience

$$\sigma,\ \text{dBsm} = 10\log(\sigma,\ \text{m}^2) \tag{1.15}$$

As indicated in Fig. 1.21, typical values of RCS range from 40 dBsm (10,000 m^2) for ships and large bombers to –30 dBsm (0.001 m^2) for insects. Modern radars are capable of detecting flocks of birds and even swarms of insects. The radar's computer will examine all detections and can discard these targets on the basis of their velocity and trajectory. However, the fact that such low RCS targets are detectable implies that the RCS design engineers have a difficult job.

Radar cross section depends on many factors. These include the *frequency* and *polarization* of the incident wave and the *target aspect* (its orientation relative to the radar). These factors are taken into consideration when a radar is designed to detect a specific target. Conversely, if a platform is to face a radar with known specifications, the target can be designed with the radar's performance in mind. For example, if an aircraft will be flying directly at a threat radar in most mission scenarios, it would be wise to put extra effort into reducing the nose-on RCS, perhaps even at the expense of raising it at broadside aspect angles. Furthermore, if the frequency of the radar is known, the RCS reduction effort need concentrate only on the threat frequency.

Example 1.4: Antenna as a Radar Target

Several interesting properties of RCS can be illustrated by considering the special case of an antenna as a radar target. Assume that the antenna mainbeam is pointed directly at the radar, as shown in Fig. 1.22. The gain and effective area in this direction are G_a and A_{ea}. Also, let the terminals

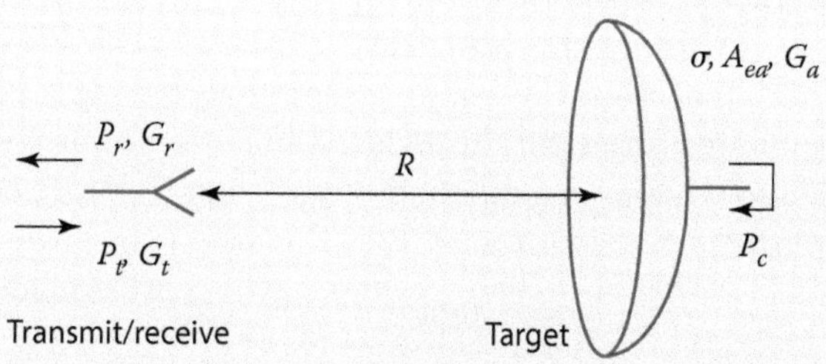

Fig. 1.22 Antenna as a radar target.

(Continued)

Example 1.4: Antenna as a Radar Target *(Continued)*

be shorted so that all the power collected is reradiated. It appears to the radar receiver that the target antenna is transmitting a power P_c, where

$$P_c = \frac{P_t G_t A_{ea}}{4\pi R^2} \tag{1.39}$$

The power scattered by the antenna will have a density back at the radar of

$$W_r = \frac{P_c G_a}{4\pi R^2} \tag{1.40}$$

which yields a received power of

$$P_r = \frac{P_t A_{ea}^2 G_t G_r}{(4\pi R^2)^2} \tag{1.41}$$

The relationship between gain and effective area of the target antenna has been used to obtain Eq. (1.41). Comparing this expression for P_r to the standard radar equation (1.6) yields a relationship between σ and A_{ea}:

$$A_{ea}^2 = \frac{\lambda^2 \sigma}{4\pi} \tag{1.42}$$

$$\sigma = \frac{4\pi A_{ea}^2}{\lambda^2} \tag{1.43}$$

For large, efficient antennas, $A_{ea} \approx A$, the physical aperture area. Equation (1.43) is frequently used to estimate the peak RCS of large, flat surfaces when viewed directly by a radar.

Some general comments on the behavior of RCS can be made based on Eq. (1.43). First, the RCS of a flat surface increases as the square of the area when the surface is viewed normally (i.e., "head on"). Second, because wavelength decreases with frequency, the RCS of a fixed surface area increases with frequency. In Example 1.4 the antenna surface was implicitly assumed to be a perfect electric conductor (PEC) with a reflection coefficient near one ($|\Gamma| \approx 1$). If the surface material were a nonconductor, intuitively one might expect that the RCS would be less than that of a PEC, based on the fact that some transmission through the material would be possible.

With regard to the antenna target, by employing Eq. (1.5), its RCS can be factored as follows:

$$\sigma = \frac{4\pi A}{\lambda^2} A = G_a A \tag{1.44}$$

Equation (1.44) shows that the surface area provides gain and, therefore, the possible square meters of RCS afforded by a surface can be much larger than its physical area in square meters.

Example 1.5: RCS of a Square Plate

Consider a square conducting plate with both edges 1 m long. At 300 MHz ($\lambda = 1$ m) the RCS is

$$\sigma = \frac{4\pi A^2}{\lambda^2} = \frac{4\pi\left[(1)^2\right]^2}{1^2} = 12.56 \text{ m}^2 = 11 \text{ dBsm}$$

At 3 GHz ($\lambda = 0.1$ m) the RCS is

$$\sigma = \frac{4\pi A^2}{\lambda^2} = \frac{4\pi\left[(1)^2\right]^2}{0.1^2} = 1256 \text{ m}^2 = 31 \text{ dBsm}$$

Frequently it is convenient to specify dimensions in terms of wavelength, rather than physical units such as meters. For example, if the square plate has edges that are 10 wavelengths long, its RCS is

$$\sigma = \frac{4\pi\left[(10\lambda)^2\right]^2}{\lambda^2} = 4\pi \times 10^4 \lambda^2 \text{ m}^2$$

Given that the wavelength is not specified, the dimensionless quantity σ/λ^2 is useful:

$$\sigma/\lambda^2 = 4\pi \times 10^4 = 125{,}600 = 51 \text{ dB}$$

Therefore, to find the value of σ at a particular frequency, it is only necessary to compute λ at that frequency and then multiply by λ^2.

1.6.2 Frequency Regions

The scattering characteristics of a target are strongly dependent on the frequency of the incident wave. There are three frequency regions in which the RCS of a target is distinctly different. They are referred to as the 1) *low-frequency*, 2) *resonance*, and 3) *high-frequency* regimes. The labels are somewhat misleading in that low and high are defined relative to the size of the target in terms of the incident wavelength rather than to a physical size. If the target is smooth, with an extent or length of L, the three frequency regimes can be defined as follows:

1. *Low-frequency region* ($kL \ll 1$): At these frequencies, the phase variation of the incident plane wave across the extent of the target is small. Thus, the induced current on the body is approximately constant in amplitude and phase. The particular shape of the body is not important. For example, both a small sphere and a small cube have essentially *isotropic* (directionally independent) scattering patterns. In general, σ vs kL is smooth and varies as $1/\lambda^4$. This region is also called the *Rayleigh region*.
2. *Resonance region* ($kL \approx 1$): For these frequencies, the phase variation of the current across the body is significant, and all parts contribute to the

scattering pattern; σ vs kL will oscillate. This region is also referred to as the *Mie region.*

3. *High-frequency region ($kL \gg 1$)*: There are many cycles in the phase variation of the current across the body and, consequently, the scattered field will be very angle dependent. The peak scattering levels are due primarily to isolated points. For example, the peak scattering from large flat plates originates from *specular* points on the surface. These are the mirrorlike reflection points that satisfy Snell's law. In this region, σ vs kL is smooth and may be independent of λ. This is also called the *optical region.*

The RCS of a sphere (Fig. 1.23) clearly illustrates the three frequency regions just defined. For $ka < 0.5$, where a is the radius, the curve is almost linear but, above 0.5, it begins to oscillate. This is the resonance region. The oscillations die out at higher values and, above $ka = 10$, the curve is essentially a constant equal to πa^2.

The cylinder is a fundamental shape that illustrates the polarization dependence of RCS. Figure 1.24 shows the RCS of a cylinder with a circular cross section of radius a. The vertical axis has been normalized by the cylinder length L and it is assumed that L is sufficiently large so that end effects are negligible. As $ka \to 0$ the cylinder becomes a thin wire, and the *azimuthal* current (the $\hat{\phi}$-component of current in a cylindrical coordinate system with z along the wire axis) is constant. The thin wire scatters only the axial component of the incident field; the transverse (perpendicular) component passes by undisturbed. As $ka \to \infty$ the cylinder approaches a flat surface, in which case both components scatter equally.

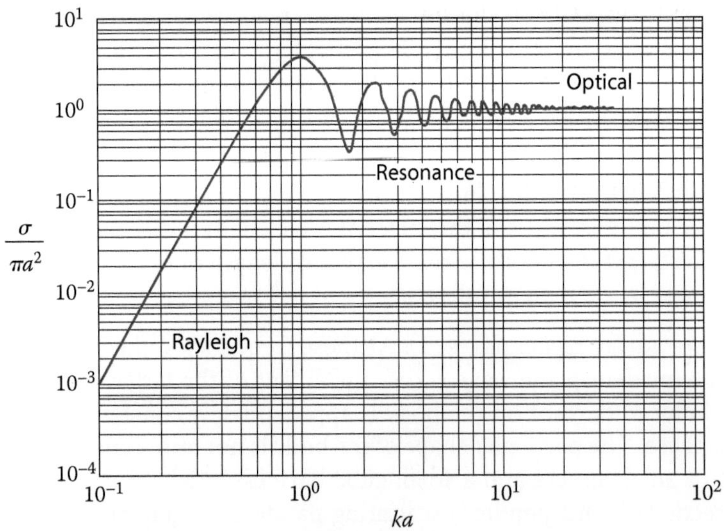

Fig. 1.23 RCS of a perfectly conducting sphere of radius a.

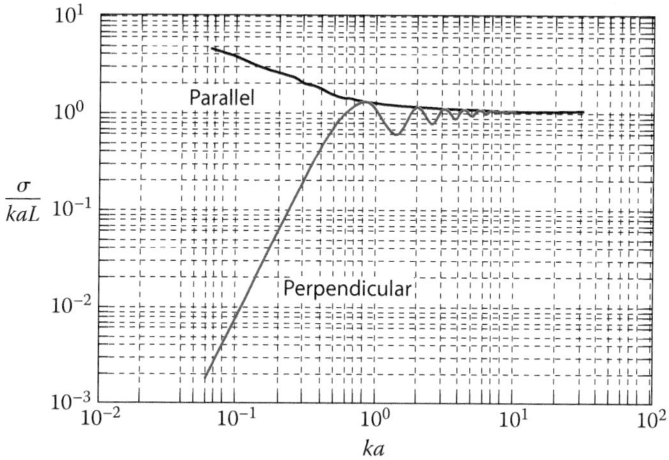

Fig. 1.24 RCS of a perfectly conducting cylinder of radius *a* and unit length.

1.7 Scattering Mechanisms

The RCS of targets encountered by most radars are more complicated than spheres or plates. There are a few exceptions, such as weather balloons and buoys. Simple shapes, however, such as plates, spheres, cylinders, and wires, are useful in studying the *phenomenology* of RCS. Furthermore, *complex targets* can be decomposed into *primitives* (basic geometrical shapes that can be assembled to form a more complex shape).

To find the RCS of any target, it is necessary to determine the total scattered electric field in Eq. (1.38). Chapter 2 explains how the scattered electric field is related to the currents induced on the target by the incident wave. For electrically large targets (i.e., in the optical region) it is possible to express the total electric field as a sum of terms, each one resulting from a scattering source on the target:

$$\left|\vec{E}_S\right| \approx \left|\sum_{n=1}^{N_S} \vec{E}_{S_n}\right| \tag{1.45}$$

where N_S is the number of scattering sources. The terms are a function of frequency and angle, and each term varies with frequency and angle in a different manner. Often individual terms can be associated with a specific physical scattering mechanism, as depicted in Fig. 1.25.

For a fixed frequency and angle, it is not unusual for one term to dominate over all others:

$$|\vec{E}_S|^2 = \left|\sum_{n=1}^{N_S} \vec{E}_{S_n}\right|^2 \approx \sum_{n=1}^{N_S} |\vec{E}_{S_n}|^2 \tag{1.46}$$

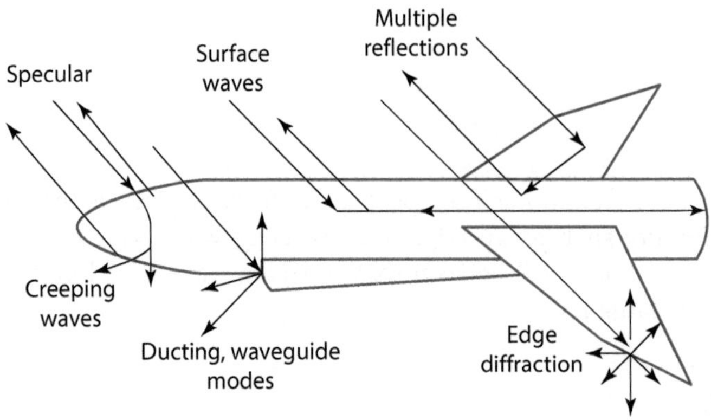

Fig. 1.25 Illustration of important scattering mechanisms.

In the approximation the cross terms in the multiplication are neglected. This is valid when one term is several orders of magnitude greater than all other terms in the summation at the frequency and angle where the sum is being evaluated. Using Eq. (1.46) in the definition of RCS gives

$$\sigma \approx \sigma_1 + \sigma_2 + \sigma_3 + \cdots + \sigma_{N_s} \tag{1.47}$$

Note that it may be the smaller terms that are the most important at the frequency of the threat radar and target aspect angle.

The important scattering mechanisms are briefly described as follows.

1.7.1 Reflections

This mechanism yields the highest RCS peaks, but these peaks are limited in number because Snell's law must be satisfied. When multiple surfaces are present, *multiple reflections* can occur. For instance, the incident plane wave could possibly reflect off the fuselage, hit a fin, and then return to the radar.

1.7.2 Diffractions

Diffracted fields are those scattered from discontinuities such as edges and tips. The waves diffracted from these shapes are less intense than reflected waves, but they can emerge over a wide range of angles. In regions of low RCS, diffracted waves can be significant.

1.7.3 Surface Waves

The term *surface wave* refers to the current traveling along a body and includes several types of waves. In general, the target acts as a transmission line, guiding the wave along its surface. If the surface is a smooth closed shape such as a sphere, the wave will circulate around the body many times. On curved bodies, the surface wave will continuously radiate. These are called *creeping waves* because they appear to creep around the back of a curved body. Radiating surface waves on flat bodies are usually called *leaky waves*. *Traveling waves* appear on slender bodies and along edges and suffer little attenuation as they propagate. If the surface is terminated with a discontinuity such as an edge, the traveling wave will be reflected back toward its origin. It will be seen that traveling wave RCS lobes can achieve surprisingly large levels.

1.7.4 Ducting

Also called *waveguide modes*, ducting occurs when a wave is trapped in a partially closed structure. An example is an air inlet cavity on a jet. Once the wave enters the cavity, many bounces can occur before a ray emerges. The ray can take many paths, and, therefore, rays will emerge at most all angles. The result is a large, broad RCS lobe. An optical analogy of this is the glowing of a cat's eye when it is illuminated by a light.

Sometimes these mechanisms interact with each other. For example, a wave reflected from a flat surface can subsequently be diffracted from an edge or enter a cavity. For a complex target, the interactions are not always obvious. In the following chapters, each of these mechanisms and their interactions will be examined in detail.

1.8 Methods of Radar Cross Section Prediction

1.8.1 General Comments

The analytical methods used to calculate RCS are the same as those used in antenna analysis. A radar target is acting essentially as a reflector antenna, albeit a poor one. The incident wave induces a current on the target, and the induced current radiates a field just as an antenna would. In this case, however, it is called a *scattered* field as opposed to a *radiated* field. The problem is, of course, that the induced current on the target is unknown and generally very difficult to determine. An *analytical* solution is possible in only a few simple cases. In other cases, it is feasible to solve Maxwell's equations or integral equations numerically. These solutions are *rigorous* because no assumptions or restrictions are imposed. The only error in the RCS is due to the numerical solution of the integral or differential equation. The alternative to a numerical solution is to take an educated guess at the current, in which case the resulting RCS is only approximate.

Electromagnetic (EM) scattering and radiation problems can be solved in either the *time domain* (TD) or *frequency domain* (FD). The RCS as a function of time is related to the RCS as a function of frequency by the Fourier

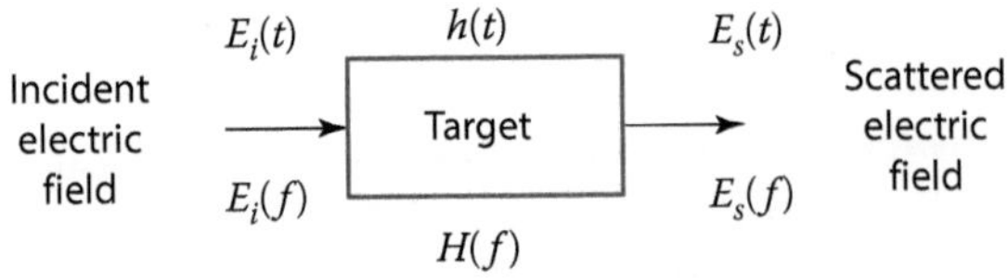

Fig. 1.26 Linear system model of a scattering target.

transform. A simple linear system model of electromagnetic scattering is shown in Fig. 1.26. If the Fourier transform pairs are denoted by a double arrow, then

$$\begin{aligned} E_i(t) &\leftrightarrow E_i(f) \\ h(t) &\leftrightarrow H(f) \\ E_s(t) &\leftrightarrow E_s(f) \end{aligned} \tag{1.48}$$

For a linear system

$$E_s(f) = H(f)E_i(f) \tag{1.49}$$

or

$$H(f) = \frac{E_s(f)}{E_i(f)} \tag{1.50}$$

From Eq. (1.38) it is apparent that the RCS is a function of the magnitude of the target's frequency response

$$\sigma \propto |H(f)|^2 \tag{1.51}$$

Clearly there are several approaches to calculating the RCS. One is to compute the *time impulse response* of the target, $h(t)$, and then take its Fourier transform. The other is to compute the *frequency transfer function*, $H(f)$, directly from the incident and scattered fields. There are advantages and disadvantages to each approach. Early measurement systems operated exclusively in the frequency domain, and the need to compare calculated and measured data dictated that calculations also be performed in the frequency domain. In a *continuous wave* (CW) measurement, the received power is plotted as the target is rotated through a range of angles while a fixed-frequency source is radiating. The frequency can be changed and another plot of RCS vs angle obtained, and so on. At any fixed angle, the RCS from a large range of frequencies can be Fourier transformed to obtain the time-domain response of the target at that particular angle. Modern measurement instrumentation is computer-controlled and capable of several operating modes. In some cases, a radar image of the target can be generated that identifies the sources of scattering and superimposes it onto an outline of the target, as shown in Chapter 8.

1.8.2 Common Methods of Prediction

Several methods of RCS prediction are described briefly in this section. The classical solution techniques are not discussed in this text because most of them are limited to one- or two-dimensional structures or simple three-dimensional shapes. The methods of interest here are those that can be applied to arbitrary three-dimensional targets. The methods most commonly encountered are *physical optics, microwave optics (ray tracing)*, the *method of moments*, the *finite element* method, and *finite difference* methods.

1.8.2.1 Physical Optics

One method of estimating the surface current induced on an arbitrary body is the physical optics (PO) approximation. On the portions of the body that are directly illuminated by the incident field, the induced current is simply proportional to the incident magnetic field intensity. On the shadowed portion of the target, the current is set to zero. The current is then used in the radiation integrals to compute the scattered field far from the target.

Physical optics is a high-frequency approximation that gives best results for electrically large bodies ($L \geq 10$). It is most accurate in the specular direction. Because PO abruptly sets the current to zero at a shadow boundary, the computed field values at wide angles and in the shadow regions are inaccurate. Furthermore, surface waves are not included. Physical optics can be used in either the time or frequency domains.

1.8.2.2 Microwave Optics

Ray-tracing methods that can be used to analyze electrically large targets of arbitrary shape are referred as microwave optics. The rules for ray tracing in a *simple* medium (linear, homogeneous, and isotropic) are similar to reflection and refraction in optics. In addition, diffracted rays are allowed that originate from the scattering of the incident wave at edges, corners, and vertices. The formulas are derived on the basis of infinite frequency ($\lambda \to 0$). This implies an electrically large target. Ray optics is frequently used in situations that severely violate this restriction and still yields surprisingly good results. The major disadvantage of ray tracing is the bookkeeping required for a complex target. It is used primarily in the frequency domain.

1.8.2.3 Method of Moments

A common technique used to solve an integral equation is the method of moments (MM). Integral equations are so named because the unknown quantity is in the integrand. In electromagnetics, integral equations are derived from Maxwell's equations and the boundary conditions. The unknown quantity can be an electric or magnetic current (either volume or surface). The MM reduces the integral equations to a set of simultaneous linear equations that can be solved using standard matrix algebra. The size of the matrix involved depends on the size of the body; computer capabilities limit the size of the targets that can be modeled.

Most MM formulations require a *discretization* (segmentation) of the body. Therefore, they are compatible with finite element methods used in structural engineering, and the two are frequently used in tandem during the design of a platform in a process called *concurrent engineering.* The MM can be used to solve both time- and frequency-domain integral equations.

1.8.2.4 Finite Element Method

The finite element method (FEM) is perhaps the oldest of all numerical methods. It is used in many engineering disciplines. The requirements for electromagnetic problems are significantly different than those for other engineering applications, and it has been only recently that FEM has achieved widespread use in EM. FEM is generally applied in the frequency domain starting with the vector wave equation. As in the case of MM, a matrix equation must be solved. Unlike MM, the resulting matrix equation for FEM is sparse (i.e., populated with a large number of zeros), which allows the application of very efficient and fast matrix solvers.

1.8.2.5 Finite Difference Methods

Finite differences are used to approximate the differential operators in Maxwell's equations in either the TD or FD. As in the MM, the target must be discretized. Maxwell's equations and the boundary conditions are enforced on the surface of the target and at the boundaries of the discretization cells. This method has found extensive use in computing the transient response of targets to various waveforms. Finite difference does not require the large matrices that the MM does because the solution is stepped in time throughout the scattering body.

Because Maxwell's equations can be written in both differential and integral form, one could numerically solve the integral forms rather than the differential forms. When starting with the integral form of Maxwell's equations, the method is known as the finite integration technique (FIT).

1.8.3 Sample RCS Pattern of an Aircraft

Figure 1.27 shows an aircraft constructed of geometrical components such as cones, plates, and spheres. A collection of basic shapes will give an acceptable RCS estimate that can be used during the initial design stages of a platform. The locations and levels of the largest RCS lobes are of most concern at this stage of the design process. The accuracy of the RCS calculation at other angles will depend on how the interactions between the various shapes are handled. These interactions are more difficult

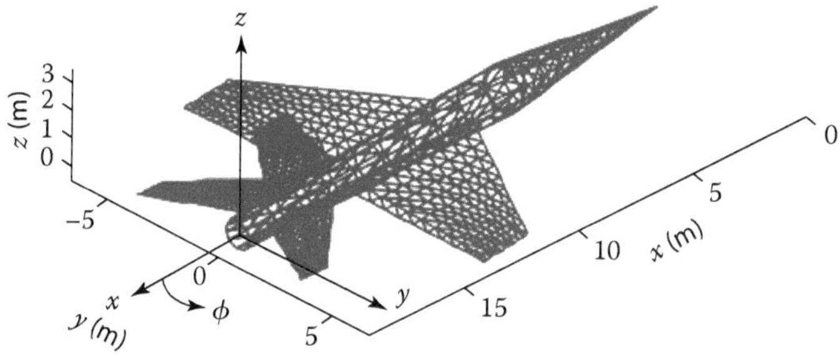

Fig. 1.27 Geometrical components model of an aircraft.

to include, as will be seen later and, even when they are included, the agreement with the measured RCS is not particularly good. For a low-observable target, more sophisticated numerical techniques must be used.

Figures 1.28 and 1.29 show the monostatic and bistatic azimuth (x–y plane) RCS patterns of the aircraft for vertical polarization, $\vec{E} = E_z\hat{z}$. In the x–y plane $\theta = 90^\circ$, so that a transformation from Cartesian to spherical coordinates gives $\vec{E} = E_\theta\hat{\theta}$. The MM was used to compute the pattern at 1 GHz. In Fig. 1.28 the lobes at 90° and 270° are due to the large reflections that occur when the sides of the fuselage and vertical tail are viewed directly. Also note the rapid fluctuation of RCS with angle. Ideally, there should be symmetry between the left and right sides of the aircraft. Symmetry is lost at low values of RCS because the discretization of the target (i.e., arrangement

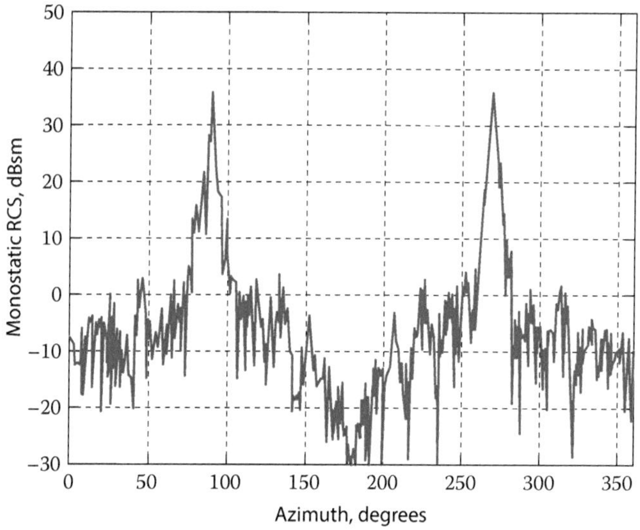

Fig. 1.28 Monostatic RCS pattern of the geometrical components aircraft in the x–y plane.

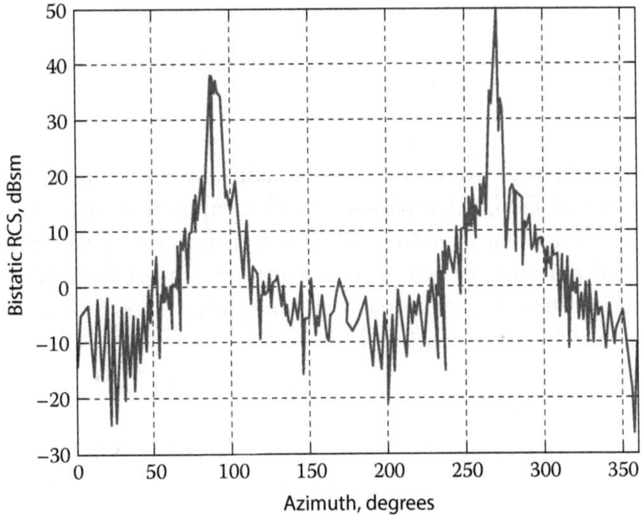

Fig. 1.29 Bistatic RCS pattern of the geometrical components aircraft in the x–y plane.

of the triangular patches) is not necessarily identical on the two halves of the aircraft.

The bistatic RCS has some significant differences when compared to the monostatic RCS. In Fig. 1.29 the incident wave (transmitter) is fixed at 90° while the receiver moves around the aircraft in the x–y plane. When the receiver is at 90° a *back-scatter* condition exists (referring to Fig. 1.1, the bistatic angle β is 0 deg) and the RCS should be the same as the monostatic case because the receiver and transmitter are colocated. When the receiver is at 270 deg, a *forward-scatter* condition exists (in Fig. 1.1, the bistatic angle is 180 deg). The RCS is greater than that for the monostatic case. For the forward-scattered case a bistatic radar has an advantage over a monostatic radar. A radar operating in this condition is called a *bistatic fence.* It is quite restrictive because it requires the target to pass between the transmitter and receiver.

Finally, note that the RCS lobe beamwidths are wider for a bistatic pattern than they are for a monostatic pattern, a fact that can help improve the probability of detecting the target. The greater the number of angles that have a high RCS (above some threshold), the higher the probability that the radar will detect the target.

1.9 Target Scattering Matrices

As discussed in Sec. 1.3, any two orthogonal components of the electric field can be used to represent any arbitrary electric field vector. Complex targets depolarize the incident electric field. For example, the scattered field can have both horizontal and vertical polarizations even if the incident field is purely vertically polarized. The many edges and curved surfaces of a complex target excite currents with many vector directions, and the resulting scattered field will no longer be a pure polarization.

The cross polarization that arises on scattering contains information about the target, and to extract the maximum amount of information from the scattered field, the cross-polarized components must be processed. The radar antenna must have the ability to measure both polarizations (i.e., a *dual-polarized* antenna) or two separate antennas must be used.

A *scattering matrix* can be used to completely specify the relationship between the incident and scattered fields:

$$\begin{bmatrix} E_{s\theta} \\ E_{s\phi} \end{bmatrix} = \begin{bmatrix} S_{\theta\theta} & S_{\theta\phi} \\ S_{\phi\theta} & S_{\phi\phi} \end{bmatrix} \begin{bmatrix} E_{i\theta} \\ E_{i\phi} \end{bmatrix} \tag{1.52}$$

The S_{pq} are the scattering parameters, and p and q can be either θ or ϕ (p, $q = \theta$, ϕ). The first index specifies the polarization of the receive antenna and the second the polarization of the incident wave. For example, $S_{\theta\theta}$ relates $E_{s\theta}$ to $E_{i\theta}$ and thus represents the response of a θ-polarized receive antenna when the target is illuminated by a pure θ-polarized incident wave.

The elements of the scattering matrix are complex quantities that can be determined by analysis or measurement. In terms of the RCS,

$$S_{pq} = \frac{\sqrt{\sigma_{pq}}}{\sqrt{4\pi R^2}} \tag{1.53}$$

In polar notation (amplitude and phase),

$$S_{pq} = \frac{|\sigma_{pq}|^{\frac{1}{2}} e^{j\psi_{pq}}}{\sqrt{4\pi R^2}} \tag{1.54}$$

$$\psi_{pq} = \arctan\left[\frac{\mathrm{Im}(\sigma_{pq})}{\mathrm{Re}(\sigma_{pq})}\right] \tag{1.55}$$

where Re and Im are the real and imaginary operators.

Example 1.6: Scattering Matrix of a Thin Wire

Consider a thin wire with its axis along the z axis as shown in Fig. 1.30. If the thin-wire approximation holds, then a current is induced on the wire only when there is a $\hat{\theta}$-component of the incident field: $\vec{E}_i = E_{\theta_i}\hat{\theta}$. Therefore,

$$E_{s\theta} = S_{\theta\theta} E_{i\theta} \tag{1.56}$$

(Continued)

Example 1.6: Scattering Matrix of a Thin Wire *(Continued)*

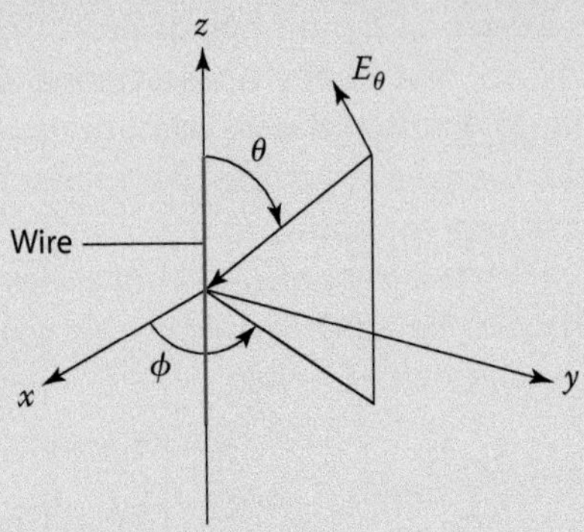

Fig. 1.30 Scattering from a thin wire.

and

$$E_{s\phi} = 0 \tag{1.57}$$

Consequently,

$$S = \begin{bmatrix} S_{\theta\theta} & S_{\theta\phi} \\ S_{\phi\theta} & S_{\phi\phi} \end{bmatrix} = \begin{bmatrix} \dfrac{\sqrt{\sigma_{\theta\theta}} e^{j\psi_{\theta\theta}}}{\sqrt{4\pi R^2}} & 0 \\ 0 & 0 \end{bmatrix} \tag{1.58}$$

References

[1] Harrington, R. F., *Time-Harmonic Electromagnetic Fields*, McGraw-Hill, New York, 1961.
[2] Ishimaru, A., *Electromagnetic Wave Propagation and Scattering*, Prentice-Hall, Englewood Cliffs, NJ, 1991.
[3] Balanis, C. A., *Advanced Engineering Electromagnetics*, Harper and Row, San Francisco, 1982.
[4] Felsen, L., and Marcuvitz, N., *Radiation and Scattering of Waves*, Prentice-Hall, Englewood Cliffs, NJ, 1973.
[5] Ruck, G., Barrick, D., Stuart, W., and Kirchbaum, C., *Radar Cross Section Handbook*, Plenum, New York, 1970.
[6] Knott, E., Shaeffer, J., and Tuley, M., *Radar Cross Section*, Artech House, Norwood, MA, 1993.
[7] Ghattacharyya, A., and Sengupta, D., *Radar Cross Section Analysis & Control*, Artech House, Norwood, MA, 1991.
[8] Sadiku, M., *Numerical Techniques in Electromagnetics*, CRC Press, Boca Raton, FL, 1992.
[9] Maffett, A., *Topics for a Statistical Description of Radar Cross Section*, Wiley Interscience, New York, 1989.
[10] Bowman, J., Senior, T., and Uslenghi, P., *Electromagnetic and Acoustic Scattering by Simple Shapes*, Hemisphere, Bristol, PA, 1969.

[11] Skolnik, M., *Introduction to Radar Systems*, 3rd ed., McGraw-Hill, NY, 2000.
[12] Phillips, C. L., and Parr, J. M., *Signals, Systems, and Transforms*, 2nd ed., Prentice-Hall, Englewood Cliffs, NJ, 1995.
[13] IEEE Standard 521–2002, *Standard Letter Designations for Radar-Frequency Bands*, https://standards.ieee.org/standard/521-2002.html
[14] Johnson, J. T., "A Study of the Four-Path Model for Scattering From an Object Above a Half-Space," *Microwave and Optical Technology Letters*, Vol. 30, No. 2, July 20, 1001, pp. 130–134.
[15] Golden, A., Jr., *Radar Electronic Warfare*, AIAA Education Series, AIAA, New York, 1987.
[16] Harger, R., "Harmonic Radar Systems for Near-Ground In-Foliage Nonlinear Scatterers" *IEEE Transactions on AES*, Vol. AES-12, No. 2, March 1976, p. 230.
[17] Kell, R., "On the Derivation of Bistatic RCS from Monostatic Measurements," *Proceedings of the IEEE*, Vol. 53, Aug. 1965, p. 983.
[18] Mack, C., and Reiffen, B. "RF Characteristics of Thin Dipoles," *Proceedings of the IEEE*, Vol. 52, May 1964, p. 533.

Problems

1.1 A target on the horizon has an RCS of 9 m^2. A monostatic radar's antenna is at a height of 60 ft above the ocean. The detection level of the radar receiver is −120 dBm at 1 GHz. What is the required gain of the antenna if 10 W is transmitted? (The distance to the horizon in miles for standard atmospheric conditions is approximately $\sqrt{2h}$, where h is the height in feet.)

1.2 (From Ref. [16].) Harmonic noise is present in all man-made objects. Its source is the imperfect contact between conducting parts. Because foliage and other natural clutter sources have no harmonic noise, a radar system that detects a harmonic of the transmitted wave will automatically eliminate return from this clutter.

In this problem, we examine the feasibility of such a system. Consider a free-space monostatic radar that transmits frequency f_0 but receives the third harmonic ($=3f_0$). (The conversion efficiency from the fundamental to the third harmonic is higher than for any other harmonics.) Define the following quantities:

P_t = transmitted power at f_0
G_t = transmit antenna gain in the direction of the target
G_r = gain of the receive antenna at the third harmonic in the direction of the target
λ_0 = wavelength at the fundamental frequency
λ_3 = wavelength at the third harmonic frequency
R = target range
P_r = received third harmonic power

a) What is the fundamental frequency power density W_i at the target (a distance R from the transmitter)?

b) The scattering cross section for the nth harmonic is

$$\sigma_n(W_i) = \sigma_h W_i^{\alpha-1} \tag{1.59}$$

where σ_h is called the harmonic scattering pseudo–cross section and α a nonlinearity parameter. These two quantities are determined by measurement, and commonly occurring values for $n = 3$ are α of about 2.5 and σ_h from –60 to –90 dB for incident power levels below 1 W/m^2. Assume $\alpha = 2.5$, and derive an expression for the received third harmonic power in terms of σ_h and the quantities defined earlier. For a standard radar system, the range dependence is R^{-4}. What is the range dependence in this case?

1.3 The radar cross section of a ship, with its bow in the $+y$ direction, at a frequency of 1 GHz is given by

$$\sigma(\phi) = \begin{cases} 1000|\text{sinc}(20.5 \sin \phi)|^{1.5}\sqrt{|\cos \phi|}, & 0 \text{ deg} \leq \phi \leq 180 \text{ deg} \\ 1000|\text{sinc}(20.5 \sin \phi)|^{1.5}, & 180 \text{ deg} \leq \phi \leq 360 \text{ deg} \end{cases}$$

where sinc(X) is defined in Eq. (1.19).

a) Generate two polar plots of the RCS as a function of ϕ, one a linear RCS scale (square meters) and the other in dB.

b) Plot the detection range of the target as a function of aspect angle. (Think of the target as being fixed and the radar circling the target.) The radar parameters are as follows:

transmitter power = 10 kW
antenna gain = 30 dB (monostatic)
minimum detectable signal (MDS) = 1 nW (10^{-9} watt)

c) Plot the detection contour if the RCS is reduced by 20 dB, that is,

$$\sigma(\phi) = \begin{cases} 10|\text{sinc}(20.5 \sin \phi)|^{1.5}\sqrt{|\cos \phi|}, & 0 \text{ deg} \leq \phi \leq 180 \text{ deg} \\ 10|\text{sinc}(20.5 \sin \phi)|^{1.5}, & 180 \text{ deg} \leq \phi \leq 360 \text{ deg} \end{cases}$$

1.4 Consider a monostatic radar tracking a target with radar cross section σ. The radar antenna main beam gain is G_0 and the main beam is on the target.

a) How much must the RCS be reduced to halve the detection range if all other parameters remain the same?

b) The target has a noise jammer on board that uses an isotropic antenna ($G_J = 0$ dB). How much can the jammer power be reduced if the RCS is reduced by 10 dB?

1.5 A 3-GHz radar antenna is mounted on a mast that is 50 m above the ocean surface. The radar detects a target whose height is 500 m at a range of 11.5 km. What is the path gain factor if the ocean reflection coefficient is −1?

1.6 a) Find the electric field attenuation constant for the atmosphere α if the loss is 0.01 dB/km.

b) A radar has a maximum detection range of 50 nautical miles (1 nautical mile = 1852 m) in clear conditions. Estimate its detection range if the atmospheric loss is 0.01 dB/m. [*Hint*: Expand the atmospheric loss factor $\exp(-4\alpha R)$ into a series and keep only the first two terms.]

1.7 Find the backscatter (monostatic radar cross section) of the following objects in square meters at 5 GHz:

a) Looking directly at a circular flat plate with a diameter of 20λ.

b) A sphere with a radius of 5λ.

c) A sphere with a radius of 0.05λ.

1.8 (From Ref. [17].) The *bistatic–monostatic equivalence* theorem states that "for perfectly conducting bodies that are sufficiently smooth, in the limit of vanishing wavelength, the bistatic cross section is equal to the monostatic cross section at the bisector of the bistatic angle."

Consider a TM_z wave incident on the following targets. Use the preceding rule to estimate the bistatic RCS at $\theta = 90$ deg:

a) A sphere with $ka = 10$.

b) A circular disk in the $z = 0$ plane for $\theta_i = 0$. [*Hint*: $E_{s\theta}(\theta = 90 \text{ deg}) \approx 0$ if the disk is very thin.]

1.9 (From Ref. [17].) The bistatic–monostatic equivalence theorem stated in Problem 1.4 can be used to derive a bistatic RCS from monostatic measurements. As a simple example, consider an arbitrary target to be composed of discrete scattering sources that behave like point scatterers. The incident wave direction is given by the unit vector $\hat{k}_i$ and the direction to the observation point by $\hat{k}_s$. The angle between the two vectors is β, as shown in Fig. P1.9. Define the z axis as the bisector between the two unit vectors. The total RCS of the target will be the *vector sum* of the individual scattered fields:

$$\sigma = |\sum_m \sqrt{\sigma_m} e^{j\psi_m}|^2$$

a) If z_m is the z coordinate of the mth scatterer, what is the two-way phase difference for a z-traveling wave?

b) Use the result of part a to describe how one could obtain the bistatic RCS by measuring the monostatic RCS at a frequency lower by an amount $\cos(\beta/2)$.

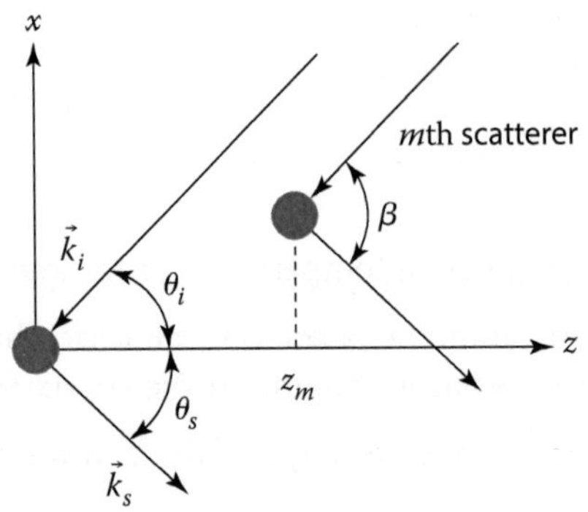

Fig. P1.9

1.10 (From Ref. [18].) A short cylindrical dipole of length L and diameter d is located along the z axis. Its RCS is given by

$$\sigma(\theta, \phi) = \sigma_{\text{max}} \sin^2(\theta)$$

where

$$\sigma_{\text{max}} = \frac{\pi^5 L^6}{16\lambda^4[\log(2L/d) - 1]^2}$$

a) Find σ_{max} for $L = 0.25\lambda$ and $L/d = 15$.

b) Find the average RCS over all angles:

$$\overline{\sigma} = \frac{1}{4\pi}\int\int \sigma(\theta, \phi) \sin\theta \, d\theta \, d\phi$$

c) A chaff cloud is composed of a number of these dipoles. Calculate the number required to achieve an average cloud cross section of 20 dBsm at a frequency of 2 GHz.

1.11 A communication system downlink from an unmanned aerial vehicle (UAV) (transmitting) to a ground station (receiving) has the following parameters:

Transmitted power = 1 W
UAV antenna gain = 3 dB
Operating frequency = 1 GHz
Receiver noise figure = 8 dB
Bandwidth = 50 MHz
Receive antenna gain = 25 dB
Losses on the transmit side (UAV) = 2 dB

Losses on the receive side (ground station) = 3 dB
Required minimum signal-to-noise ratio, $\text{SNR}_{\text{min}} = 15\ dB$

Assume that all antennas are polarization matched and there are no antenna mismatches.

a) What is the effective temperature of the receiver?

b) The ground antenna is looking at a sky background such that $T_A = 300$ K. What is the total noise power in the receiver?

c) What is the maximum range of the communication system?

1.12 A noise jammer is operating against the communication system defined in problem 1.11 (see Fig. 1.16). The jammer transmits noise with a ERP_J of 1 W and located in a direction where the ground station antenna has a gain, G_s, 23 dB below the peak gain ($G_0 = 25$ dB as given in problem 1.11). Find the burnthrough range. You can assume that the bandwidths of the jammer and communication system are the same.

Chapter 2 Basic Theorems, Concepts, and Methods

2.1 Introduction

Before discussing the various aspects of radar cross section prediction and control, it is helpful to introduce a few fundamental concepts. Many of these methods will be familiar from antenna studies, in which case this chapter serves as a review. However, the discussions and examples in this chapter will be directed toward RCS problems. The application of these methods to RCS can be substantially different from their corresponding antenna applications.

As in antenna theory, we consider fields to be set up by a combination of *impressed* electric and (in general) magnetic currents. Currents are *induced* such that the fields satisfy the boundary conditions and Maxwell's equations. The *radiation integrals* provide a means of calculating the fields due to a prescribed set of currents. Unfortunately, the currents on a body are usually not known. It is necessary to find the currents first or, at least, to provide an estimate of them. Finding the current involves solving a *differential* or *integral equation*; these methods will be discussed in Chapters 3 and 4. A good guess for the current is provided by the *physical optics* approximation presented in Sec. 2.10. First, several theorems are introduced that will be used to advantage in subsequent sections to formulate and solve scattering problems.

2.2 Uniqueness Theorem

In general, the *uniqueness theorem* states that a solution to Maxwell's equations that satisfies the boundary conditions is unique. More specifically of interest to us is the following result: For a linear, dissipative region V, enclosed by a surface S, that contains impressed current sources $\vec{J}$ and $\vec{J}_m$, the fields inside are determined uniquely by the tangential component of the electric (or magnetic) field intensity on the surface and the impressed currents inside. The situation is illustrated in Fig. 2.1, where double-tipped arrows represent magnetic currents.

The uniqueness theorem allows us to replace the impressed sources outside a closed surface with currents flowing on the boundary S. According to the theorem, the problem can be solved by using an electric current everywhere on the surface, a magnetic current everywhere on the surface, or a

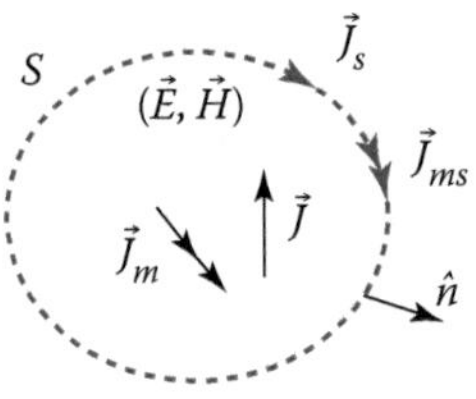

Fig. 2.1 Uniqueness theorem.

combination of electric and magnetic currents on discrete parts of the surface.

The theorem is actually more general than it is as stated here. It can be applied to lossless media and open surfaces as well. Two related theorems are of practical interest:

1. If a source-free region is completely occupied with *dissipative* media and if $\vec{E}_{\text{tan}}$ and $\vec{H}_{\text{tan}}$ are zero everywhere on the surface that encloses the region, then, the electromagnetic field vanishes everywhere in the region.
2. If a source-free region is completely occupied with *lossless* media and if $\vec{E}_{\text{tan}}$ and $\vec{H}_{\text{tan}}$ are zero everywhere on the surface that encloses the region, then, the electromagnetic field in the region is either zero or a combination of resonant mode fields. (A resonant mode field is one in which the time-averaged energy stored in the electric field is equal to the time-averaged energy stored in the magnetic field.)

The beauty of the uniqueness theorem is that it assures us that there is only one answer to a problem; the method of solution is inconsequential, even if it is a guess.

2.3 Reciprocity Theorem

Let there be two sets of currents such that $(\vec{E}_1, \vec{H}_1)$ are due to $(\vec{J}_1, \vec{J}_{m1})$ and $(\vec{E}_2, \vec{H}_2)$ are due to $(\vec{J}_2, \vec{J}_{m2})$. From Maxwell's second equation, assuming a linear and isotropic medium,

$$-\nabla \cdot (\vec{E}_1 \times \vec{H}_2 - \vec{E}_2 \times \vec{H}_1) = \vec{E}_1 \cdot \vec{J}_2 + \vec{H}_2 \cdot \vec{J}_{m1} - \vec{E}_2 \cdot \vec{J}_1 - \vec{H}_1 \cdot \vec{J}_{m2}. \quad (2.1)$$

This [Eq. (2.1)] is the *Lorentz reciprocity theorem.*

If the sources are enclosed in a finite volume V and the fields are observed in the far field (no $\hat{r}$ components), the divergences are zero and the left-hand side of Eq. (2.1) vanishes. Rearranging the right-hand side and integrating give

$$\iiint_V \left(\vec{E}_1 \cdot \vec{J}_2 - \vec{H}_1 \cdot \vec{J}_{m2}\right) \mathrm{d}v = \iiint_V \left(\vec{E}_2 \cdot \vec{J}_1 - \vec{H}_2 \cdot \vec{J}_{m1}\right) \mathrm{d}v \quad (2.2)$$

The preceding equation still holds if the subscripts 1 and 2 are interchanged. This implies that, in general, the *response of a system to a source is unchanged if the source and observer are interchanged.* As applied to antennas, the receiving pattern of an antenna constructed of linear isotropic materials is identical to its transmitting pattern.

2.4 Duality Theorem

Symmetries exist between electric and magnetic field quantities and equations. This allows us to apply the solution of an electric (or magnetic) problem to its *dual problem.* Dual quantities are listed in Table 2.1.

Table 2.1 Dual Quantities

Electric Sources	Magnetic Sources
$\vec{E}$	$\vec{H}$
$\vec{H}$	$-\vec{E}$
$\vec{J}$	$\vec{J}_m$
ϵ	μ
μ	ϵ
k	k
η	$1/\eta$

Example 2.1: Duality Applied to Maxwell's Second Equation

Maxwell's second equation is

$$\nabla \times \vec{H} = \vec{J} + j\omega\epsilon\vec{E} \tag{2.3}$$

After substitution of dual quantities into Eq. (2.3), Maxwell's first equation is obtained:

$$-\nabla \times \vec{E} = \vec{J}_m + j\omega\mu\vec{H} \tag{2.4}$$

A less trivial application of the duality theorem is *Babinet's principle,* which is illustrated in Fig. 2.2. A slot of width w in an infinite ground plane and a strip of width w in free space are *complementary structures.* A joining of the two surfaces results in a continuous ground plane. Babinet's principle gives a relationship between the scattering from complementary structures. Thus, it is only necessary to solve one of two problems.

In Fig. 2.2, a transverse electromagnetic (TEM) plane wave with fields $(\vec{E}_i, \vec{H}_i)$ is incident on a slot of width w and length l. The transmitted fields are $(\vec{E}_t, \vec{H}_t)$. Next, the same plane wave with $(\vec{E}_i, \vec{H}_i)$ is incident on a strip of width w and length l. Note that the strip is rotated 90 deg relative to its cutout in the ground plane (which is equivalent to rotating the polarization of the incident wave by 90 deg). In this case, the transmitted fields are $(\vec{E}'_t, \vec{H}'_t)$. Babinet's principle relates the two transmitted fields and the

(Continued)

Example 2.1: Duality Applied to Maxwell's Second Equation *(Continued)*

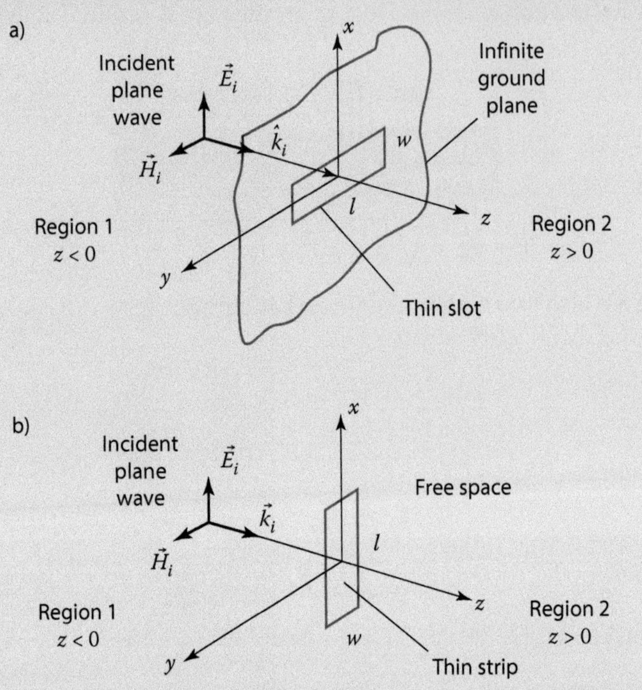

Fig. 2.2 Illustration of Babinet's principle for complementary structures. (a) A plane wave illuminating a thin aperture in an infinite conducting plane. (b) A plane wave incident on a thin strip.

incident field:

$$|\vec{E}_t| + |\vec{E}_t'| = |\vec{E}_i| \tag{2.5}$$

Thus, if the scattered field from the strip ($\vec{E}_t'$) is known, it is straightforward to compute $\vec{E}_t$.

2.5 Radiation Integrals

2.5.1 General Form

The *radiation integrals* or *Stratton–Chu integrals* are integral solutions to Maxwell's equations. They can be derived directly by taking the curl of Maxwell's first two equations, using the vector Green's theorem, and then integrating [1]. In this section, they will be derived via *potentials*.

Recall that the *magnetic vector potential* is given by

$$\vec{A}(x, y, z) = \frac{\mu}{4\pi} \iiint_V \vec{J}(x', y', z') \frac{e^{-jkR}}{R} \, dv' \tag{2.6}$$

and its dual, the *electric vector potential,* is

$$\vec{F}(x, y, z) = \frac{\epsilon}{4\pi} \iiint_V \vec{J}_m(x', y', z') \frac{e^{-jkR}}{R} \, dv' \tag{2.7}$$

where

$$R = |\vec{r} - \vec{r}'| = \sqrt{(x - x')^2 + (y - y')^2 + (z - z')^2} \tag{2.8}$$

There are also two scalar potentials that contribute to the field at (x, y, z). They are the *scalar electric potential,*

$$\Phi_e(x, y, z) = \frac{1}{4\pi\epsilon} \iiint_V \rho_v(x', y', z') \frac{e^{-jkR}}{R} \, dv' \tag{2.9}$$

and the *scalar magnetic potential,*

$$\Phi_m(x, y, z) = \frac{1}{4\pi\mu} \iiint_V \rho_{mv}(x', y', z') \frac{e^{-jkR}}{R} \, dv' \tag{2.10}$$

The charges and currents satisfy the continuity equations

$$\nabla \cdot \vec{J} = -j\omega\rho_v \tag{2.11}$$

and

$$\nabla \cdot \vec{J}_m = -j\omega\rho_{mv} \tag{2.12}$$

The vector and scalar potentials are related through the *Lorentz gauge*:

$$\nabla \cdot \vec{A} = -j\omega\mu\epsilon\Phi_e \tag{2.13}$$

and

$$\nabla \cdot \vec{F} = -j\omega\mu\epsilon\Phi_m \tag{2.14}$$

By combining Eqs. (2.6–2.14), the electric and magnetic fields at an observation point in an unbounded homogeneous medium can be expressed entirely in terms of the currents:

$$\vec{E} = -j\omega\vec{A} - \frac{j}{\omega\mu\epsilon}\nabla(\nabla \cdot \vec{A}) - \frac{1}{\epsilon}\nabla \times \vec{F} \tag{2.15}$$

$$\vec{H} = -j\omega\vec{F} - \frac{j}{\omega\mu\epsilon}\nabla(\nabla \cdot \vec{F}) + \frac{1}{\mu}\nabla \times \vec{A} \tag{2.16}$$

Note that the ∇ operator operates on the observation coordinates (x, y, z).

2.5.2 Far-Zone Scattered Fields

For RCS calculations, the observation point will be in the far zone of the target which is located at the origin of the coordinate system. In this case, the vectors $\vec{R}$ and $\vec{r}$ are approximately parallel, as shown in Fig. 2.3

$$R = |\vec{R}| = |\vec{r} - \vec{r}'| \approx |\vec{r}| - \vec{r}' \cdot \hat{r} \tag{2.17}$$

Using this approximation in the integrals for the potentials $\vec{A}$ and $\vec{F}$, and inserting them into Eq. (2.15) yield

$$\vec{E}(x, y, z) = \frac{-j\omega\mu}{4\pi r} e^{-jkr} \iiint_V \left[\vec{J} - (\vec{J} \cdot \hat{r})\hat{r} + \frac{1}{\eta}\vec{J}_m \times \hat{r} \right] e^{jk\vec{r}' \cdot \hat{r}} \mathrm{d}v' + O\left(\frac{1}{r^2}\right) \tag{2.18}$$

In Eq. (2.18), the position vector to a source point is

$$\vec{r}' = \hat{x}x' + \hat{y}y' + \hat{z}z' \tag{2.19}$$

and the unit vector in the direction of the observation point is

$$\hat{r} = \hat{x}u + \hat{y}v + \hat{z}w \tag{2.20}$$

where u, v, w are the direction cosines of the observation point (see Appendix B):

$$\begin{aligned} u &= \sin\theta\cos\phi \\ v &= \sin\theta\sin\phi \\ w &= \cos\theta \end{aligned} \tag{2.21}$$

The components of the electric field tangential to a sphere at radius r are

$$E_\theta(x, y, z) = \frac{-jk\eta}{4\pi r} e^{-jkr} \iiint_V \left[\vec{J} \cdot \hat{\theta} + \frac{1}{\eta}\vec{J}_m \cdot \hat{\phi} \right] e^{jk\vec{r}' \cdot \hat{r}} \mathrm{d}v' \tag{2.22}$$

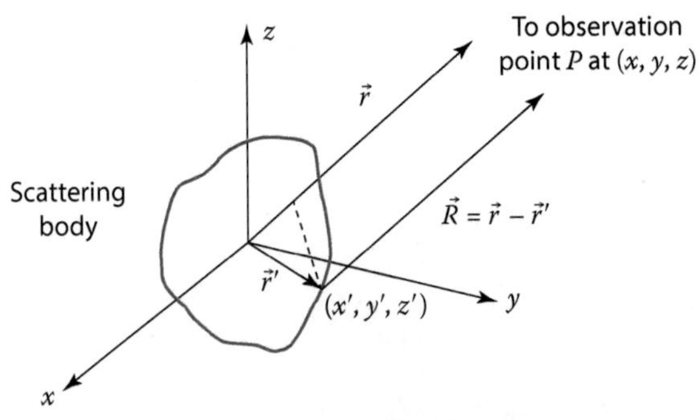

Fig. 2.3 Target far-field approximation.

$$E_\phi(x, y, z) = \frac{-jk\eta}{4\pi r} e^{-jkr} \iiint_V \left[\vec{J} \cdot \hat{\phi} + \frac{1}{\eta} \vec{J}_m \cdot \hat{\theta}\right] e^{jk\vec{r}' \cdot \hat{r}} \mathrm{d}v' \tag{2.23}$$

Because the observation point is assumed to be in the far field, the spherical wave front scattered from the target can be considered planar in the vicinity of the observation point. Thus, the wave is approximately TEM, and the vectors $\vec{E}$, $\vec{H}$, and $\hat{r}$ are orthogonal, so that

$$\vec{H}(x, y, z) \approx \frac{\hat{r} \times \vec{E}(x, y, z)}{\eta} \tag{2.24}$$

If $\vec{J}_m = 0$ then Eqs. (2.22) and (2.23) can be written concisely as

$$\vec{E}(r, \theta, \phi) = \frac{-jk\eta}{4\pi r} e^{-jkr} \iiint_V \vec{J} e^{jkg} \mathrm{d}v' \tag{2.25}$$

where

$$g = \vec{r}' \cdot \hat{r} = x'u + y'v + z'w \tag{2.26}$$

has been defined for convenience, and it is understood that the component E_r is ignored.

2.6 Superposition Theorem

The superposition theorem states that, if a medium is linear, the field intensity due to two or more sources turned on simultaneously is equal to the sum of the field intensities of the sources energized separately. For example, if $(\vec{E}_1, \vec{H}_1)$ are due to the impressed currents $(\vec{J}_1, \vec{J}_{m1})$ and if $(\vec{E}_2, \vec{H}_2)$ are due to $(\vec{J}_2, \vec{J}_{m2})$, then,

$$\nabla \times \vec{H}_1 = \vec{J}_1 + j\omega\epsilon\vec{E}_1$$

$$\nabla \times \vec{H}_2 = \vec{J}_2 + j\omega\epsilon\vec{E}_2$$

$$\nabla \times \vec{H} = \vec{J} + j\omega\epsilon\vec{E}$$

where $\vec{E} = \vec{E}_1 + \vec{E}_2$, $\vec{H} = \vec{H}_1 + \vec{H}_2$, and $\vec{J} = \vec{J}_1 + \vec{J}_2$. This is a consequence of the linearity of Maxwell's equations.

It should be noted that scaled combinations of the two solutions are included (e.g., $\vec{E} = c_1\vec{E}_1 + c_2\vec{E}_2$). All physically valid solutions are constrained by conservation of energy and the boundary conditions.

2.7 Theorem of Similitude

The theorem of similitude [2] provides scaling relationships between the various quantities that occur in Maxwell's equations. This is important in

Table 2.2 Scaling Factors for Electromagnetic Quantities

Quantity	General Case	Geometrical Model
Time	$t' = (t/q)$	$t' = (t/p)$
Length	$L' = (L/p)$	$L' = (L/p)$
Wavelength	$\lambda' = (\lambda/q)$	$\lambda' = (\lambda/p)$
Frequency	$f' = qf$	$f' = pf$
Permittivity	$\epsilon' = (p\alpha\epsilon/\beta q)$	$\epsilon' = \epsilon$
Permeability	$\mu' = (p\beta\mu/\alpha q)$	$\mu' = \mu$
Conductivity	$\sigma_c' = (p\alpha\sigma_c/\beta)$	$\sigma_c' = p\sigma_c$
Current density	$\vec{J}' = (p\vec{J}/\beta)$	
Power density	$\vec{W}' = (\vec{W}/\alpha\beta)$	
Phase velocity	$u_p' = (qu_p/p)$	$u_p' = u_p$
Antenna gain	$G' = G$	
Propagation constant	$\gamma' = p\gamma$	
Impedance	$\eta' = (\beta\eta/\alpha)$	$\eta' = \eta$
Radar cross section	$\sigma' = (\sigma/p^2)$	$\sigma' = (\sigma/p^2)$

predicting the RCS of *full-scale* targets from measurements on *scaled models.* Target models are used in the early stages of the design process to verify critical design concepts. Obviously, it is more cost-effective to build a small model and modify it than to build a full-scale model.

Let the full-scale quantities for length and time be (x, y, z, t) and the corresponding quantities in the scaled system be (x', y', z', t'). Define scale factors for time and length as q $[t' = (t/q)]$ and p $[x' = (x/p), y' = (y/p), z' = (z/p)]$. Scale factors for the field intensities are also defined:

$$\vec{E}'(x', y', z', t') = \frac{1}{\alpha}\vec{E}(x, y, z, t)$$

$$\vec{H}'(x', y', z', t') = \frac{1}{\beta}\vec{H}(x, y, z, t)$$

Because both sets of quantities must satisfy Maxwell's equations, the relationships listed in Table 2.2 must exist.

The third column in the table occurs when $\alpha = \beta$ and $p = q$. This is referred to as the *geometrical model.* The constitutive parameters ϵ and μ are the same in both systems, and most other quantities are scaled by p. This allows essentially the same materials to be used in the model as in the full-scale system; only the dimensions are scaled. An important exception to the similar materials statement is conductivity σ, which must be scaled by the length factor. This implies that, if the full-scale material is a good conductor, the scaled model must be an even better conductor.

Example 2.2: Frequency Scaling

The RCS of a 5-m square plate is needed at 2 GHz, but our measurement system operates at 10 GHz. What size should the model plate be? What RCS will be measured?

Assuming the geometrical model, the ratio of frequencies can be used to obtain the length scale factor: $p = (f'/f) = 5$. Thus, the model dimension is $L' = (L/p) = 1\,m$. This is referred to as a "one-fifth" scaled model. From Example 1.5, at 10 GHz, the measured RCS is

$$\sigma' = \frac{4\pi A^2}{\lambda^2} = \frac{4\pi(1^2)^2}{0.03^2} = 13{,}956\,\text{m}^2 = 41.44\,\text{dBsm} \tag{2.27}$$

The RCS of the full-scale plate is

$$\sigma = \sigma' p^2 = (13{,}956)(25) = 348{,}890\,\text{m}^2 = 55.4\,\text{dBsm} \tag{2.28}$$

As a check, we can calculate the RCS of the 5-m plate at 2 GHz directly.

The RCS has its largest value looking directly at the plate, and this is why it is important to avoid orienting large flat surfaces in the direction of a threat radar.

2.8 Method of Images

The *method of images* provides a means of obtaining a simple equivalent problem that can be solved in place of an original problem that is more difficult to solve. The uniqueness theorem assures us that, if the equivalent problem satisfies the same boundary conditions as the original one, the same solution will be obtained. It is usually not possible to find an equivalent problem that satisfies the original boundary conditions everywhere. The equivalent problem is valid only in restricted regions, but this is sufficient in most practical cases.

A simple example is a vertical *Hertzian dipole* (current element) a distance h above an infinite perfect electric ground plane, shown in Fig. 2.4. The electric field components at the conductor are broken into normal and tangential components relative to the surface. The boundary condition at the surface requires that the total tangential component be zero. In the original problem, a current is induced on the conductor that radiates a *scattered field*. If the induced current were known, it could

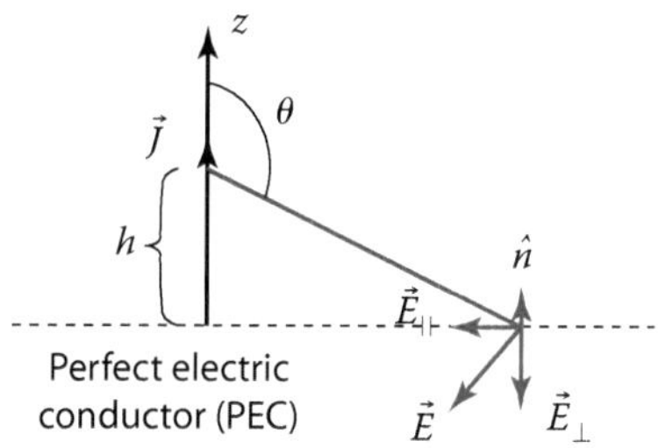

Fig. 2.4 Vertical dipole above a perfect electric ground plane.

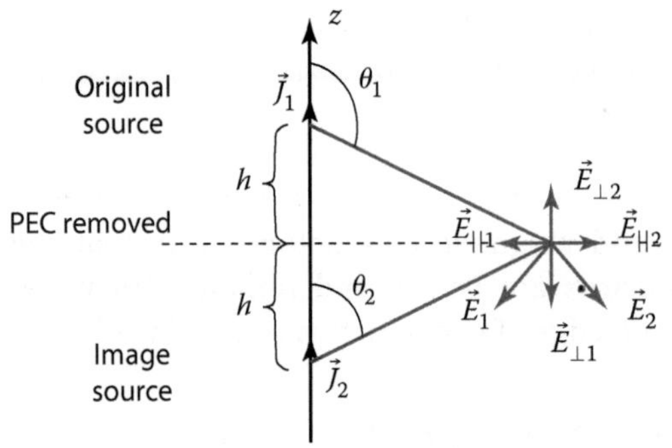

Fig. 2.5 Image equivalent for a vertical dipole above a PEC ground plane.

be substituted into the radiation integrals and evaluated on the surface. The resulting scattered field $\vec{E}_s$ would exactly cancel the incident field on the surface, thereby satisfying the boundary conditions.

A clever observer might note that the total electric field can be forced to zero at the surface by introducing a second electric field term on the surface that exactly cancels with the dipole's tangential component. Another vertical dipole a distance h below the conductor/air interface is such a source. Thus, the conducting material can be removed and replaced with an *image dipole* in free space. The equivalent problem is shown in Fig. 2.5. Note that the image source is located inside the volume containing the original conducting material and, because it is now filled with free space,

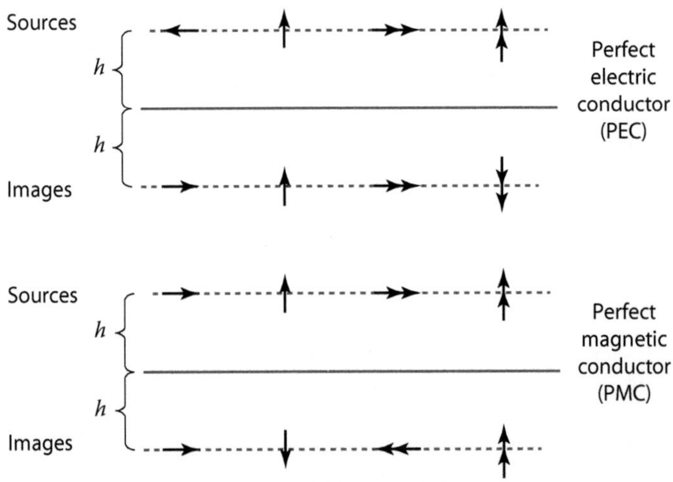

Fig. 2.6 Image equivalent for electric and magnetic current elements above perfect electric and magnetic ground planes.

an electric field exists. Therefore, the equivalent problem is valid only in the region of space outside the original conductor.

Images exist for both electric and magnetic currents. Their orientation varies with the type of material and the geometry of the conductor surface. Figure 2.6 shows electric and magnetic current elements over infinite, flat perfect electric and magnetic conductors. Note that these are valid only when the conductors are *perfect* and *infinite*. When the surface is finite, there will be some edge effect, which is not included.

2.9 Equivalence Principles

Equivalence principles are powerful tools for the analysis of scattering from bodies of arbitrary shapes and materials. The *volume equivalence principle* deals with a volume of space filled with linear isotropic material (μ, ϵ). Electric and magnetic sources radiating in unbounded space filled with this material set up the fields $(\vec{E}, \vec{H})$. If the same sources radiate in free space rather than in the material, assume that the fields $(\vec{E}_0, \vec{H}_0)$ are set up. Define the *scattered fields* as the difference:

$$\vec{E}_s = \vec{E} - \vec{E}_0$$

$$\vec{H}_s = \vec{H} - \vec{H}_0$$

Now, equivalent currents can be defined that set up the scattered fields:

$$\vec{J}_{\text{eq}} = j\omega(\epsilon - \epsilon_0)\vec{E} \tag{2.29}$$

$$\vec{J}_{\text{meq}} = j\omega(\mu - \mu_0)\vec{H} \tag{2.30}$$

Note that $\vec{J}_{\text{eq}}$ and $\vec{J}_{\text{meq}}$ will exist only in regions of space where $\mu \neq \mu_0$ and $\epsilon \neq \epsilon_0$. This version of the equivalence principle is not of much practical use because the fields $(\vec{E}, \vec{H})$ are unknown. However, Eqs. (2.29) and (2.30) can form the basis of integral equations that can be solved numerically for the currents.

The *surface equivalence principle* is of much more practical use. Consider electric and magnetic fields $(\vec{E}, \vec{H})$ set up by the impressed current sources $\vec{J}$ and $\vec{J}_m$. The medium is composed of material with parameters μ and ϵ, as depicted in Fig. 2.7. If a surface S encloses a volume V in which there are no impressed sources, the same fields can be maintained in V with the sources turned off if surface currents flow on S that satisfy

$$\vec{J}_s = \hat{n} \times \vec{H} \tag{2.31}$$

$$\vec{J}_{ms} = -\hat{n} \times \vec{E} \tag{2.32}$$

A problem equivalent to the original one is also shown in Fig. 2.7. Note that there are an infinite number of possible currents that will maintain the correct field inside S but that each yields different values for the field

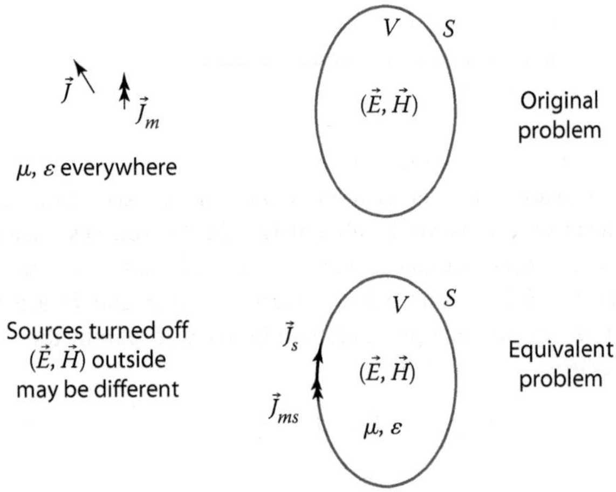

Fig. 2.7 Surface equivalence principle applied to an exterior region.

outside S. Usually $\vec{J}_s$ and $\vec{J}_{ms}$ are chosen to give zero in the exterior region. In this case, because there are no fields external to S, the values of μ and ϵ outside V are not important.

It is possible to reverse the situation just described to the one shown in Fig. 2.8; the fields could be maintained outside S and a null field specified inside S. The region inside S could be an antenna, in which case we need concern ourselves only with the equivalent currents (and therefore fields) on a surface that encloses the antenna. The choice of $\vec{E} = \vec{H} = 0$ in the interior region is referred to as *Love's equivalence principle.*

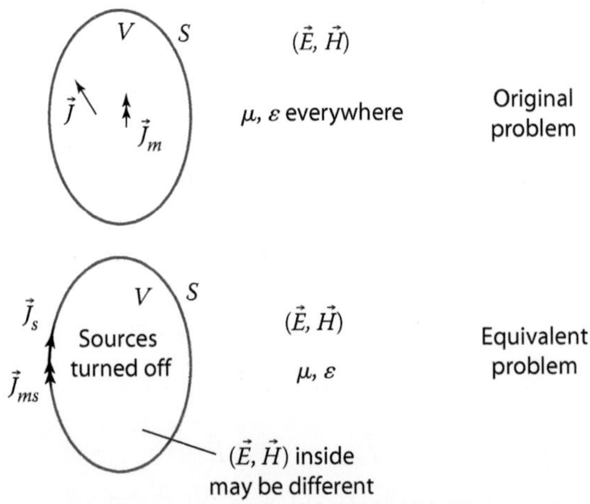

Fig. 2.8 Surface equivalence principle applied to an interior region.

Example 2.3: Diffraction by an Aperture Using Love's Equivalence Principle

A rectangular aperture is located in an infinite ground plane, as shown in Fig. 2.9. A plane wave is incident along the z axis from the negative direction. Because the wave is obviously due to sources located in the region $z < 0$, a closed surface S can be defined that includes this entire region. Thus, $z > 0$ is a source-free region and, if $\vec{E}$ and $\vec{H}$ are to be zero inside S, Love's equivalence principle can be used to obtain the currents on the $z = 0$ plane:

$$\vec{J}_s = \hat{n} \times \vec{H}_a \tag{2.33}$$

$$\vec{J}_{ms} = -\hat{n} \times \vec{E}_a \tag{2.34}$$

The subscript a refers to the fields in the aperture.

Because the fields are zero inside S, the material is arbitrary. The $z = 0$ plane can be replaced by a continuous electric or magnetic conductor. If a perfect electric conductor is added, the electric current is shorted out. This can be seen by examining a differential element of $\vec{J}_s$ and its image. Because these are oppositely directed, they cancel as they approach the ground plane at $z = 0$. On the other hand, the magnetic current image is in the same direction as the magnetic current element and they add. Thus, the perfect electric ground plane can be removed and the magnetic current,

$$\vec{J}_{ms} = -2\hat{n} \times \vec{E}_a \tag{2.35}$$

is considered to radiate in free space, as shown in Fig. 2.10.

It is equally valid to insert a perfect magnetic conductor behind the $z = 0$ plane. In this case, the magnetic current is shorted and the electric

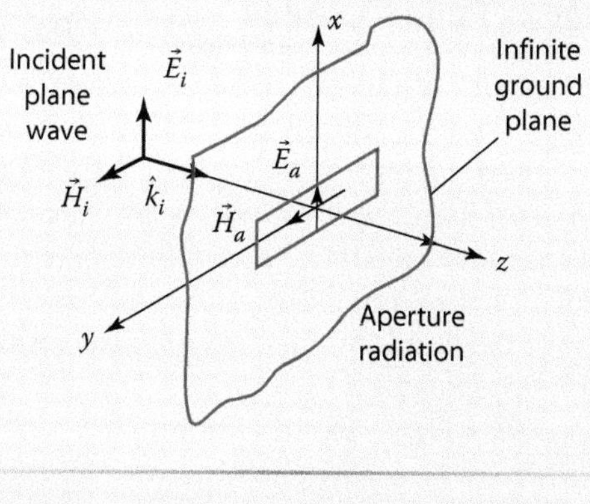

Fig. 2.9 Diffraction by an aperture–original problem.

(Continued)

Example 2.3: Diffraction by an Aperture Using Love's Equivalence Principle *(Continued)*

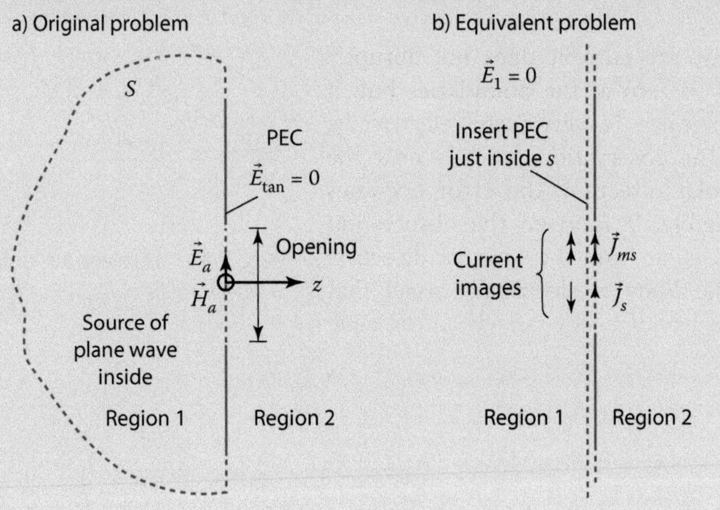

Fig. 2.10 Application of Love's Equivalence Principle to the aperture problem.

current doubled, yielding

$$\vec{J}_s = 2\hat{n} \times \vec{H}_a \tag{2.36}$$

which flows over the original aperture area.

2.10 Physical Optics Approximation

The radiation integrals are a means of determining the fields due to a prescribed set of currents $\vec{J}$ and $\vec{J}_m$. In practice, the currents are unknown; only the excitation (impressed current or voltage) and the boundary conditions are known. The *physical optics approximation* provides an estimate for the unknown currents that gives relatively accurate results for electrically large bodies. The current is approximated by the *geometrical optics* current on the illuminated part of the body, but no current exists on shadowed portions:

$$\vec{J}_s \approx \begin{cases} 2\hat{n} \times \vec{H}_i, & \text{for the illuminated portion} \\ 0, & \text{for the shadowed portion} \end{cases} \tag{2.37}$$

$\vec{H}_i$ is the incident magnetic field intensity at the surface and $\hat{n}$ is the local surface normal unit vector.

Figure 2.11 illustrates how the physical optics approximation is applied for plane wave incidence. The current flowing on the surface is only nonzero between the shadow boundaries. In reality, the current does not abruptly drop to zero at the boundaries but, if the surface is very large electrically, and the observation point is near the specular direction, the error becomes negligible. It is when the observation point is far from the specular direction or the body is electrically small that the behavior of the current at the edge significantly affects the scattered field.

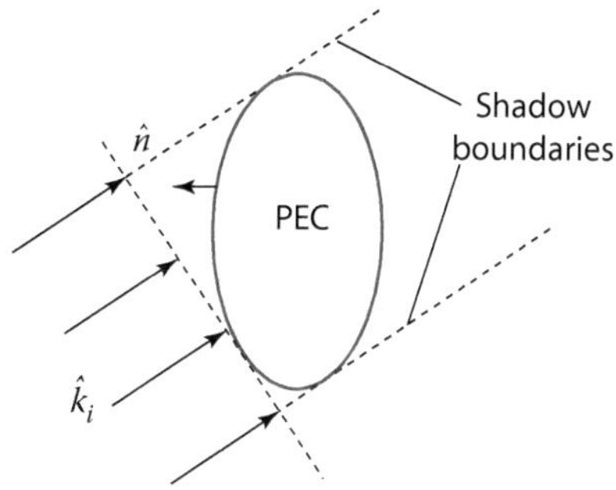

Fig. 2.11 Illuminated and shadowed parts of a surface for application of the physical optics approximation.

Example 2.4: Scattering by a Rectangular Plate Using Physical Optics

Consider the rectangular plate with dimensions a and b shown in Fig. 2.12. A plane wave of arbitrary polarization is incident from an angle (θ, ϕ). The wave polarization is determined by the constants $E_{0\theta}$ and $E_{0\phi}$ in the expression

$$\vec{E}_i = (E_{0\theta}\hat{\theta} + E_{0\phi}\hat{\phi})e^{-j\vec{k}_i\cdot\vec{r}} \tag{2.38}$$

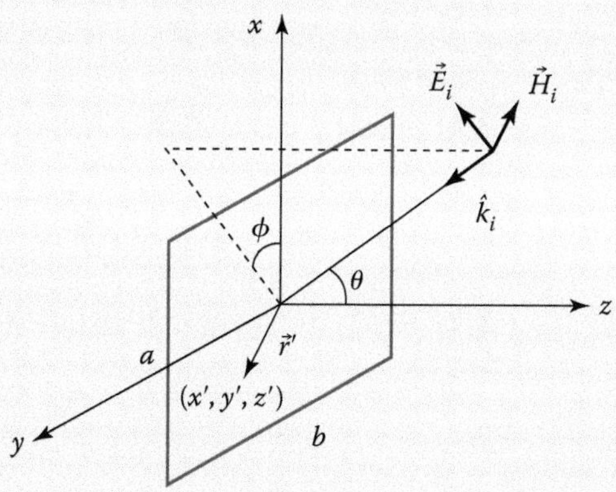

Fig. 2.12 Plane wave scattering by a rectangular plate.

(Continued)

Example 2.4: Scattering by a Rectangular Plate Using Physical Optics *(Continued)*

If the wave is propagating toward the origin, $\hat{k}_i = -\hat{r}$ and, because it is a plane wave, the magnetic field intensity is

$$\vec{H}_i = \frac{-\hat{r} \times \vec{E}_i}{\eta} = -(E_{0\theta}\hat{\phi} - E_{0\phi}\hat{\theta})\frac{e^{-j\vec{k}_i \cdot \vec{r}}}{\eta} \tag{2.39}$$

It has been assumed that the plate is located in a medium with intrinsic impedance η.

Now, the current on the plate is approximated by

$$\vec{J}_s \approx -2\hat{z} \times (E_{0\theta}\hat{\phi} - E_{0\phi}\hat{\theta})\frac{e^{-j\vec{k}_i \cdot \vec{r}}}{\eta} \tag{2.40}$$

The vectors and products required to evaluate Eq. (2.40) are

$$\vec{r}' = \text{position vector for a point } (x', y', z') \text{ on the surface} = \hat{x}x' + \hat{y}y'$$

$$-\hat{k}_i = \hat{x}\sin\theta\cos\phi + \hat{y}\sin\theta\sin\phi + \hat{z}\cos\theta$$

$$-\vec{k}_i \cdot \vec{r} = k(x'\sin\theta\cos\phi + y'\sin\theta\sin\phi)$$

$$\hat{z} \times \hat{\theta} = -\hat{x}\cos\theta\sin\phi + \hat{y}\cos\theta\cos\phi$$

$$\hat{z} \times \hat{\phi} = -\hat{x}\cos\phi - \hat{y}\sin\phi$$

Thus the approximation for the current is

$$\vec{J}_s \approx \frac{2e^{jkh}}{\eta}\left[\hat{x}(E_{0\theta}\cos\phi - E_{0\phi}\cos\theta\sin\phi) + \hat{y}(E_{0\theta}\sin\phi + E_{0\phi}\cos\theta\cos\phi)\right] \tag{2.41}$$

where $h = x'\sin\theta\cos\phi + y'\sin\theta\sin\phi$. Using this current in the radiation integral (2.22) with $\vec{J}_m = 0$ gives

$$E_\theta(P) = \frac{-jk\eta}{4\pi r}e^{-jkr}\iint_S \frac{2e^{jkh}}{\eta}[(E_{0\theta}\cos\phi - E_{0\phi}\cos\theta\sin\phi)\cos\theta_P\cos\phi_P$$
$$+ (E_{0\theta}\sin\phi + E_{0\phi}\cos\theta\cos\phi)\cos\theta_P\sin\phi_P]e^{jkg}\,dx'dy' \tag{2.42}$$

where the subscript P denotes observation point quantities and

$$g = x'\sin\theta_P\cos\phi_P + y'\sin\theta_P\sin\phi_P = \vec{r}' \cdot \hat{r} \tag{2.43}$$

For monostatic scattering,

$$\theta_P = \theta$$
$$\phi_P = \phi$$
$$g = h$$

(Continued)

Example 2.4: Scattering by a Rectangular Plate Using Physical Optics *(Continued)*

Furthermore, assume that the incident wave is θ -polarized ($E_{0\phi} = 0$):

$$E_\theta(r, \theta, \phi) = \frac{-jk}{2\pi r} e^{-jkr} \int_{-(a/2)}^{a/2} \int_{-(b/2)}^{b/2} E_{0\theta} \cos\theta e^{j2kh} \mathrm{d}y' \mathrm{d}x' \tag{2.44}$$

$$E_\theta(r, \theta, \phi) = \frac{-jk}{2\pi r} e^{-jkr} E_{0\theta} \cos\theta \int_{-(a/2)}^{a/2} e^{j2kx'u} \mathrm{d}x' \int_{-(b/2)}^{b/2} e^{j2ky'v} \mathrm{d}y' \tag{2.45}$$

where u and v are the x and y direction cosines. Evaluating the integrals yields

$$E_\theta(r, \theta, \phi) = \frac{-jk}{2\pi r} e^{-jkr} E_{0\theta} \cos\theta\, ab\, \mathrm{sinc}(kau)\mathrm{sinc}(kbv) \tag{2.46}$$

with the sinc function defined in Eq (1.19).

The RCS is obtained from Eq. (1.38). Denoting the plate area by A ($=ab$) gives

$$\sigma = \lim_{r\to\infty} 4\pi r^2 \frac{|\vec{E}_s|^2}{|\vec{E}_i|^2} = \lim_{r\to\infty} 4\pi r^2 \frac{|\vec{E}(r, \theta, \phi)|^2}{|E_{0\theta}|^2} \tag{2.47}$$

$$\sigma_{\theta\theta}(\theta, \phi) = 4\pi r^2 \left[\frac{A^2 k^2}{4\pi^2 r^2} \cos^2\theta\right] \mathrm{sinc}^2(kau)\mathrm{sinc}^2(kbv) \tag{2.48}$$

$$\sigma_{\theta\theta}(\theta, \phi) = \frac{4\pi A^2}{\lambda^2} \cos^2\theta\, \mathrm{sinc}^2(kau)\mathrm{sinc}^2(kbv) \tag{2.49}$$

Figure 2.13 shows the monostatic scattering pattern for plates with edge lengths of 5λ and 10λ. Usually the frequency-independent quantity σ/λ^2 is plotted rather than σ.

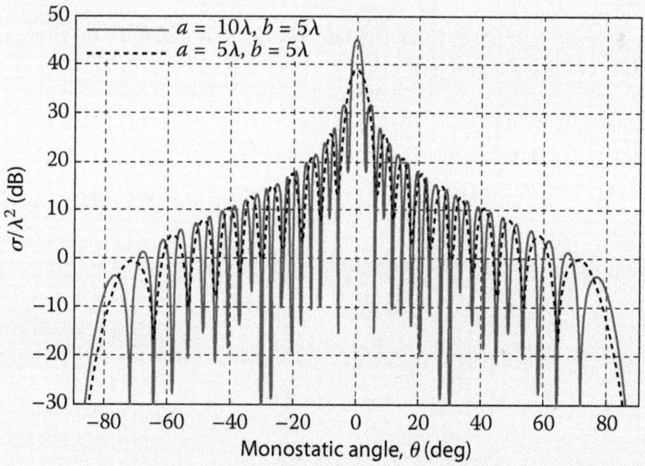

Fig. 2.13 Radar cross section of rectangular plates ($\phi = 0$ deg).

An examination of Eq. (2.49) reveals several characteristics of the physical optics approximation and scattering in general:

1. The maximum value of σ occurs when $\theta = 0$ and has a value

$$\sigma_{\max} = \frac{4\pi A^2}{\lambda^2} \tag{2.50}$$

2. The scattering pattern has the same functional dependence on angle as the radiation pattern from a rectangular aperture of the same area, [i.e., sinc(X)] except for a factor of 2 in the argument. This arises from the path difference between a point on the plate and the origin, which is a two-way distance for scattering. For example, the term

$$\text{sinc}(kau/2)$$

occurs for radiation from a rectangular aperture, whereas

$$\text{sinc}(kau)$$

occurs for scattering. Because $k = (2\pi)/\lambda = (2\pi f)/c$, the scattering pattern looks like the radiation pattern of a larger aperture with dimension $2a$. Alternately, the scattering pattern is equivalent to radiation from the same aperture but at twice the frequency.
3. The scattering pattern is *separable* (factorable) in the direction cosines u and v. This arises from the fact that the plate edges conform to the Cartesian coordinate axes and is not true in general.
4. The phase of the scattered field does not affect the RCS. Thus, the phase reference used in the scattering calculation is arbitrary and, frequently, the leading exponential e^{-jkr} is dropped.

When planar surfaces with edges are involved, the polarization designations TM and TE are frequently used. The definitions are illustrated in Fig 1.12.

In light of the similarity between radiation and scattering patterns for a given aperture, many of the same analysis and synthesis methods used for antennas can be applied to RCS. Because RCS engineers are concerned with reducing scattering, amplitude tapering of the kind used to lower antenna sidelobes is of great interest.

Example 2.5: Radar Cross Section Reduction Using a Linear Aperture Taper

Again, consider the rectangular plate shown in Fig. 2.12. Now, assume that the conductivity (or some other property of the plate surface) can be varied such that the current density is linearly distributed, as shown in Fig. 2.14.

(Continued)

Example 2.5: Radar Cross Section Reduction Using a Linear Aperture Taper *(Continued)*

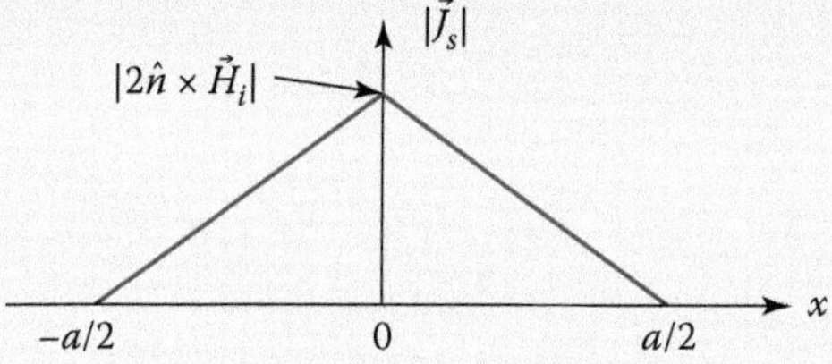

Fig. 2.14 Linearly tapered current density.

The incident wave is TM_z-polarized and, for simplicity, the conductivity is considered constant in y. Furthermore, the wave is restricted to being incident in the $x-z$ plane. Thus, the incident wave is of the form

$$\vec{H}_i = \hat{y}H_0 e^{-j\vec{k}_i \cdot \vec{r}} \tag{2.51}$$

The assumed current will have the PO value at the center ($x = 0$) and taper to zero at the edges:

$$\vec{J}_s \approx -2H_0\hat{x}\left(1 - \frac{2|x'|}{a}\right)e^{jk(x'u+y'v)} \tag{2.52}$$

Using this current in the radiation integral (2.25) gives

$$\vec{E}(r, \theta, \phi) = \hat{x}\frac{jk\eta H_0}{2\pi r}e^{-jkr}\int_{-(a/2)}^{a/2}\left(1 - \frac{2|x'|}{a}\right)e^{j2kx'u}\mathrm{d}x' \times \int_{-(b/2)}^{b/2} e^{j2ky'v}\mathrm{d}y' \tag{2.53}$$

Performing the integrations and using $\hat{\theta} \cdot \hat{x} = \cos\theta$ when $\phi = 0$ ($v = 0$) yields

$$E_\theta(r, \theta, 0) = \frac{jk\eta H_0}{2\pi r}e^{-jkr}\cos\theta\frac{ab}{2}\operatorname{sinc}^2(kau/2) \tag{2.54}$$

Finally,

$$\sigma(\theta, 0) = \frac{4\pi A^2}{\lambda^2}\frac{1}{4}\cos^2\theta\operatorname{sinc}^4(kau/2) \tag{2.55}$$

A comparison of scattering from uniformly illuminated and linearly tapered plates is shown in Fig. 2.15 for a 5λ square plate.

(Continued)

Example 2.5: Radar Cross Section Reduction Using a Linear Aperture Taper *(Continued)*

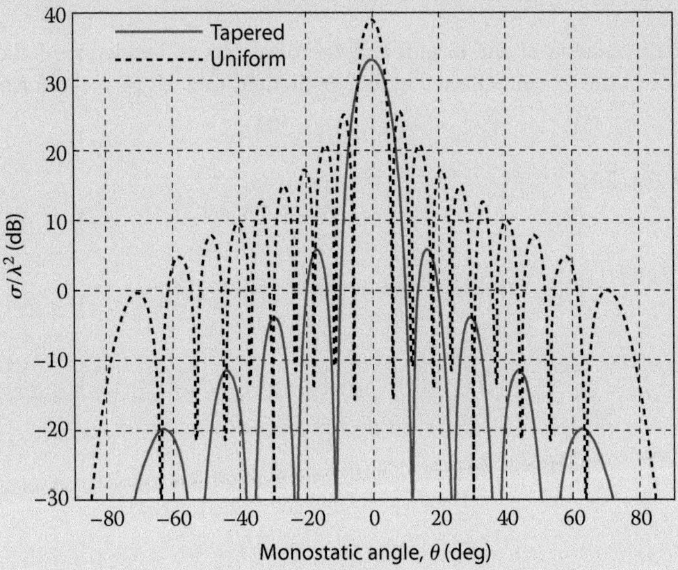

Fig. 2.15 RCS of uniformly illuminated and linearly tapered 5λ by 5λ square plate.

Note the change in pattern characteristics for the linearly tapered pattern relative to the uniform case:

1. In the plane of the taper ($x - z$ or $\phi = 0$), the RCS sidelobes have been reduced from −13.2 dB to approximately −26 dB relative to the peak.
2. The mainlobe beam width has increased.
3. The peak RCS value at $\theta = 0$ has been reduced by a factor of 4 (6 dB).
4. The nulls have shifted outward.

Tapering the current distribution is potentially a powerful means of reducing RCS. Unfortunately, this is difficult to do in practice. Attenuating the current requires that the loss be controlled precisely from point to point on the surface over a wide range of values. Few candidate materials exist that satisfy these requirements, and those that do are expensive and may not satisfy other operational requirements (mechanical, weight, etc.). Chapter 7 will address this problem in more detail.

The following example illustrates the application of the physical optics approximation to a curved surface.

Example 2.6: Radar Cross Section of a Sphere Using Physical Optics

A plane wave is incident on a sphere of radius a as shown in Fig. 2.16. The sphere is located at the origin, and the direction of incidence of the plane wave is in the $-z$ direction. The magnetic field intensity has the form

$$\vec{H}_i = \hat{x}\, H_0 e^{jkz} \tag{2.56}$$

From Fig. 2.16

$$\hat{n} = \hat{r}' = \hat{x}u' + \hat{y}v' + \hat{z}w'$$

$$u' = \sin\theta' \cos\phi'$$

$$v' = \sin\theta' \sin\phi'$$

$$w' = \cos\theta'$$

$$z' = a\cos\theta' = aw'$$

$$\Delta = a - z' = a(1 - w')$$

where the primed quantities refer to a point on the surface of the sphere. Using these in the PO approximation gives

$$\vec{J}_s = 2H_0 e^{-jk\Delta}(\hat{y}w' - \hat{z}v') \tag{2.57}$$

Plugging $\vec{J}_s$ into the radiation integral and integrating over the illuminated part of the sphere (a hemisphere) give

$$\vec{E}(r, \theta, \phi) = \frac{-jk\eta H_0}{2\pi r} e^{-jkr} \int_0^{2\pi} \int_0^{\pi/2} (\hat{y}w' - \hat{z}v') e^{-j2k\Delta} a^2 \sin\theta' \mathrm{d}\theta' \mathrm{d}\phi' \tag{2.58}$$

The z component vanishes when the ϕ' integration is performed because

$$\int_0^{2\pi} \sin\phi' \mathrm{d}\phi' = 0$$

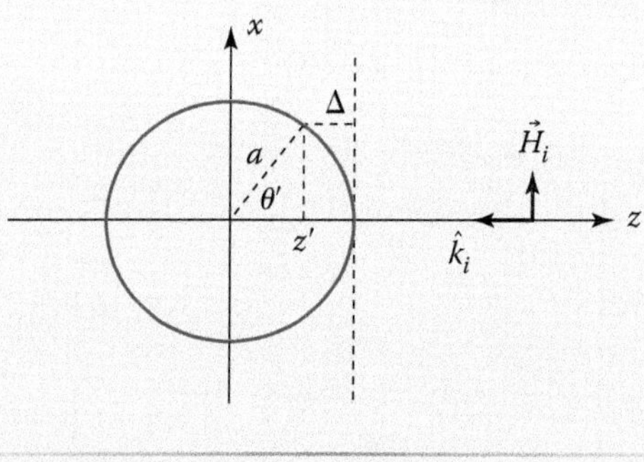

Fig. 2.16 Plane wave incident on a sphere.

(Continued)

Example 2.6: Radar Cross Section of a Sphere Using Physical Optics *(Continued)*

Thus,

$$\vec{E}(r,\theta,\phi) = \frac{-jk\eta H_0 a^2}{r} e^{-jkr}\hat{y}\int_0^{\pi/2} \cos\theta' \sin\theta' \exp[-j2ka(1-\cos\theta')]\mathrm{d}\theta' \tag{2.59}$$

To evaluate the integral, work in terms of $w' = \cos\theta'$. Therefore, $\mathrm{d}w' = -\sin\theta'\ \mathrm{d}\theta'$ and

$$\vec{E}(r,\theta,\phi) = \frac{-jk\eta H_0 a^2}{r} e^{-jk(r+2a)}\hat{y}\int_0^{1} w' e^{j2kaw'}(-\mathrm{d}w') \tag{2.60}$$

But

$$\int_0^1 w' e^{j2kaw'}\mathrm{d}w' = \frac{e^{jka}}{j2ka}[1-\mathrm{sinc}(ka)] \tag{2.61}$$

Thus,

$$\vec{E}(r,\theta,\phi) = \frac{-j\eta H_0 a}{2r} e^{-jk(r+2a)}\hat{y}[1-\mathrm{sinc}(ka)] \tag{2.62}$$

The θ and ϕ components are (given that $\phi = 0$)

$$\begin{aligned} E_\theta(r,\theta,0) &= \cos\theta\sin\phi E(r,\theta,0) = 0 \\ E_\phi(r,\theta,0) &= \cos\phi E(r,\theta,0) = E_y \end{aligned} \tag{2.63}$$

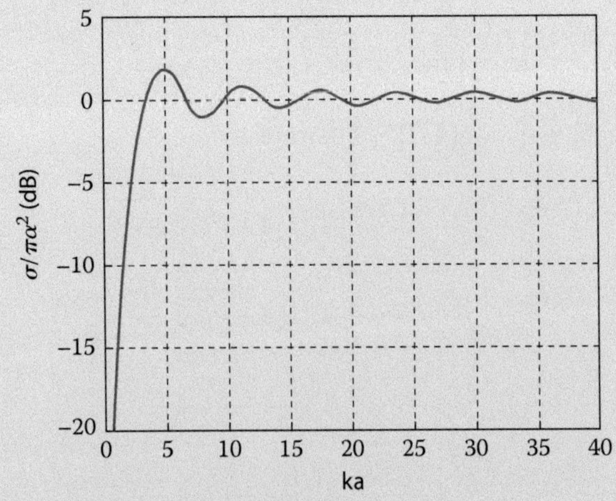

Fig. 2.17 Backscatter from a sphere using PO.

(Continued)

Example 2.6: Radar Cross Section of a Sphere Using Physical Optics *(Continued)*

The RCS is computed using Eqs. (2.62) and (2.63) and $E_0 = \eta H_0$

$$\sigma(\theta, 0) = \pi a^2 [1 - \mathrm{sinc}(ka)]^2 \tag{2.64}$$

Note that, when the frequency becomes very high ($\lambda \to 0$), the sinc term goes to zero, and the RCS is simply

$$\sigma(\lambda \to 0) = \pi a^2$$

The backscatter from a sphere as a function of radius computed using the physical optics approximation is shown in Fig. 2.17.

The curve of Fig. 2.17 bears some similarity to Fig. 1.23, which is a plot obtained from the Mie series. Both curves approach 1 as $ka \to \infty$, but the period and amplitude of the oscillations are much different. The Mie solution is exact and, therefore, includes all scattering components. The PO approximation does not include surface waves, nor does it correctly model the surface current distribution near the shadow boundaries. The oscillations in Fig. 2.17 are due to the radiation from current on different parts of the surface adding and canceling. The oscillations in Fig. 1.6 are due primarily to creeping waves (or surface waves) that travel around the shadowed back side of the sphere.

Radar cross section estimates for complex targets that consist of a collection of basic shapes are frequently obtained by superposition. The RCS of each component is computed using the physical optics approximation, and the total RCS is determined from the vector sum of the scattered fields. It is important, however, to include the effect of shadowing due to the target on itself, as Example 2.7 illustrates.

Example 2.7: Radar Cross Section of a Finite Cylinder with Triangular Cross Section

Figure 2.18 shows a cylinder of length b with a triangular cross section. Each side consists of a plate with dimensions a by b. The origin is located at the center of the bottom plate and a TM_z plane wave is incident in the $\phi = 0$ plane. The monostatic RCS of this structure can be computed using Eq (2.46) for each plate with the proper values of u. This is because TM_z is also TM_{z1} and TM_{z2}. The contribution from a plate is only included in the total RCS when it is illuminated by the incident wave.

(Continued)

Example 2.7: Radar Cross Section of a Finite Cylinder with Triangular Cross Section *(Continued)*

A *local coordinate system* is defined for each plate as shown in Fig. 2.18. Because $\phi = 0$, $v = 0$ and the factor $(kvb) \to 1$. The angle θ is known and, from geometry

$$\begin{aligned} \theta_2 &= \theta - 60\,\text{deg} \\ \theta_1 &= \theta + 60\,\text{deg} \end{aligned} \tag{2.65}$$

where positive values are toward the x_1 and x_2 axes. A rotation of coordinates gives

$$\begin{aligned} \hat{x}_1 &= \hat{x}\cos(60\,\text{deg}) + \hat{z}\sin(60\,\text{deg}) \\ \hat{x}_2 &= \hat{x}\cos(-60\,\text{deg}) + \hat{z}\sin(-60\,\text{deg}) \end{aligned} \tag{2.66}$$

and, because direction cosines obey the same transformation rules (see Appendix B),

$$\begin{aligned} u_1 &= \sin\theta_1 = u\cos(60\,\text{deg}) + w\sin(60\,\text{deg}) \\ u_2 &= \sin\theta_2 = u\cos(-60\,\text{deg}) + w\sin(-60\,\text{deg}) \end{aligned} \tag{2.67}$$

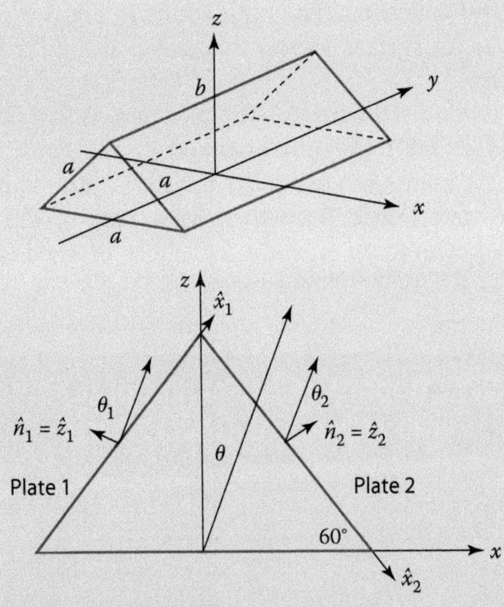

Fig. 2.18 Finite cylinder with a triangular cross section (tent).

(Continued)

Example 2.7: Radar Cross Section of a Finite Cylinder with Triangular Cross Section *(Continued)*

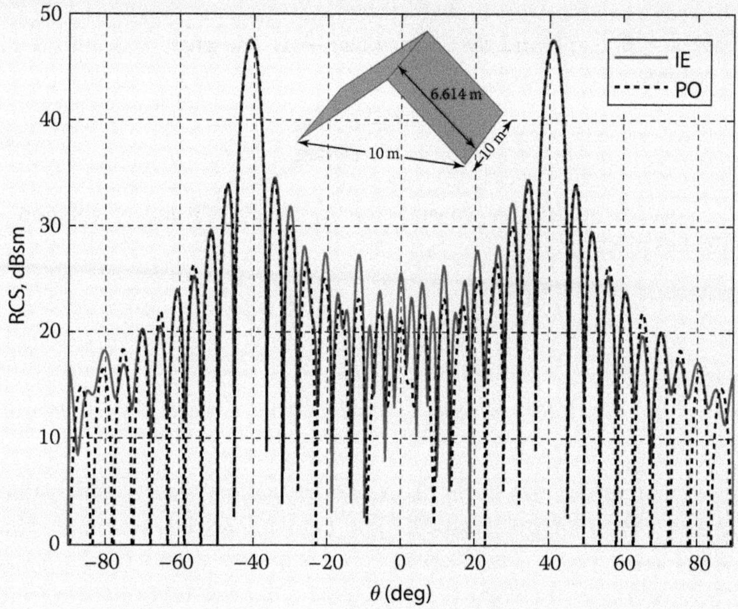

Fig. 2.19 Principal plane RCS of the triangular tent structure.

Thus, the total RCS is given by

$$\sigma_{\theta\theta}(\theta, \phi = 0) = \frac{4\pi A^2}{\lambda^2} |\cos\theta_1 \,\mathrm{sinc}(ku_1 a)\exp\{jka[-u/2 + 0.866w]\} + \cos\theta_2 \,\mathrm{sinc}(ku_2 a)\exp\{jka[u/2 + 0.866w]\}|^2 \quad (2.68)$$

The exponential terms are required to reference the phase of each plate back to the *global coordinate system* origin. The quantities in the exponents are twice the path difference between the plate centers and the phase reference. Note also that, if $|\theta_1|$, $|\theta_2| > 90$ deg, the plate is not illuminated and the corresponding term in Eq (2.69) must be set to zero. Figure 2.19 shows the RCS when $a = 5\lambda$ and $b = 10\lambda$. The PO result is compared to an integral equation (IE) result that is rigorous and includes all scattering mechanisms.

2.11 Huygen's Principle

Huygen's principle states that each point on a wave front can be considered as a new source of secondary spherical waves and that a new wave front can be constructed from the envelope of these secondary waves.

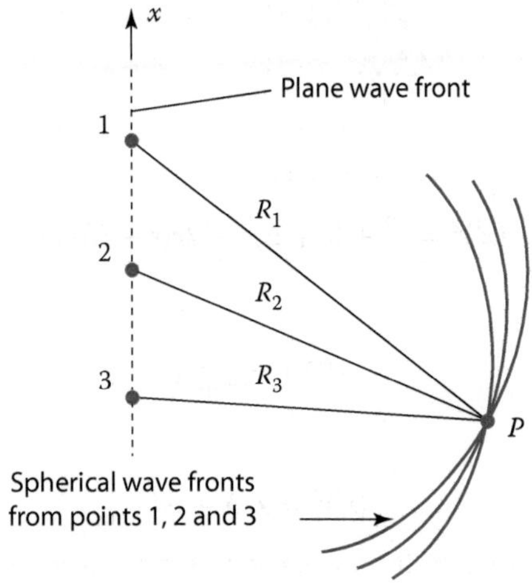

Fig. 2.20 Huygen's principle applied to a plane wave.

Huygen's principle is the basis of *scalar diffraction theory*, which does not include the effects of polarization. The principle is illustrated in Fig. 2.20 for a plane wave front. The field at P is a superposition of spherical waves from each point on the original wave front. For a discrete number of points,

$$|\vec{E}(P)| \sim \left| \sum_{n=-\infty}^{\infty} \frac{e^{-jkR_n}}{R_n} \right| \tag{2.69}$$

and, as the distance between the sources is reduced,

$$|\vec{E}(P)| \sim \left| \int_{-\infty}^{\infty} \frac{e^{-jkR(x')}}{R(x')} \mathrm{d}x' \right| \tag{2.70}$$

Example 2.8: Diffraction by a Half-Plane

A plane wave is incident on a half-plane located at $z = 0$ and extending from $x = -\infty$ to $x = a$ as shown in Fig. 2.21. To determine the electric field at a point $P(0, 0, z)$ in the shadow, the contributions from all Huygen's wavelets for $x \geq a$ must be included:

$$|\vec{E}(P)| = \left| E_0 \int_{a}^{\infty} \frac{e^{-jkR(x')}}{R(x')} \mathrm{d}x' \right| \tag{2.71}$$

(Continued)

Example 2.8: Diffraction by a Half-Plane *(Continued)*

E_0 is the amplitude of the incident wave. For any point on the wave front, the distance to P is $R = r + \Delta$. Using the law of cosines for the right triangle gives

$$(x')^2 = r^2 + (r+\Delta)^2 - 2r(r+\Delta)\cos\alpha$$

But

$$\cos\alpha = \frac{r}{r+\Delta}$$

so that

$$(x')^2 = 2r\Delta + \Delta^2$$

The contribution to $E(P)$ from each source diminishes as its distance from the observation point increases. Because P is in the shadow, points in the vicinity of $x = a$ will contribute more strongly than points toward ∞; therefore, it can be assumed that $\Delta \ll r$ in Eq. (2.72):

$$\Delta \approx \frac{(x')^2}{2r} \tag{2.72}$$

Inserting Eq. (2.73) in Eq. (2.72) gives

$$|\vec{E}(P)| = \left| E_0 \frac{e^{-jkr}}{r} \int_a^{\infty} e^{-jk(x')^2/2r} \mathrm{d}x' \right| \tag{2.73}$$

This integral can be cast into the form of standard Fresnel sine and cosine integrals using the substitutions $\xi^2 = 2/(r\lambda)$ and $\zeta = \xi x$. Using these in Eq. (2.73) gives

$$|\vec{E}(P)| = \left| \frac{E_0 e^{-jkr}}{\xi r} \int_{\xi a}^{\infty} e^{-j\pi\zeta^2/2} \mathrm{d}\zeta \right|$$

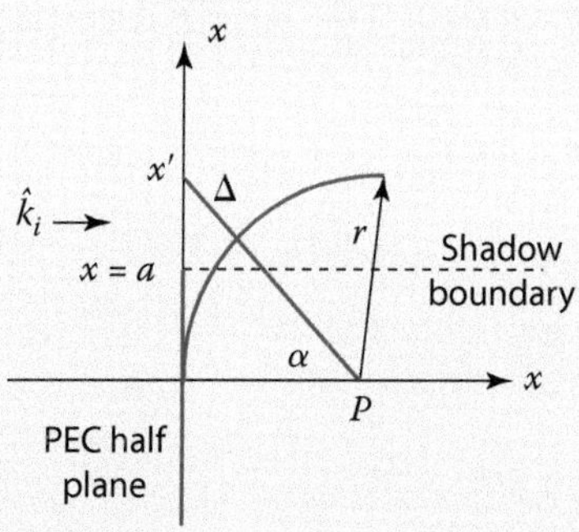

Fig. 2.21 Diffraction by a half-plane using Huygen's principle.

(Continued)

Example 2.8: Diffraction by a Half-Plane *(Continued)*

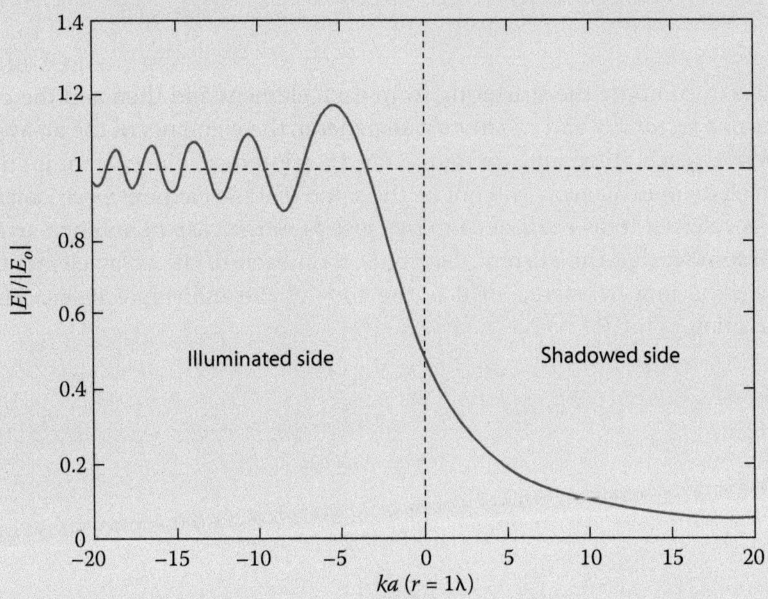

Fig. 2.22 Field strength as the observation point is moved from the shadow to the lit region.

$$|\vec{E}(P)| = \left|\frac{E_0 e^{-jkr}}{\xi r}\left[\int_0^{\infty} e^{-j\pi\zeta^2/2}\mathrm{d}\zeta - \int_0^{\xi a} e^{-j\pi\zeta^2/2}\mathrm{d}\zeta\right]\right|$$

The Fresnel sine and cosine integrals (which are tabulated in most math handbooks and available in many computer utilities) are defined as

$$C(x) = \int_0^x \cos\left(\frac{\pi\xi^2}{2}\right)\mathrm{d}\xi \tag{2.74}$$

$$S(x) = \int_0^x \sin\left(\frac{\pi\xi^2}{2}\right)\mathrm{d}\xi \tag{2.75}$$

Thus, in terms of Eqs. (2.74) and (2.75),

$$|\vec{E}(P)| = \left|\frac{E_0 e^{-jkr}}{\xi r}\left[\frac{1+j}{2} - C(\xi a) - jS(\xi a)\right]\right| \tag{2.76}$$

Figure 2.22 shows the relative electric field strength as the edge is moved parallel to the x axis at a constant distance r. (This is equivalent to P approaching from $+\infty$.) The oscillations grow in the vicinity of $x = a$. At the shadow boundary, the electric field is one-half of its free space value (-6 dB power) and, when P is well into the shadow region, the electric field is essentially zero.

2.12 Arrays of Scatterers

2.12.1 General

An array of scatterers simply refers to a collection of targets that may or may not be identical. The simplest approach to calculating the RCS of the array is to compute the scattering from each element and then add the contributions *vectorally* and *coherently*. In general, the elements of the array can be close to each other, and, consequently, they interact. The current induced on an element in an array will not be the same as if the element were isolated. This is referred to as *mutual coupling*, just as in the case of antenna arrays.

Assuming that the current distribution on each of the array elements is known, the total scattered field is the sum of the individual fields. At an observation point P,

$$\vec{E}(P) = \sum_{n=1}^{N} \vec{E}_n(P) \tag{2.77}$$

where N is the number of elements. If the observation point is in the far zone of the array, as shown in Fig. 2.23, and the nth element has a current density $\vec{J}_n$, the radiation integral for the electric field becomes

$$\vec{E}(x, y, z) = \frac{-jk\eta e^{-jkr}}{4\pi r} \sum_{n=1}^{N} e^{jkg_n} \iiint_{V_n} \vec{J}_n e^{jkg} \mathrm{d}v' \tag{2.78}$$

where it is understood that E_r is ignored. The coordinates (x_n, y_n, z_n) specify the location of the *local* coordinate system for element n in terms of the *global* coordinate system. If the observation point P is at (r, θ, ϕ), then,

$$g_n = x_n u + y_n v + z_n w \tag{2.79}$$

$$g = x' u + y' v + z' w \tag{2.80}$$

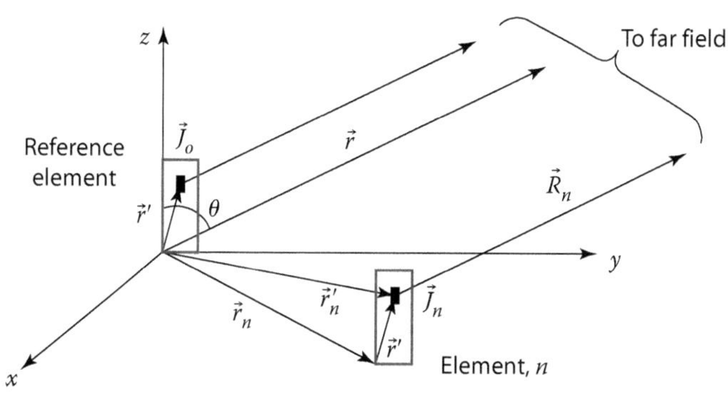

Fig. 2.23 Array of scatterers with the observation point in the far field.

where u, v, and w are the direction cosines and (x', y', z') are the integration coordinates on element n.

2.12.2 Identical Scatterers

For the special case of identical scatterers, the current density on all elements is the same. One of the elements, usually the one located at the origin of the global coordinate system, is defined as the reference. With the current density on the reference element denoted $\vec{J}_0$ for monostatic scattering, Eq. (2.78) can be written

$$\vec{E}(x, y, z) = \left[\sum_{n=1}^{N} e^{j2kg_n}\right]\left[\frac{-jk\eta e^{-jkr}}{4\pi r}\iiint_{V_0} \vec{J}_0 e^{jkg} \mathrm{d}v'\right] \tag{2.81}$$

where $g_n = \vec{r}_n \cdot \hat{r}$ and $g = \vec{r}' \cdot \hat{r}$.

This has a form similar to the field radiated from an array of antennas, and the same terminology can be adopted. The first term in brackets is an array factor; it depends only on the geometry and dimensions of the array and is independent of the element characteristics. The second term is essentially an element factor that is determined entirely by the scattering properties of the reference body. Equation (2.81) is the scattering equivalent of the *principle of pattern multiplication* encountered in antenna theory.

Example 2.9: Scattering from an Array of Thin Strips

Consider a TE_z plane wave incident on an array of thin strips, as shown in Fig. 2.24. The incident field is of the form

$$\vec{E}_i = \hat{\phi} E_0 e^{-j\vec{k}_i \cdot \vec{r}}$$

The spacing is d and the dimensions of each strip are a in x and b in y. It is assumed that $a \ll \lambda$, and we are interested in finding the monostatic RCS

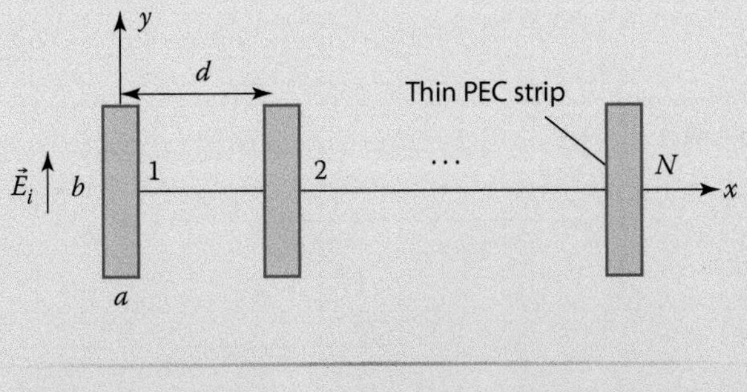

Fig. 2.24 TE_z-polarized wave incident onto an array of thin PEC strips.

(Continued)

Example 2.9: Scattering from an Array of Thin Strips (Continued)

when the wave is incident in the $x-z$ plane ($\phi = 0$). The current induced on all strips is assumed to be identical and given by the physical optics approximation

$$\vec{J}_s \approx 2\hat{n} \times \vec{H}_i = 2\hat{z} \times \hat{\theta}\,\frac{E_0}{\eta}e^{-j\vec{k}_i\cdot\vec{r}} \tag{2.82}$$

In general,

$$\hat{z} \times \hat{\theta} = \hat{y}\cos\theta\cos\phi - \hat{x}\cos\theta\sin\phi \tag{2.83}$$

but, since $\phi = 0$, Eq. (2.83) reduces to

$$\hat{z} \times \hat{\theta} = \hat{y}\cos\theta \tag{2.84}$$

The radiation integral for a single element is

$$E_y = \frac{-jk\eta}{4\pi r}e^{-jkr}\frac{2E_0}{\eta}\cos\theta\int_{-(a/2)}^{a/2} e^{j2kx'u}\mathrm{d}x'\int_{-(b/2)}^{b/2} e^{j2ky'v}\mathrm{d}y' \tag{2.85}$$

The integral with respect to y simply gives b because $v = 0$. Using the result of Example 2.3 and the fact that $\hat{\phi}\cdot\hat{y} = \cos\phi = 1$,

$$E_\phi = \frac{-jk}{2\pi r}e^{-jkr}E_0\cos\theta\; ab\;\mathrm{sinc}(kau) \tag{2.86}$$

If $a \ll \lambda$ (even though this is a violation of PO), then $\mathrm{sinc}(kau) \to 1$ and

$$E_\phi = \frac{-jk}{2\pi r}e^{-jkr}E_0 ab\cos\theta \tag{2.87}$$

Thus, when viewed by an observer in the $x-z$ plane, the strips are approximately isotropic scatterers except for the projected aperture factor $\cos\theta$. The array factor (summation term) is

$$\sum_{n=1}^{N} e^{j2kg_n} = \sum_{n=1}^{N} e^{j2kx_n u} = \sum_{n=1}^{N} e^{j2kd(n-1)\sin\theta} \tag{2.88}$$

Shifting the summation index and factoring out n in the exponent yields a form of a geometric series

$$\sum_{n=0}^{N-1}\exp[(j2kd\sin\theta)]^n = \exp[j(N-1)kd\sin\theta]\frac{\sin(Nkd\sin\theta)}{\sin(kd\sin\theta)} \tag{2.89}$$

The final expression for E_ϕ (neglecting the leading phase term) is

$$E_\phi = \frac{-jk}{2\pi r}e^{-jkr}E_0 ab\cos\theta\frac{\sin(Nkd\sin\theta)}{\sin(kd\sin\theta)} \tag{2.90}$$

(Continued)

Example 2.9: Scattering from an Array of Thin Strips (Continued)

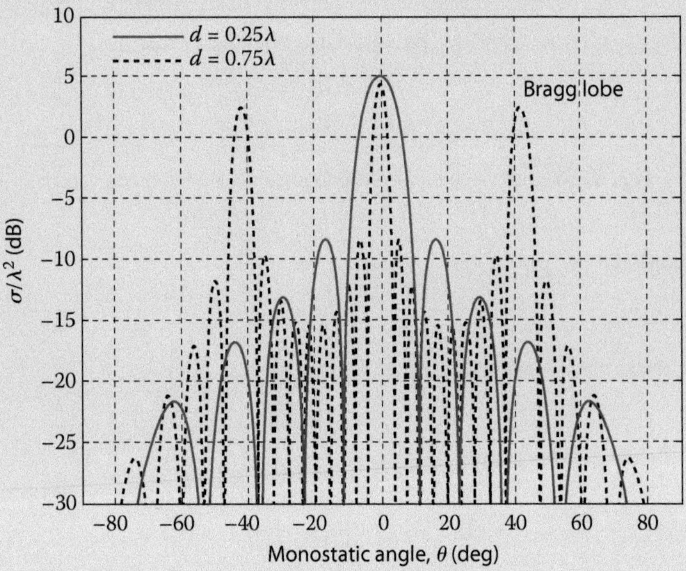

Fig. 2.25 Radar cross section of a 10-element array of thin strips.

resulting in an RCS of

$$\sigma_{\phi\phi} = \frac{4\pi A^2}{\lambda^2}\cos^2\theta\left[\frac{\sin(Nkd\sin\theta)}{\sin(kd\sin\theta)}\right]^2 \tag{2.91}$$

where A is the area of a single strip. Figure 2.25 shows the monostatic scattering pattern from an array of 10 strips, each $0.1\lambda \times 0.5\lambda$.

2.12.3 Bragg Diffraction

For the largest element spacing in Fig. 2.25, secondary major maxima occur. These secondary peaks are called *Bragg lobes* and are analogous to grating lobes in the radiation case. The location and spacing of Bragg lobes depend only on the geometry of the array *lattice*, not on the size of the array. The peaks occur when the round-trip path difference (2Δ) between adjacent elements in the array is a multiple of 2π, as depicted in Fig. 2.26. From Eq. (2.91), this condition is satisfied whenever the denominator is zero:

$$kd\sin\theta = m\pi, \quad \text{where} \quad m = \pm 1, \pm 2, \ldots, \tag{2.92}$$

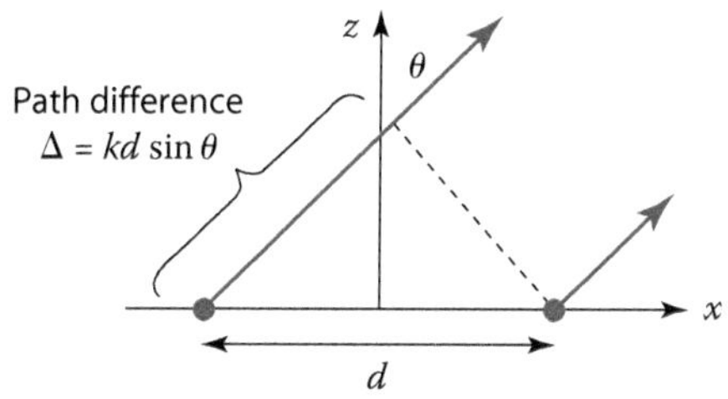

Fig. 2.26 Path length difference for determining the Bragg condition.

which leads to

$$\theta_m = \sin^{-1}\frac{m\lambda}{2d} \tag{2.93}$$

The index m is limited to values that yield $0 \le |\theta| \le \pi/2$ (for $0 \le \phi \le 2\pi$), which is the *visible region.*

By induction, Eq. (2.91) can be extended to a two-dimensional array with a rectangular grid (d_x, d_y):

$$\sigma_{\phi\phi} \sim \left[\frac{\sin(N_x k d_x u)}{\sin(k d_x u)}\right]^2 \left[\frac{\sin(N_y k d_y v)}{\sin(k d_y v)}\right]^2 \tag{2.94}$$

where the subscripts x and y refer to quantities related to the x and y directions. The Bragg condition is satisfied when both factors are a maximum:

$$\begin{aligned} k d_x u_m &= m\pi, \quad \text{where} \quad m = \pm 1, \pm 2, \ldots \\ k d_y v_\ell &= \ell\pi, \quad \text{where} \quad \ell = \pm 1, \pm 2, \ldots \end{aligned} \tag{2.95}$$

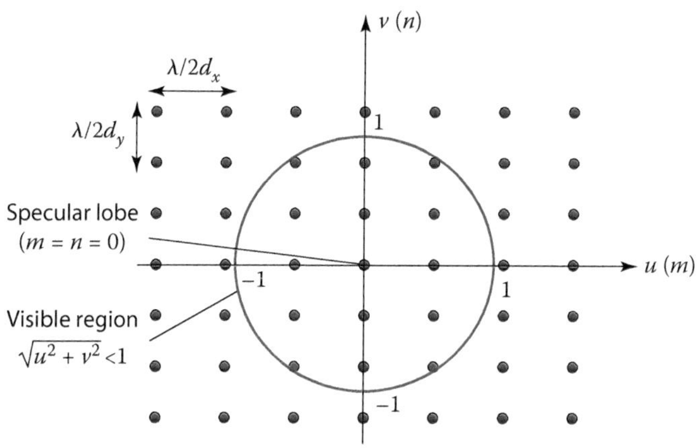

Fig. 2.27 Bragg diagram for a planar array with a rectangular grid.

yielding

$$u_m = \frac{m\lambda}{2d_x}, \quad v_\ell = \frac{\ell\lambda}{2d_y} \tag{2.96}$$

A plot of Bragg lobe locations in direction cosine space is called a *Bragg diagram*. An example is shown in Fig. 2.27. The unit circle represents a hemisphere of the visible region. If the element grid is rectangular, the spacing of lobes is also rectangular. A diagram such as this can be used to determine the required spacing of an array to restrict the location and number of Bragg lobes.

Example 2.10: Bragg Diffraction from a Rectangular Grid

A rectangular grid of 40×40 isotropic elements has a spacing of $\lambda/2$ in both x and y. Bragg lobes will occur at

$$u_1 = v_1 = \pm 1$$

The Bragg diagram for the array factor in direction cosine space is shown in Fig. 2.28. In both principal planes, the lobes are just entering the visible

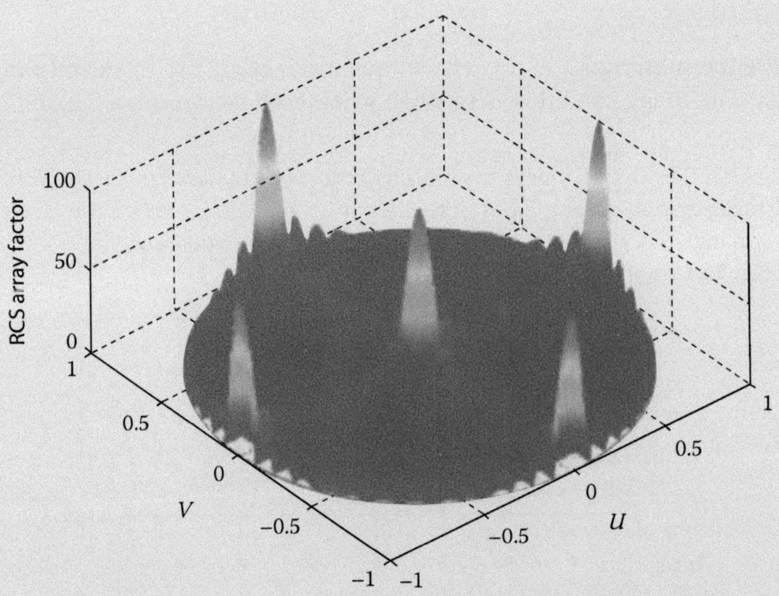

Fig. 2.28 Bragg scattering pattern in direction cosine space for an array with $N_x = N_y = 10$ and $d_x = d_y = 0.5\lambda$.

(Continued)

Example 2.10: Bragg Diffraction from a Rectangular Grid *(Continued)*

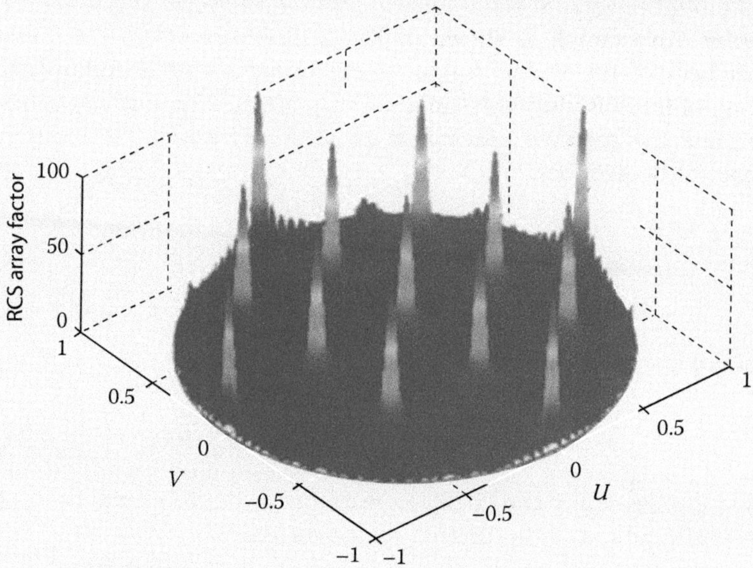

Fig. 2.29 Bragg scattering pattern in direction cosine space for an array with $N_x = N_y = 10$ and $d_x = d_y = 1\lambda$.

region. The central lobe is the specular reflection at normal incidence. (Recall that radiation grating lobes first enter the visible region when the element spacing is λ.) If the spacing is increased to 1λ, the Bragg diagram of Fig. 2.29 results.

2.13 Impedance Boundary Conditions

2.13.1 Definition of Surface Impedance

The equivalence principle can be used to obtain the currents flowing on a surface in terms of the total fields $\vec{E}$ and $\vec{H}$. The total fields are a sum of the incident and scattered fields:

$$\begin{aligned}\vec{E} &= \vec{E}_i + \vec{E}_s \\ \vec{H} &= \vec{H}_i + \vec{H}_s\end{aligned} \tag{2.97}$$

Unfortunately, the scattered fields are unknown because the current induced throughout the scattering body is unknown. An estimate for the current that depends only on the incident field is desirable. The physical optics approximation is one such estimate for perfect electric or magnetic conductors. A similar approximation for bodies of more complex composition is based on the *impedance boundary condition* (IBC).

The impedance boundary concept was conceived in the late 1940s by Leontovich [3] and is frequently referred to as the *Leontovich boundary condition.* It relates the tangential components of the electric and magnetic fields at every point on the surface:

$$\eta_s = \frac{E_{\tan}}{H_{\tan}} \quad \text{on } S \tag{2.98}$$

where η_s is the *surface impedance* in ohms. In vector form, Eq. (2.98) can be written

$$\vec{E} - \hat{n}(\hat{n} \cdot \vec{E}) = \eta_s \hat{n} \times \vec{H} = \eta_s \vec{J}_s \tag{2.99}$$

There is also a dual to this equation:

$$\vec{H} - \hat{n}\left(\hat{n} \cdot \vec{H}\right) = -\frac{1}{\eta_s} \hat{n} \times \vec{E} = \frac{1}{\eta_s} \vec{J}_{ms} \tag{2.100}$$

Comparing Eqs. (2.99) and (2.100) reveals that the surface impedance relates the electric and magnetic currents on the surface:

$$-\vec{J}_{ms} = \eta_s \hat{n} \times \vec{J}_s \tag{2.101}$$

Frequently, the *normalized surface impedance* Z_s is used. It is defined by

$$Z_s = \frac{\eta_s}{\eta_0} \tag{2.102}$$

where η_0 is the impedance of free space. Note that $\eta_s = 0$ corresponds to a perfect electric conductor.

2.13.2 Equivalent Surface Impedance of a Boundary

An example of how one might apply the impedance boundary condition to a scattering problem is illustrated in Fig. 2.30. The target, which can be composed of multiple types of materials, is enclosed by a surface *S*. To solve this problem rigorously requires that the current be determined throughout the volume. Solving for the current is an extremely difficult problem for targets of arbitrary shape and composition. Given that our objective is to find the scattered fields outside the body, the surface equivalence principle can be invoked to find the surface currents that will provide the same fields outside *S* but null fields inside *S*. These currents will depend on the *total* fields on the surface

$$\begin{aligned} \vec{J}_s &= \hat{n} \times (\vec{H}_i + \vec{H}_s) \\ \vec{J}_{ms} &= -\hat{n} \times (\vec{E}_i + \vec{E}_s) \end{aligned} \tag{2.103}$$

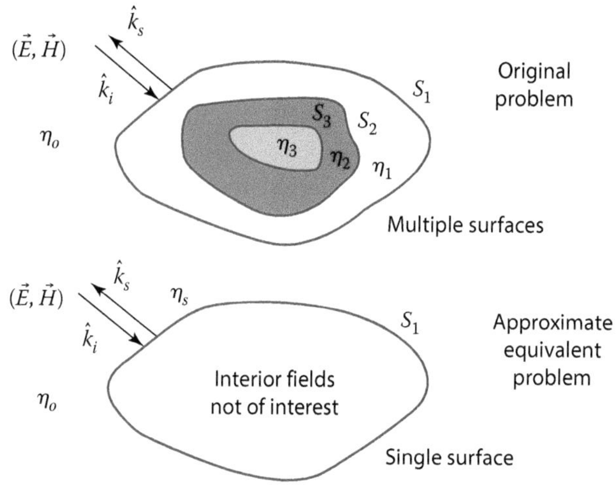

Fig. 2.30 Using the impedance boundary condition to simplify a scattering problem.

In general, both electric and magnetic currents are present. However, the impedance boundary condition relates the two and thus eliminates one unknown quantity (either $\vec{J}_s$ or $\vec{J}_{ms}$). The remaining unknown must be determined analytically or numerically, or an estimate must be provided as in the case of the physical optics approximation.

Example 2.11: Equivalent Surface Impedance of a Planar Interface

A simple example of a surface impedance calculation is one for a planar interface between air and a dielectric with relative permittivity ϵ_r. If a TEM wave is normally incident, as shown in Fig. 2.31, the wave impedance in the dielectric is the intrinsic impedance because the angle of incidence is zero:

$$\eta = \sqrt{\frac{\mu_0}{\epsilon}} = \frac{\eta_0}{\sqrt{\epsilon_r}} \tag{2.104}$$

Using a ray-based approach the scattered field consists of reflected and transmitted terms. The reflected field exists in region 1 and the transmitted field exists in region 2. The tangential fields must satisfy the boundary conditions,

$$|\vec{E}_{\text{tan}}| = E_i + E_r = E_t \rightarrow 1 + \Gamma = \tau \tag{2.105}$$

$$|\vec{H}_{\text{tan}}| = H_i + H_r = H_t \rightarrow \frac{1}{\eta_0}(1 - \Gamma) = \frac{\tau}{\eta} \tag{2.106}$$

(Continued)

Example 2.11: Equivalent Surface Impedance of a Planar Interface *(Continued)*

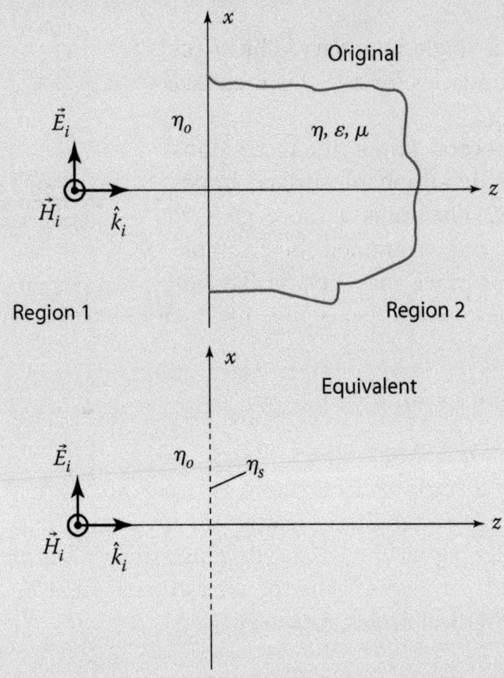

Fig. 2.31 Impedance boundary conditions applied to a planar interface.

where the subscripts *i*, *r*, and *t* refer to incident, reflected, and transmitted terms of the tangential fields. Taking the ratio of Eqs. (2.105) and (2.106) and comparing the result to the definition of surface impedance in Eq. (2.98) give

$$\eta_s = \frac{E_{\tan}}{H_{\tan}} = \eta \tag{2.107}$$

For an observer in free space (region 1), a problem equivalent to the original one is also shown in Fig. 2.31. Medium 2 is enclosed by a surface *S* that extends to infinity with surface impedance $\eta_s = \eta$. The same boundary conditions are satisfied by the tangential components of the fields as in the original problem.

From Example 2.11, it appears that there may not be any advantage to using a surface impedance because a knowledge of the reflection coefficient was required to obtain η_s. However, it is possible to estimate η_s directly from the material parameters in region 2 and their arrangement. For example, the multilayered half-space can also be reduced to the same equivalent problem as the one in Fig. 2.31. The surface impedance will be determined by the wave impedance looking into the material at $z = 0$. When calculating the scattering for the medium using the surface impedance approach, one need only deal with a single boundary. The effect of all other boundaries for $z > 0$ is included in the value of η_s.

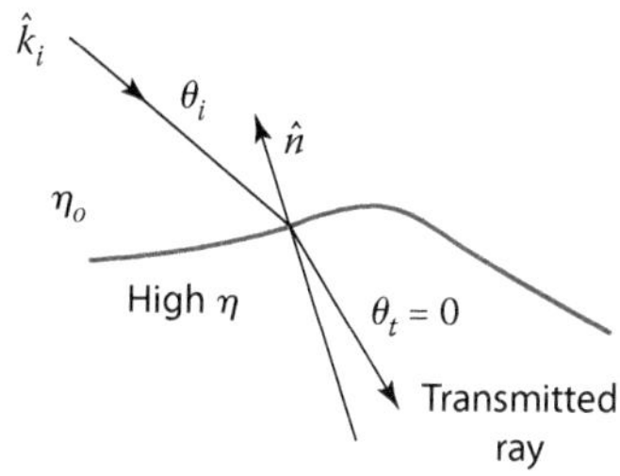

Fig. 2.32 Surface impedance approximation restrictions.

There are several important restrictions implied in the definition of surface impedance. Fig. 2.32 illustrates a more general case than the one examined in Example 2.11. First, the surface may not be flat but may have some curvature. Second, the incident wave will not always be normal to the surface and, consequently, the field vectors will have a normal component in addition to the tangential components. In light of these possibilities, the surface impedance approximation is expected to be accurate only for

1. Near-normal incidence angles, $\theta_i \approx 0$.
2. High-dielectric interfaces, $\left|\sqrt{\epsilon_r}\right| \gg 1$, or $Z_s = (\eta_s/\eta_0) \ll 1$.
3. "Locally flat surfaces" (radius of curvature $\gg \lambda$).

Even though the restrictions appear to be severe, the surface impedance approximation is frequently used when the conditions are grossly violated; nevertheless, the results are useful. The scattering from an impedance-loaded plate illustrates the procedure for computing the RCS using the IBC.

Example 2.12: Scattering from an Impedance-Loaded Plate

The same plate geometry as in Example 2.4 (Fig. 2.12) is assumed. The plate has a constant surface impedance η_s, and a TM$_z$-polarized plane wave is incident [Eq. (2.38) with $E_{0\phi} = 0$]. Let the current on the plate be given by the physical optics approximation of Eq. (2.108), again assuming the plate is located in a material with intrinsic impedance η

$$\vec{J}_s \approx -2\hat{z} \times \hat{\phi} E_{0\theta} \frac{e^{-j\vec{k}_i \cdot \vec{r}}}{\eta} \tag{2.108}$$

or

$$\vec{J}_s \approx \frac{2e^{jkh}}{\eta} E_{0\theta}(\hat{x}\cos\phi + \hat{y}\sin\phi) = \frac{2e^{jkh}}{\eta} E_{0\theta}\hat{\rho} \tag{2.109}$$

(Continued)

Example 2.12: Scattering from an Impedance-Loaded Plate *(Continued)*

where $h = x' \sin\theta\cos\phi + y' \sin\theta\sin\phi$. The magnetic current on the surface is obtained using the IBC (2.101)

$$\vec{J}_{ms} = -\eta_s \hat{n} \times \vec{J}_s = \frac{-2\eta_s E_{0\theta} e^{jkh}}{\eta} \hat{z} \times \hat{\rho} \tag{2.110}$$

When this current is used in the radiation integral (2.22),

$$E_\theta(x, y, z) = \frac{-jk}{2\pi r} E_{0\theta} e^{-jkr} \int_{-(a/2)}^{a/2} \int_{-(b/2)}^{b/2} \left[\hat{\rho} \cdot \hat{\theta} - \frac{\eta_s}{\eta} (\hat{z} \times \hat{\rho}) \cdot \hat{\phi} \right] e^{jkh} e^{jkg} \mathrm{d}y' \, \mathrm{d}x' \tag{2.111}$$

For monostatic scattering, $h = g$, and using

$$\begin{aligned} \hat{\rho} \cdot \hat{\theta} &= \cos\theta \\ (\hat{z} \times \hat{\rho}) \cdot \hat{\phi} &= 1 \end{aligned} \tag{2.112}$$

in Eq. (2.111) yields

$$E_\theta(x, y, z) = \frac{-jk}{2\pi r} E_{0\theta} e^{-jkr} (\cos\theta - Z_s) \int_{-(a/2)}^{a/2} \int_{-(b/2)}^{b/2} e^{j2kg} \mathrm{d}y' \, \mathrm{d}x' \tag{2.113}$$

Performing the integration gives

$$E_\theta(x, y, z) = \frac{-jkab}{2\pi r} E_{0\theta} e^{-jkr} (\cos\theta - Z_s) \operatorname{sinc}(kua)\operatorname{sinc}(kbv) \tag{2.114}$$

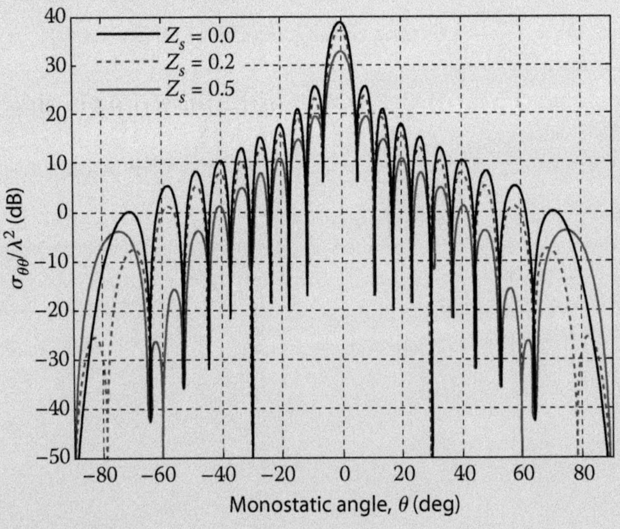

Fig. 2.33 Principal plane RCS of a 5λ by 5λ plate (TM polarization).

(Continued)

Example 2.12: Scattering from an Impedance-Loaded Plate *(Continued)*

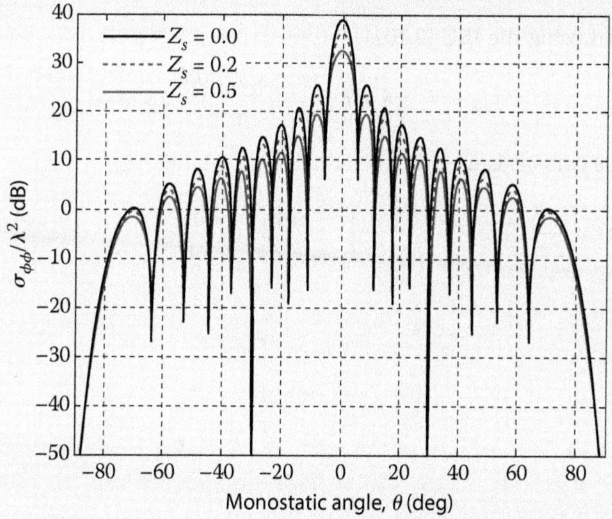

Fig. 2.34 Principal plane RCS of a 5λ by 5λ plate (TE polarization).

and a RCS of

$$\sigma_{\theta\theta}(\theta,\phi) = \frac{4\pi A^2}{\lambda^2}|\cos\theta\ \mathrm{sinc}(kua)\mathrm{sinc}(kbv)(1 - Z_s/\cos\theta)|^2 \tag{2.115}$$

A similar calculation for the TE_z polarization (with $E_{0\theta} = 0$) yields

$$\sigma_{\phi\phi}(\theta,\phi) = \frac{4\pi A^2}{\lambda^2}|\cos\theta\ \mathrm{sinc}(kua)\mathrm{sinc}(kbv)(1 - Z_s\cos\theta)|^2 \tag{2.116}$$

Some typical scattering patterns for a 5λ square plate with various surface impedances are shown in Figs. 2.33–2.34. In all cases, $\phi = 0$.

2.13.3 Reflection Coefficient of an Impedance-Coated Surface

An examination of Eqs. (2.115) and (2.116) reveals that

1. $Z_s = 0$ reduces to the PEC case as expected.
2. $Z_s = 1$ forces the RCS to zero and thus corresponds to a surface matched to free space for normal incidence.
3. For $Z_s > 1$, the RCS increases without bound. This is physically not possible and is attributed to the fact that the restriction $\sqrt{\epsilon_r} \gg 1$ is violated.

Recalling that the plane wave reflection coefficient for a perpendicularly polarized wave is (see Appendix A)

$$\Gamma_{\perp} = \Gamma_{\mathrm{TE}} = \frac{\eta_2 \cos\theta_i - \eta_1 \cos\theta_t}{\eta_2 \cos\theta_i + \eta_1 \cos\theta_t} = \frac{Z_s \cos\theta - 1}{Z_s \cos\theta + 1} \approx Z_s \cos\theta - 1 \quad (2.117)$$

for $|Z_s| \ll \cos\theta$ and $\cos\theta_t \approx 1$. Similarly, for parallel polarization,

$$\Gamma_{\parallel} = \Gamma_{\mathrm{TM}} = \frac{\eta_2 \cos\theta_t - \eta_1 \cos\theta_i}{\eta_2 \cos\theta_t + \eta_1 \cos\theta_i} = \frac{Z_s - \cos\theta}{Z_s + \cos\theta} \approx Z_s - \cos\theta \quad (2.118)$$

From these, one obtains

$$|\Gamma_{\mathrm{TM}}|^2 \approx |\cos\theta - Z_s|^2 \quad (2.119)$$

$$|\Gamma_{\mathrm{TE}}|^2 \approx |1 - Z_s \cos\theta|^2 \quad (2.120)$$

which leads to the result

$$\sigma_{\theta\theta} = \sigma_{\mathrm{PEC}} |\Gamma_{\mathrm{TM}}|^2 \quad (2.121)$$

$$\sigma_{\phi\phi} = \sigma_{\mathrm{PEC}} |\Gamma_{\mathrm{TE}}|^2 \quad (2.122)$$

σ_{PEC} is the RCS of a plate of the same dimensions made of perfect electric conducting material. Thus, the surface impedance can be interpreted as reducing the reflection from the plate by an amount determined by the value of $|\Gamma|^2$.

As a final comment, it should be recalled that the physical optics approximation was intuitively explained on the basis of image theory. The tangential components of $\vec{H}$ over a ground plane are just twice those from the same source in unbounded free space. However, this assumes a perfect ground plane, which is not the case when $Z_s > 0$. Thus, when Z_s is substantially different from zero, a better estimate of the current might be

$$\vec{J}_s \approx (1 - \Gamma)\hat{n} \times \vec{H}_i, \quad \text{where} \quad \Gamma = \frac{Z_s - 1}{Z_s + 1} \quad (2.123)$$

The factor $1 - \Gamma$ accounts for the fact that the intensity of the image is reduced when the surface is not a perfect conductor. When $\Gamma = -1$, Eq. (2.123) reduces to the usual PO expression.

2.14 Discontinuity Boundary Conditions

Discontinuity boundary conditions are used to represent thin films. They are referred to as the *resistive sheet* and *conductive sheet* boundary conditions. They provide a relationship between the current on the surface and the tangential field components and are illustrated in Fig. 2.35.

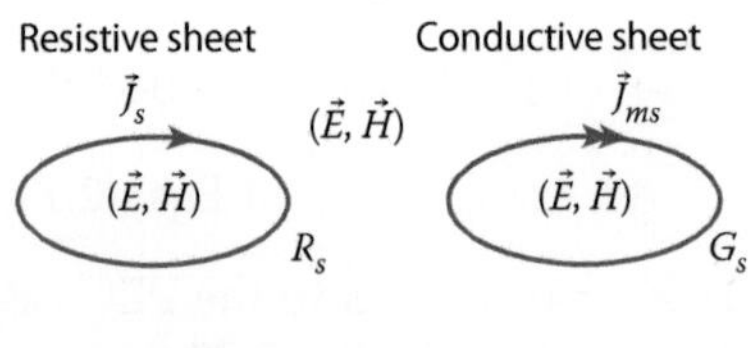

Fig. 2.35 Resistive and conductive sheet boundary conditions.

2.14.1 Resistive Sheet

A *resistive sheet* is an infinitely thin imperfect electric conductor. Its conductivity is finite ($\sigma_c < \infty$), and it does not support a magnetic current ($\vec{J}_{ms} = 0$). The surface current is related to the tangential electric field through the surface resistivity as

$$\vec{E} - \hat{n}(\hat{n} \cdot \vec{E}) = R_s \vec{J}_s \tag{2.124}$$

Taking $\hat{n} \times$ both sides of Eq. (2.124) and solving for $\vec{J}_s$ give

$$\vec{J}_s = -\frac{1}{R_s} \hat{n} \times (\hat{n} \times \vec{E}) \tag{2.125}$$

The common unit of surface resistance is *ohms per square*, Ω/sq. This is derived from the basic definition of resistance for the special case of length equal to width ($\ell = w$):

$$R_s = \frac{\ell}{\sigma_c t w} = \frac{1}{\sigma_c t} \tag{2.126}$$

For a very thin film of material that may be only a few atoms thick, the *film conductivity* σ_f should be used in place of the *bulk conductivity* $\sigma_f \leq \sigma_c$ because the mean free path of electrons is reduced in the film.

Example 2.13: Reflection Coefficient of a Resistive Sheet

Figure 2.36 shows a plane wave incident on an infinite resistive sheet of thickness t and surface resistivity R_s. A transmission line equivalent is also shown. The characteristic impedance of the section of line representing the sheet is

$$Z_c = \sqrt{\frac{j\omega\mu}{\sigma_c + j\omega\epsilon}} \tag{2.127}$$

The wave impedance looking into the film at the surface is

$$Z_{\text{in}} = Z_c \frac{Z_0 + Z_c \tanh \gamma t}{Z_c + Z_0 \tanh \gamma t} \tag{2.128}$$

where Z_0 is the wave impedance of the section of transmission line that represents free space on the back side ($= \eta_0$ for normal incidence). The propagation constant is

$$\gamma = \alpha + j\beta = \sqrt{j\omega\mu(\sigma_c + j\omega\epsilon)} \approx \frac{(1+j)}{\delta} \tag{2.129}$$

(Continued)

Example 2.13: Reflection Coefficient of a Resistive Sheet *(Continued)*

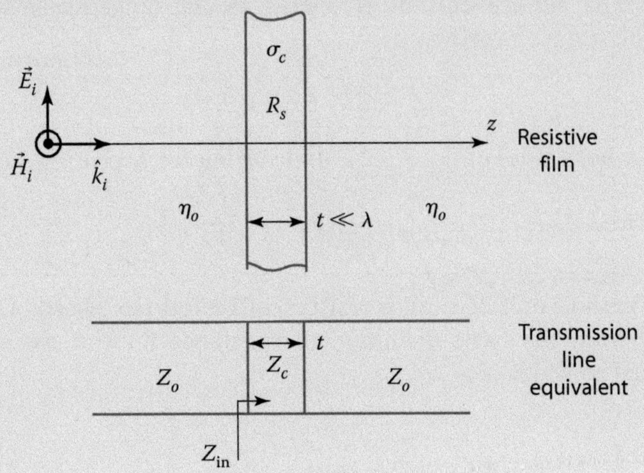

Fig. 2.36 Plane wave incident on a resistive film and its transmission line equivalent circuit.

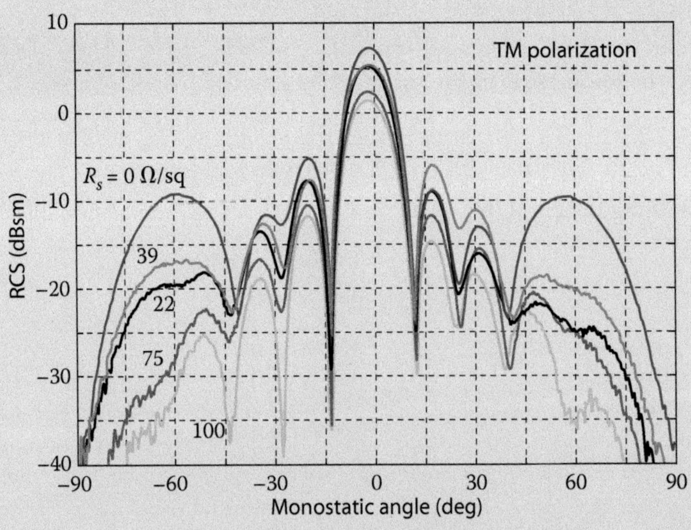

Fig. 2.37 Measured principal plane RCS of 2.2λ by 2.2λ resistive sheets (TE polarization).

(Continued)

Example 2.13: Reflection Coefficient of a Resistive Sheet *(Continued)*

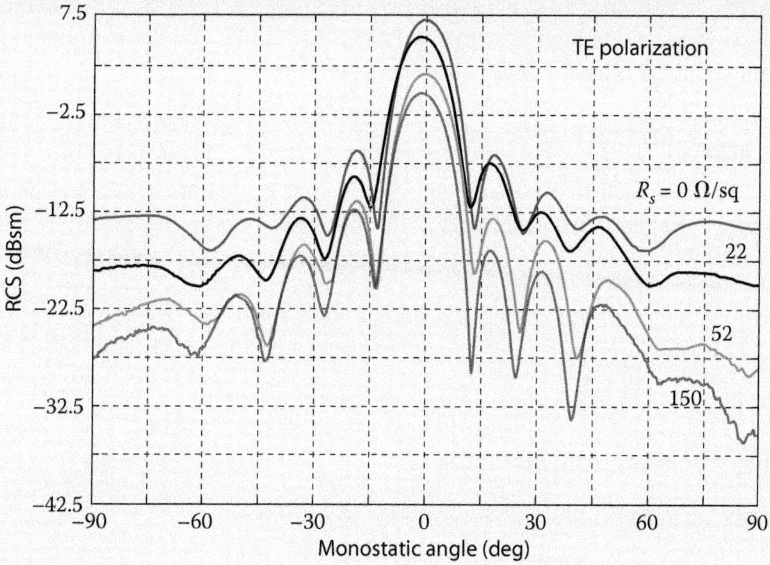

Fig. 2.38 Measured principal plane RCS of 2.2λ by 2.2λ resistive sheets (TM polarization).

and, if the thickness is much less than the skin depth ($t \ll \delta$), then,

$$\tanh \gamma t = \tanh\left[\frac{t(1+j)}{\delta}\right] \approx \gamma t \tag{2.130}$$

Substituting in $R_s = 1/(\sigma_c\, t)$ and $\delta = t$, the input impedance becomes

$$Z_{\text{in}} \approx Z_0 \frac{1 + j\beta t}{1 + Z_0 \sigma_c t} = \frac{R_s Z_0}{R_s + Z_0} \tag{2.131}$$

Finally, the reflection coefficient is

$$\Gamma = \frac{Z_{\text{in}} - Z_0}{Z_{\text{in}} + Z_0} = \frac{-Z_0}{2R_s + Z_0} \tag{2.132}$$

Figure 2.37 and Fig. 2.38 show the measured RCS of 2.2λ square sheets with various resistivities.

The two limiting values of surface resistance are $R_s = 0$ ($\Gamma = -1$), which represents a perfect electric conductor, and $R_s = \infty$ ($\Gamma = 0$), which represents a surface matched to free space.

Resistive films are an effective means of controlling the reflection coefficient of a surface. They consist of thin metallic deposits on low dielectric films such as Mylar. A common example of a commercially available resistive film is window tinting of the type used for automobiles. Note that the vast majority of the energy that is not reflected at the sheet is transmitted. In spite of the designation *resistive surface*, very little energy is absorbed by the material, as would be the case for a resistor in a circuit.

2.14.2 Conductive Sheet

There is a dual to the resistive sheet called the *conductive sheet*. It is an imperfect magnetic conductor ($\vec{J}_s = 0$), and the magnetic current is related to the total magnetic field at the surface by

$$\vec{J}_{ms} = -\frac{1}{G_s}\hat{n} \times (\hat{n} \times \vec{H}) \tag{2.133}$$

2.15 Surface Waves

2.15.1 General Comments

The term *surface wave* is used to describe several related propagation mechanisms that occur at an interface between two different media. For example, in the case of plane wave reflection at an air/glass interface, a surface wave can propagate relatively unattenuated along the boundary when certain requirements are satisfied (see Appendix A). The boundary acts like an efficient transmission line, guiding the wave along until a discontinuity is encountered, at which point a reflection or radiation occurs.

Surface waves are a very important consideration in RCS design. Although their scattered signal levels are generally low compared to specular reflections, surface waves are frequently troublesome because they can be difficult to control. Surface wave control has been the driving factor in the development of new radar absorbing materials (RAM) as well as in the shaping of edges and other discontinuities.

Surface waves that propagate with very little or no attenuation are referred to as *traveling waves*. In most cases, a surface wave will radiate as it propagates. These are called *leaky waves* and are commonly found on curved bodies. A *creeping wave* is one that travels around the back (shadowed) side of a curved body such as a sphere and reemerges on the illuminated side. A creeping wave is continually radiating as it propagates and is therefore losing energy. For a good conducting body, the rate of energy loss depends on the curvature of the surface: the smaller the radius of curvature in terms of wavelength, the stronger the radiation.

2.15.2 Surface Wave Solutions to the Wave Equation

The behavior of surface waves can be determined by considering the interface to be a waveguide and computing its propagation constant.

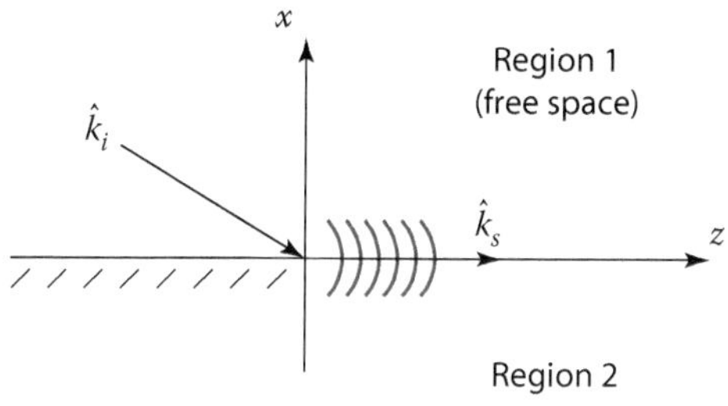

Fig. 2.39 Coordinate systems for surface wave propagation.

The propagation constant, in turn, depends on the surface impedance of the interface. For waveguide analysis, the z axis is traditionally chosen as the direction of propagation and, therefore, the coordinate system defined in Fig. 2.39 will be used in this section. The figure shows an infinite planar surface in the $x = 0$ plane. Waves existing in the source-free region $x > 0$ must satisfy the scalar Helmholtz equation

$$(\nabla^2 + k^2)\Psi(x, z) = 0 \qquad (2.134)$$

where ψ is a scalar wave function related to either the electric or magnetic field. TM_z and TE_z waves for incidence in the x–z plane can be defined by

$$\text{TM: } H_y = \Psi(x, z)H_0 \qquad (2.135)$$

$$\text{TE: } E_y = \Psi(x, z)E_0 \qquad (2.136)$$

With a separation of variables approach [4], the wave function can be written as the product

$$\Psi(x, z) = e^{-\gamma_x x}e^{-\gamma_z z} \qquad (2.137)$$

and using this in Eq. (2.134) gives the *separation equation*

$$\gamma_x^2 + \gamma_z^2 + k^2 = 0 \qquad (2.138)$$

Because γ_x and γ_z are propagation constants and, in general, can be complex, each can be written in terms of attenuation and phase constants:

$$\gamma_x = \alpha_x + j\beta_x \qquad (2.139)$$

$$\gamma_z = \alpha_z + j\beta_z \qquad (2.140)$$

A vector propagation constant can be defined using Eqs. (2.139) and (2.140):

$$\vec{\gamma} = \gamma_x\hat{x} + \gamma_z\hat{z} = \vec{\alpha} + j\vec{\beta} \qquad (2.141)$$

Because a position vector is defined by $\vec{r} = x\hat{x} + y\hat{y} + z\hat{z}$, the wave function can be written

$$\Psi(x, z) = e^{-\vec{\gamma} \cdot \vec{r}} \tag{2.142}$$

Returning to the separation equation, expanding

$$(\alpha_x + j\beta_x)^2 + (\alpha_z + j\beta_z)^2 + k^2 = 0 \tag{2.143}$$

and separating into two equations based on the real and imaginary parts give

$$\text{Real part:} \quad \alpha_x^2 + \alpha_z^2 - \beta_x^2 - \beta_z^2 + k^2 = 0 \tag{2.144}$$

$$\text{Imaginary part:} \quad \alpha_x\beta_x + \alpha_z\beta_z = 0 \tag{2.145}$$

The last equation implies that $\vec{\alpha} \cdot \vec{\beta} = 0$; therefore, these two vectors are perpendicular. The planes of constant phase can be determined from the imaginary part,

$$\vec{\beta} \cdot \vec{r} = \text{constant}$$

whereas the planes of constant amplitude are determined from the real part,

$$\vec{\alpha} \cdot \vec{r} = \text{constant}$$

Now, consider a traveling wave propagating in the z direction ($\beta_z > 0$ and $\alpha_z = 0$). The amplitude dependence of the wave in the x direction is

$$|e^{-\gamma_x x}| = e^{-\alpha_x x} \tag{2.146}$$

If $\alpha_x < 0$, there is an exponential increase in the wave amplitude with distance from the surface (this is an *improper wave*). On the other hand, if $\alpha_x > 0$, the wave decays exponentially with increasing x (a *proper wave*).

The characteristics of surface waves and, consequently, their classification are dependent on the values of the attenuation and phase constants. The relationships between all of these cases are summarized in Table 2.3 (from Ref. [5]) for a surface wave propagating in the $+z$ direction ($\beta_z > 0$).

Table 2.3 Attenuation and Phase Constants for Surface Waves (After Ref. [5])

Type	α_x	β_x	α_z	Common Name
Proper	0	>0	0	Fast wave (waveguide mode)
	>0	>0	<0	Backward leaky wave
	>0	0	0	Trapped surface wave
	>0	<0	>0	Zenneck wave
	0	>0	0	Plane wave incidence
Improper	<0	<0	<0	
	<0	0	0	Untrapped surface wave
	<0	>0	>0	Forward leaky wave

Example 2.14: Trapped Surface Wave in an Optical Fiber

For a trapped surface wave, the following conditions hold. 1) The velocity is less than that of free space (slow wave): $\beta_z > \beta_0 = (2\pi)/\lambda = k$. 2) Propagation in the z direction is unattenuated: $\alpha_z = 0$.3) There is no propagation in the x direction: $\beta_x = 0$.

Given that these hold, the wave function reduces to

$$\Psi(x, z) = e^{-\alpha_x x} e^{-j\beta_z z} \tag{2.147}$$

and the separation equation for this case is

$$(j\beta_z)^2 + \alpha_x^2 + k^2 = 0 \tag{2.148}$$

yielding $\alpha_x = \sqrt{\beta_z^2 - k^2}$. Therefore, the "slower" the wave, the greater is β_z $(=\omega/u_p)$ and the greater is α_x. A large α_x implies that the wave decays rapidly with distance from the interface. This is called a *tightly bound* surface wave because it is concentrated near the interface.

The conditions listed in Example 2.14 are frequently satisfied by phased array apertures that consist of periodic structures (e.g., dipoles). When the wave is incident at a critical angle, it will be trapped at the array surface and will propagate along it with little loss. None of the energy enters the antenna, and the received signal drops to zero. This is called *scan blindness* and is avoided by proper design of the radiating elements and judicious choice of the geometry for the frequency band of operation.

Leaky waves are surface waves that radiate as they propagate along a surface. They have the following properties:

1. The velocity is greater than that of free space: $\beta_z < \beta_0 = (2\pi)/\lambda = k$.
2. Attenuation is present in the z direction: $\alpha_z > 0$.
3. The wave propagates away from the surface in the $+x$ direction: $\beta_x > 0$.

From Property 3 and Eq. (2.145), $\alpha_x < 0$. Thus, the amplitude of the wave increases with distance from the interface. This seems intuitively impossible, but leaky waves can exist in limited regions of space. An example is the slotted waveguide, where the amplitude decreases as the wave propagates in the positive z direction [5].

2.15.3 Surface Impedance and Surface Waves

The next task is to obtain a relationship between the surface wave propagation constants and the surface impedance. Surface wave behavior was encountered at a planar interface when the index of refraction in region 1 was greater than the index of refraction in region 2 (see Appendix A). With the equations in Appendix A rearranged to conform to the coordinate

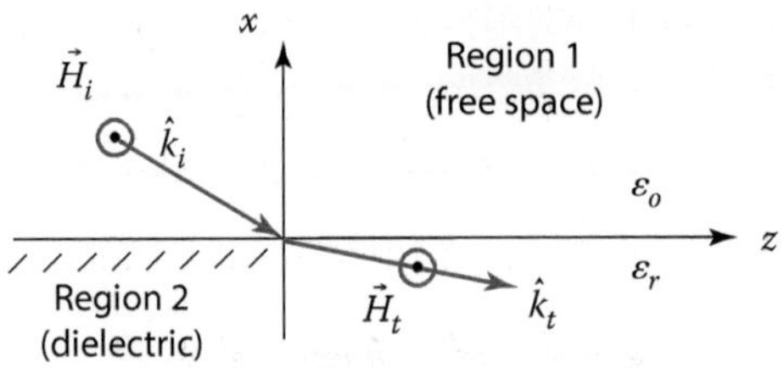

Fig. 2.40 TM$_x$-polarized plane wave incident on a dielectric/air interface.

system in Fig. 2.40 (i.e., the x and z axes interchanged), the transmitted field at the Brewster's angle is [6]

$$H_{yi} = \begin{cases} H_0 \exp\left[-jk_0 \dfrac{\sqrt{\epsilon_r} z - x}{\sqrt{\epsilon_r + 1}}\right], & x > 0 \\ H_0 \exp\left[-jk_0 \dfrac{\sqrt{\epsilon_r} z - \epsilon_r x}{\sqrt{\epsilon_r + 1}}\right], & x < 0 \end{cases} \tag{2.149}$$

It is anticipated that a surface wave traveling along the interface will have the same form as Eq. (2.149). The incident wave is arriving from region 1 and traveling in the $-x$ direction. Because the reflected wave is zero ($H_{ry} = 0$), the total field in region 1 will be

$$H_{y1} = H_{yi} = H_0 \exp[(\gamma_{x1} x) - (\gamma_z z)] \tag{2.150}$$

From Maxwell's equations,

$$E_{z1} = \frac{\gamma_{x1}}{j\omega\epsilon_0} H_{y1} \tag{2.151}$$

which gives

$$Z_1 = \frac{E_{z1}}{H_{y1}} = \frac{-j\gamma_{x1}\eta_0}{k_0} \tag{2.152}$$

Free space has been assumed for region 1 and, therefore, the subscript 0 is used on k and η.

Below the surface ($x < 0$), the total field consists solely of the transmitted component,

$$H_{y2} = H_{yt} = H_0 \exp[(\gamma_{x2} x) - (\gamma_z z)] \tag{2.153}$$

Again, from Maxwell's equations,

$$E_{z2} = \frac{\gamma_{x2}}{j\omega\epsilon} H_{y2} \tag{2.154}$$

which gives

$$Z_2 = \frac{E_{z2}}{H_{y2}} = \frac{-j\gamma_{x2}\eta}{k} \tag{2.155}$$

Example 2.15: Propagation Constants at an Air/Dielectric Interface

In this example, the special case of a low-loss dielectric is examined:

$$\epsilon = \epsilon' - j\epsilon'' \quad \text{where} \quad \epsilon' \gg \epsilon'' \tag{2.156}$$

Because the separation equation must hold at the interface ($x = 0$),

$$\gamma_z|_{x>0} = \gamma_z|_{x<0} \tag{2.157}$$

$$k_0^2 + \gamma_{x1}^2 = k^2 + \gamma_{x2}^2 \tag{2.158}$$

Solving for γ_{x1} gives

$$\gamma_{x1} = \sqrt{\gamma_{x2}^2 + k^2 - k_0^2} = \sqrt{\gamma_{x2}^2 - k_0^2(1 - \epsilon_r)} \tag{2.159}$$

An additional relationship between the propagation constants in regions 1 and 2 is obtained by enforcing continuity of the tangential component of the fields at the boundary. From Eqs. (2.152) and (2.155),

$$Z_1 = Z_2 \quad \text{or} \quad \frac{\gamma_{x1}}{\epsilon_0} = \frac{\gamma_{x2}}{\epsilon} \tag{2.160}$$

yielding

$$\gamma_{x1} = \frac{\gamma_{x2}}{\epsilon_r} \tag{2.161}$$

Equations (2.159) and (2.161) provide enough information to solve for the two propagation constants,

$$\gamma_{x2} = \frac{jk_0\epsilon_r}{\sqrt{\epsilon_r + 1}} \tag{2.162}$$

$$\gamma_{x1} = \frac{k_0}{\sqrt{\epsilon_r + 1}} \tag{2.163}$$

Substituting in the complex representation for ϵ gives

$$\gamma_{x2} = \frac{jk_0(\epsilon_r' - j\epsilon_r'')}{\sqrt{\epsilon_r' + 1}\sqrt{1 - [j\epsilon_r''/(1 + \epsilon_r')]}} \tag{2.164}$$

The square root term in the denominator that contains ϵ_r'' can be expanded in a binomial series and all but the first two terms dropped. The results are

$$\gamma_{x2} \approx \frac{jk_0}{\sqrt{\epsilon_r' + 1}}\left[\epsilon_r' - \frac{j\epsilon_r''}{2}\frac{\epsilon_r' + 2}{\epsilon_r' + 1}\right] \tag{2.165}$$

(Continued)

Example 2.15: Propagation Constants at an Air/Dielectric Interface *(Continued)*

and, similarly, for γ_{x1},

$$\gamma_{x1} \approx \frac{jk_0}{\sqrt{\epsilon'_r + 1}}\left[1 + \frac{j\epsilon''_r}{2(\epsilon'_r + 1)}\right] \tag{2.166}$$

When the real parts of Eqs. (2.165) and (2.166) are examined, it can be seen that $\alpha_{x1} < 0$ and $\alpha_{x2} > 0$. Thus, the wave attenuates with distance from $x = 0$ as expected.

Referring again to Fig. 2.40 and Eq. (2.152), we have

$$Z_1 = \frac{E_z}{H_y} = -\frac{j\gamma_{x1}}{k_0}\eta_0 = Z_s\eta_0 \tag{2.167}$$

Therefore,

$$Z_s = -j\frac{\alpha_{x1} + j\beta_{x1}}{k_0} = \frac{1}{k_0}(\beta_{x1} - j\alpha_{x1}) \equiv R'_s + jX'_s \tag{2.168}$$

from which the normalized resistance and reactance are obtained:

$$R'_s = \frac{\beta_{x1}}{k_0} \quad \text{and} \quad X'_s = -\frac{\alpha_{x1}}{k_0} \tag{2.169}$$

Because H_{y1} varies as $e^{\gamma_{x1}x}$, there will be an exponential decay away from the surface (in the positive x direction) only when $\alpha_{x1} < 0$. Consequently, for TM-polarized incidence, a surface wave will be supported by the interface only when

$$X'_s > 0 \Rightarrow \text{inductive surface impedance} \tag{2.170}$$

From the separation equation, at the boundary,

$$\begin{aligned} \gamma_z^2 &= -k_0^2 - \gamma_{x1}^2 = -k_0^2 - (\alpha_{x1} + j\beta_{x1})^2 \\ \gamma_z &= k_0\sqrt{(R'_s)^2 - [1 + (X'_s)^2] + 2jX'_sR'_s} \end{aligned} \tag{2.171}$$

Equation (2.171) shows that, for small attenuation in the z direction, the product $X'_sR'_s$ must be small. For a closely bound wave that attenuates rapidly with distance from the surface, X'_s must be large. These conditions are satisfied by periodic structures and coated conductors such as those shown in Fig. 2.41.

For TE polarization the roles of the electric and magnetic fields are interchanged. A similar analysis for TE polarization would show that a surface wave is supported only when there is a capacitive component of the

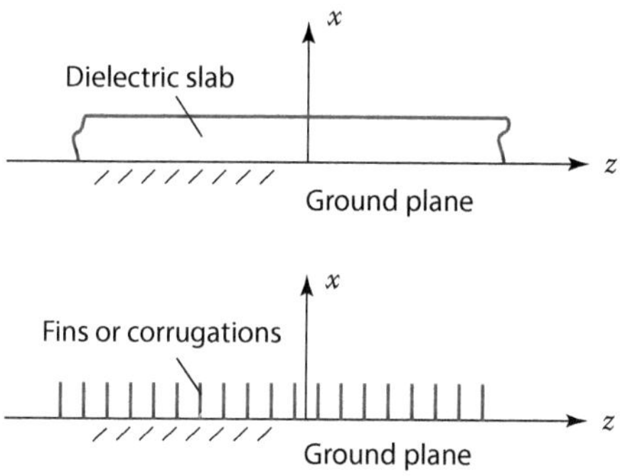

Fig. 2.41 Structures that support surface waves.

surface impedance. One situation where this occurs is when the permeabilities of the two materials forming the interface are different. Recently, RCS treatments have incorporated more magnetic materials because they tend to maintain their performance over a wider bandwidth than nonmagnetic materials do. Thus, surface waves for TE incidence will be a growing concern.

Example 2.16: Traveling Wave Along a Thin Strip

Figure 2.42 shows a TM-polarized plane wave incident on a thin strip (after Ref. [6]). If the material is an imperfect conductor ($\sigma_c < \infty$), the normalized surface impedance can be computed by (see Appendix A)

$$Z_s = (1 + j)\sqrt{\frac{k_0}{2\sigma_c \eta_0}} = R'_s + jX'_s \tag{2.172}$$

Using Eq. (2.171) yields

$$\gamma_z = jk_0\sqrt{1 - \frac{jk_0}{\sigma_c \eta_0}} \tag{2.173}$$

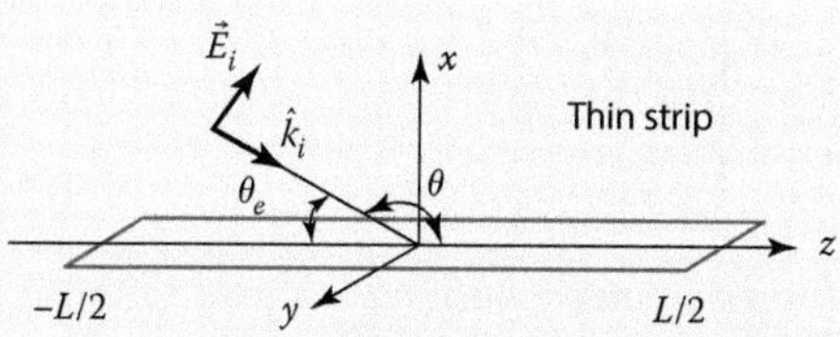

Fig. 2.42 TM wave incident on a thin strip.

(Continued)

Example 2.16: Traveling Wave Along a Thin Strip *(Continued)*

and, because σ_C is very large,

$$\gamma_z \approx jk_0 + \frac{k_0^2}{2\sigma_c\eta_0} \tag{2.174}$$

Typical values are $\sigma_c = 10^7$ S/m, $\eta_0 = 377$ Ω, $\lambda = 3.14$ cm, and $k_0 = 200$ rad/m, which give

$$\alpha_z = 5.3 \times 10^{-6} \text{ Np/m}$$

$$\alpha_{x1} = 3.2 \times 10^{-2} \text{ Np/m}$$

It is apparent that there is essentially no attenuation as the wave travels along the interface, and the attenuation away from the surface is negligible for several thousand wavelengths.

The monostatic scattering characteristics of a traveling wave are clearly illustrated in Fig. 2.43 for $L = 5\lambda$. When the wave is TM-polarized and incident from an angle θ_e, the surface impedance is inductive and the incident wave is captured by the surface. It propagates unattenuated, just as if

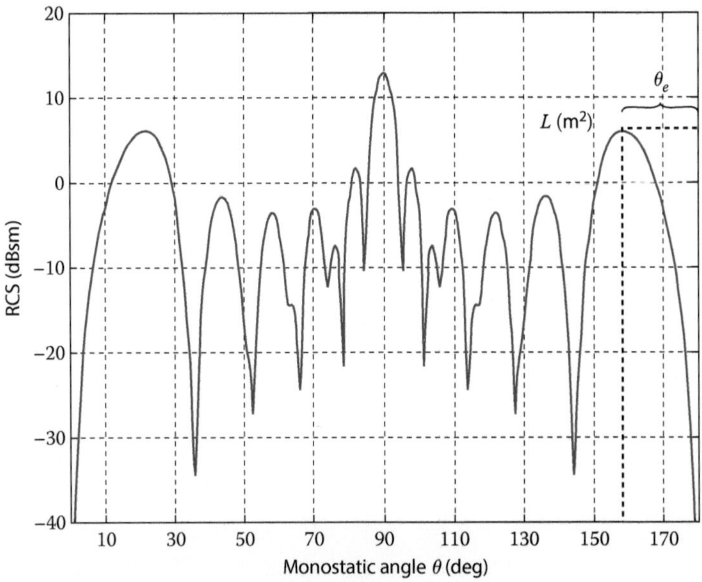

Fig. 2.43 RCS of a 5-wavelength long thin strip showing the traveling wave lobe location and height.

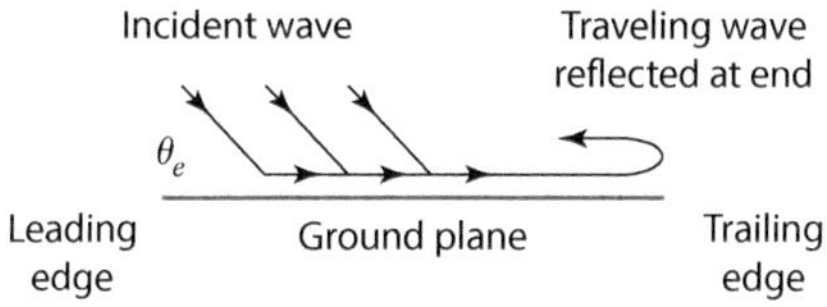

Fig. 2.44 Traveling wave reflection at the trailing edge of a strip.

it were a low-loss transmission line, until it encounters an impedance mismatch at the trailing edge of the strip. This causes most of the surface wave to be reflected and, as it travels back toward the leading edge, it radiates back in the direction θ_e. The process is illustrated in Fig. 2.44. The location and height of the traveling wave lobe depend on the length of the structure L and its surface resistivity. As Fig. 2.37 illustrates, as $R_s \to \infty$, the reflection coefficient at the trailing edge becomes smaller. Consequently, more energy is scattered in the forward direction and less in the direction of θ_e. For good conductors, the lobe height is approximately L^2 m^2, and the edge angle is given by $\theta_e \approx 49.35\sqrt{\lambda/L}$ deg [7].

References

[1] Silver, S., (ed), *Microwave Antenna Theory and Design,* McGraw-Hill, New York, 1949, pp. 80–90.

[2] Sinclair, G., "Theory of Models of Electromagnetic Systems," *Proceedings of the IEE,* Nov. 1948, p. 1364.

[3] Leontovich, M. A., "On the Approximate Boundary Conditions for the Electromagnetic Field on the Surface of Well Conducting Bodies," in: B.A. Vvedenskii, Ed., *Investigations on Propagation of Radio Waves,* AN SSSR, Moscow, p. 5–10, (1948).

[4] Harrington, R. F., *Time-Harmonic Electromagnetic Fields,* McGraw-Hill, New York, 1961.

[5] Ishimaru, A., *Electromagnetic Wave Propagation, Radiation, and Scattering,* Prentice-Hall, Englewood Cliffs, NJ, 1991.

[6] Collin, R. E., *Field Theory of Guided Waves,* McGraw-Hill, New York, 1960.

[7] Peters, L., "End-Fire Echo Area of Long, Thin Bodies," *IRE Transactions on Antennas and Propagation,* Jan. 1958, p. 133.

[8] Astrakhan, M., "Reflecting and Screening Properties of Plane Wire Grids," *Telecommunications and Radio Engineering* (English Ed.), Vol. 23, No. 1, Jan. 1968, p. 76.

[9] Moreira, F., and Prata, A., Jr., "A Self-Checking Predictor-Corrector Algorithm for Efficient Evaluation of Reflector Antenna Radiation Integrals," *IEEE Transactions on Antennas and Propagation,* Vol. 42, No. 2, Feb. 1994, pp. 246–254.

[10] Andreasen, M. G., "Scattering from Cylinders with Arbitrary Surface Impedance," *Proceedings of the IEEE,* Aug. 1965, p. 812.

[11] Mei, K. K., and Van Bladel, J. G., "Scattering by Perfectly Conducting Rectangular Cylinders," *IEEE Transactions on Antennas and Propagation,* Vol. 11, March 1963, pp. 185–192.

Problems

2.1 A center-fed half-wave dipole in free space has a radius of $a = \lambda/1000$ and an impedance of $Z = 80 + j42\ \Omega$. A scale model of this dipole is embedded in a large block of lossless dielectric with μ, ϵ, and its dimensions are $L' = \lambda'/2$ and $a' = \lambda'/1000$.

a) Relate the impedance of the original and scale-model antennas, z and Z'.

b) Obtain an equation that relates the constitutive parameters in the two cases if the original medium is lossless but the scale-model medium is lossy (with conductivity σ) by using a complex frequency ω_c (time dependence $e^{j\omega_c t}$) with the original model and a real frequency with the scale model. *Hint:* Dimensionless quantities such as γL and Z/η have a scale factor of unity.

2.2 A TE_z plane wave is incident on a rectangular plate like the one shown in Fig. 2.12.

a) Using the physical optics approximation, what is the expression for the current on the plate surface $\vec{J}_s$?

b) Show that the monostatic scattered field using the PO approximation is

$$E_\phi(\theta, \phi) = \frac{jkabE_{0\phi}}{2\pi r} e^{-jkr} \cos\theta\, \text{sinc}(kua)\text{sinc}(kvb)$$

2.3 A square plate with edge length a has a square hole with edge length b cut into it. As shown in Fig. P2.3, the hole is centered at (x_0, y_0) and is rotated an angle δ with respect to the x axis. Find the monostatic RCS for a TM_z-polarized incident plane wave using the physical optics approximation. Plot the result for $\phi = 45°$ and $\phi = 0°$. Verify your result using POFACETS with $x_0 = y_0 = 3\lambda$, $b = 2\lambda$, and $a = 10\lambda$.

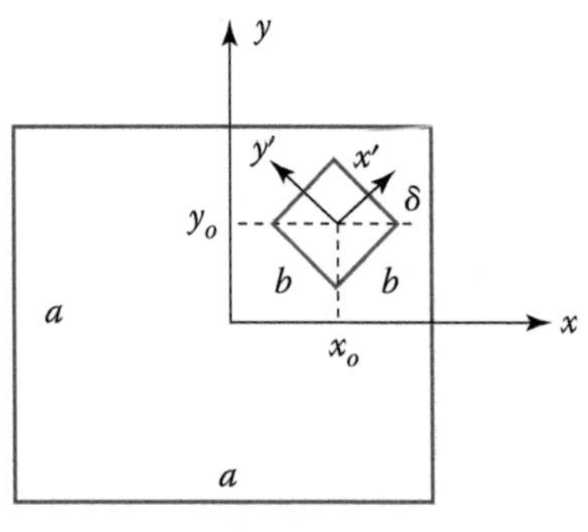

Fig. P2.3

2.4 A circular plate of radius a lies in the $x-y$ plane. A source point is given by the cylindrical coordinates (ρ', ϕ') as shown in Fig. P2.4. A TM-polarized plane wave is incident on the plate from a direction (θ, ϕ),

$$\vec{H}_i = \hat{\phi} H_0 e^{-j\vec{k}_i \cdot \vec{r}}$$

a) Show that the path difference relative to the origin in terms of cylindrical coordinates is $-\hat{k} \cdot \vec{r}' = \rho' \sin\theta \cos(\phi - \phi')$.

b) Using the physical optics approximation, find the monostatic radar cross section of the plate for any direction, $\sigma(\theta)$. *Hint*: Make use of the following integrals:

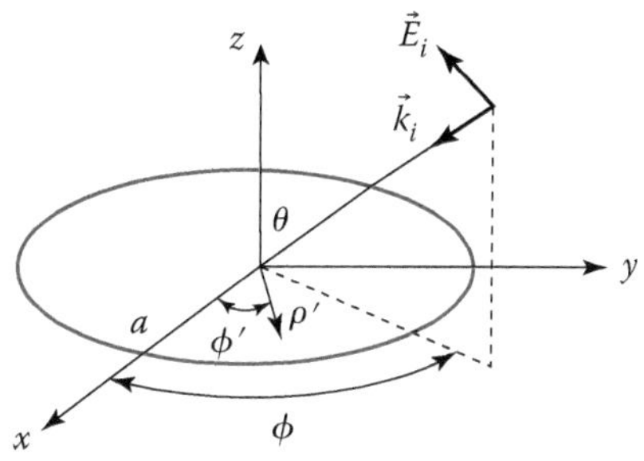

Fig. P2.4

$$\int_0^{2\pi} \exp[j\alpha\cos(\phi - \phi')]\mathrm{d}\phi' = 2\pi J_0(\alpha) \text{ and } \int \alpha J_0(\alpha)\mathrm{d}\alpha = \alpha J_1(\alpha)$$

2.5 Edge tapering is often used to reduce the wide-angle RCS of flat objects. We have already seen this for the triangular taper (Example 2.5). In this problem, we derive a closed-form expression for the RCS of a rectangular plate with a linear edge taper (see Fig. P2.5). A simple approach to the problem results when the current distribution is broken into an array of overlapping triangles and pattern multiplication is used.

a) Each subarray has a width a and a triangular current distribution. What is the subarray factor (or element factor if you consider this an element)?

b) What is the array factor if there are N triangles?

c) Assume that the current distribution is uniform in the y direction with length b and that the length of the plate in x is L. What is the peak value of the RCS?

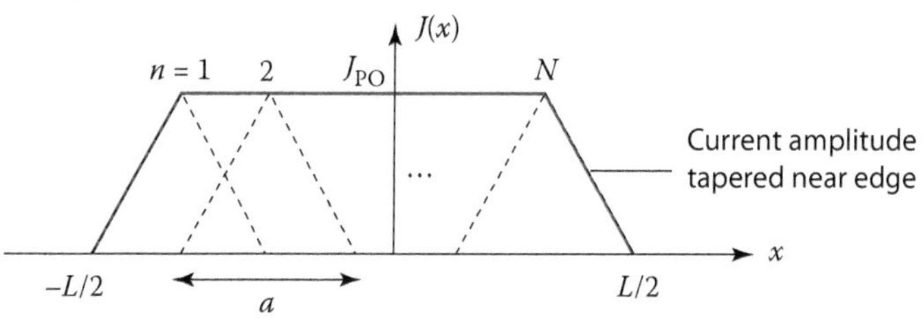

Fig. P2.5

2.6 A rectangular plate lies in the $z = 0$ plane, and a second plate is centered directly above it at height h, as shown in Fig. P2.6. The top plate has width $2a$, the bottom plate has width $2b$ ($a < b$), and both have length L. Find the total RCS of the configuration for $\theta < (\pi/2)$, $\phi = 0$ using the physical optics approximation. Assume a TM$_z$-polarized incident wave, and include shadowing by setting $\vec{J}_s = 0$ in the shadow region. (This is referred to as the *null-field hypothesis*.)

Select a value for a, b, and L. Compute the patterns using your formulas and compare to POFACETS. (Note that POFACETS does not include shadowing.)

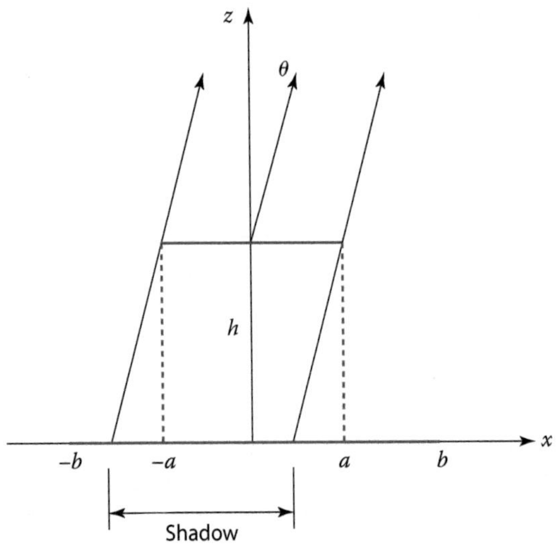

Fig. P2.6

2.7 The elements of an array antenna are laid out in a square grid with rows and columns parallel to the x and y axes and spaced $d = 0.25\lambda_0$ at the array's operating frequency f_0. At what frequency f relative to f_0 will Bragg lobes first appear at 60 deg from the normal to the array aperture in the monostatic RCS pattern? In what ϕ planes will these lobes first occur? Show this on a Bragg diagram.

2.8 Isotropic elements are arranged in a triangular grid as shown in Fig. P2.8 (dark circles). Find an expression for the RCS when a TM$_z$ wave is incident from an angle (θ, ϕ). Plot the Bragg diagram for $d_x = 0.3\lambda$ and $d_y = 0.8\lambda$. *Hint*: Consider the triangular array to be a rectangular array with every other element removed. Use the principle of pattern multiplication.

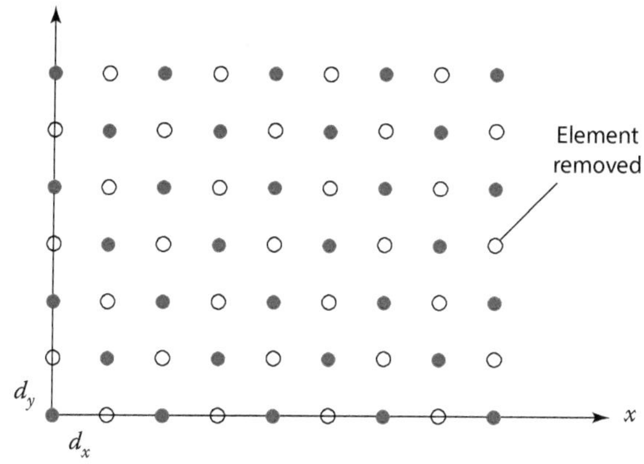

Fig. P2.8

2.9 A large square conducting plate with sides of length a is located in the $z = 0$ plane and has a small circular hole of radius b cut in it. The hole is centered at a point (x_0, y_0). Determine the monostatic RCS of the plate with the hole for a TM_z-polarized incident plane wave using the physical optics approximation.

2.10 A right-triangular conducting plate has an arbitrary orientation with respect to the coordinate system (x, y, z) as shown in Fig. P2.10. Only the coordinates of the vertices are known: $P_1(x_1, y_1, z_1)$, $P_2(x_2, y_2, z_2)$, and $P_3(x_3, y_3, z_3)$.

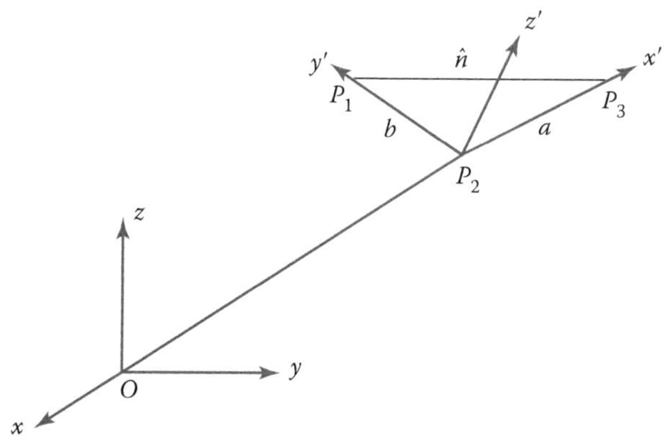

Fig. P2.10

a) Write expressions for the edge vectors $\vec{a}$ and $\vec{b}$ and the unit vector normal to the plate $\hat{n}$:

$$\vec{a} = a_x\hat{x} + a_y\hat{y} + a_z\hat{z}$$

$$\vec{b} = b_x\hat{x} + b_y\hat{y} + b_z\hat{z}$$

$$\vec{n} = n_x\hat{x} + n_y\hat{y} + n_z\hat{z}$$

Find the components of these three vectors in terms of the vertex points.

b) Write expressions for the edge vectors $\vec{a}$ and $\vec{b}$ in terms of the local plate (primed) coordinate system (x', y', z').

c) By equating the expressions for $\vec{a}$ and $\vec{b}$ in parts a and b, find a transformation matrix **T** that relates the Cartesian coordinates in these two systems.

d) A plane wave is TM to z and incident from a direction $(\theta, \phi = 0)$:

$$\vec{H}_i = \hat{y}H_0 e^{-j\vec{k}_i\cdot\vec{r}}$$

Using the physical optics approximation, determine the primed components of the surface current density $J_{sx'}$ and $J_{sy'}$ at a point on the plate (x', y'). Reference the phase of the current to the origin of the unprimed coordinate system, O.

2.11 Wire grids are often used to approximate conducting surfaces. An example is a window in a low-observable platform, where the glass or plastic has a wire mesh embedded in it. This prevents waves from penetrating into the interior of the platform and also reduces scattering from the edges of the window opening.

The reflection coefficient for wire grids can be obtained by several different methods. One is to use the method of *averaged boundary conditions* developed by Astrakhan [8]. Depending on the properties of the grid and the degree of electrical contact at the junctions, depolarization may occur on reflection. For example, if a pure TM wave is incident, the reflected wave may contain both TM and TE components. Consequently, four reflection coefficients are required. For a wire mesh with dimensions a in x and b in y located in the $z = 0$ plane, the most general equations for the reflection coefficients follow.

TM (parallel) polarization incident:

$$\Gamma_{\parallel}^{e} = \frac{k}{I_0}\cos\theta\{1 - k\cos\theta[\gamma_2\cos^2\phi + (\delta_2 - \delta_1)\sin\phi\cos\phi - \gamma_1\sin^2\phi]\}$$

$$\Gamma_{\perp}^{e} = \frac{k^2}{I_0} \cos\theta \left[\delta_1 \sin^2\phi - (\gamma_1 + \gamma_2) \sin\phi \cos\phi + \delta_2 \cos^2\phi \right]$$

TE (perpendicular) polarization incident:

$$\Gamma_{\perp}^{h} = \frac{k}{I_0} \left\{ \cos\theta + k\left[\gamma_1 \cos^2\phi + (\delta_2 - \delta_1) \sin\phi \cos\phi - \gamma_2 \sin^2\phi \right] \right\}$$

$$\Gamma_{\parallel}^{h} = \frac{k^2}{I_0} \cos\theta \left[\delta_1 \cos^2\phi + (\gamma_1 + \gamma_2) \sin\phi \cos\phi + \delta_2 \cos^2\phi \right]$$

where

$$\begin{aligned} \frac{I_0}{k} = {} & \cos\theta (1 + k^2 \delta_1 \delta_2 - k^2 \gamma_1 \gamma_2) \\ & + k \sin^2\theta [\gamma_2 \cos^2\phi + (\delta_2 - \delta_1) \sin\phi \cos\phi - \gamma_1 \sin^2\phi] \\ & + k(\gamma_1 - \gamma_2) \end{aligned}$$

$$\gamma_1 = \alpha_1 \left[1 + F_x - \left[\frac{(a/b) + \chi_x}{1 + (a/b) + \chi_x} \right] \sin^2\theta \cos^2\phi \right]$$

$$\gamma_2 = -\alpha_2 \left[1 + F_y - \left[\frac{(b/a) + \chi_y}{1 + (b/a) + \chi_y} \right] \sin^2\theta \sin^2\phi \right]$$

$$\delta_1 = \alpha_1 \left[\frac{(a/b)}{1 + (a/b) + \chi_x} \right] \sin^2\theta \cos\phi \sin\phi$$

$$\delta_2 = -\alpha_2 \left[\frac{(b/a)}{1 + (b/a) + \chi_y} \right] \sin^2\theta \sin\phi \cos\phi$$

$$\alpha_1 = \frac{jb}{\pi} \ell n \left(\frac{b}{2\pi r_0} \right) \quad \text{and} \quad \alpha_2 = \frac{ja}{\pi} \ell n \left(\frac{a}{2\pi r_0} \right)$$

The parameters F_x and F_y allow for imperfect conductors ($F_x = F_y = 0$ for a perfect conductor), and χ_x and χ_y characterize the contact between the wires at the junctions ($\chi_x = \chi_y = 0$ for a soldered grid). All wires are assumed to have a radius of r_0.

a) Consider a square grid with spacing d and a plane wave normally incident ($\theta = 0$). What is the reflection coefficient if the wires are perfectly conducting and there is perfect electrical contact at the wire nodes?

b) What is the surface resistivity of the grid in Ω/sq?

2.12 If the impedance boundary condition holds on a surface, derive the relationship between the electric and magnetic surface currents $\vec{J}_{ms}$ and $\vec{J}_s$ from the IBC equations.

2.13 A time-harmonic field in a source-free region has the following magnetic field intensity:

$$\vec{H} = \hat{y}H_0 e^{-\alpha x} e^{-j\beta x}$$

where α and β are real. Determine $\vec{E}(x, y, z)$ and the time-averaged Poynting vector $\vec{W}_{av}(x, y, z)$.

2.14 A time-harmonic field in a source-free region with parameters (μ, ϵ) has the following magnetic field intensity:

$$\vec{H} = \hat{z}H_0[\cos(hx) + \cos(hy)]$$

a) Determine h in terms of ω, μ, and ϵ.

b) Determine the values of hd such that the field can exist in a triangular waveguide with perfectly conducting walls, with corners located at $(x, y) = (0, 0)$, (d, d), and $(d, -d)$. The axis of the waveguide is parallel to the z axis.

2.15 As shown in Fig. P2.14, a plane wave has oblique incidence on a slab with thickness t, permeability μ, and permittivity ϵ. At $z = 0$, the bottom surface of the slab is coated with a perfectly conducting ground plane. The time-harmonic incident field is

$$\vec{E}_i = \hat{x}\, E_0 \exp(-jky \sin\theta) \exp(jkz \cos\theta)$$

where $k = \omega\sqrt{\mu_0 \epsilon_0}$ and the total field in the slab is

$$\vec{E}_2 = \hat{x}\, E_0 e^{-jgy}(Ae^{jhz} + Be^{-jhz})$$

a) Write an expression for the reflected field outside the slab, using Γ as the reflection coefficient at $z = t$.

b) Determine the constants g and h.

c) Develop enough equations to permit solution of the constants A, B, and Γ.

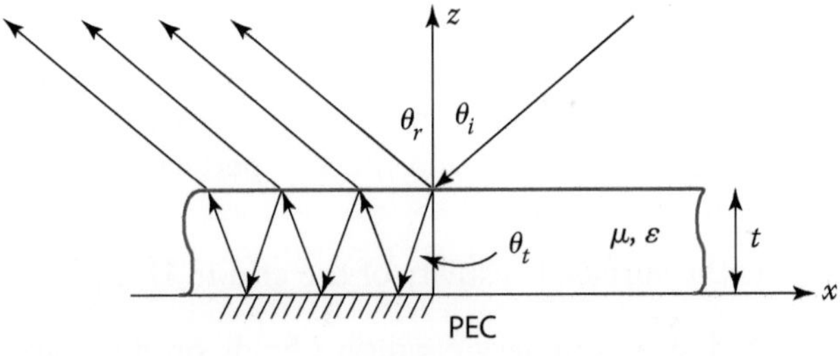

Fig. P2.14.

2.16 A conductor is coated with a material of dielectric constant ϵ_r and thickness t as in Problem 2.15. Find an expression for the surface impedance of the coated conductor for a TM_z-polarized plane wave incident at an angle θ.

2.17 A TM-polarized wave is incident on a right-triangular plate with sides of lengths a and b as shown in Fig. P2.17. Show that the monostatic RCS using the physical optics approximation is given by

$$\sigma_{\theta\theta} = \frac{4\pi A^2}{\lambda^2}\cos^2\theta \left|\frac{e^{j\beta}\text{sinc}(\beta) - e^{j\alpha}\text{sinc}(\alpha)}{\alpha - \beta}\right|^2$$

where $\alpha = kau$ and $\beta = kbv$ and $\alpha - \beta \neq 0$. More general formulas are given in Ref. 9. Plot the RCS pattern and compare to POFACETS. (Note that any triangle can be decomposed into a collection of right triangles.)

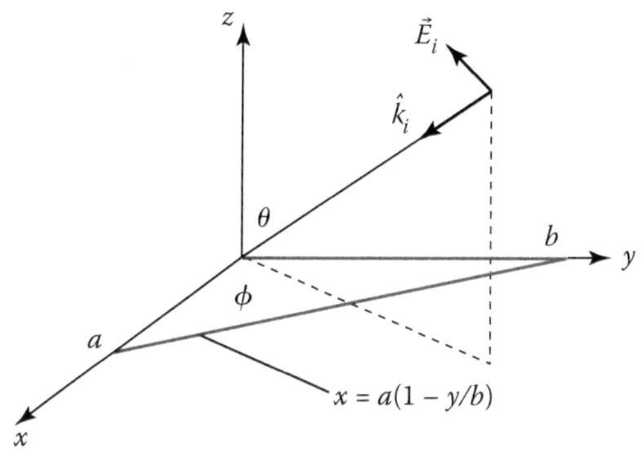

Fig. P2.17

2.18 A plane wave is normally incident on a resistive sheet with surface resistivity of R_s (see Fig. P2.18):

$$\vec{E}_i = \hat{x}E_0 e^{-jkz}$$

At the sheet ($z = 0$), the following boundary conditions must be satisfied:

a) $\vec{E}_{\tan}$ is continuous ($\vec{J}_{ms} = 0$).

b) $\vec{H}_{\tan}$ is discontinuous by an amount $\vec{J}_s$.

c) $\vec{E}_{\tan} = \vec{J}_s R_s$.

Apply the preceding conditions at $z = 0$ to obtain a reflection coefficient for the sheet in terms of R_s.

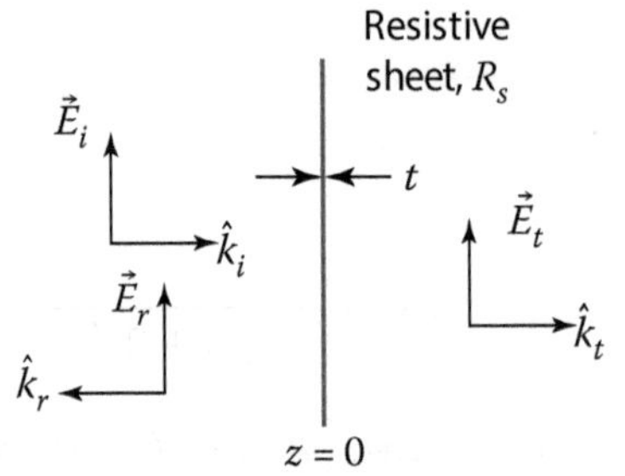

Fig. P2.18

2.19 A time-harmonic field is incident on a small dielectric sphere of radius a, permittivity ϵ, and permeability μ_0. The sphere is centered at the origin, and the incident plane wave is

$$\vec{E}_i = \hat{z}E_0 e^{-jkx}$$

At each point in space, the electric field intensity can be expressed as the sum of an incident field (the field that exists when the sphere is absent) plus a scattered field (the disturbance in the field associated with the dielectric sphere):

$$\vec{E} = \vec{E}_i + \vec{E}_s$$

a) The scattered field can be considered to be due to an equivalent volume current density $\vec{J}$ radiating in free space. Using the volume equivalence principle, write an expression for $\vec{J}$ in terms of ϵ, ϵ_0, and $\vec{E}$.

b) If a is much less than a wavelength, $\vec{E}$ is approximately constant inside the sphere: $\vec{E} = \hat{z}E_1$. Solve for E_1 in terms of E_i, $\epsilon_r(=\epsilon/\epsilon_0)$, and $\vec{J}$ by enforcing $\vec{E} = \vec{E}_i + \vec{E}_s$ at the center of the sphere. *Hint:* When a uniform current density $\vec{J}$ exists in a tiny spherical region and radiates in free space, it generates the field

$$\vec{E}_s = \frac{j\vec{J}}{3\omega\epsilon_0}$$

at its center.

c) Using the results of parts a and b show that

$$J_z = \frac{j3\omega\epsilon_0(\epsilon_r - 1)}{(\epsilon_r + 2)} E_0$$

d) Find the far field scattered from the dielectric sphere by using $\vec{J}$ in the radiation integral.

e) Show that the backscattered field of the tiny sphere is given by

$$\sigma_b = \frac{4\pi k^4 a^6 (\epsilon_r - 1)^2}{(\epsilon_r + 2)^2}$$

This is a form of the *Rayleigh law of scattering.*

2.20 An aircraft is approaching a radar as shown in the Fig. P2.20. The edge of the wing is straight and 1.5 m long. For $f = 2$ GHz, at what angle θ will a traveling wave lobe be observed? Estimate its level in dBsm.

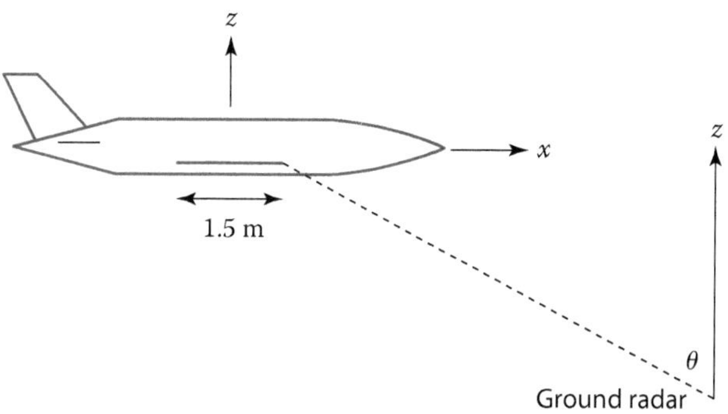

Fig. P2.20

2.21 An infinitely long circular cylinder is coaxial with the z axis and has a radius a and normalized surface impedance Z_S. Assume bistatic scattering, and denote the bistatic angle as ϕ', that is,

$$\hat{k}_i \cdot \hat{k}_s = \cos \phi'$$

Using the method of separation of variables, show that, for parallel polarization [10],

$$\sigma_{\|}/\lambda = \frac{2}{\pi} \left| \sum_{n=0}^{\infty} \epsilon_n \frac{J_n(ka) + jZ_s J_n'(ka)}{H_n^{(2)}(ka) + jZ_s H_n^{(2)'}(ka)} \cos(n\phi') \right|^2$$

and, for perpendicular polarization,

$$\sigma_{\perp}/\lambda = \frac{2}{\pi}\left|\sum_{n=0}^{\infty} \epsilon_n \frac{J_n'(ka) - jZ_s J_n(ka)}{H_n^{(2)'}(ka) - jZ_s H_n^{(2)}(ka)} \cos(n\phi')\right|^2$$

where

$$\epsilon_n = \begin{cases} 1, & n = 0 \\ 2, & n > 0 \end{cases}$$

$J_n(ka)$ is the Bessel function of the first kind, order n, argument ka, and $H_n^{(2)}(ka)$ is the Hankel function of the second kind, order n, argument ka. The primes denote differentiation with respect to the argument.

2.22 Reference 11 gives some bistatic RCS patterns for infinitely long cylinders with rectangular cross sections. For a square cylinder of side length a, where $ka = 5$, plot the bistatic RCS for a plane wave incident 45 deg off of a face normal. Use POFACETS and choose a cylinder length much greater than the side ($L \gg a$). Note that the two-dimensional scattering cross section of an infinitely long cylinder is approximately related to the three-dimensional RCS by

$$\sigma_{3D} \approx \sigma_{2D}\frac{2L^2}{\lambda}$$

Chapter 3 Frequency-Domain Numerical Methods

3.1 Introduction

The radiation integrals introduced in Chapter 2 provide a means of calculating the scattered field from a target once the induced current on the surface is known. Except for the simplest shapes, the current distribution on the target is not known accurately a priori. For electrically large bodies that are smooth and relatively flat, the physical optics approximation provides a good estimate at points far from *discontinuities* and shadow boundaries. A discontinuity is any abrupt change in the surface contour or composition. Examples are edges, steps, cracks, and material joints. It is expected that the smaller the discontinuity, the less impact it will have on RCS. This statement is true when the maximum RCS level of the target is used as a reference. However, small scattering sources tend to have broad patterns and, therefore, scattering from a discontinuity can dominate in directions in which other contributors are negligible.

For an accurate calculation of RCS, all significant current contributions must be included in the radiation integrals. For objects whose surfaces are metals (good conductors), the small skin depth justifies the use of an idealized surface current $\vec{J}_s$. If the object has dielectric or magnetic regions, then polarization and magnetization currents may arise, respectively, in those materials. If so, then volume electric $\vec{J}$ and magnetic $\vec{J}_m$ currents must be determined along with the surface currents.

In this chapter, *integral* and *differential* equations will be formulated from Maxwell's equations and the boundary conditions on the target. Most engineers are familiar with differential equations but may not have encountered integral equations. An integral equation (IE) has the unknown quantity, which is usually the current, in the integrand and possibly in other terms outside of the integral. Equations of this type occur in all branches of physics and engineering but, in spite of their importance, analytical solutions exist for only a few cases. In most problems of practical interest, the integral equation must be solved numerically on a computer. Similarly, mathematicians and engineers have been working with differential equations for hundreds of years, and many closed-form solutions exist. Yet, in just about all cases, the equations that arise from scattering by complex objects must be solved numerically. Differential equation (DE) formulations of scattering generally solve for the fields E or H rather than the currents.

A popular solution technique for surface integral equations is the *method of moments* (MM), which consists of expanding the unknown quantity (usually the surface current) into a series with unknown coefficients. The expansion is inserted into the IE and a testing procedure is performed. The properties of the expansion functions are used to reduce the integral equation to a set of simultaneous linear equations that can be solved for the coefficients. Once the series expression for the current is known, it can be inserted into the radiation integrals and used to calculate the far-scattered field.

The finite element method (FEM) is closely related to the method of moments. FEM is one of the oldest numerical techniques applied to engineering problems. Traditional applications in the disciplines of aeronautical and mechanical engineering employ *scalar node* basis functions. For spatial time-varying electromagnetic applications, *vector edge* basis functions are desirable. It has been only recently that efficient edge-based functions have been developed. The starting point for the FEM solution is typically a vector wave equation (DE) for the fields in a bounded region of space around the scatterer (the *computational domain*). The fields are approximated by a series expansion, and, as in the case of MM, a testing procedure is performed to obtain expressions for the expansion coefficients. Once the expansion coefficients are determined, E and H over the surface of the computational domain are known and the equivalent electric and magnetic currents on the computational boundary can be used to calculate the scattered field.

Generally, the target interior [if it is not a perfect electric conductor (PEC)] and a small region of free space around the target comprise the FEM computational domain. For RCS calculations, both the source and observer are outside of the domain. This is referred to as an *open problem.* In the case of RCS, the boundary between open space and the discretized "gridded" computational domain must be transparent to the incoming incident and outgoing scattered waves. Boundary conditions must be enforced at the grid edges to ensure that the fields are continuous and that no artificial reflections arise.

Before discussing the MM and FEM techniques, several fundamental integral and differential equations will be derived.

3.2 Electric Field Integral Equation

The E-field integral equation (EFIE) is the most commonly encountered integral equation. It can be derived from the radiation integrals by letting the observation point be located on the surface of the body. The target is assumed to be a PEC, and the coordinate system is defined in Fig. 3.1. Because the surface is a PEC, $\vec{J}_{ms} = 0$. and, from Sec. 2.5, the scattered field due to the electric current is

$$\vec{E} = -j\omega\vec{A} - \frac{j}{\omega\mu\epsilon}\nabla(\nabla\cdot\vec{A}) \tag{3.1}$$

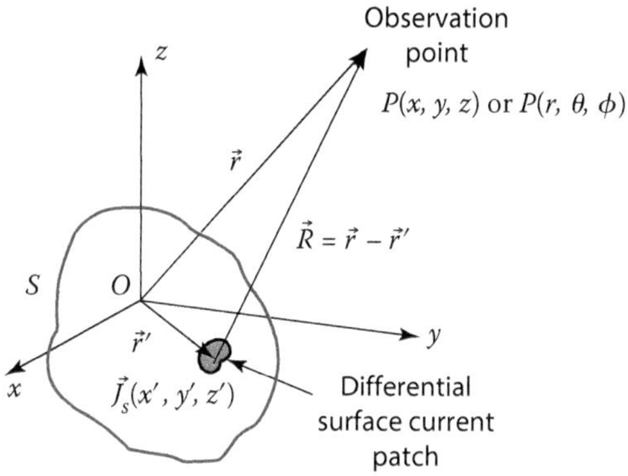

Fig. 3.1 Geometry for derivation of the EFIE.

$$\vec{A}(\vec{r}) = \frac{\mu}{4\pi}\iint_S \vec{J}_s(\vec{r}\,')\frac{e^{-jkR}}{R}\,\mathrm{d}s'. \tag{3.2}$$

and, as usual,

$$R = \left|\vec{r} - \vec{r}\,'\right| = \sqrt{(x-x')^2 + (y-y')^2 + (z-z')^2} \tag{3.3}$$

For notational convenience, the free-space Green's function will be denoted by G:

$$G(\vec{r}, \vec{r}\,') = \frac{e^{-jkR}}{4\pi R} \tag{3.4}$$

The tangential component of the total field on the surface must be zero to satisfy the boundary conditions. The total field consists of incident and scattered parts, so that the boundary conditions can be expressed as

$$[\vec{E}_i(P) + \vec{E}_s(P)]_{\tan} = 0, \;\; P \; on \; S \tag{3.5}$$

Substituting Eq. (3.1) in Eq. (3.5) gives

$$\vec{E}_i(P)|_{\tan} = -\vec{E}_s(P)|_{\tan} = \left[-j\omega\vec{A} - \frac{j}{\omega\mu\epsilon}\nabla(\nabla\cdot\vec{A})\right]_{\tan} \tag{3.6}$$

Writing Eq. (3.6) explicitly in terms of the surface current yields the EFIE

$$\vec{E}_i(\vec{r})|_{\tan} = \left\{ j\omega\mu\iint_S \vec{J}_s(\vec{r}\,')G(\vec{r},\vec{r}\,') + \frac{j}{\omega\epsilon}\nabla\left[\nabla\cdot\iint_S \vec{J}_s(\vec{r}\,')G(\vec{r},\vec{r}\,')\mathrm{d}s'\right]\right\}_{\tan} \tag{3.7}$$

The only unknown quantity is the surface current $\vec{J}_s$. The quantity in curly brackets is an *integrodifferential operator* $\mathcal{L}$, which operates on $\vec{J}_s$. Thus, using operator notation,

$$\mathcal{L}(\vec{J}_s) = \left\{ j\omega\mu \iint_S \vec{J}_s(\vec{r}')G(\vec{r},\vec{r}')\mathrm{d}s' + \frac{j}{\omega\epsilon}\nabla\left[\nabla \cdot \iint_S \vec{J}_s(\vec{r}')G(\vec{r},\vec{r}')\mathrm{d}s'\right]\right\}_{\tan} \tag{3.8}$$

The last term in Eq. (3.8) has mixed operations on vectors and scalars that depend on both the primed and unprimed coordinates. It is more convenient to write all quantities and operators in terms of the integration variables (primed coordinates). To this end, first interchange the del operator and the integral,

$$\nabla \cdot \iint_S \vec{J}_s(\vec{r}')G(\vec{r},\vec{r}')\mathrm{d}s' = \iint_S \nabla \cdot \left[\vec{J}_s(\vec{r}')G(\vec{r},\vec{r}')\right]\mathrm{d}s' \tag{3.9}$$

By vector identity,

$$\nabla \cdot (\vec{J}_s G) = \nabla G \cdot \vec{J}_s + G(\nabla \cdot \vec{J}_s) \tag{3.10}$$

The last term is zero because ∇ operates on (x, y, z) and $\vec{J}_s$ is a function of (x', y', z'). Because of the symmetry between primed and unprimed coordinates, the gradient term can be rewritten

$$\begin{aligned}\nabla G &= \left(\frac{\partial}{\partial x}\hat{x} + \frac{\partial}{\partial y}\hat{y} + \frac{\partial}{\partial z}\hat{z}\right)\frac{e^{-jkR}}{4\pi R} \\ &= -\left[jk + \frac{1}{R}\right]\frac{e^{-jkR}}{4\pi R}\hat{R} = -\nabla' G\end{aligned} \tag{3.11}$$

Using Eqs. (3.10) and (3.11), the last term in Eq. (3.8) becomes

$$\nabla \cdot \iint_S \vec{J}_s(\vec{r}')G(\vec{r},\vec{r}')\mathrm{d}s' = -\iint_S \nabla' G(\vec{r},\vec{r}') \cdot \vec{J}_s(\vec{r}')\mathrm{d}s' \tag{3.12}$$

Now, the vector identity of Eq. (3.10) is applied again:

$$\nabla' \cdot (\vec{J}_s G) = \nabla' G \cdot \vec{J}_s + G(\nabla' \cdot \vec{J}_s) \tag{3.13}$$

The *surface divergence theorem* [1, 2] can be used to show that

$$\iint_S \nabla' \cdot \left[G(\vec{r},\vec{r}')\vec{J}_s(\vec{r}')\right]\mathrm{d}s' = 0 \tag{3.14}$$

The final form suitable for solution by a numerical technique is

$$\mathcal{L}(\vec{J}_s) = \iint_S \left\{ j\omega\mu\vec{J}_s(\vec{r}')G(\vec{r},\vec{r}') - \frac{j}{\omega\epsilon}\left[\nabla' \cdot \vec{J}_s(\vec{r}')\right]\nabla' G(\vec{r},\vec{r}')\right\}_{\tan}\mathrm{d}s' \tag{3.15}$$

The EFIE can be written concisely as

$$\vec{E}_i\big|_{\tan} = \mathcal{L}(\vec{J}_s) \tag{3.16}$$

Equation (3.16) is a general form of the EFIE that can be applied to any arbitrary PEC surface. The incident field is known, and so the only unknown quantity is the surface current density. The incident field could be due to a potential across points on the surface (such as the feed points of a dipole) or to an incident wave from a radar.

If the surface S happens to be a straight wire with radius a that is thin compared to a wavelength, some further simplification of Eq. (3.15) is possible. For a wire along the z axis, the integral and differential operations can be replaced as follows:

$$\begin{aligned} \nabla &\Rightarrow \frac{\partial}{\partial z}\hat{z} \\ \iint_S \mathrm{d}s' &\Rightarrow 2\pi a \int_L \mathrm{d}z' \\ \vec{J}_s(\vec{r}') &\Rightarrow \frac{I(z')}{2\pi a}\hat{z} \end{aligned} \tag{3.17}$$

The last condition implies that no current is flowing around the circumference of the wire; there is only an axial component of current. This is referred to as the *thin-wire approximation* and is valid when $a \ll \lambda$ and $a \ll L$ (L is the length of the wire). Incorporating the foregoing conditions into the EFIE yields

$$\hat{z}E_{iz} = \hat{z}\frac{j}{\omega\epsilon}\int_{-(L/2)}^{L/2} I(z')\left[k^2 + \frac{\partial^2}{\partial z^2}\right] G(z, z')\,\mathrm{d}z' \tag{3.18}$$

This form of the EFIE is known as *Pocklington's equation.* With some further manipulation, it can be shown that

$$\left[k^2 + \frac{\partial^2}{\partial z^2}\right] G(z, z') = \frac{G(z, z')}{R^4}\left[(1 + jkR)(2R^2 - 3a^2) + (kaR)^2\right] \tag{3.19}$$

where $R = \sqrt{(z - z')^2 + a^2}$.

3.3 Magnetic Field Integral Equation

A second integral equation for the electric current density on the target can be derived by imposing the boundary conditions on the total magnetic field at the surface. For a PEC, the scattered magnetic field intensity is given by

$$\vec{H}_s(P) = \frac{1}{\mu}\nabla \times \vec{A} = \nabla \times \iint_S \vec{J}_s(\vec{r}')G(\vec{r}, \vec{r}')\,\mathrm{d}s' = \iint_S \nabla \times \left[\vec{J}_s(\vec{r}')G(\vec{r}, \vec{r}')\right]\mathrm{d}s' \tag{3.20}$$

Using the vector identity (3.10) and noting that $\nabla \times \vec{J}_s = 0$ give

$$\vec{H}_s(P) = \iint_S \left[\vec{J}_s(\vec{r}') \times \nabla G(\vec{r},\vec{r}')\right] ds' \tag{3.21}$$

Moving the observation point onto the surface and enforcing the boundary condition

$$\vec{J}_s = \hat{n} \times \left[\vec{H}_i + \vec{H}_s\right] \quad \text{for } P \text{ on } S \tag{3.22}$$

yield

$$\hat{n} \times \vec{H}_i(\vec{r}') = \vec{J}_s(\vec{r}') - \hat{n} \times \iint_S \left[\vec{J}_s(\vec{r}') \times \nabla G(\vec{r},\vec{r}')\right] ds' \tag{3.23}$$

This is the general form of the magnetic field integral equation (MFIE) and is valid only for closed surfaces. It is sometimes called the H-field integral equation (HFIE).

3.4 Vector Wave Equation

Maxwell's first two equations are coupled differential equations involving the electric and magnetic fields

$$\nabla \times \vec{E} = -j\omega\mu\vec{H} - \vec{J}_m \tag{3.24}$$

and

$$\nabla \times \vec{H} = j\omega\epsilon\vec{E} + \vec{J} \tag{3.25}$$

To eliminate H, first solve Eq. (3.24) for $\vec{H}$

$$\vec{H} = \frac{\nabla \times \vec{E} + \vec{J}_m}{-j\omega\mu} \tag{3.26}$$

and then substitute Eq. (3.25) into Eq. (3.24). Arrange the equation so that terms involving the sources (currents) are on the right-hand side:

$$\underbrace{\nabla \times \left(\frac{\nabla \times \vec{E}}{\mu_r}\right) - k_o^2\epsilon_r\vec{E}}_{\equiv \mathcal{L}_E(\vec{E})} = \underbrace{-jk_o\eta_o\vec{J} - \nabla \times \left(\frac{\vec{J}_m}{\mu_r}\right)}_{\equiv \vec{f}_E} \tag{3.27}$$

where $k_o = \omega\sqrt{\mu_o\epsilon_o}$. This is the *inhomogeneous* form of the vector wave equation because it has not been assumed that μ_r is independent of position. The symbol $\mathcal{L}_E$ is a differential operator and $\vec{f}_E$ the forcing function (or excitation), so that a shorthand form of the equation is

$$\mathcal{L}_E(\vec{E}) = \vec{f}_E \tag{3.28}$$

If the medium is homogeneous, then μ_r can be pulled out of the curl operator

$$\nabla \times \nabla \times \vec{E} - k^2 \vec{E} = -jk\eta\vec{J} - \nabla \times \vec{J}_m \tag{3.29}$$

Clearly, there is a dual to this equation involving the magnetic field intensity:

$$\underbrace{\nabla \times \left(\frac{\nabla \times \vec{H}}{\epsilon_r}\right) - k_o^2 \mu_r \vec{H}}_{\equiv \mathcal{L}_H(\vec{H})} = \underbrace{-j\omega\epsilon_o \vec{J}_m + \nabla \times \left(\frac{\vec{J}}{\epsilon_r}\right)}_{\equiv \vec{f}_H} \tag{3.30}$$

If the medium is homogeneous, then

$$\nabla \times \nabla \times \vec{H} - k^2 \vec{H} = -j\omega\epsilon\vec{J}_m + \nabla \times \vec{J} \tag{3.31}$$

Furthermore, if there are no impressed sources and the volume of interest V is a source free region (i.e., this is not an antenna problem), then $\vec{J} = \vec{J}_m = 0$. Under this condition, the double curl can be replaced by the Laplacian because the divergences of both $\vec{E}$ and $\vec{H}$ are zero in a source free region. The familiar forms of the wave equations result:

$$\nabla^2 \vec{E} + k^2 \vec{E} = 0 \tag{3.32}$$

and

$$\nabla^2 \vec{H} + k^2 \vec{H} = 0 \tag{3.33}$$

The wave equations are differential equations that relate the field to the sources at a point in space, subject to any boundary conditions. Equations (3.27) and (3.30) will be the starting point for formulation of the FEM.

3.5 Method of Moments Technique

3.5.1 Testing Procedure

Both the EFIE and MFIE have $\vec{J}_s$ as the unknown quantity. In mathematics, they are classified as *Fredholm equations* because the limits of the integrals are fixed. In Fredholm equations of the first kind, the unknown appears only in the integrand; in Fredholm equations of the second kind, the unknown appears outside as well as in the integrand. Therefore, the EFIE is a Fredholm equation of the first kind and the MFIE is a Fredholm equation of the second kind.

There are many ways to solve integral equations, but most are applicable only to very specific coordinate systems and surface shapes. The MM is a numerical technique that has become very popular because it can be applied to arbitrary bodies. Although the MM is conceptually simple, the mathematics and bookkeeping can become overwhelming. In a nutshell, it reduces the integral equation to a matrix problem. The size of the matrix is related directly to the electrical size of the scattering body in terms of the incident wavelength. Large bodies result in large matrices and, therefore, require large computers with fast processing units. Fortunately, the rapid increase in computer capability in recent years has allowed larger targets to be analyzed as well as a reduction in computation time.

The first step in applying the method of moments is to represent the current by a series with unknown expansion coefficients I_n:

$$\vec{J}_s = \sum_{n=1}^{N} I_n \vec{J}_n \tag{3.34}$$

The vectors $\vec{J}_n$ are the *expansion* or *basis functions*, and both I_n and $\vec{J}_n$ can be complex. The series representation for the current density is inserted into the integral equation. To solve the EFIE, Eq. (3.16) is used:

$$\vec{E}_i(\vec{r})\big|_{\tan} = \sum_{n=1}^{N} I_n \iint_S \left\{ j\omega\mu \vec{J}_n(\vec{r}\,')G(\vec{r},\vec{r}\,') - \frac{j}{\omega\epsilon}\left[\nabla' \cdot \vec{J}_n(\vec{r}\,')\right]\nabla' G(\vec{r},\vec{r}\,') \right\}_{\tan} ds' \tag{3.35}$$

The selection of the basis functions is an important consideration in the MM solution. Basis functions should be mathematically convenient (easy to integrate and differentiate) and should also be consistent with the behavior of the current. Typical expansion functions (shown in Fig. 3.2) are pulses, triangles, sinusoids, and δ functions. When each expansion function extends over the entire body, the functions are referred to as *entire-domain* basis functions. If each $\vec{J}_n$ is nonzero on only a portion of the surface, the basis functions are *subsectional*. A comparison of the two is shown in Fig. 3.3. Subdomain functions are used most frequently because of their flexibility in conforming to disjoint shapes.

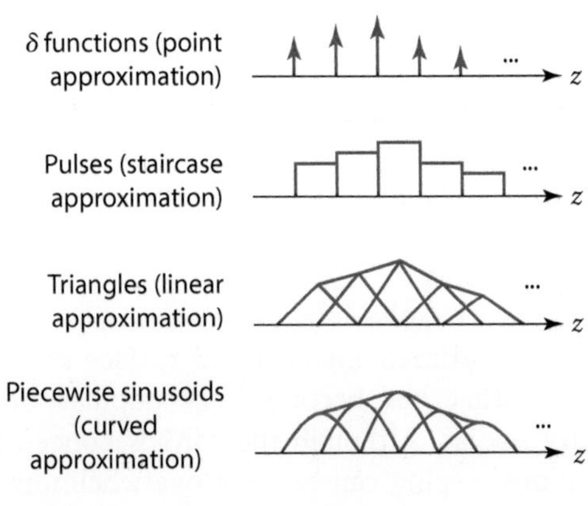

Fig. 3.2 Examples of subsectional basis functions.

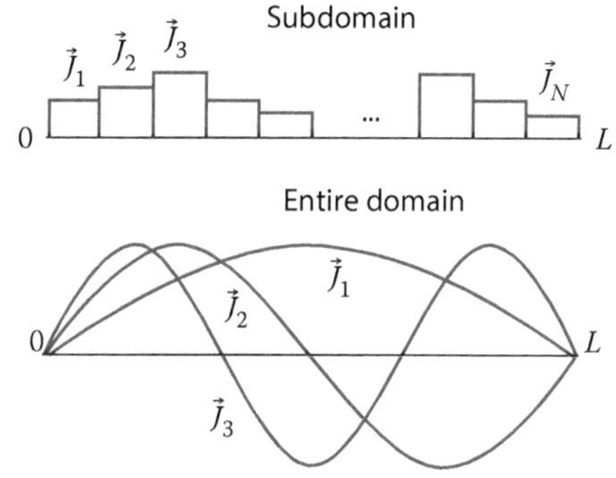

Fig. 3.3 Comparison of subsectional (top) and entire domain (bottom) basis functions.

3.5.2 Method of Moments Matrix Equation

Returning to Eq. (3.35), we wish to solve for the series coefficients by taking advantage of the properties of the expansion functions. Define a set of *testing* or *weighting functions*, $\vec{W}_m$, $m = 1, 2, \ldots, N$. There are no restrictions on the weighting functions although they are generally chosen to be the complex conjugates of the basis functions, $\vec{W}_m = \vec{J}_m^*$. This choice is referred to as *Galerkin's method*. Now, multiply both sides of Eq. (3.35) by each of the weighting functions and integrate over its domain. The result is N equations of the form

$$\iint_{S_m} \vec{W}_m(\vec{r}) \cdot \vec{E}_i(\vec{r})|_{\tan} \mathrm{d}s = \iint_{S_m} \vec{W}_m(\vec{r}) \cdot \left\{ \sum_{n=1}^{N} I_n \times \iint_{S_n} [j\omega\mu \vec{J}_n(\vec{r}') G(\vec{r}, \vec{r}') - \frac{j}{\omega\epsilon} \nabla' \cdot \vec{J}_n(\vec{r}') \nabla' G(\vec{r}, \vec{r}')]_{\tan} \mathrm{d}s' \right\} \mathrm{d}s \tag{3.36}$$

for $m = 1, 2, \ldots, N$. Now, define the following quantity:

$$Z_{mn} \equiv \iint_{S_m} \mathrm{d}s \iint_{S_n} \mathrm{d}s' [j\omega\mu \vec{W}_m(\vec{r}) \cdot \vec{J}_n(\vec{r}') - \frac{j}{\omega\epsilon} (\nabla' \cdot \vec{J}_n(\vec{r}'))(\nabla \cdot \vec{W}_m(\vec{r}))] G(\vec{r}, \vec{r}') \tag{3.37}$$

where the surface divergence theorem has been applied to obtain

$$\iint_S \vec{W}_m \cdot \nabla' G \, \mathrm{d}s = \iint_S G(\nabla \cdot \vec{W}_m) \, \mathrm{d}s \tag{3.38}$$

Let the right-hand side of Eq. (3.36) be defined as

$$V_m \equiv \iint_{S_m} \vec{W}_m(\vec{r}) \cdot \vec{E}_i(\vec{r}) \, \mathrm{d}s \tag{3.39}$$

The subscript tan has been dropped because the dot product of a vector with $\vec{W}_m$ pulls out the tangential component. Equation (3.36) can now be written as

$$V_m = \sum_{n=1}^{N} I_n Z_{mn} \quad \text{for } m = 1, 2, \ldots, N \tag{3.40}$$

or, in matrix form,

$$\mathbf{V} = \mathbf{ZI} \tag{3.41}$$

The matrix **Z** has units of ohms and is called the *impedance matrix*. **V** is the *excitation vector* and has units of volts. **I** is a vector containing the unknown current expansion coefficients, which are determined by solving Eq. (3.31):

$$\mathbf{I} = \mathbf{Z}^{-1}\mathbf{V} \tag{3.42}$$

3.5.3 Measurement Vector

Once the I_n are known, the series representation for the current, Eq. (3.34), is defined, and it can be used in the radiation integral to obtain $\vec{E}(P)$. Using the Lorentz reciprocity theorem (see Problem 3.15) it is possible to define a *measurement matrix* **R** with elements

$$R_n = \iint_{S_n} \vec{J}_n(\vec{r}\,') \cdot \vec{E}_r(\vec{r})\, ds' \tag{3.43}$$

where $\vec{E}_r$ is a unit radiated plane wave. As an example, consider a θ-polarized wave:

$$\vec{E}_r = \hat{\theta} e^{-j\vec{k}_r \cdot \vec{r}} = \hat{\theta} e^{-jkr} e^{jkg} \tag{3.44}$$

Equation (3.43) can now be written with the superscript θ added to denote the polarization of $\vec{E}_r$:

$$R_n^{\theta} = \iint_{S_n} \left[\hat{\theta} \cdot \vec{J}_n(\vec{r})\right] \exp[jk(xu + yv + zw)]\, ds \tag{3.45}$$

A similar analysis for a ϕ-polarized radiated plane wave gives

$$R_n^{\phi} = \iint_{S_n} \left[\hat{\phi} \cdot \vec{J}_n(\vec{r})\right] \exp[jk(xu + yv + zw)]\, ds \tag{3.46}$$

The components of the electric field will be

$$E_\theta = \frac{-jk\eta}{4\pi r} e^{-jkr} \sum_{n=1}^{N} R_n^\theta I_n \tag{3.47}$$

$$E_\phi = \frac{-jk\eta}{4\pi r} e^{-jkr} \sum_{n=1}^{N} R_n^\phi I_n \tag{3.48}$$

The MM is a powerful tool that provides a *rigorous* solution for the induced current density on a body. Various proofs and theorems assure us that the MM solution is rigorous under "realistic" conditions [3]. In general, this statement can be made: *If the method of moments is properly applied, the series representation for the current will converge to the actual current as the number of basis functions is increased.*

Example 3.1: Scattering from a Thin Wire Using Pulse Basis Functions

As an example of how the MM is applied, consider the thin wire of radius a and length L shown in Fig. 3.4. Pulse basis functions will be used to obtain a step approximation to the current on the wire. Each pulse exists over a segment of length Δ and is centered at the point z_n, as shown in Fig. 3.5:

$$p_n(z') = \begin{cases} 1, & z_n - \frac{\Delta}{2} \le z' \le z_n + \frac{\Delta}{2} \\ 0, & \text{otherwise} \end{cases} \tag{3.49}$$

If the thin-wire approximation holds, the surface current can be written as

$$\vec{J}_s = \frac{I(z')}{2\pi a}\hat{z} = \frac{\hat{z}}{2\pi a} \sum_{n=1}^{N} I_n p_n(z') \tag{3.50}$$

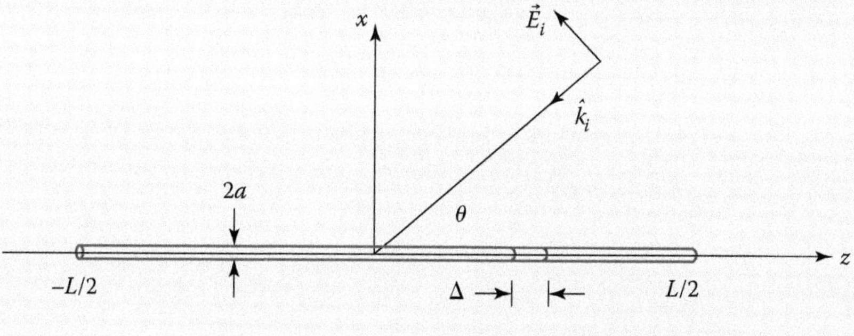

Fig. 3.4 MM applied to a thin wire, TM_z polarization.

(Continued)

Example 3.1: Scattering from a Thin Wire Using Pulse Basis Functions *(Continued)*

For a thin wire, the EFIE reduces to

$$\hat{z}E_{iz} = \hat{z}\frac{j}{\omega\epsilon}\int_{-(L/2)}^{L/2} I(z')\left[k^2 + \frac{\partial^2}{\partial z^2}\right] G(z,z')\,dz' \tag{3.51}$$

where

$$G(z,z') = \frac{e^{-jk|z-z'|}}{4\pi|z-z'|} \tag{3.52}$$

The second term in Eq. (3.51) contains the derivative of the pulse function. Using a δ function approximation for the derivative, as illustrated in Fig. 3.5:

$$p'_n(z') \approx \delta\left[z' - \left(z_n - \frac{\Delta}{2}\right)\right] - \delta\left[z' - \left(z_n + \frac{\Delta}{2}\right)\right] \equiv \delta_n^- - \delta_n^+ \tag{3.53}$$

The prime on p denotes the derivative with respect to its argument. The thin-wire approximation has allowed us to replace the two-dimensional integral over the surface of the wire with a one-dimensional integral along the surface parallel to the wire axis. [The original integral around the circumference of the wire simply yields a factor of $2\pi a$, which cancels with the denominator in Eq. (3.50).] Problems will occur with the *self term* ($m = n$) because there is a singularity in the integrand where $z = z'$. The easiest way to get around this problem is to move either the test point (z) or the source point (z') from the surface to the axis, as indicated in Fig. 3.6. Thus, the current flows along the wire axis, but testing is performed on the wire surface.

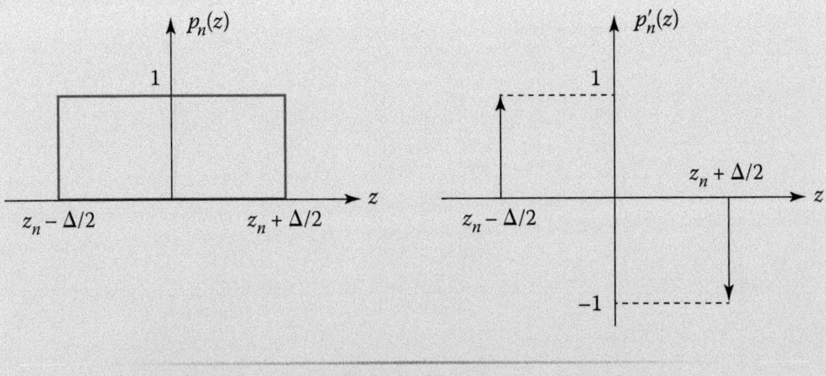

Fig. 3.5 Pulse basis function (left) and its derivative (right).

(Continued)

Example 3.1: Scattering from a Thin Wire Using Pulse Basis Functions *(Continued)*

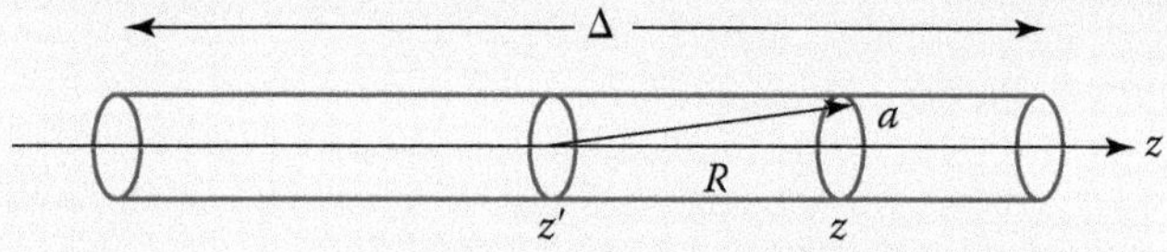

Fig. 3.6 Displacement of the test point to avoid the singularity.

The distance between source and test points is

$$R=\sqrt{(z-z')^2+a^2} \tag{3.54}$$

For Galerkin's method, the impedance elements are given by

$$\begin{aligned} Z_{mn} &= jk\eta\int_{z_m-(\Delta/2)}^{z_m+(\Delta/2)} \mathrm{d}z\int_{z_n-(\Delta/2)}^{z_n+(\Delta/2)} \mathrm{d}z'\frac{e^{-jkR}}{4\pi R} \\ &\quad -\frac{j\eta}{k^2}\int_{-(L/2)}^{(L/2)} \mathrm{d}z\int_{-(L/2)}^{(L/2)} \mathrm{d}z'(\delta_n^- - \delta_n^+)(\delta_m^- - \delta_m^+)\frac{e^{-jkR}}{4\pi R} \end{aligned} \tag{3.55}$$

Both of these integrals could be evaluated numerically but, for a long wire, this could take a large amount of computer time. A closed-form approximation for one of the integrations in each term would reduce the total number of integration steps without significantly degrading the final result. Assuming that Δ is small and that the integrand is slowly varying over the interval, an approximation for one of the integrals can be obtained using the rectangular rule

$$\int_{z_m-(\Delta/2)}^{z_m+(\Delta/2)} f(z)\,\mathrm{d}z \approx \Delta f(z_m)$$

which gives

$$\text{First term} \approx jk\eta\Delta\int_{z_n-(\Delta/2)}^{z_n+(\Delta/2)} \mathrm{d}z'\frac{e^{-jkR_m}}{4\pi R_m} \tag{3.56}$$

R_m is R evaluated at $z=z_m$. This remaining integral can be evaluated numerically.

(Continued)

Example 3.1: Scattering from a Thin Wire Using Pulse Basis Functions *(Continued)*

The properties of the δ functions can be used to reduce the second term to:

$$\begin{aligned}\text{Second term} &\approx \frac{j\eta}{k^2}\left\{G\left(z_m-\frac{\Delta}{2}, z_n-\frac{\Delta}{2}\right)-G\left(z_m-\frac{\Delta}{2}, z_n+\frac{\Delta}{2}\right)\right.\\ &\left.-G\left(z_m+\frac{\Delta}{2}, z_n-\frac{\Delta}{2}\right)+G\left(z_m+\frac{\Delta}{2}, z_n+\frac{\Delta}{2}\right)\right\} \\ &= \frac{j\eta}{k^2}\left\{\frac{2e^{-jkR_1}}{4\pi R_1}-\frac{e^{-jkR_2}}{4\pi R_2}-\frac{e^{-jkR_3}}{4\pi R_3}\right\}\end{aligned} \tag{3.57}$$

where

$$\begin{aligned}R_1 &= \sqrt{(z_m-z_n)^2+a^2}\\ R_2 &= \sqrt{(z_m-z_n-\Delta)^2+a^2}\\ R_3 &= \sqrt{(z_m-z_n+\Delta)^2+a^2}\end{aligned} \tag{3.58}$$

This completes the evaluation of the impedance elements.

Next, the excitation and measurement vectors must be evaluated. For a TM incident wave,

$$\vec{E}_i = \hat{\theta}e^{-j\vec{k}_i\cdot\vec{r}} \tag{3.59}$$

Inserting this into Eq. (3.39) gives

$$V_m^\theta = \int_{-L/2}^{L/2}\int_0^{2\pi}\left(\hat{\theta}e^{jkz\cos\theta}\right)\cdot\left(\frac{I(z)}{2\pi a}\hat{z}\right)ad\phi dz = \sin\theta\int_{z_m-(\Delta/2)}^{z_m+(\Delta/2)} e^{jkz\cos\theta}dz \tag{3.60}$$

Performing the integration yields

$$V_m^\theta = \Delta\sin\theta\,\mathrm{sinc}\left(\frac{k\Delta}{2}\cos\theta\right)e^{jkz_m\cos\theta} \tag{3.61}$$

because Galerkin's method is used and the basis functions are real, and $\vec{W}_m = \vec{J}_m$ and $R_m = V_m$. Finally, the electric field in the direction of θ is

$$E_\theta(\theta) = \frac{-jk\eta}{4\pi r}e^{-jkr}\mathbf{R}^T\mathbf{Z}^{-1}\mathbf{V} \tag{3.62}$$

where T signifies transpose and the RCS

$$\sigma_{\theta\theta}(\theta) = \frac{k^2\eta^2}{4\pi}|\mathbf{R}^T\mathbf{Z}^{-1}\mathbf{V}|^2 \tag{3.63}$$

(Continued)

Example 3.1: Scattering from a Thin Wire Using Pulse Basis Functions *(Continued)*

Figure 3.7 shows the backscatter from a thin wire as a function of frequency.

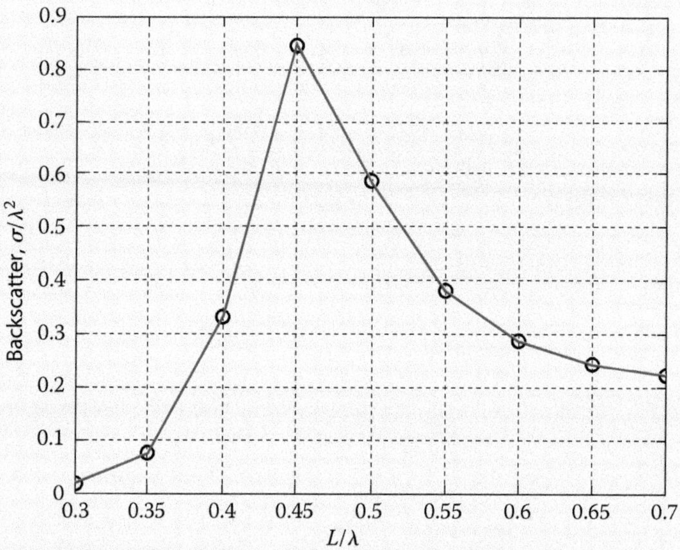

Fig. 3.7 Radar cross section of a thin wire using MM with pulse basis functions.

The backscatter from thin wires has peak values at odd integer multiples of 0.5λ ($\approx 0.5\lambda$, 1.5λ, ...). These peaks are much higher than the surrounding values and are called *resonances.* At these frequencies, the elements of the impedance matrix are phased such that large currents result. The strongest resonance is generally the first one.

3.5.4 Matrix Properties and Convergence

The pulse basis functions are among the simplest and easiest to work with, even though, as Example 3.1 illustrates, reducing the expressions for the impedance elements to the point at which they can be computer programmed takes a substantial amount of work. More complex basis functions (i.e., those that overlap) require fewer terms in the sum for convergence but yield more complicated expressions for the matrix elements. Nevertheless, if a large number of scattering computations are to be performed, it would pay

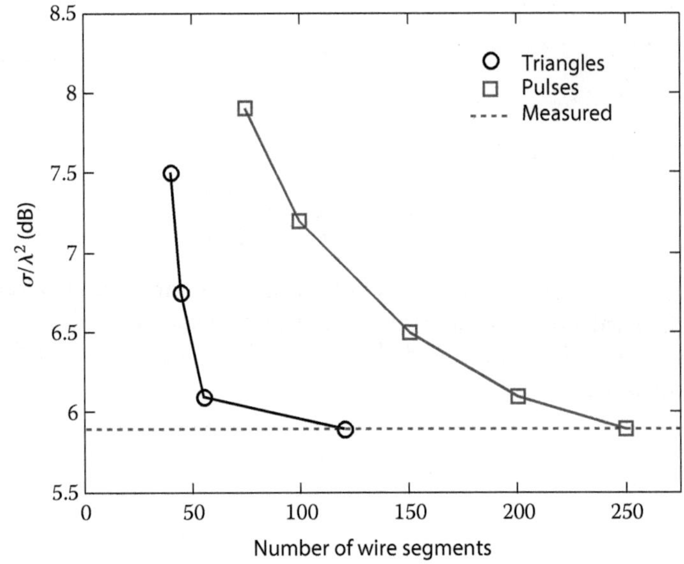

Fig. 3.8 Comparison of the solution convergence for pulses and overlapping triangles.

to employ overlapping basis functions. Figure 3.8 presents a comparison of pulses (step approximation) and overlapping triangles (piecewise linear approximation) for a 5.2λ long wire. More than four times as many pulses are necessary for convergence, and this has a significant effect on computer run time because impedance matrix size increases as N [2]. A typical rule of thumb for convergence is $\Delta \lesssim 0.1\lambda$.

In Example 3.1, symmetry exists in the impedance elements Z_{mn} and, therefore, symmetry is present in the matrix **Z**. In general, there is reciprocity between basis functions m and n so that $Z_{mn} = Z_{nm}$. For the special case of equal segment lengths, the elements of Z depend only on the relative values of m and n. For example, $Z_{11} = Z_{22} = Z_{33} = \cdots = Z_{NN}$ and, similarly, $Z_{12} = Z_{23} = Z_{34} = \cdots = Z_{N-1N}$, etc. Only N impedance values are needed to define the matrix completely. A *Toeplitz* matrix results for which efficient inversion algorithms exist [4].

Example 3.2: Scattering from a Thin-Wire Loop Using Exponential Basis Functions

Figure 3.9 shows a thin-wire loop of radius b located in the $x-y$ plane [3]. One method of solving the EFIE is to approximate the loop by a series of straight-line segments and use the formulas of Example 3.1. An alternative is to use entire-domain functions such as exponentials. In the latter case,

(Continued)

Example 3.2: Scattering from a Thin-Wire Loop Using Exponential Basis Functions *(Continued)*

the current is approximated by

$$\vec{J}_s(\phi) = \hat{\phi} \sum_{n=-\infty}^{\infty} I_n \frac{e^{jn\phi}}{2\pi a} \tag{3.64}$$

which is recognized as a Fourier series. Each term corresponds to a value of n and is called a *mode*. (The term ϕ is usually an azimuth variable so that the term *azimuthal modes* is used.) When Galerkin's method is applied the test functions are

$$\vec{W}_n(\phi) = \hat{\phi}\frac{e^{-jn\phi}}{2\pi a} \tag{3.65}$$

In cylindrical coordinates, only the ϕ variable is changing and, thus, the del operator becomes

$$\nabla \equiv \frac{1}{b}\frac{\partial}{\partial \phi}$$

Noting that

$$\hat{\phi} \cdot \hat{\phi}' = \cos(\phi - \phi')$$

we can reduce the expression for the impedance elements to

$$Z_{mn} = \int_0^{2\pi} b d\phi \int_0^{2\pi} b d\phi' \left[j\omega\mu \cos(\phi - \phi') - \frac{jmn}{\omega\epsilon b^2} \right] \times \exp[j(m\phi - n\phi')] \frac{e^{-jkR}}{4\pi R} \tag{3.66}$$

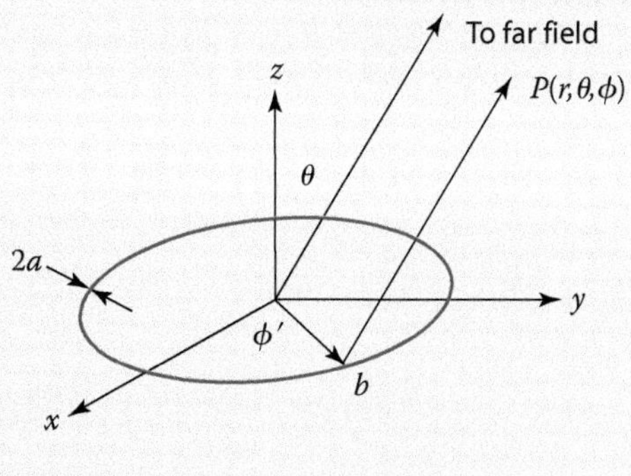

Fig. 3.9 Scattering from a thin wire circular loop.

(Continued)

Example 3.2: Scattering from a Thin-Wire Loop Using Exponential Basis Functions *(Continued)*

The distance between the source and test points is

$$R = b\sqrt{4\sin^2\left(\frac{\phi-\phi'}{2}\right)+\left(\frac{a}{b}\right)^2} \tag{3.67}$$

As in Example 3.1, the source point is considered to lie on the wire axis, whereas the test point lies on the surface.

The orthogonality property of the exponential functions can be used to eliminate one of the integrals. The integral is only nonzero when $m = n$, and replacing $\phi - \phi'$ with α yields

$$Z_{nn} = \frac{1}{2\pi}\int_0^{2\pi} d\alpha\left[j\omega\mu b\cos\alpha - \frac{jn^2}{\omega\epsilon b}e^{jn\alpha}\right]\frac{e^{-jkR}}{R} \tag{3.68}$$

where R is now given by

$$R = b\sqrt{4\sin^2\left(\frac{\alpha}{2}\right)+\left(\frac{a}{b}\right)^2} \tag{3.69}$$

Note that the limits of integration have been kept the same in spite of the transformation of variables to α. This is allowed because the integrand is still periodic with period 2π. Because the limits are over one complete period, the start and stop values are arbitrary. Equation (3.68) can be evaluated numerically and, because $m = n$, the resulting impedance matrix is diagonal.

The excitation and receive elements will be the complex conjugates of each other. As an example, the unit radiated plane wave with ϕ polarization is

$$\vec{E}_r = \hat{\phi}e^{-j\vec{k}_r\cdot\vec{r}} = \hat{\phi}\exp[jkb\sin\theta\cos(\phi-\phi')] \tag{3.70}$$

The values (θ, ϕ) specify the direction of the observation point, and ϕ' is the location of the source point on the wire (see Problem 2.4). The receive elements are given by

$$R_n^{\phi} = \int_0^{2\pi}\vec{E}_r\cdot\vec{J}_n b\,d\phi' = \int_0^{2\pi}\exp\left[jkb\sin\theta\cos(\phi-\phi')\right]e^{jn\phi'}b\,d\phi' \tag{3.71}$$

Using the identity provided in Problem 2.4, we obtain

$$R_n^{\phi} = \pi b j^{n+1}e^{jn\phi}[J_{n+1}(kb\sin\theta) - J_{n-1}(kb\sin\theta)] \tag{3.72}$$

When the wave is axially incident $(\theta = 0)$, only the $n = \pm 1$ modes are required; all other elements of $\mathbf{R}$ (and $\mathbf{V}$) will be zero. As the incidence angle as measured from the loop axis increases, more modes are required for convergence.

3.5.5 Selection of Basis Functions and Convergence

Subsectional basis functions are the best choice when the scattering body contains discontinuities. On a smooth body, the current density away from edges will be approximately a constant equal to the geometrical optics current. Near edges and discontinuities, the current varies more rapidly. When subsectional basis functions are used, localized variations in the current can be represented more accurately than with entire-domain functions. A common method of improving convergence is to use unequal subdomains. A small Δ is used where the current is expected to vary rapidly, but a large Δ is used away from discontinuities.

The characteristics of MM solutions for wires have been studied extensively. The subjects of study include the effect of different basis and testing functions [5], how the wire thickness affects the solution [6], and various methods of treating the singularity [7, 8]. As a rule of thumb, there is no advantage to using basis functions more sophisticated than overlapping triangles. Piecewise sinusoids are frequently used in place of triangles because, in some cases, they yield more concise expressions. Both overlapping triangles and piecewise sinusoids adequately represent the change in current amplitude across a segment. A problem that may occur is that the phase can vary significantly over a segment if Δ becomes too large, even if the amplitude is essentially constant. As a result, the magnitude of $\vec{J}_s$ is converged when the phase is not.

3.6 Method of Moments for Surfaces

3.6.1 Rectangular Subdomains

In the MM solution, we are free to choose the basis functions used in the expansion and testing. As in the case of the one-dimensional wire problem, the basis functions should be mathematically simple, yet be capable of efficiently representing the current so that rapid convergence is achieved. To represent a surface current, a minimum of two orthogonal components are required. More than two are needed if the basis functions are not orthogonal.

A simple example where two orthogonal components are sufficient is in computing the RCS of a flat rectangular plate, as shown in Fig. 3.10. The surface is discretized into small rectangular *patches* (the subdomains). Two orthogonal vector basis functions span each patch. The basis functions $\vec{J}^x_{mn}$ and $\vec{J}^y_{mn}$ shown in Fig. 3.10 are *pedestals*; they are the two-dimensional extensions of the pulses used in the one-dimensional case. To satisfy the boundary conditions the components should be parallel and perpendicular to the edges of the plate. For the rectangular plate with edges aligned along the x and y axes, an appropriate current expansion is

$$\vec{J}_s(x, y) = \sum_{n=1}^{N_x} \sum_{m=1}^{N_y} \left(I^x_{mn} \vec{J}^x_{mn} + I^y_{mn} \vec{J}^y_{mn} \right) \tag{3.73}$$

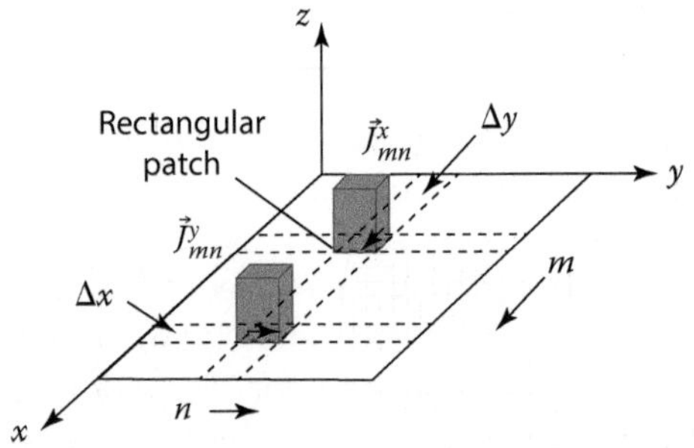

Fig. 3.10 Pulse basis functions for a rectangular plate.

where N_x and N_y are the number of patches in the x and y directions, respectively.

As in the one-dimensional case, overlapping basis functions provide a smoother approximation to the current and therefore converge faster. The two-dimensional version of the triangle is the rooftop function shown in Fig. 3.11. Each rooftop spans two patches. Again, the typical rule of thumb for convergence down to the -30 dB level is that the patches should be no greater than $\lambda/10$ by $\lambda/10$.

Rectangular patches are not an efficient means of approximating curved edges, in which case there is a *quantization* or step error introduced, as shown in Fig. 3.12a. To adequately represent the components of current tangential and normal to the edge, the patches must be very small, and therefore the resulting impedance matrix becomes large and convergence is very slow.

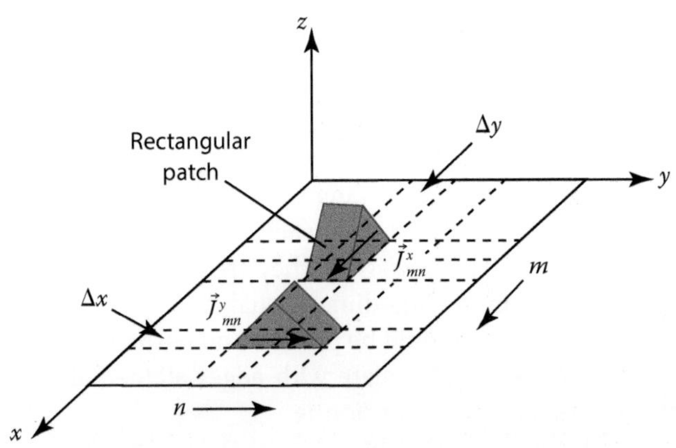

Fig. 3.11 Overlapping triangular basis functions for a rectangular plate.

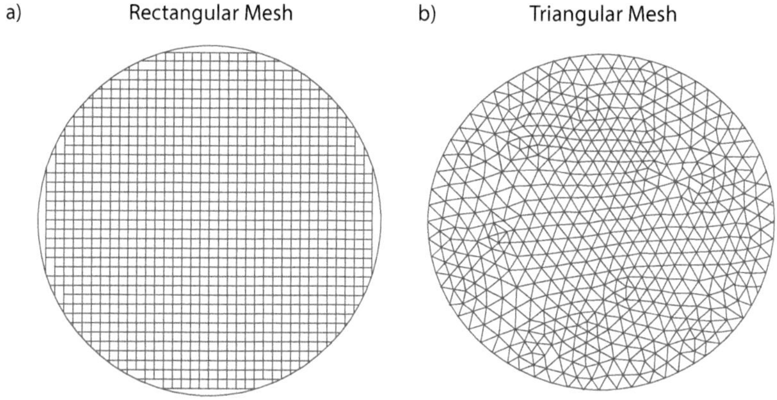

Fig. 3.12 Comparison of a) rectangular and b) triangular mesh approximations for a circle.

3.6.2 Triangular Subdomains

A more efficient basis function for curved edges is the *triangular subdomain* proposed by Rao, Wilton, and Glisson (RWG) [9]. Figure 3.12b shows a curved surface represented with a triangular mesh or *triangular facets*. The current expansion has the form

$$\vec{J}_s(\vec{r}) = \sum_{n=1}^{N} I_n \vec{J}_n \tag{3.74}$$

where N is the number of basis functions.

As illustrated in Fig. 3.13, each basis function spans two adjacent subdomains denoted as T_n^+ and T_n^-. The common edge is L_n and $\vec{\rho}_m^{\,c\pm}$ are vectors from the free vertices to the centroids of the two patches that form subdomain n. The basis functions have the form

$$\vec{J}_n(\vec{r}) = \begin{cases} \dfrac{L_n \rho_n^+}{2A_n^+} & \vec{r} \text{ in } T_n^+ \\ \dfrac{L_n \rho_n^-}{2A_n^-} & \vec{r} \text{ in } T_n^- \end{cases} \tag{3.75}$$

An example of a vector basis function plotted at various points in T_n^+ is shown in Fig. 3.14 for a triangle with a base and height of 1 unit. An example of how the current on an interior triangle is represented as the sum of the three basis functions is depicted in Fig. 3.15.

The desirable characteristics of the basis functions are as follows.

1. For triangle edges that represent body edges, there is no normal component of the surface current.

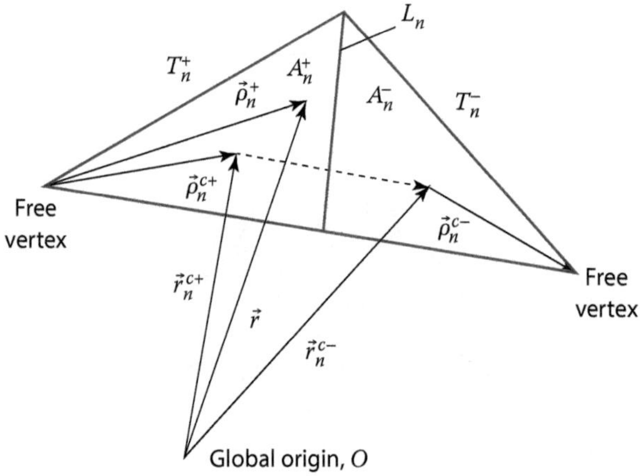

Fig. 3.13 Triangular subdomains and definition of quantities for the RWG basis function.

2. The current normal to an internal edge n is constant and continuous across the edge. Thus, the continuity equation is satisfied.
3. The divergence (first derivative) of $\vec{J}_n$ exists and has a simple form:

$$\nabla \cdot \vec{J}_n = \begin{cases} \dfrac{L_n}{A_n^+} & \vec{r} \text{ in } T_n^+ \\ -\dfrac{L_n}{A_n^-} & \vec{r} \text{ in } T_n^- \end{cases} \tag{3.76}$$

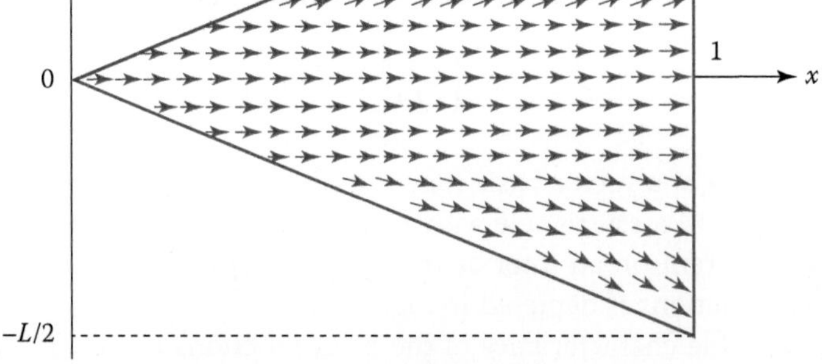

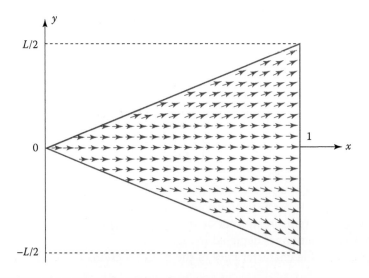

Fig. 3.14 Example of the vector plot of the part of basis function $\vec{J}_n$ on T_n^+.

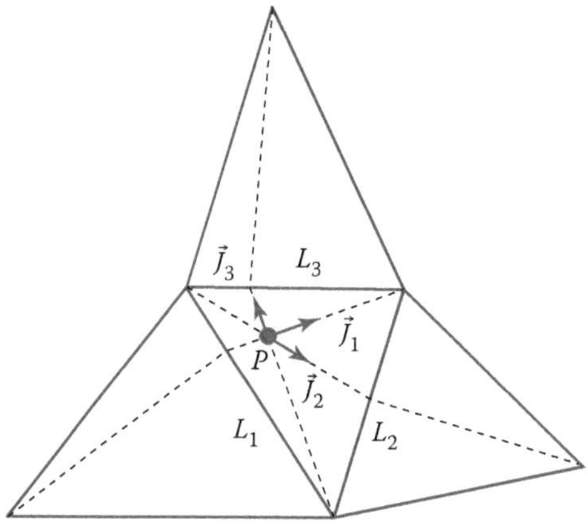

Fig. 3.15 Example of how the current at point P can be represented by three non-orthogonal components.

The equations for the impedance and excitation elements have been derived elsewhere [10]. The resulting equations are relatively simple when expressed in terms of vector quantities. The impedance elements are

$$Z_{mn} = L_m \left[\frac{j\omega}{2} \left(\vec{A}^{+}_{mn} \cdot \vec{\rho}^{c+}_{m} + \vec{A}^{-}_{mn} \cdot \vec{\rho}^{c-}_{m} \right) + \Phi^{-}_{mn} - \Phi^{+}_{mn} \right] \tag{3.77}$$

The vector and scalar potentials are

$$\vec{A}^{\pm}_{mn} = \frac{\mu}{4\pi} \iint_{S_n} \vec{J}_n(\vec{r}\,') \frac{e^{-jkR^{\pm}_{m}}}{R^{\pm}_{m}} \, ds' \tag{3.78}$$

$$\Phi^{\pm}_{mn} = \frac{j}{4\pi\omega\epsilon} \iint_{S_n} \nabla' \cdot \vec{J}_n(\vec{r}\,') \frac{e^{-jkR^{\pm}_{m}}}{R^{\pm}_{m}} \, ds' \tag{3.79}$$

and $R^{\pm}_{m} = \left| \vec{r}^{\,c\pm}_{m} - \vec{r}\,' \right|$. The excitation vector elements are

$$V_m = \frac{L_m}{2} \left[\vec{E}_i(\vec{\rho}^{c+}_{m}) \cdot \vec{\rho}^{c+}_{m} + \vec{E}_i(\vec{\rho}^{c-}_{m}) \cdot \vec{\rho}^{c-}_{m} \right] \tag{3.80}$$

The expressions for the impedance elements contain integrals that must be evaluated numerically. For triangular subdomains, this is most easily done by transforming to a set of localized coordinates on each triangle and then working in a set of normalized area coordinates. Finally, care must be taken with the self-term to avoid integration errors due to the singularity.

Computer coding the equations is straightforward and need only be done once. A more complicated task is generating the target and meshing the surfaces. Much effort has gone into developing mesh and grid generation software [11]. Several packages are available commercially and some are even offered for free on the Internet [12].

After the current coefficients have been determined, the scattered fields can be obtained by substituting the current series expansion into the radiation integrals and performing the required integration. A simpler approach that is accurate as long as the triangles are much smaller than a wavelength is to consider each subdomain (triangle pair) as a small dipole with a moment directed between the two triangle centroids

$$\vec{m}_n = L_n I_n \left(\vec{r}_n^{\,c-} - \vec{r}_n^{\,c+}\right) \tag{3.81}$$

The electric field due to this dipole, if it is located at the origin, is

$$\vec{E}_n(\vec{r}) = \frac{\eta_o e^{-jkr}}{4\pi} \left\{ \left[\left(\frac{\vec{r} \cdot \vec{m}_n}{r^2} \right) - \vec{m}_n \right] \left(\frac{jk}{r} + \frac{1}{r^2} \left[1 + \frac{1}{jkr} \right] \right) \right.$$
$$\left. + 2 \left(\frac{\vec{r} \cdot \vec{m}_n}{r^2} \right) \frac{1}{r^2} \left[1 + \frac{1}{jkr} \right] \right. \tag{3.82}$$

The contributions from all dipoles can be summed and the appropriate simplifications made for observation points in the far field. Of course, care must be taken to use a common phase reference for all of the dipoles.

Example 3.3: Monostatic RCS of a Plate Computed by the Method of Moments

A triangular meshed three-wavelength square plate is shown in Fig. 3.16 for two mesh configurations. One is a regularized mesh and the other

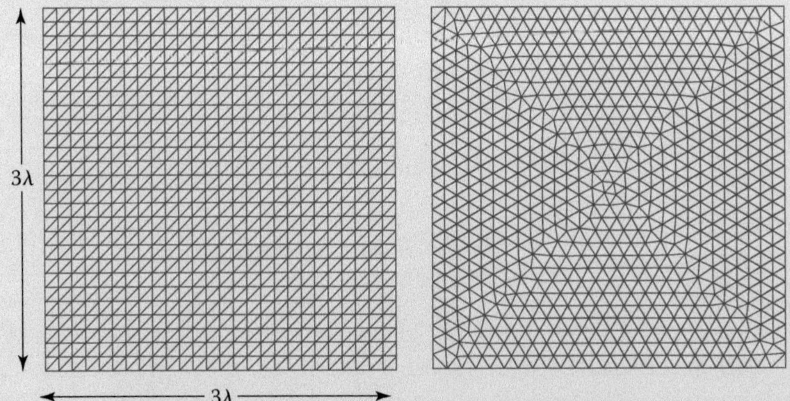

Fig. 3.16 Examples of meshed square plates.

(Continued)

Example 3.3: Monostatic RCS of a Plate Computed by the Method of Moments *(Continued)*

an equalized mesh. Modern meshing software gives users a range of options for mesh generation. The monostatic RCS for TM-polarized incident plane wave is shown in Fig. 3.17. The plate mesh contains 2760 edges.

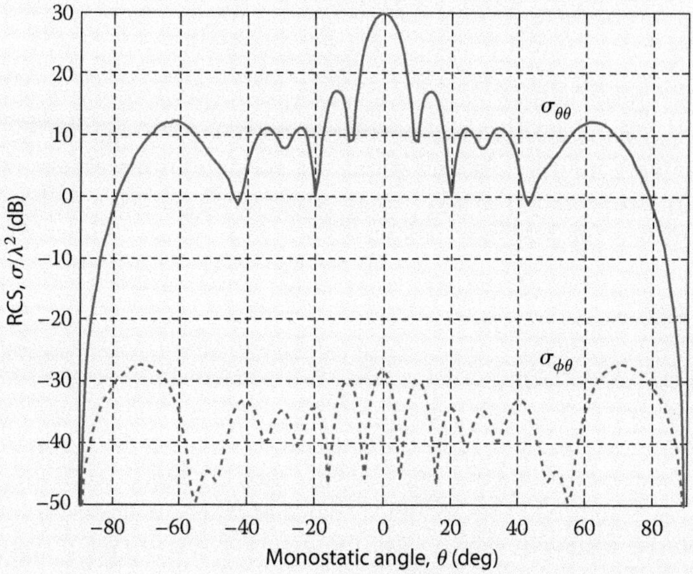

Fig. 3.17 Monostatic RCS of the three wavelength square plate, TM_z polarized incidence.

3.6.3 Other Basis Functions

There are other types of basis functions that have been developed for specific target types. For example, the use of hybrid basis functions is an efficient approach for a body of revolution (BOR) [1, 13]. A BOR is generated by rotating a plane curve around an axis of symmetry. Examples are disks, cylinders, cones, ogives, and spheres. Clearly a triangular mesh can accurately represent these shapes; however, a more computationally efficient approach is to use a combination of the basis functions introduced in Examples 3.1 and 3.2. The BOR generating curve is approximated by straight-line segments and subsectional basis functions used radially (along the curve, whose distance from the axis of rotation is a function of radius, ρ). In the azimuthal direction (around the body), a series of exponentials is used to approximate the current. In effect, on each segment of the BOR curve, a complex Fourier series ($\sim\exp[\pm jn\phi]$) accounts for the ϕ variation of the current. This approach yields an impedance matrix for each azimuthal mode: $n = 1, 2, \ldots, \infty$. However, each of the matrices is much smaller than the one that would occur using RWG basis functions. Furthermore, the mode impedance matrices are independent, and they can be inverted separately.

The maximum number of azimuthal modes required for convergence, that is, $n = 1, 2, \ldots, N_{\max}$, increases with the diameter of the BOR in wavelengths and as the incidence angle θ_i off of the axis of symmetry grows. A commonly used guideline for convergence is

$$N_{\max} \approx k\rho_{\max} \sin\theta_i \tag{3.83}$$

Thus, given limited computer resources, a much larger body can be handled with the BOR formulation. However, given the complexities of real-world targets and the relatively rapid and continuous increase in computer processing capability, the RWG basis functions have become the standard basis function for MM calculations.

3.6.4 Convergence Issues

Numerical techniques such as the MM are appealing because they can be shown mathematically to converge to the correct solution, assuming that they have been formulated and implemented properly [3]. Convergence is checked by increasing the number of terms in the expansion until the result has stabilized. This assumes that computer roundoff errors do not corrupt the result, in which case the solution may never converge. The following five important attributes of the solution technique affect the convergence.

3.6.4.1 Number of Terms in the Current Expansion

This fact is illustrated in Fig. 3.8 for the thin wire. Both curves (pulses and triangles) approach the measured "true" value. Not only measurements but other analytical and numerical techniques can also be used to establish convergence when a new MM formulation is being validated.

Generally the magnitudes of the current coefficients $\{I_n\}$ converge faster than the phase. For example, the current in the center of a 10λ square metal plate is approximately the physical optics value, $2\hat{n} \times \vec{H}_i$. The magnitude is constant, but the phase varies significantly over the scale of a wavelength. Therefore it is important to check the convergence of both the magnitude and phase. On the other hand, complete convergence of the phase is not absolutely necessary for a good RCS calculation. It is possible to have a good result for the far-scattered field even though the current expansion itself is not completely converged. This is not surprising given that the far field is determined from integration of the current, and integration is a smoothing or averaging process.

3.6.4.2 Type of Basis Functions

The fact that the type of basis function affects the rate of convergence has already been discussed in terms of pulses versus triangles for the thin wire. Similarly, for surfaces, the rooftop functions shown in Fig. 3.11 converge with larger patch sizes than do the pedestals shown in Fig. 3.10. The general rule of thumb, which has already been stated, is that the subdomains should be less than approximately 0.1λ for convergence down to the -30 dB level.

3.6.4.3 Method of Numerical Integration

Numerical integration must be performed to determine the impedance, excitation, and measurement (receive) elements. There are many types of integration algorithms. The most common are the simple *rectangular rule, Simpsons rule,* and *Gaussian quadrature* [4, 14]. If the approximate result for an integral is significantly in error, then the rest of the matrix calculation will also be in error.

In general, the impedance element integrands contain oscillatory functions (complex exponentials) and singularities, and are therefore sensitive to the number of integration points used. The method of integration must also be considered carefully. For example, let overlapping triangular basis functions be employed, but a rectangular rule approximation is used to evaluate the resulting integrals. This combination is essentially the same as using pulses because the integrand is sampled at only one point in the integration process. Improving accuracy in this case would require using smaller segments and, therefore, the advantage of using the triangular basis functions would be lost.

3.6.4.4 Method of Treating the Singularity

The self-term Z_{nn} can pose a problem in the numerical evaluation of an integral because of the singularity in the integrand. The simplest approach is to avoid the singularity by keeping the expansion and test points from occurring at the same location, as in Example 3.1. For surfaces, a variation of this approach is to continue to partition the subdomain into smaller patches of the same shape. For example, as shown in Fig. 3.18, a triangle can be broken into a number of similar smaller triangles, with the integrand approximately constant over each one [10]. The test point remains at the center of the original triangle.

The EFIE contains the free-space Green's function in the integrand. As pointed out earlier, the integrand can become singular for the self-term because both the source and test points lie on the same segment. Even if the source and test points are not colocated but are slightly separated, R can be so small that numerical inaccuracies result in a large integration error. This problem can be avoided by *subtracting out the singularity.*

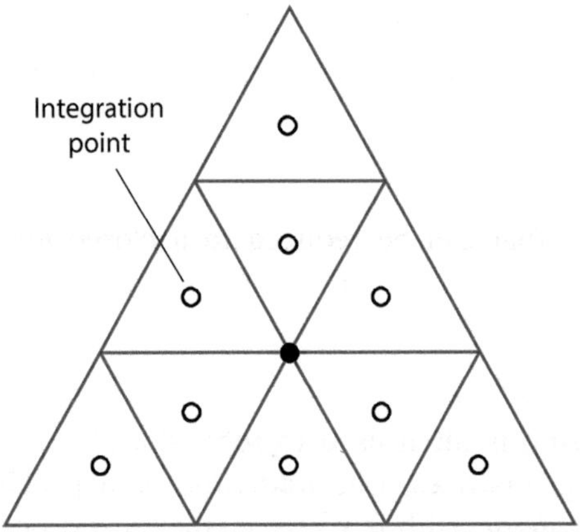

Fig. 3.18 Integration subdivision scheme for a triangle that avoids the singularity.

Consider a thin wire as described earlier in Fig. 3.6. If both the source and test points lie on the wire axis, the impedance matrix self-terms have the following form:

$$Z_{nn} \approx \int_{S_n}\int_{S_n} [\cdots] \frac{e^{-jk|z-z'|}}{4\pi|z-z'|} dz dz' \tag{3.84}$$

where we ignore the specific form of the term in brackets but assume that it is approximately constant in the region of the singularity. Near $z = z'$,

$$\frac{e^{-jk|z-z'|}}{4\pi|z-z'|} \approx \frac{1}{4\pi|z-z'|}$$

Thus, Eq. (3.84) is modified by adding and subtracting the approximation to R:

$$Z_{nn} \approx \int_{S_n}\int_{S_n} [\cdots] \left\{ \frac{e^{-jk|z-z'|}}{4\pi|z-z'|} + \frac{1}{4\pi|z-z'|} - \frac{1}{4\pi|z-z'|} \right\} dz\, dz' \tag{3.85}$$

$$Z_{nn} \approx \int_{S_n}\int_{S_n} [\cdots] \frac{e^{-jk|z-z'|}}{4\pi|z-z'|} dz\, dz' + \int_{S_n}\int_{S_n} [\cdots] \frac{1}{4\pi|z-z'|} dz\, dz' \tag{3.86}$$

The first term in Eq. (3.86) is now well behaved in the vicinity of $z = z'$ and can therefore be integrated numerically with good stability. The second integral can be reduced to closed form if [...] is constant over the limits of integration.

Singularity subtraction for two-dimensional integrals can be applied in several ways. If the integration variables are x and y, for instance, the elimination of the singularity can be applied to either of the variables individually or to the double integral. The latter method is generally preferable for complete elimination of the singularity, but it is more difficult to find an approximation for the integrand that can be reduced to a closed-form function of both variables.

3.6.4.5 Matrix Inversion

Matrix inversion is often used to solve Eq. (3.41). For solutions of the EFIE having large matrices, *ill-conditioning* is a problem. Ill-conditioned matrices are very *dense,* with few zero elements and nonzero elements that span several orders of magnitude.

The EFIE has the undesirable property that, as the segment size is reduced, the diagonal elements (self-terms) as well as off-diagonal elements are reduced in size. Thus, all elements of the impedance matrix are small but still important. This may not seem intuitively correct because elements of **Z** that are far apart correspond to source and test points that are far apart on the structure. People have hypothesized that these interactions could be neglected, resulting in a *sparse* matrix that is more easily inverted. Sometimes this "matrix-thinning" approach works and sometimes it does not. Evidently, even though the individual impedance terms are small, their collective effect is significant.

Most matrix inversion routines provide a *condition number* along with the inverted matrix to estimate the accuracy of the returned data.

3.6.4.6 Final Comments

There are a few cases when MM gives erroneous results for closed conducting bodies. This is most severe for the EFIE at frequencies where there is an internal resonance. Similar problems occur when MM is applied to the HFIE, however, the resonant frequencies are different than those for the EFIE. This has led to a combined field solution that uses a weighted combination of the EFIE and HFIE [15].

Finally, it is important to note that although measured data are often used to check calculations, they are not necessarily the "true" or "exact" values. Measurement errors are always present to some degree, and in fact, computer simulations are often used to validate a measurement.

3.6.5 Wire-Grid Models

Wire grids are frequently used to represent solid conducting surfaces. Problem 2.10 shows that when the spacing between wires is small in terms of the incident field wavelength, the reflection coefficient of the grid is essentially 1. Wire-grid representations of scatterers were used a great deal in the early stages of numerical electromagnetics for several reasons:

1. The frequencies of interest were predominately in the UHF and L bands and, therefore, a large body could be represented by just a few wires.
2. If the wires satisfy the thin-wire conditions, the expressions for the impedance elements require only a one-dimensional integration.
3. The segment size can be electrically very short so that approximate integration methods can be used. In some cases, the elements of **Z** can be reduced to closed-form expressions.

For efficient wire-grid modeling, it is important that the basis functions satisfy the boundary conditions at the wire end; that is, the current must vanish at a free end. Piecewise sinusoidal functions are a good choice. A wire is mathematically modeled as a series of overlapping dipoles. If all wire segments have a length Δ, the basis functions are of the form [16]

$$\vec{J}_i = \hat{t}\frac{S_i(t)}{2\pi a} \tag{3.87}$$

where t is an arclength variable along the wire axis, a its radius, and $S_i(t)$ the "dipole mode" function:

$$S_i(t) = \begin{cases} \dfrac{\sin k[\Delta - (t - t_i)]}{\sin k\Delta} = \cos\left[\dfrac{\pi(t - t_i)}{2\Delta}\right], & t_i - \Delta \le t \le t_i + \Delta \\ 0, & \text{otherwise} \end{cases} \tag{3.88}$$

These basis functions can also be extended to two dimensions and used as surface patch modes [17]. A final requirement is that current be continuous at the junction of the wires.

Figure 3.19 shows a crude low-frequency wire-grid approximation of an aircraft. Generally, the width of the apertures in the grid (w) should not exceed 0.25λ and preferably be closer to $\lambda/10$. If the wires are too close, however, the current begins to vary circumferentially, which then violates the thin-wire approximation [18]. The radius of the wires should satisfy $a \approx w/25$ and $a < 0.02\lambda$.

A host of limitations are involved in wire-grid modeling, especially for closed bodies [19]. Wire-grid models are rarely used to model complex high-frequency targets because surface patch codes are widely available and more accurate. The fundamental difference between the two is that, for a wire grid, the current is confined to the wires. A continuous surface current density does not exist on the entire target body as it does for a patch model.

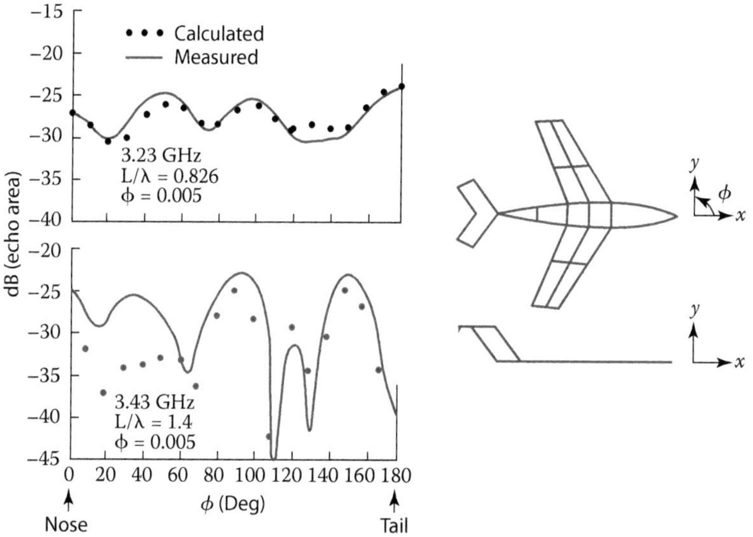

Fig. 3.19 Wiregrid model of an aircraft and MM computed patterns (from Ref. [16], courtesy IEEE).

Example 3.4: Monostatic RCS of a Wiregrid Aircraft

Figure 3.20 shows a crude model of an aircraft that can be used at low frequencies. The monostatic RCS versus frequency is plotted in Fig. 3.21 for a

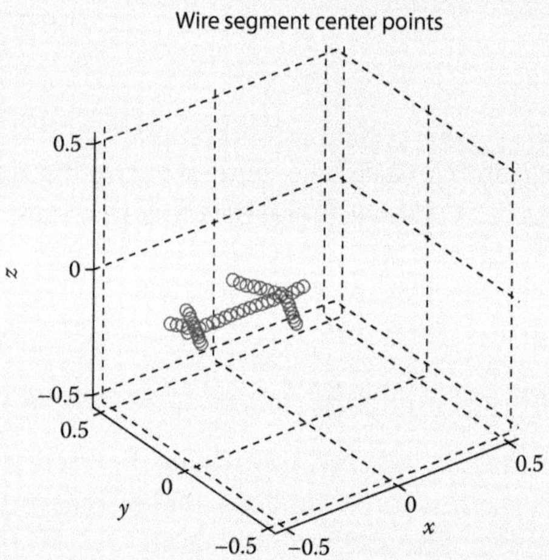

Fig. 3.20 Approximate low frequency model of an aircraft (circles denote the centers of subdomains).

(Continued)

Example 3.4: Monostatic RCS of a Wiregrid Aircraft (Continued)

TE-polarized incident plane wave. The peak in the curve around 270 MHz is due to a resonance condition for the fuselage. Resonances occur at several frequencies due to various aircraft parts and are most pronounced for slender metallic shapes. Resonances can be used to identify an aircraft, or the radar can operate at a target resonance to take advantage of the higher RCS.

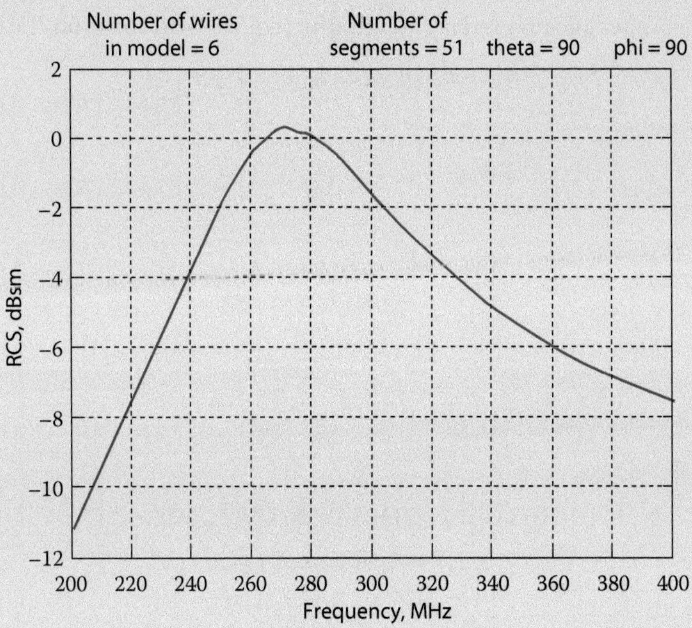

Fig. 3.21 Broadside RCS of the aircraft as a function of frequency using MM with pulse basis functions.

3.7 Other Integral Equations

3.7.1 Integral Equations for Impedance Boundary Conditions

In Sec. 2.13, it was found that impedance boundary conditions were a convenient approximation for complex surfaces that support both electric and magnetic currents. The surface impedance relates the tangential components of the electric and magnetic fields on S and, consequently, it also relates the currents,

$$\vec{J}_{ms} = -Z_s \eta_0 \hat{n} \times \vec{J}_s \tag{3.89}$$

where $\hat{n}$ is an outward unit vector.

At an observation point P, the scattered electric field that arises from the currents $\vec{J}_s$ and $\vec{J}_{ms}$ radiating in free space is given in terms of the electric and magnetic vector potentials:

$$\vec{E}_s(P) = -j\omega\vec{A} - \frac{j}{\omega\mu\epsilon}\nabla(\nabla\cdot\vec{A}) - \frac{1}{\epsilon}\nabla\times\vec{F} \tag{3.90}$$

The magnetic field intensity is given by

$$\vec{H}_s(P) = \frac{\nabla\times\vec{A}}{\mu} - j\omega\vec{F} - \frac{j}{\omega\mu\epsilon}\nabla(\nabla\cdot\vec{F}) \tag{3.91}$$

The impedance boundary conditions require that, for P on the surface,

$$[\vec{E}_i + \vec{E}_s]_{\tan} = Z_s\eta_0\hat{n}\times\left[\vec{H}_i + \vec{H}_s\right] \tag{3.92}$$

or, equivalently,

$$\vec{E}_i\Big|_{\tan} - Z_s\eta_0\hat{n}\times\vec{H}_i = Z_s\eta_0\hat{n}\times\left\{\left[\frac{1}{\mu}\nabla\times\vec{A} - j\omega\vec{F} - \frac{j}{\omega\epsilon\mu}\nabla\left(\nabla\cdot\vec{F}\right)\right]\right.$$
$$\left.+\left[j\omega\vec{A} + \frac{j}{\omega\epsilon\mu}\nabla\left(\nabla\cdot\vec{A}\right) + \frac{1}{\epsilon}\nabla\times\vec{F}\right]\right\}\Bigg|_{\tan} \tag{3.93}$$

Now, substitute in the integral expressions for the potentials and make use of the following relationships:

$$G(\vec{r},\vec{r}') = \frac{e^{-jk|\vec{r}-\vec{r}'|}}{4\pi|\vec{r}-\vec{r}'|}$$
$$\nabla[\nabla\cdot\textstyle\iint\vec{J}_{ms}Gds'] = \iint(\nabla'\cdot\vec{J}_{ms})\nabla Gds'$$
$$\nabla'\cdot(\hat{n}'\times\vec{J}_s) = \vec{J}_s\cdot(\nabla'\times\hat{n}') - \hat{n}'\cdot(\nabla'\times\vec{J}_s)$$
$$\nabla\times(\vec{J}_sG) = -\vec{J}_s\times\nabla G$$
$$\omega\mu = k\eta$$
$$\omega\epsilon = \frac{k}{\eta}$$

Combining all of the relationships in the listed equations yields a rather formidable integral equation for the surface current density $\vec{J}_s$:

$$\frac{\vec{E}_i(\vec{r})|_{\tan}}{\eta_0} - Z_s(\vec{r})\hat{n}(\vec{r})\times\vec{H}_i(\vec{r}) = Z_s(\vec{r})\hat{n}(\vec{r})\times\left[\nabla\times\iint_S\vec{J}_sGds'\right.$$
$$\left.+\,jk\iint_S Z_s(\hat{n}'\times\vec{J}_s)Gds' + \frac{j}{k}\iint_S\nabla'\cdot[Z_s\hat{n}'\times\vec{J}_s]\nabla Gds'\right]$$
$$+\,[jk\iint_S\vec{J}_sGds' + \frac{j}{k}\iint_S(\nabla'\cdot\vec{J}_s)\nabla Gds'$$
$$-\,\nabla\times\iint_S Z_s(\hat{n}'\times\vec{J}_s)Gds']_{\tan} \tag{3.94}$$

The $\vec{r}$ dependence has been explicitly included, specifying the location of an observation point on S. The quantities $\hat{n}'$, Z_s, G, and $\vec{J}_s$ in the integrands are functions of the source point coordinates, even though the $\vec{r}'$ dependence has not been explicitly written.

The MM can be used to solve for the current density. For an arbitrary surface, the current is approximated by the series

$$\vec{J}_s(x_1', x_2') = \sum_{n=1}^{N} \sum_{m=1}^{M} [I_{mn}^{x_1} \vec{J}_{mn}^{x_1}(x_1', x_2') + I_{mn}^{x_2} \vec{J}_{mn}^{x_2}(x_1', x_2')] \tag{3.95}$$

where x_1' and x_2' refer to the two chosen orthogonal coordinate variables used to describe $\vec{r}'$. If Galerkin's method is used, the testing functions will be of the form $\vec{W}_{mn}^{x_{1,2}}(x_1, x_2) = \vec{J}_{mn}^{x_{1,2}}(x_1, x_2)^*$, where x_1 and x_2 are the orthogonal coordinate variables used to describe $\vec{r}$. Testing the left side of Eq. (3.94) yields an excitation vector; testing the right side provides the impedance elements.

3.7.2 Resistive Sheet Integral Equation

Resistive sheets only support electric current and satisfy the condition

$$\vec{E}_{\tan} = R_s \vec{J}_s \tag{3.96}$$

or, in terms of the incident and scattered fields,

$$\vec{E}_i|_{\tan} = R_s \vec{J}_s - \vec{E}_s|_{\tan} \tag{3.97}$$

Using the radiation integrals to obtain an expression for the scattered field in terms of the electric current and moving the observation point onto the surface give

$$\vec{E}_i(\vec{r})|_{\tan} = R_s(\vec{r})\vec{J}_s(\vec{r}) + \left[jk\eta_0 \iint_S \vec{J}_s G ds' + \frac{jk}{\eta_0} \nabla\nabla \cdot \iint_S \vec{J}_s G ds' \right]_{\tan} \tag{3.98}$$

Equation (3.98) is simply the EFIE for a PEC with the term $R_s(\vec{r})\vec{J}_s(\vec{r})$ added to the right-hand side. The MM testing procedure will result in an additional term for each element of the impedance matrix,

$$Z_{mn} = (Z_{MM} + Z_L)_{mn}$$

Z_{MM} is given by Eq. (3.37), and the *load impedance matrix* elements are

$$(Z_L)_{mn} = \int_{S_m} \int_{S_n} \vec{W}_m \cdot [R_s(\vec{r})\vec{J}_s(\vec{r})] ds \tag{3.99}$$

Obtaining the expansion coefficients requires solving the matrix equation

$$\mathbf{I} = [\mathbf{Z}_{MM} + \mathbf{Z}_L]^{-1}\mathbf{V} \tag{3.100}$$

Example 3.5: Scattering from a Thin Resistive Wire

In Example 3.1, scattering from a PEC strip was calculated using the MM. For this example, assume that the strip is no longer a perfect conductor but has a surface resistivity R_s. Pulse basis functions and Galerkin's method are used as in Example 3.1. The load impedance matrix elements are determined from

$$(Z_L)_{mn} = R_s(2\pi a)\int_{-(L/2)}^{L/2}\left[\frac{p_m(z)}{2\pi a}\hat{z}\right]\cdot\left[\hat{z}\frac{p_n(z)}{2\pi a}\right]dz \tag{3.101}$$

Because of the nonoverlapping property of the pulses, the matrix correction term is

$$(Z_L)_{mn} = \begin{cases} \dfrac{R_s\Delta}{2\pi a}, & m = n \\ 0, & m \neq n \end{cases} \tag{3.102}$$

The backscatter was computed for a wire of length 5.22λ with a radius of 0.0025λ. Triangular and pulse functions were compared for two surface resistivities. The results are summarized in Table 3.1. The entries are σ/λ^2 in dB, and the number of segments for convergence is given in parentheses.

Table 3.1 Results of Example 3.5

R_s/η_0	Pulses, dB (N)	Triangles, dB (N)
0	5.57(250)	5.75(100)
0.2	−4.12(100)	−4.29(70)

3.7.3 Volume Integral Equations

Penetrable bodies are those for which a field and current can exist internally. For example, in a dielectric, the response to an impressed field is due to the bound charges, that is, a *polarization current*. For a magnetic body ($\mu \neq \mu_0$), the response is due to the *magnetization current*. Integral equations for these volume current densities can be derived based on the equivalence theorem of Sec. 2.9 [3].

3.7.3.1 Dielectric Bodies

A *dielectric body* is a body composed of material for which $\epsilon_r > 1$ and. $\mu \neq \mu_0$. The polarization current density is given by Eq. (2.29):

$$\vec{J} = j\omega\Delta\epsilon\left(\vec{E}_i + \vec{E}_s\right) \tag{3.103}$$

where $\Delta\epsilon = \epsilon - \epsilon_0$. The scattered field is given in operator notation by

$$\mathcal{L}(\vec{J}) = -\vec{E}_s = j\omega\vec{A} + \nabla\Phi_e \tag{3.104}$$

The potentials $\vec{A}$ and Φ_e are defined in Eqs. (2.6) and (2.9). Combining Eqs. (3.103) and (3.104) yields

$$\mathcal{L}(\vec{J}) + \frac{\vec{J}}{j\omega\Delta\epsilon} = \vec{E}_i \tag{3.105}$$

This is recognized as a PEC integral equation modified by the term $\vec{J}/(j\omega\Delta\epsilon)$. An MM procedure with the current represented by

$$\vec{J} = \sum_n I_n\vec{J}_n \tag{3.106}$$

would yield the following matrix equation:

$$[\mathbf{Z}_{MM} + \mathbf{Z}_L]\mathbf{I} = \mathbf{V}$$

The elements of $\mathbf{Z}_{MM}$ are the same as in the PEC case, and

$$(Z_L)_{mn} = \iiint_V \vec{W}_m \cdot \frac{\vec{J}_n}{j\omega\Delta\epsilon}\,\mathrm{d}v' \tag{3.107}$$

3.7.3.2 Magnetic Bodies

A similar procedure can be applied to *magnetic materials,* starting with Eq. (2.30):

$$\vec{J}_m = j\omega\Delta\mu(\vec{H}_i + \vec{H}_s) \tag{3.108}$$

where $\Delta\mu = \mu - \mu_0$ and

$$\mathcal{L}(\vec{J}_m) = -\vec{H}_s = j\omega\vec{F} + \nabla\Phi_m \tag{3.109}$$

with the potentials given by Eqs. (2.7) and (2.10). Rearranging gives

$$\mathcal{L}(\vec{J}_m) + \frac{\vec{J}_m}{j\omega\Delta\mu} = \vec{H}_i \tag{3.110}$$

When the MM is applied to the solution of magnetic currents, the unknown expansion coefficients play the role of voltages. Therefore, the magnetic current is approximated by

$$\vec{J}_m = \sum_n V_n\vec{J}_{mn} \tag{3.111}$$

Note that the subscript m refers to magnetic current and is not an index.

Testing the right side of Eq. (3.110) provides the excitation vector, which now has units of current,

$$I_n = \iiint_V \vec{W}_n \cdot \vec{H}_i\mathrm{d}v \tag{3.112}$$

Testing the left side yields an *admittance matrix* **Y**. The resulting MM matrix equation will have the form

$$[\mathbf{Y}_{MM} + \mathbf{Y}_L]\mathbf{V} = \mathbf{I}$$

where the matrix $\mathbf{Y}_{MM}$ is the dual of $\mathbf{Z}_{MM}$ in the PEC case and

$$(Y_L)_{ln} = \iiint_V \vec{W}_l \cdot \frac{\vec{J}_n}{j\omega\Delta\mu} dv' \tag{3.113}$$

3.7.3.3 Dielectric and Magnetic Bodies

For bodies that are both dielectric and magnetic, the interaction between the polarization and magnetization currents must be considered. This interaction is accounted for by the curl terms in Eqs. (2.15) and (2.16). Define a new operator $\mathcal{K}$ such that

$$\begin{aligned} \mathcal{K}(\vec{J}) &\equiv \frac{1}{\mu}\nabla \times \vec{A} \\ \mathcal{K}(\vec{J}_m) &\equiv \frac{1}{\epsilon}\nabla \times \vec{F} \end{aligned} \tag{3.114}$$

A set of *coupled integral equations* must be solved:

$$\begin{aligned} \mathcal{L}(\vec{J}) + \mathcal{K}(-\vec{J}_m) + \frac{\vec{J}}{j\omega\Delta\epsilon} &= \vec{E}_i \\ \mathcal{K}(\vec{J}) - \mathcal{L}(-\vec{J}_m) + \frac{\vec{J}_m}{j\omega\Delta\mu} &= \vec{H}_i \end{aligned} \tag{3.115}$$

In matrix notation,

$$\begin{bmatrix} \mathcal{L} & \mathcal{K} \\ \mathcal{K} & -\mathcal{L} \end{bmatrix} \begin{bmatrix} \vec{J} \\ -\vec{J}_m \end{bmatrix} + \begin{bmatrix} \dfrac{\vec{J}}{j\omega\Delta\epsilon} \\ \dfrac{\vec{J}_m}{j\omega\Delta\mu} \end{bmatrix} = \begin{bmatrix} \vec{E}_i \\ \vec{H}_i \end{bmatrix} \tag{3.116}$$

Using the current expansion in Eqs. (3.106) and (3.111) and testing with all weighting functions result in the following matrix equation:

$$\begin{bmatrix} \mathbf{Z}_{MM} + \mathbf{Z}_L & \mathbf{B} \\ \mathbf{C} & \mathbf{Y}_{MM} + \mathbf{Y}_L \end{bmatrix} \begin{bmatrix} \mathbf{I} \\ \mathbf{V} \end{bmatrix} = \begin{bmatrix} \mathbf{V}_i \\ \mathbf{I}_i \end{bmatrix} \tag{3.117}$$

The subscript i has been added to the excitation vectors to distinguish them from the vector of expansion coefficients. The matrices **B** and **C** characterize the interaction between the polarization and magnetization currents.

Example 3.6: Thin-Shell Approximation

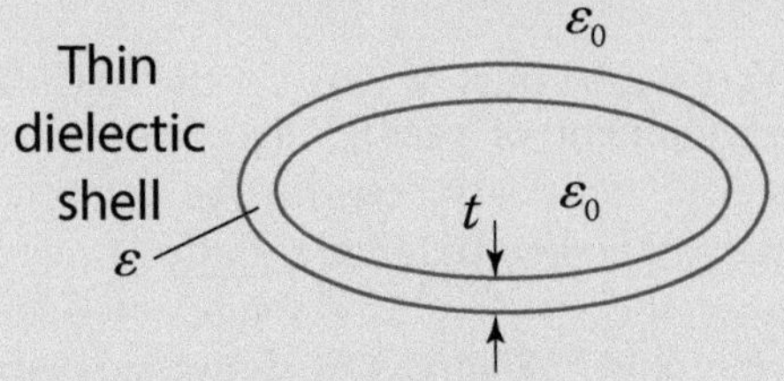

Fig. 3.22 Thin-shell approximation for dielectrics.

If a dielectric body resembles a sheet that is very thin in terms of wavelength, the volume polarization current density can be approximated by a surface current density [20]. The thickness of the shell is t and the dielectric constant is ϵ, as illustrated in Fig. 3.22. The surface current density is approximated by

$$\vec{J}_s \approx \vec{J} \cdot t = j\omega\Delta\epsilon t\vec{E} \tag{3.118}$$

and the integral equation for the surface current follows directly from Eq. (3.105):

$$\mathcal{L}(\vec{J}_s) + \frac{\vec{J}_s}{j\omega\Delta\epsilon t} = \vec{E}_i \tag{3.119}$$

The impedance matrix load terms are

$$(Z_L)_{mn} = \iint_S \vec{W}_m \cdot \frac{\vec{J}_n}{j\omega\Delta\epsilon t} \mathrm{d}s \tag{3.120}$$

Consider the resistive wire of Example 3.3 with radius a and surface resistivity R_s. If we imagine the wire to be a sheet of resistive film that is wrapped like a straw, we see that it could be cut in the axial direction and flattened into a strip. The width of the strip is equal to the circumference of the wire, $2\pi\ a$. If the film conductivity is σ_f, the complex dielectric constant is

$$\epsilon = \epsilon' - j\epsilon'' \approx \epsilon_0 - j\frac{\sigma_f}{\omega}$$

The scalar quantity in the integrand becomes

$$\frac{1}{j\omega\Delta\epsilon t} \approx \frac{1}{\sigma_f t} = R_s \tag{3.121}$$

This is identical to the load impedance term that appears in the resistive sheet integral equation (3.99). Thus, a resistive sheet is equivalent to a thin dielectric shell.

3.8 Finite Element Method

3.8.1 Wave Equations

In recent years the finite element method (FEM) has emerged as an efficient method for solving the vector wave equation [2, 21, 22]. This was because of the development of vector edge basis functions and efficient methods of handing the boundary between the computational domain and unbounded free space. The computational domain refers to the penetrable (non-PEC) volume of the target plus a region of free space immediately around the scattering object. Figure 3.23 shows an example of the domain for a complex target with several discrete pieces. If the electric and magnetic fields can be determined on the surface S enclosing the computational domain, then equivalent electric and magnetic currents can be computed. They, in turn, can be used in the radiation integrals of Chapter 2 to calculate the far-scattered field and RCS.

There are many formulations of FEM that start from a variety of differential equations. The finite element method can be applied directly to Maxwell's curl equations,

$$\begin{aligned} \nabla \times \vec{E} &= -j\omega\mu\vec{H} \\ \nabla \times \vec{H} &= j\omega\epsilon\vec{E} \end{aligned} \tag{3.122}$$

to the vector Helmholtz equations,

$$\begin{aligned} \nabla^2\vec{E} + k^2\vec{E} &= 0 \\ \nabla^2\vec{H} + k^2\vec{H} &= 0 \end{aligned} \tag{3.123}$$

or, more commonly, to the vector wave equations [Eqs. (3.27) or (3.30)], also referred to as the "double-curl" equations. The quantities in the equations are the total fields, which are sums of the incident and scattered fields: $\vec{E} = \vec{E}_s + \vec{E}_i$ and $\vec{H} = \vec{H}_s + \vec{H}_i$. It is assumed that the incident fields are known. For the RCS problem they are plane wave fields.

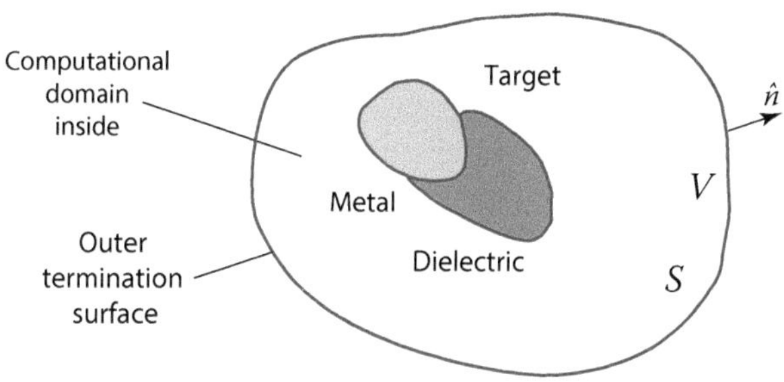

Fig. 3.23 Scattering object and its accompanying computational domain.

The starting point for the FEM formulation is the vector wave equation in a source free region, Eqs. (3.27) and (3.30),

$$\nabla \times \left(\frac{\nabla \times \vec{E}}{\mu_r} \right) - k_o^2 \epsilon_r \vec{E} = 0 \tag{3.124}$$

$$\nabla \times \left(\frac{\nabla \times \vec{H}}{\epsilon_r} \right) - k_o^2 \mu_r \vec{H} = 0 \tag{3.125}$$

The incident fields satisfy

$$\nabla \times \nabla \times \vec{E}_i - k_o^2 \vec{E}_i = 0 \tag{3.126}$$

$$\nabla \times \nabla \times \vec{H}_i - k_o^2 \vec{H}_i = 0 \tag{3.127}$$

Subtracting Eq. (3.127) from (3.125) gives

$$\nabla \times \left(\frac{\nabla \times \vec{H}_s}{\epsilon_r} \right) - k_o^2 \mu_r \vec{H}_s + \nabla \times \left[\left(\frac{1}{\epsilon_r} - 1 \right) \nabla \times \vec{H}_i \right] - k_o^2 (\mu_r - 1) \vec{H}_i = 0 \tag{3.128}$$

and, similarly, for the double-curl equation for the electric field

$$\nabla \times \left(\frac{\nabla \times \vec{E}_s}{\mu_r} \right) - k_o^2 \epsilon_r \vec{E}_s + \nabla \times \left[\left(\frac{1}{\mu_r} - 1 \right) \nabla \times \vec{E}_i \right] - k_o^2 (\epsilon_r - 1) \vec{E}_i = 0 \tag{3.129}$$

As a first step, the scattered electric or magnetic field is expanded in terms of basis functions, $\vec{W}_n$. The types of basis functions and their selection will be discussed later. For example, if the wave equation for $\vec{H}_s$ is to be solved, the expansion takes the form

$$\vec{H}_s = \sum_{n=1}^{N} H_n \vec{W}_n \tag{3.130}$$

where $\{H_n\}$ are unknown coefficients. To solve for the expansion coefficients, a testing procedure is applied to the vector wave equation. Assuming real basis functions and that Galerkin's method is used, testing Eq. (3.128) with $\vec{W}$ gives

$$\iiint_V \vec{W} \cdot \left\{ \nabla \times \left(\frac{\nabla \times \vec{H}_s}{\epsilon_r} \right) - k_0^2 \mu_r \vec{H}_s + \nabla \times \left[\left(\frac{1}{\epsilon_r} - 1 \right) \nabla \times \vec{H}_i \right] - k_0^2 (\mu_r - 1) \vec{H}_i \right\} dv = 0 \tag{3.131}$$

Because of the double curl in Eq. (3.131), the selected expansion and testing functions must have a second derivative. This is highly restrictive, and it also turns out that the approximations in the numerical evaluation of the second derivatives limit the accuracy of the results and lead to spurious solutions or *parasites* [23]. To relax the demands on the selection of $\vec{W}$, Green's identity can be applied. Green's first vector identity is [2]

$$\oint\!\!\oint_S \left(\vec{A} \times \nabla \times \vec{B}\right) \cdot \hat{n} ds = \iiint_V \left(\nabla \times \vec{A} \cdot \nabla \times \vec{B} - \vec{A} \cdot \nabla \times \nabla \times \vec{B}\right) dv \tag{3.132}$$

where S is a closed surface surrounding the volume V. In terms of the total field, the resulting so-called *weak form* is

$$\begin{aligned}
&\iiint_V \left(\nabla \times \vec{W} \cdot \frac{\nabla \times \vec{H}_s}{\epsilon_r} - k_0^2 \mu_r \vec{W} \cdot \vec{H}_s\right) dv \\
&+ \iiint_V \left(\nabla \times \vec{W} \cdot \left[\left(\frac{1}{\epsilon_r} - 1\right) \nabla \times \vec{H}_i\right] - k_0^2 (\mu_r - 1) \vec{W} \cdot \vec{H}_i\right) dv \\
&- \oint\!\!\oint_S \vec{W} \cdot \left(\hat{n} \times \left[\frac{1}{\epsilon_r} \nabla \times (\vec{H}_i + \vec{H}_s) - \nabla \times \vec{H}_i\right]\right) ds = 0
\end{aligned} \tag{3.133}$$

As shown in Figure 3.23, $\hat{n}$ is the outward unit normal from the closed volume V. The exterior part of S is always in free space, so that $\mu_r = \epsilon_r = 1$ and therefore

$$\begin{aligned}
&\iiint_V \left(\nabla \times \vec{W} \cdot \frac{\nabla \times \vec{H}_s}{\epsilon_r} - k_0^2 \mu_r \vec{W} \cdot \vec{H}_s\right) dv - \oint\!\!\oint_S \vec{W} \cdot (\hat{n} \times \nabla \times \vec{H}_s) ds \\
&= \iiint_V \left(\nabla \times \vec{W} \cdot \left[\left(\frac{1}{\epsilon_r} - 1\right) \nabla \times \vec{H}_i\right] - k_0^2 (\mu_r - 1) \vec{W} \cdot \vec{H}_i\right) dv
\end{aligned} \tag{3.134}$$

Likewise, there is a dual equation for the electric field:

$$\begin{aligned}
&\iiint_V \left(\nabla \times \vec{W} \cdot \frac{\nabla \times \vec{E}_s}{\mu_r} - k_0^2 \epsilon_r \vec{W} \cdot \vec{E}_s\right) dv - \oint\!\!\oint_S \vec{W} \cdot \left(\hat{n} \times \nabla \times \vec{E}_s\right) ds \\
&= \iiint_V \left(\nabla \times \vec{W} \cdot \left[\left(\frac{1}{\mu_r} - 1\right) \nabla \times \vec{E}_i\right] - k_0^2 (\epsilon_r - 1) \vec{W} \cdot \vec{E}_i\right) dv
\end{aligned} \tag{3.135}$$

To continue the FEM solution, the basis functions must be specified. A two-dimensional problem will be used to illustrate the procedure.

3.8.2 Basis Functions for Two and Three Dimensions

For a two-dimensional problem the domain is an open surface S that represents the geometry cross section, surrounded by a closed contour C. The normal to the contour $\hat{n}$ points outward on the surface as shown in Fig. 3.24. For a two-dimensional formulation, Eq. (3.135) becomes [21]

$$\iint_S \left[\nabla \times \vec{W} \cdot \left(\frac{1}{\mu_r} \nabla \times \vec{E}_s\right) - k_0^2 \epsilon_r \vec{W} \cdot \vec{E}_s\right] \mathrm{d}s$$
$$= \oint_C \left(\hat{n} \times \frac{1}{\mu_r} \nabla \times \vec{E}_i\right) \cdot \vec{W} \mathrm{d}l - \iint_S \left[\nabla \times \left(\frac{\nabla \times \vec{E}_i}{\mu_r}\right) - k_0^2 \epsilon_r \vec{E}_i\right] \cdot \vec{W} \mathrm{d}s \tag{3.136}$$

Note that in two dimensions ∇ becomes a transverse operator. For example, if the geometry is invariant to z, then $\partial/\partial_z = 0$ and $\nabla \equiv \hat{x}\, \partial/\partial x + \hat{y}\, \partial/\partial y$.

Now consider the infinitely long PEC cylinder shown in Fig. 3.25. The cylinder cross section is indicated in the figure, along with a triangular mesh of the computational domain. As in the case of MM subdomains, a triangle is a convenient and flexible shape that can be used to approximate curved contours. To represent the field inside of a triangle, three vector edge basis functions are used, as described in Fig. 3.26. The numbering convention for the edges and vertices is shown in the figure. For triangle number e ($e = 1, 2, \ldots, N_e$, where N_e is the total number of triangles in the mesh), the three basis functions are

$$\begin{aligned} \vec{W}_1^e &= \ell_1^e \left(L_1^e \nabla L_2^e - L_2^e \nabla L_1^e\right) \\ \vec{W}_2^e &= \ell_2^e \left(L_2^e \nabla L_3^e - L_3^e \nabla L_2^e\right) \\ \vec{W}_3^e &= \ell_3^e \left(L_3^e \nabla L_1^e - L_1^e \nabla L_3^e\right) \end{aligned} \tag{3.137}$$

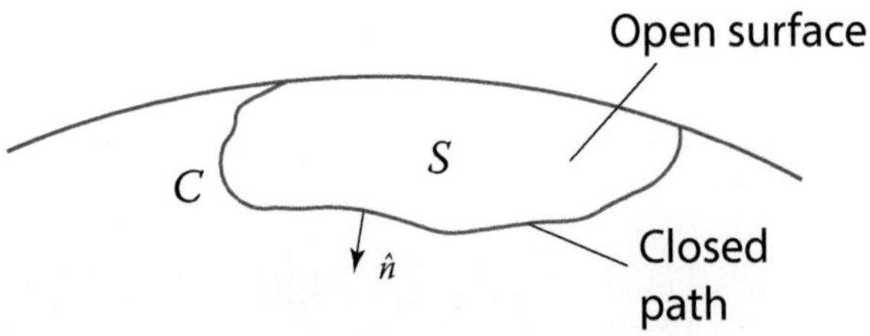

Fig. 3.24 Computational domain for a two-dimensional problem (open surface S and closed path C).

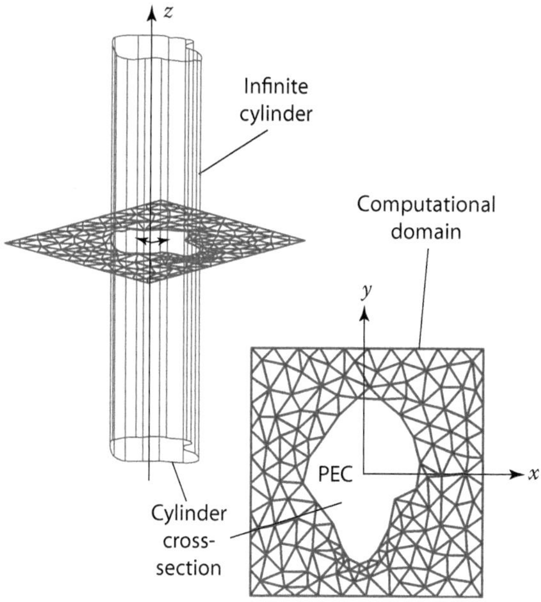

Fig. 3.25 Infinitely long cylinder with meshed computational domain for two-dimensional scattering calculation.

Thus, for a point P(x, y) inside of triangle e,

$$
\begin{aligned}
L_1^e &= \frac{\text{AREA } P23}{\text{AREA } 123} \\
L_2^e &= \frac{\text{AREA } P31}{\text{AREA } 123} \\
L_3^e &= \frac{\text{AREA } P12}{\text{AREA } 123}
\end{aligned}
\tag{3.138}
$$

The areas are easily computed using the cross product of two edge vectors.

$$\vec{W}_k^e = \ell_k (L_i^e \Delta L_j^e - L_j^e \Delta L_i^e)$$

Edge k	Node 1 i	Node 2 j
1	1	2
2	2	3
3	3	1

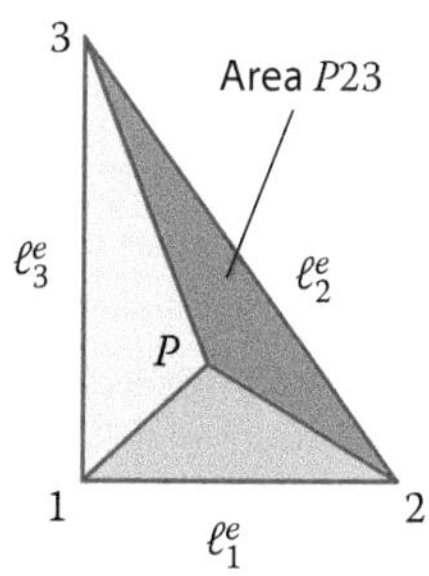

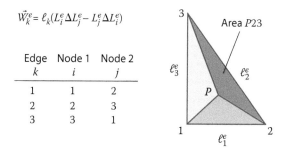

Fig. 3.26 Edge-based vector basis functions for a triangle of the mesh.

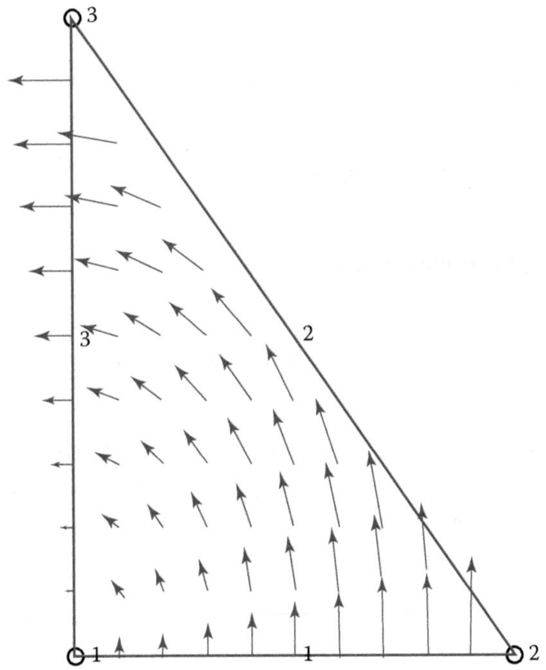

Fig. 3.27 Plot of the vector function $\vec{W}_2^e$; for a triangular subdomain.

Each of the three functions described in Eq. (3.137) rotates about one of the vertices of the triangle, as shown in Fig. 3.27 for $\vec{W}_2^e$. The vector is normal to the two edges that intersect at the node about which the basis function rotates (node 1), and the tangential component is constant along the edge opposite this node (edge 2). The direction of the electric field at point $P(x, y, z)$ is the vector sum of the three basis functions evaluated at (x, y, z), each weighted by the appropriate expansion coefficient. Thus for triangle e

$$\vec{E}_s^e(x, y, z) = \sum_{n=1}^{3} E_n^e \vec{W}_n^e(x, y, z) \tag{3.139}$$

In the actual implementation of the equations it will be more convenient to have a single index that runs over all edges, rather than separate indices for edges and triangles.

Now that the basis and weighting functions have been defined, the testing process can be completed. Testing Eq. (3.136) with each basis function gives coefficients of the form

$$A_{mn}^e = \iint_S \left[\nabla \times \vec{W}_m^e \cdot \left(\frac{1}{\mu_r} \nabla \times \vec{W}_n^e \right) - k_o^2 \epsilon_r \, \vec{W}_m^e \cdot \vec{W}_n^e \right] ds \tag{3.140}$$

$$B_m^e = \oint_C \left(\hat{n} \times \frac{1}{\mu_r} \nabla \times \vec{E}_i \right) \cdot \vec{W}_m^e \mathrm{d}l - \iint_S \left[\nabla \times \left(\frac{\nabla \times \vec{E}_i}{\mu_r} \right) - k_o^2 \epsilon_r \vec{E}_i \right] \cdot \vec{W}_m^e \, \mathrm{d}s \tag{3.141}$$

which can be put in matrix form

$$\mathbf{A}\,\mathbf{E} = \mathbf{B} \tag{3.142}$$

The matrix equation is solved for the vector **E**. Once the coefficients E_n^e are determined, the electric field is known everywhere in the computational domain. An important property of **A** is that it is *sparse* (many zero elements) due to the fact that a single edge basis function is shared by only a few triangles. Highly efficient matrix solution techniques exist for sparse matrices [24] (see http://www.netlib.org/).

In three dimensions the most flexible basis function is the tetrahedron shown in Fig. 3.28. There is a vector field associated with each of the six edges. Figure 3.29 shows the field for edge $k = 1$, which is defined by the expression

$$\vec{W}_k^e = \ell_k \left(L_i^e \nabla L_j^e - L_j^e \nabla L_i^e \right) \tag{3.143}$$

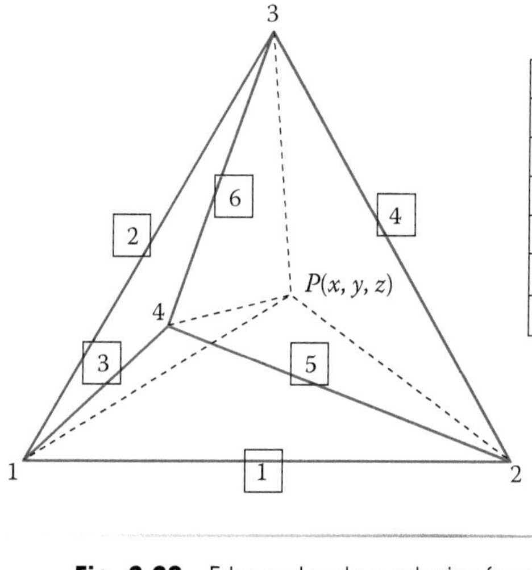

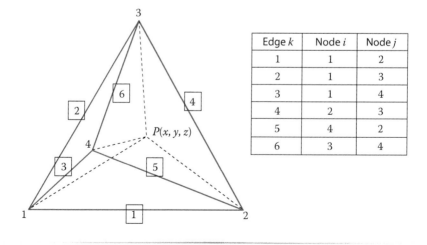

Edge k	Node i	Node j
1	1	2
2	1	3
3	1	4
4	2	3
5	4	2
6	3	4

Fig. 3.28 Edge and node numbering for a tetrahedral subdomain with internal observation point P.

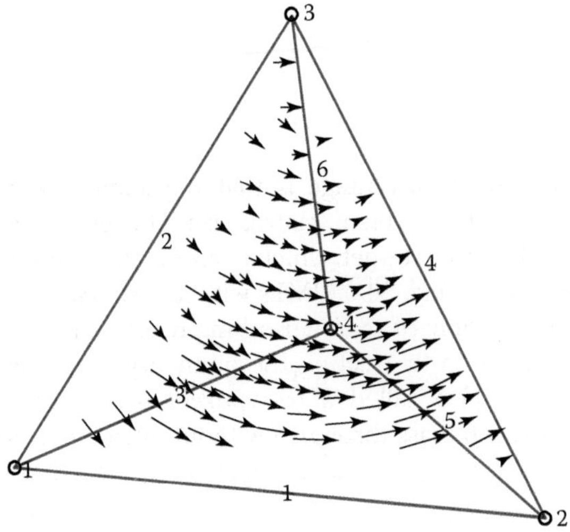

Fig. 3.29 Plot of the vector function $\vec{W}_1^e$; for the tetrahedral subdomain.

where

$$
\begin{aligned}
L_1^e &= \frac{\text{VOLUME } P234}{\text{VOLUME } 1234} \\
L_2^e &= \frac{\text{VOLUME } P341}{\text{VOLUME } 1234} \\
L_3^e &= \frac{\text{VOLUME } P412}{\text{VOLUME } 1234} \\
L_4^e &= \frac{\text{VOLUME } P123}{\text{VOLUME } 1234}
\end{aligned}
\tag{3.144}
$$

and l_k is the edge length, having endpoints i and j as defined in Fig. 3.28. The volumes are easily computed from a triple product of the edge vectors.

A specific case of Eq. (3.143) is shown in Fig. 3.29, for the field $\vec{W}_1^e$. The vector field rotates around edge 6 defined by nodes 3 and 4 and is normal to planes containing those nodes. The circulation is along edge 1 (nodes 1 and 2). At any point $P(x, y, z)$ inside of the tetrahedron the total field is the vector sum of the six edge basis vectors at (x, y, z).

3.8.3 Grid Termination

An important issue in the application of the FEM to open problems is the method of terminating the computational domain. Two approaches are: 1) a *perfectly matched layer* (PML) or 2) an *absorbing boundary condition* (ABC). A PML is an artificial material that is reflectionless, no matter what the type, angle, frequency, or polarization of the wave incident on the layer. A thin layer of PML (generally about $0.1\lambda_0$ where λ_0 is the free space wavelength) is added just inside of the boundary, with a PEC on S. An ideal PML absorbs all of the field incident on it, allowing outward propagating waves to appear to exit the domain, thereby simulating a boundary with free space. This method is appealing because it simply requires adding a thin layer of material to the original problem. The disadvantage is that the additional basis functions of the PML increase the problem matrix size. However, a good PML can be made conformal with the target, which decreases the number of additional nodes and is depicted in Fig. 3.30. PMLs will be discussed in more detail in Chapter 7, where the topic of radar absorbing material is introduced.

The second approach in treating the boundary is to use an absorbing boundary condition (also called a *transparent boundary condition*) that allows outgoing waves to pass through S without introducing artificial reflections [24, 25]. They are imposed locally and therefore are approximate but retain the sparse nature of the matrix $\mathbf{A}$ in Eq. (3.142). The boundary condition is incorporated into the surface terms in Eq. (3.134) or (3.135). Let $\vec{\psi}$ be either $\vec{E}_s$ or $\vec{H}_s$ so that the vector wave equations can be written as

$$\nabla \times \nabla \times \vec{\psi} - k^2 \vec{\psi} = 0 \tag{3.145}$$

All physically realizable fields must decay to zero at infinity. This is the *Sommerfeld radiation condition*, which can be expressed as

$$\lim_{r \to \infty} r \left\{ \hat{r} \times \left(\nabla \times \vec{\psi} \right) - jk\vec{\psi} \right\} = 0 \tag{3.146}$$

Equation (3.146) can be considered an ABC. When applied to a surface S in the far field of the scatterer, it is exact because an outgoing spherical wave exists on both sides of the surface. At an arbitrary distance from the scatterer, the fields can be expressed as a series of terms in powers of $1/r$ (known as the *Wilcox expansion theorem*) [26]

$$\vec{\psi}(r, \theta, \phi) = \frac{e^{-jkr}}{4\pi r} \sum_{n=0}^{\infty} \left(\frac{\vec{\psi}_n(\theta, \phi)}{r^n} \right) = \frac{e^{-jkr}}{4\pi r} \left\{ \vec{\psi}_0 + \frac{\vec{\psi}_1}{r} + \frac{\vec{\psi}_2}{r^2} + \cdots \right\} \tag{3.147}$$

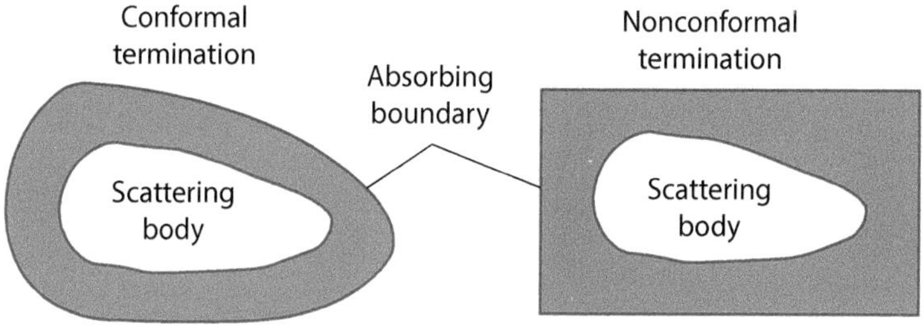

Fig. 3.30 Conformal (left) versus nonconformal (right) boundary surfaces.

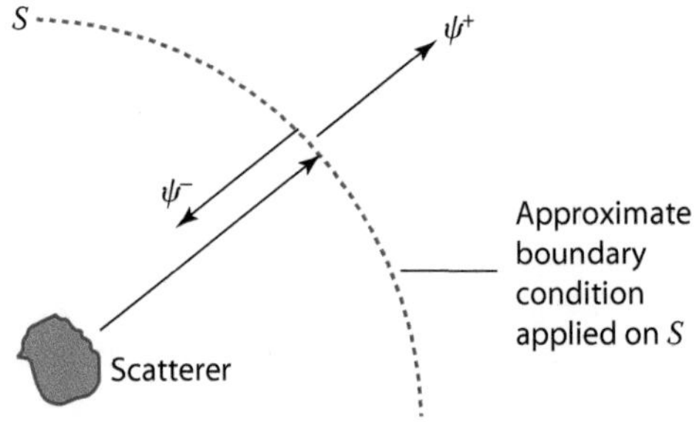

Fig. 3.31 Illustration of how nonphysical waves are set up at an artificial boundary.

Using the series in the radiation condition gives

$$\hat{r}\times\left(\nabla\times\vec{\psi}\right)-jk\vec{\psi}=\frac{e^{-jkr}}{4\pi r}\sum_{n=0}^{\infty}\left(-jk\frac{\hat{r}\left(\hat{r}\cdot\vec{\psi}\right)}{r^{n}}+\frac{r\nabla\left(\hat{r}\cdot\vec{\psi}\right)-n\hat{r}\times\left(\hat{r}\times\vec{\psi}\right)}{r^{n+1}}\right) \tag{3.148}$$

Equation (3.148) forms the basis of an ABC that can be applied at an arbitrary distance r from the scatterer. As illustrated in Fig. 3.31, if the boundary condition is applied at a distance r that is in the near field, then a sufficient number of higher order terms must be kept so that the significant number of terms in the physical field matches the number of significant terms in the boundary condition. If not, then nonphysical waves will be set up to match the error terms at the boundary.

Higher-order boundary conditions can be derived that give good results for surfaces near to the scatterer; however, they are more complicated and difficult to apply.

Example 3.7: Second Order Absorbing Boundary Condition [24]

For a second-order condition the error is $O(r^{-5})$. The exact scattered field at distance r has an infinite number of terms:

$$\vec{\psi}_{\text{exact}}=\frac{e^{-jkr}}{4\pi r}\left(\vec{\psi}_0=\frac{\vec{\psi}_1}{r}+\frac{\vec{\psi}_2}{r^2}+\cdots\right)$$

(Continued)

Example 3.7: Second Order Absorbing Boundary Condition [24] *(Continued)*

The boundary condition allows the retention of terms up to r^{-4}

$$\vec{\psi}_{\text{approx}} = \frac{e^{-jkr}}{4\pi r}\left(\vec{\psi}_0 + \frac{\vec{\psi}_1}{r} + \frac{\vec{\psi}_2}{r^2} + \frac{\vec{\psi}_3}{r^3}\right) + \underbrace{\frac{e^{-jkr}}{4\pi r}\left(\frac{\vec{A}_4}{r^4} + \frac{\vec{A}_5}{r^5} + \cdots\right)}_{\text{Outgoing waves, } \psi^+} + \underbrace{\frac{e^{jkr}}{4\pi r}\left(\frac{\vec{B}_4}{r^4} + \frac{\vec{B}_5}{r^5} + \cdots\right)}_{\text{Fictitious incoming waves, } \psi^-}$$

Therefore, if a higher-order outgoing term is not sufficiently small at the boundary, then a fictitious incoming wave (i.e., reflection) is set up to cancel it on S.

Several points worth noting are the following:

1. Low-order boundary conditions must be used at large r or significant errors occur. However, this requires a larger computational domain (i.e., more basis functions).
2. Higher-order boundary conditions can be applied closer to the surface but are more complex and difficult to implement.
3. An increasing number of derivatives are required as the order increases.

There are many forms of absorbing boundary conditions, but in general they are all derived from the radiation conditions as $r \to \infty$, and their particular form depends on the distance from the scattering surface at which they are applied. Relatively simple equations (first order) can be used when applied far from the target ($1\lambda_o - 2\lambda_o$), but this is usually not acceptable because of the large number of additional basis functions that must be added to the domain between S and the scatterer's surface. Therefore, even though ABCs applied near to the scatterer on conformal surfaces are more complex than those applied on coordinate variable constant planes far away, overall, it usually pays to employ conformal ABCs close to the boundary.

Example 3.8: Bistatic RCS of a Square Plate Using FEM

Several FEM software packages are commercially available. In this example the bistatic RCS of a 5λ by 5λ PEC plate is computed for an incidence angle of 45 deg. The software package Ansys HFSS (High Frequency Structures Simulator) was used. (See www.ansys.com for information on the HFSS software

(Continued)

Example 3.8: Bistatic RCS of a Square Plate Using FEM (*Continued*)

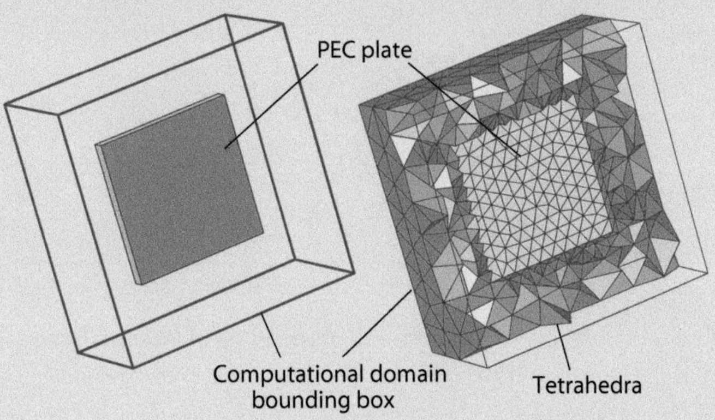

Fig. 3.32 FEM model for square plate bistatic RCS calculation. Right shows a close-up of the FEM meshing of the radiation box (from CST Microwave Studio).

package.) Figure 3.32 shows the plate and computational domain. The outer box is the surface over which the absorbing boundary conditions are automatically applied. The principal plane RCS of the plate is shown in Fig. 3.33. The electric and magnetic fields on the bounding box are used to compute the far-scattered field based on equivalent electric and magnetic currents.

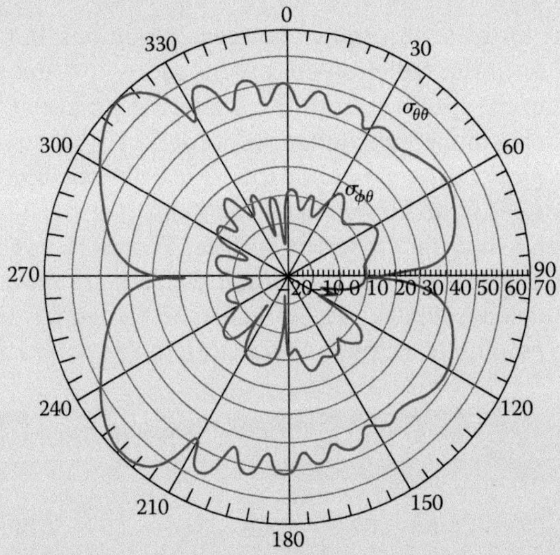

Fig. 3.33 Bistatic RCS of the 5-wavelength square plate, for TM incidence at 45 deg and 7.5 MHz.

Example 3.9: Bistatic RCS of a Gripen Aircraft Using FEM

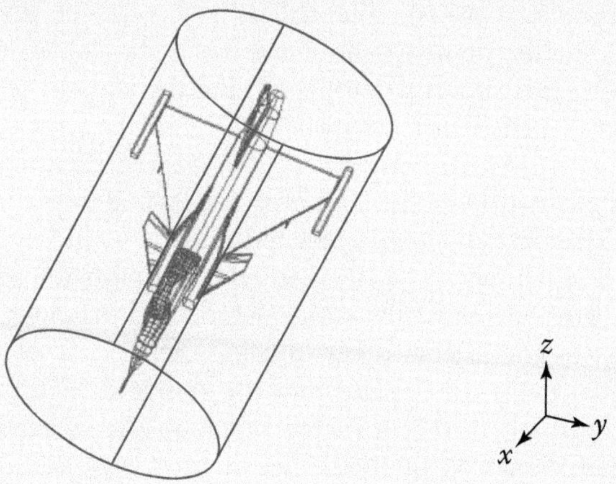

Fig. 3.34 Approximate model of a Gripen fighter aircraft with cylindrical radiation boundary (left).

Figure 3.34 shows an approximate model of a Gripen jet aircraft, with its computed bistatic RCS at 1500 MHz shown in Fig. 3.35. The highest frequency is limited by the available computer memory to restrict the tetrahedra dimension to approximately 0.1λ.

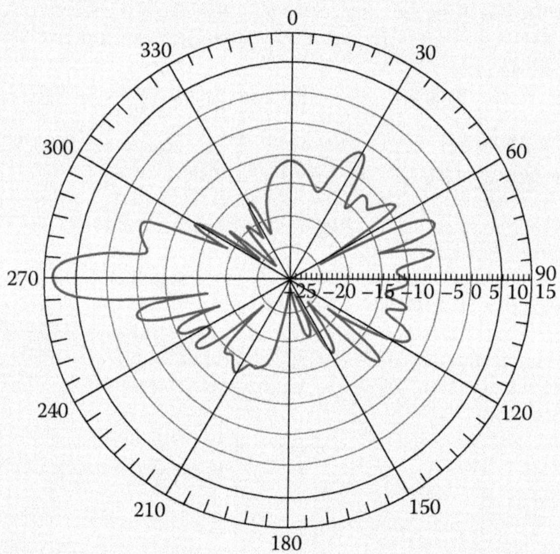

Fig. 3.35 Bistatic RCS of the Gripen fighter aircraft ($\sigma_{\phi\phi}$ for $\theta = \phi = 90°$ incidence).

FEM has become one of the most used computational EM methods. Tetrahedral subdomains are flexible and can be applied to arbitrary three-dimensional shapes and materials. However, other types of three-dimensional basis functions are possible, and as in the case of MM, special basis functions for bodies of revolution have been developed. Over the years, efficient ABCs have also been proposed, and the sources of spurious solutions have been identified and controlled [27]. Software packages such as *Ansys HFSS, Altair FEKO, CST Microwave Studio,* and *Comsol FEMLAB* [28] have powerful geometry builders and post-processing tools.

There is a third approach to handling the terminating boundary called the finite-element boundary-integral (FE-BI) method [2]. An integral equation is formulated for the surface *S* that relates the electric and magnetic fields on the boundary. The method is equivalent to using the method of moments to find the tangential fields on *S*. Following the standard MM procedure, the integral equation is reduced to a matrix equation that is solved simultaneously with the FEM matrix equation. This method is appealing because it is rigorous in principle. However, the surface portion of the matrix is dense, and the sparse nature of the FEM solution is destroyed.

Although FEM can be applied in the time domain [29], to date, it has been used almost exclusively in the frequency domain.

References

[1] Mautz, J. R., and Harrington, R. F., "Radiation and Scattering From Bodies of Revolution," *Applied Science Research*, Vol. 20, June 1969, p. 405.

[2] Jin, J., *The Finite Element Method in Electromagnetics*, 2nd ed. Wiley-Interscience, New York, 2002.

[3] Harrington, R. F., *Field Computation by Moment Methods*, Macmillan, New York, 1961.

[4] Press, W. H., Fannery, B. P., Teukolsky, S. A., and Vetterling, W. T., *Numerical Recipes*, Cambridge Univ. Press, New York, 1986.

[5] Klein, C. A., and Mittra, R., "The Effect of Different Testing Functions in the Moment Method Solution of Thin-Wire Antenna Problems," *IEEE Transaction on Antennas and Propagation*, March 1975, p. 258.

[6] Imbraile, W. A., and Ingerson, P. G., "On Numerical Convergence of Moment Method Solutions of Moderately Thick Wire Antennas Using Sinusoidal Basis Functions," *IEEE Transactions on Antennas and Propagation*, May 1973, p. 363.

[7] Butler, C. M., "Evaluation of Potential Integral at Singularity of Exact Kernel in the Thin-Wire Calculations," *IEEE Transactions on Antennas and Propagation*, March 1975, p. 293.

[8] Butler, C. M., and Wilton, D. R., "Analysis of Various Numerical Techniques Applied to Thin-Wire Scatterers," *IEEE Transactions on Antennas and Propagation*, July 1975, p. 534.

[9] Rao, S. M., Wilton, D. R., and Glisson, A. W., "Electromagnetic Scattering by Surfaces of Arbitrary Shape," *IEEE Transactions on Antennas and Propagation*, Vol. AP-30, No. 3, May 1982, p. 409.

[10] Makarov, S. N., *Antenna and E M Modeling with MATLAB*, Wiley, New York, 2002.

[11] Knupp, P., and Steinberg, S., *Fundamentals of Grid Generation*, CRC Press, London, 1994.

[12] Awadhiya, A., Barba, P., and Kempel, L., "Finite Element Method Programming Made Easy???," *IEEE Antennas and Propagation Magazine*, Vol. 45, No. 4, 2003, pp. 73–79.
[13] Mautz, J. R., and Harrington, R. F., "An Improved E-Field Solution for a Conducting Body of Revolution," Syracuse Univ. Syracuse, NY, Tech. Rept. TR-80-1, 1980.
[14] Hombeck, R. W., *Numerical Methods*, Quantum Publishers, New York, 1975.
[15] Mautz, J. R., and Harrington, R. F., "H-Field, E-Field, and Combined Field Solutions for Bodies of Revolution," Syracuse Univ., Syracuse, NY, Tech. Rept. TR-77-2, 1977.
[16] Lin, Y. T., and Richmond, J. H., "EM Modeling of Aircraft at Low Frequencies," *IEEE Transactions on Antennas and Propagation*, Vol. AP-23, Jan. 1975, p. 53.
[17] Newman, E. H., and Tulyathan, P., "A Surface Patch Model for Polygonal Plates," *IEEE Transactions on Antennas and Propagation*, Vol. AP-30, No. 4, July 1982, p. 588.
[18] Tulyathan, P., and Newman, E. H., "The Circumferential Variation of the Axial Component of Current in Closely Spaced Thin-Wire Scatterers," *IEEE Transactions on Antennas and Propagation*, Vol. AP-27, Jan. 1979, p. 46.
[19] Lee, K. S. H., Marin, L., and Castillo, J. P., "Limitations of Wire-Grid Modeling of a Closed Surface," *IEEE Transactions on Electromagnetic Compatibility*, Vol. EMC-18, No. 3, Aug. 1976, p. 123.
[20] Harrington, R. F., and Mautz, J. R., "An Impedance Sheet Approximation for Thin Dielectric Shells," *IEEE Transactions on Antennas and Propagation*, Vol. AP-29, No. 4, July 1975, p. 531.
[21] Volakis, J. L., Chatteijee, A., and Kempel, L. C., "Review of the Finite-Element Method for Three-Dimensional Electromagnetic Scattering," *Journal of the Optical Society of America, A*, Vol. 11, No. 4, 1994, pp. 1422–1433.
[22] Volakis, J. L., Chatteijee, A., and Kempel, L. C., *Finite Element Method for Electromagnetics*, IEEE Press, New York, 1998.
[24] Lynch, D. R., and Paulsen, K. D., "Origin of Vector Parasites in Numerical Maxwell's Solutions," *IEEE Transactions on Microwave Theory and Technology*, Vol. 39, No. 3, 1991, pp. 383–394.
[24] Pederson, A., "Absorbing Boundary Conditions for the Vector Wave Equation," *Microwave and Optical Technology Letters*, Vol. 1, No. 2, April 1988, pp. 62–64.
[25] Webb, J. P., and Kanellopoulos, V. N., "Absorbing Boundary Conditions for the Finite Element Solution of the Vector Wave Equation," *Microwave and Optical Technology Letters*, Vol. 2, 1998, pp. 370–372.
[26] Wilcox, C. H., "An Expansion Theorem for Electromagnetic Fields," *Communications on Pure and Applied Mathematics*, Vol. 9, 1956, pp. 115–134.
[27] Khebir, A., and D'Angelo, J., "A New Finite Element Formulation for RF Scattering by Complex Bodies of Revolution," *IEEE Transactions on Antennas and Propagation*, Vol. 41, No. 5, May 1993, pp. 534–541.
[28] Paulsen, K. D., and Lynch, D. R., "Elimination of Vector Parasites in Finite Element Maxwell Solutions," *IEEE Transactions on Microwave Theory and Technology*, Vol. 39, No. 3, 1991, pp. 395–404.
[29] Lee, J.-F., Lee, R., and Cangellaris, A., "Time-Domain Finite Element Methods," *IEEE Transactions on Antennas and Propagation*, Vol. 45, No. 3, March 1997, pp. 430–442.
[30] Wang, N. N., Richmond, J. H., and Gilreath, M. C., "Sinusiodal Reaction Formulation for Radiation and Scattering from Conducting Surfaces," *IEEE Transactions on Antennas and Propagation*, Vol. AP-23, No. 3, May 1975, p. 376.
[31] Butler, C. M., "The Equivalent Radius of a Narrow Conducting Strip," *IEEE Transactions on Antennas and Propagation*, Vol. AP-30, No. 4, July 1982, p. 755.
[32] Tsai, L. L., Dudley, D. G., and Wilton, D. R., "Electromagnetic Scattering by a Three-Dimensional Conducting Rectangular Box," *Journal of Applied Physics*, Vol. 45, No. 10, Oct. 1974, p. 4393.

Problems

3.1 A TM-polarized plane wave is incident on a thin wire with endpoints at $z = 0$ and $z = L$. Using the MM procedure, derive expressions for the impedance, excitation, and receive elements for sinusoidal basis functions,

$$\vec{J}_n(z) = \hat{z}\frac{\sin(n\pi z/L)}{2\pi a}$$

where $n = 1, 2, \ldots, N$. Solve the matrix equation and plot the RCS for a 1.25λ wire. Compare the results with those from pulbf.m.

3.2 A thin wire of length $L = 0.5\lambda$ is conductively attached to a body. The MM is used to compute a series representation for the current over the entire body. For the wire basis functions and the resulting expansion coefficients given in each of the following parts, sketch (to scale) the current along the wire.

a) Pulse basis functions (all subsections are the same length):

$$I(x) = \sum_{i=1}^{N} I_i p_i(x) \qquad i = 1, \ \ldots, \ 5$$

where $I_1 = 0.634$, $I_2 = 0.634$, $I_3 = 0.250$, $I_4 = -0.135$, and $I_5 = -0.135$.

b) Triangular basis functions (all subsections are the same length):

$$I(x) = \sum_{i=1}^{N} I_i T_i(x) \qquad i = 1, \ \ldots, 4$$

where $I_1 = 0.690$, $I_2 = 0.474$, $I_3 = 0.026$, and $I_4 = -0.190$.

c) Exponential basis functions:

$$I(x) = \sum_{n=-N}^{N} I_n e^{jn\pi x/L} \qquad n = 0, \ \pm 1, \ \pm 2$$

where $I_0 = 0.25$, $I_1 = I_{-1} = 0.125$, and $I_2 = -I_{-2} = -0.125j$.

d) Assume that the exponential basis functions give the correct current distribution. Which end of the wire is attached to the body: $x = 0$ or $x = L$? Discuss the advantages and disadvantages of using pulses and triangles in this case.

3.3 For the wire in Problem 3.1, derive the load impedance elements if the wire has a resistivity of R_s.

3.4 For the loop in Example 3.2, show that the receive vector elements for a θ-polarized incident field are given by

$$R_n^\theta = \pi b j^n e^{jn\phi} \cos\phi [J_{n+1}(kb\sin\theta) + J_{n-1}(kb\sin\theta)]$$

3.5 In this problem, the *reaction integral equation* will be derived [30]. As shown in Fig. P3.5, sources $\vec{J}_i$, $\vec{J}_{mi}$ set up electric and magnetic fields $\vec{E}$, $\vec{H}$ in the presence of a scatter. The surface of the scatterer is denoted S, and it may have a surface impedance Z_S.

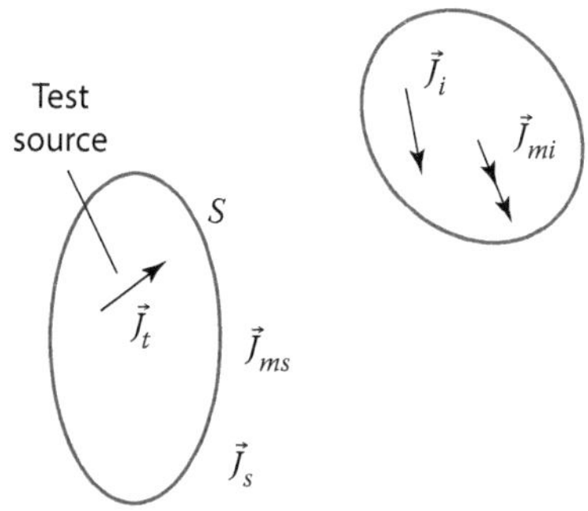

Fig. P3.5

a) Using Love's equivalence principle, what surface currents $\vec{J}_s$, $\vec{J}_{ms}$ should flow on S to give a null field inside?

b) An electric test source $\vec{J}_t$ is placed in the interior. The fields due to this source are $\vec{E}_t$, $\vec{H}_t$. Use the reciprocity theorem to show that

$$\oint_S \left(\vec{J}_s \cdot \vec{E}_t - \vec{J}_{ms} \cdot \vec{H}_t\right) \mathrm{d}s + \iiint_V \left(\vec{J}_i \cdot \vec{E}_t - \vec{J}_{mi} \cdot \vec{H}_t\right) \mathrm{d}v = 0$$

c) Use the impedance boundary condition to obtain the following integral equation for the electric current $\vec{J}_s$:

$$-\iint_S \vec{J}_s \cdot \left[\vec{E}_t - (\hat{n} \times \vec{H}_t) Z_s\right] \mathrm{d}s = \iint_S \vec{J}_i \cdot \vec{E}_t \mathrm{d}s$$

3.6 The loop of Example 3.2 is made of resistive film with surface resistivity R_S. Find an expression for the MM load impedance elements $(Z_L)_n$, where n is the azimuthal mode index.

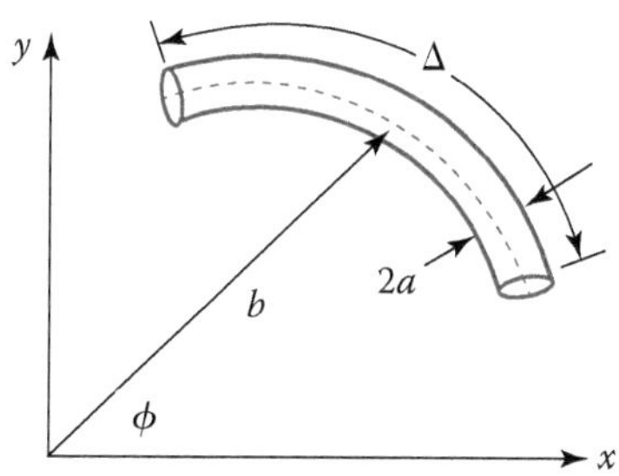

Fig. P3.7

3.7 Scattering from the loop of Example 3.2 is to be computed using *conformal* subdomains as shown in Fig. P3.7. The loop is discretized into N equal curved segments of arclength Δ. The arclength variable is $t = b\phi$. Thus, for pulse-type basis functions,

$$\vec{J}_s(t) = \sum_n I_n \hat{t} \frac{p_n(t)}{2\pi a}$$

a) Find an expression for Z_{mn} using Galerkin's method and displacing the test and source points to avoid the singularity as was done in Examples 3.1 and 3.2.

b) Find an expression for the plane wave excitation vector for θ and ϕ polarizations.

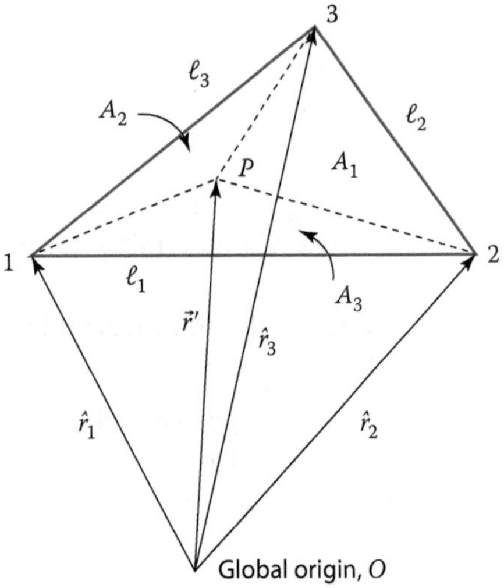

Fig. P3.10

3.8 The equations developed in Examples 3.1 and 3.4 can be used to compute the RCS of a flat resistive strip of width w by using an equivalent radius $a_e \approx 0.225\, w$[31]. Compare the RCS of a strip with $Z_s = 0.5$ to the RCS of a perfectly conducting strip for the parameters $L = 3\lambda$, $w = 0.1\lambda$, and the number of segments $N = 150$.

3.9 A perfectly conducting cube with sides of length L is placed a height h over an infinite PEC ground plane. Formulate an MM solution for this configuration based on the method of images. Specify how the structure must be excited to simulate arbitrary incidence angle and polarization [32].

3.10 Triangles are used as subdomains in PO, MM, and two-dimensional FEM. When calculations deal with surface areas it is convenient to work in normalized area coordinates, also called *natural* or *simplex* coordinates:

$$\zeta = \frac{A_1}{A}, \quad \xi = \frac{A_2}{A}, \quad \eta = \frac{A_3}{A} \quad \text{where} \quad \zeta + \xi + \eta = 1.$$

a) Referring to Fig. P3.10, write the edge vectors $\vec{l}_i$ ($i = 1, 2, 3$) in terms of the position vectors $\vec{r}_i$

b) Express the total area of the triangle in terms of a cross product of edge vectors.

c) Express $\int_A ds$ in terms of the simplex variables (ζ, ξ, η).

d) Express the one-way phase at a point P inside of the triangle relative to a point at infinity in the direction (θ, ϕ) in terms of the simplex variables. (In other words, find $kg = k\hat{r} \cdot \vec{r}'$.)

e) Suggest a simple test involving the total area A and the subareas A_i that will determine if P is located inside or outside of the triangle.

3.11 Consider the tetrahedron in Fig. 3.28 with edge vectors $\vec{l}_i$ ($i = 1, \ldots, 6$).

a) Write equations (3.144) in terms of triple scalar products of the edge vectors.

b) Show that $\nabla \cdot \vec{W}_1 = 0$ and $\vec{\ell}_1 \cdot \vec{W}_1 = \text{constant}$.

3.12 A plane wave propagating in the $+x$ -direction

$$E_z = E_o e^{-j(k_x x + k_y y)}$$

is incident at an angle ϕ onto an infinite boundary at $x = 0$ ($k_x = k \cos\phi$, $k_y = k \sin\phi$). The exact boundary condition at $x = 0$ that gives zero reflection is

$$\mathrm{d}E_z/\mathrm{d}x = -jk \cos \phi E_z$$

Because the boundary condition must be angle independent, use the approximate boundary condition $\mathrm{d}E_z/\mathrm{d}x \approx -jkE_z$ and find the resulting reflection coefficient.

3.13 Two adjacent triangle subdomains for a two-dimensional FEM problem are shown in Fig. P3.13: (124) and (234). They share a common edge (24) andtogether the two triangles comprise a unit square in free space. In computing the matrix elements we are confronted with the integrals

$$P^e_{mn} = \iint_S \left[\nabla \times \vec{W}^e_m \cdot \nabla \times \vec{W}^e_n\right] ds$$

$$\text{and} \quad Q^e_{mn} = \iint_S \left[\vec{W}^e_m \cdot \vec{W}^e_n\right] ds$$

Evaluate the two integrals for all combinations of m, n, and e.

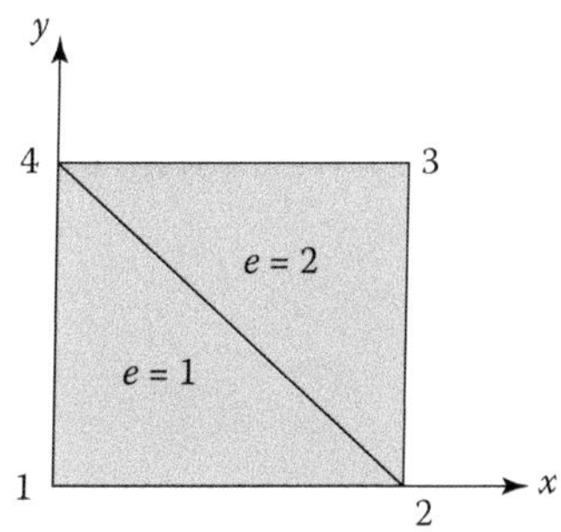

Fig. P3.13

3.14 Modify Eq. (3.135) to include a resistive surface S_r inside of the domain V that satisfies the resistive sheet boundary condition

$$\hat{n}_+ \times \hat{n}_+ \times \vec{E} = -R_s \hat{n}_+ \times \left(\vec{H}_+ - \vec{H}_-\right)$$

where $\vec{H}_\pm$ is the magnetic field on the top and bottom of the sheet and $\hat{n}_+$ is a unit vector normal to the sheet pointing toward the region above.

3.15 Consider a single subdomain with current basis function $\vec{J}_n(\vec{r}')$ and expansion coefficient I_n. Assume that a Hertzian dipole is located at a far-field point P at $\vec{r}_n$. Using the Lorentz reciprocity theorem, Eq. (2.2), and the definition of the measurement element R_n in Eq. (3.43), verify Eqs. (3.47) and (3.48) for a single subdomain, n.

3.16 Starting with Maxwell's first two "curl" equations, derive Eqs. (3.27) and (3.30).

Chapter 4 Time-Domain Numerical Methods

Jovan Lebaric and David Jenn

4.1 Introduction

The analysis of electromagnetic problems in the time domain has become more common in recent decades. In this approach, the differential or integral equations for the particular problem under consideration are typically solved numerically using a short time-duration waveform as excitation. By far the most common of these techniques is the so-called *finite difference time-domain* (FDTD) method [1,2]. There are other useful time-domain techniques such as time-domain physical optics (TDPO) [3], the finite integration technique (FIT), and time-domain integral equations (TDIE) [4–6], but FDTD is the most common and, consequently, it is the focus of the material in this chapter.

In the FDTD method, the target is first discretized in a convenient grid coordinate system. The differential operators in Maxwell's equations are approximated by finite differences. The selected waveform illuminates the target, and the fields at the grid nodes are computed at discrete time steps $n\Delta t$, where n is an integer. This process is referred to *marching in time.* As described in the following sections, the field at a particular node at time t can be determined from the fields at the same and adjacent nodes at the previous time step.

In some cases, the time-domain scattered field provides the desired target information. An example is the resolution of two individual scatterers on a complex target (see Chapter 8 for further discussion). If radar cross section (RCS) is the quantity of interest, however, the time-domain scattered field must be Fourier-transformed to obtain the frequency response of the target. Thus, the final RCS value obtained via the FDTD method is the result of a multistepped process, and accuracy is affected by the approximations and limitations introduced at each step. These include the grid size, time step, length of time observation, and the excitation waveform, to name a few.

There are several appealing advantages to the FDTD method as compared to frequency-domain approaches. First, it is simple to implement for complex scatterers because of the discretization. Each cell can be assigned arbitrary electric and magnetic properties. Although the update equations appear complicated, they contain only additions, subtractions, and

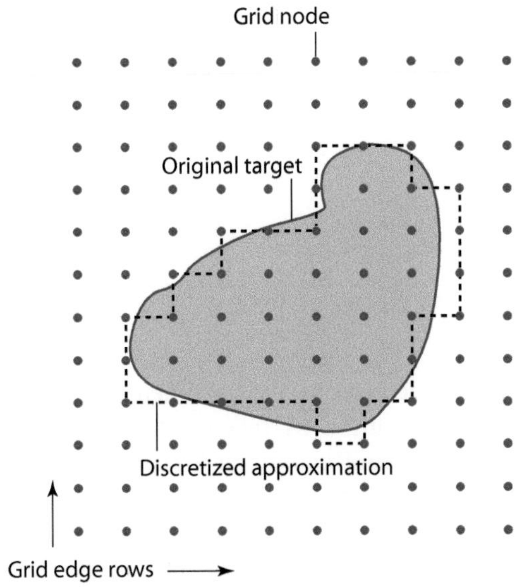

Fig. 4.1 Application of FDTD discretization to a two-dimensional target with an arbitrary cross section.

multiplications; there are no integrations to perform. Second, the memory requirements are usually less demanding than those for the method of moments.

An important consideration in the FDTD method is how to terminate the grid. In principle, all space surrounding the target, as well as the target itself, should be discretized. If the target is in free space, however, the propagation characteristics of the incident and scattered waves are known. Therefore, it is not necessary to discretize free space far from the scatterer; typically, only the region near the target is gridded as illustrated in Fig. 4.1. The FDTD solution is initiated by "injecting" the incident field into the grid. This must be performed in such a way that the field entering the grid is identical (or acceptably close) to what it would be if the grid were infinite. Similarly, if the wave passes through the target, the wave must exit the grid without adversely affecting the solution. This is referred to as *transparent grid termination* (TGT).

The Fourier transformation to get RCS from the time function is crucial in determining the efficiency of the FDTD for a particular problem. Early radars were narrowband and some even single frequency (i.e., CW radars). Only the RCS at a single frequency was of interest and, therefore, frequency-domain analysis methods were the obvious choice. To solve a single-frequency problem in the time domain requires an infinitely long time observation. (Recall that the Fourier transform of a delta function is a constant.) In practice, the observation window must be finite, and this introduces error into the RCS calculation.

Modern multifunction radars operate over much larger frequency ranges (typically 10%), and some proposed ultrawideband radars would operate over several decades. (Each decade is an order of magnitude change in frequency, e.g., 1–10 GHz). Wide frequency ranges reduce the observation time requirement, whereas the calculation of RCS at many frequencies using the MM can be extremely time consuming for a large target. In this case, FDTD techniques become more appealing.

The emphasis in this chapter deviates slightly from that of other chapters. Here, only a one-dimensional application of the FDTD is discussed in detail. This allows the subtleties of the solution to be explained without the notation becoming too overwhelming. It sufficiently illustrates the trade-offs involved in choosing calculation parameters and some of the limitations of the FDTD method.

The extension to two and three dimensions is treated briefly. The two-dimensional case starts with the problem setup and then proceeds directly to the final update equations for scatterers that are independent of the z coordinate. Finally, for the extension to three dimensions, only the update equations are presented. Other aspects of the two- and three-dimensional FDTD solution can be found in the references provided [7,8].

The important steps of the FDTD solution discussed in the following sections include 1) definition of the computational grid; 2) discretizing Maxwell's equations in time and space and derivation of the TM and TE *update equations*; 3) criteria for determining the permissible grid spacing, extent of the computational domain, and limitations on the time-stepping process; 4) how to handle the grid edges to minimize artificial reflections from the computational boundary; 5) waveform requirements and limitations; 6) how to couple (or inject) the incident field into the computational grid; and 7) transforming the near fields to the far fields and calculation of RCS. This chapter includes a brief introduction to the finite integration technique (FIT). In this method, Maxwell's equations in their integral form are solved using a time-stepping process similar to the FDTD. Finally, we conclude with an overview of the transmission line matrix (TLM) method.

4.2 Relationship Between Time and Frequency

4.2.1 Impulse and Frequency Responses of Linear Systems

Consider a two-port model of a linear time-invariant system shown in Fig. 4.2. A signal $x(t)$ is applied to the input, and the response $y(t)$ is observed

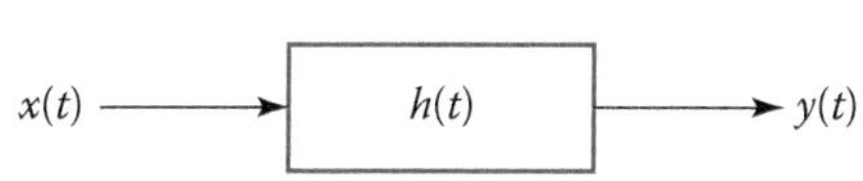

Fig. 4.2 Two-port model of a linear system.

at the output. The response is obtained as the convolution of the input $x(t)$ and the *system impulse response* $h(t)$ [9]:

$$y(t) = x(t) * h(t) = \int_{-\infty}^{\infty} x(t)h(t-\tau)\mathrm{d}\tau \tag{4.1}$$

If the signal $x(t)$ is zero for $t < 0$ and the system is *causal*, that is, the impulse response $h(t)$ is zero for $t < 0$, the convolution integral can be written as

$$y(t) = \int_{0}^{t} x(t)h(t-\tau)\mathrm{d}\tau \tag{4.2}$$

The system impulse response is the system's response to a Dirac delta function $\delta(t)$, which is nonzero only at $t = 0$. A causal system does not generate a response that precedes the excitation and, thus, the response of a causal system must be zero for $t < 0$. Using the property that the Fourier transform of a convolution is the product of Fourier transforms [9], gives the input-output relationship in the frequency domain:

$$\mathcal{F}\{y(t)\} = \mathcal{F}\{x(t)\}\mathcal{F}\{h(t)\} \tag{4.3}$$

where the operator $\mathcal{F}$ denotes a Fourier transform. Equation (4.3) can be written as

$$Y(\omega) = X(\omega)H(\omega) \tag{4.4}$$

where the upper-case letters denote the Fourier transforms of the lower-case time functions (e.g., $X(\omega) = \mathcal{F}\{x(t)\}$). The function $H(\omega)$ is the system's *frequency response.*

The linear system concept just discussed can be extended to linear time-invariant systems with more than one input and output. Such a system can be represented as a multiport, which can have N inputs and M outputs as illustrated in Fig. 4.3. The signal at the nth input is denoted $x_n(t)$, and the signal

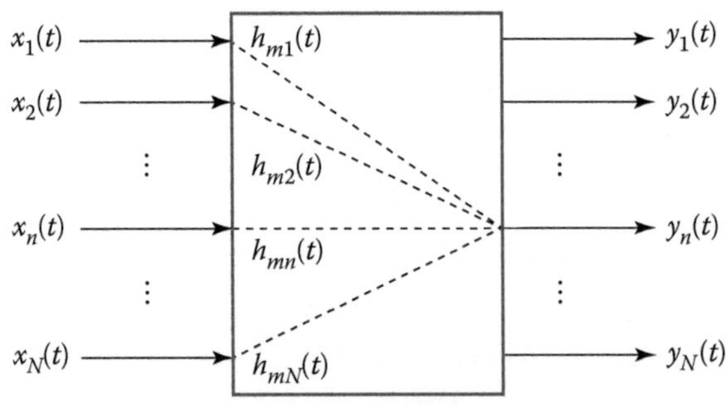

Fig. 4.3 Multiport model of a linear system.

observed at the mth output is denoted $y_m(t)$. The system is described by a matrix of impulse responses with elements $h_{mn}(t)$. Each $h_{mn}(t)$ represents the response observed at the mth output due to the Dirac delta function $\delta(t)$ applied to the nth input with all other inputs at zero.

The response $y_m(t)$ at the mth output of a multiport linear system can be determined as a sum of convolutions:

$$y_m(t) = \sum_{n=1}^{N} x_n(t) * h_{mn}(t) \tag{4.5}$$

The corresponding expression in the frequency domain is

$$Y_m(\omega) = \sum_{n=1}^{N} X_n(\omega) H_{mn}(\omega) \tag{4.6}$$

A multiport can be used to model the plane wave scattering process, with the input ports representing different directions of incidence of a plane wave onto the scatterer, and the output ports different observation points on a sphere of very large radius with its origin at the center of the scatterer.

4.2.2 Multiport Linear System Model of a Scattering Problem

Consider a scatterer illuminated by a plane wave. The incident electric field $E_i = |\vec{E}_i|$ (not accounting for polarization) on the surface of the scatterer is a function of direction of incidence specified by (θ_i, ϕ_i). Each direction of incidence can be thought of as one input port of a linear multiport. Because there are infinitely many possible directions of incidence, the multiport has, theoretically, infinitely many ports. However, one normally discretizes the continuous space by sampling the possible directions of incidence as finely as needed. Thus, the number of input ports is, in practice, determined by the resolution of the directions of incidence.

The scattered electric field $\vec{E}_s$ at a distant (far-field) observation point is a function of the direction of the observation point specified by (θ, ϕ) as well as the distance r between the observation point and the scatterer local coordinate origin. Selecting all observation points to be on a sphere of radius r centered at the scatterer allows one to uniquely specify an observation point by its direction (θ, ϕ).

Each observation point can now be thought of as an output port of a linear multiport. The response at an observation port corresponding to the mth direction, $E_s(\theta_m, \phi_m) = E_{sm}$, can now be found from a sum of convolutions, just as for a multiport:

$$E_{sm}(t) = \sum_{n=1}^{N} E_{in}(t) * h_{mn}(t) \tag{4.7}$$

or, in the frequency domain:

$$E_{sm}(\omega) = \sum_{n=1}^{N} E_{in}(\omega)H_{mn}(\omega) \tag{4.8}$$

In the case of illumination by a single plane wave from a direction (θ_n, ϕ_n), Eq. (4.8) simplifies to

$$E_{sm}(\omega) = E_{in}(\omega)H_{mn}(\omega) \tag{4.9}$$

The magnitude squared of the scattered field at the mth observation point, specified by angles (θ_m, ϕ_m), normalized to the magnitude squared of the incident field, is

$$\frac{|E_{sm}(\omega)|^2}{|E_{in}(\omega)|^2} = |H_{mn}(\omega)|^2 = |\mathcal{F}\{h_{mn}(t)\}|^2 \tag{4.10}$$

Multiplying Eq. (4.10) by $4\pi r^2$, where r is the distance to the observation point, we get

$$4\pi r^2 \frac{|E_{sm}(\omega)|^2}{|E_{in}(\omega)|^2} = 4\pi r^2 |H_{mn}(\omega)|^2 = 4\pi r^2 |\mathcal{F}\{h_{mn}(t)\}|^2 \tag{4.11}$$

The left-hand side is the definition of RCS if we assume that the distance r tends to infinity. (In practice, this means that r is very large relative to the scatterer dimensions and the operating wavelength such that the scatterer appears as a point source to the observer.) Thus, the RCS for a particular incidence (index n) and observation (index m) directions can be determined from the scatterer's impulse response:

$$\sigma(\omega, \theta_m, \phi_m, \theta_n, \phi_n) = 4\pi r^2 |H_{mn}(\omega)|^2 = 4\pi r^2 |\mathcal{F}\{h_{mn}(t)\}|^2 \tag{4.12}$$

Equation (4.12) allows us to establish a relationship between the RCS, which is defined in the frequency domain, and a time-domain quantity that represents the scatterer's impulse response. Note that to determine the scatterer's impulse response, we can theoretically use any waveform whose spectrum extends to infinity and is free of zeros, not just the Dirac delta function. For example, we can use a unit step function, $u_n(t)$, and determine its response, $a_{mn}(t)$. Because the unit step is the integral of the Dirac delta function and the system is linear, the impulse response $h_{mn}(t)$ will be the derivative of the unit step response $a_{mn}(t)$. Furthermore, we can use an incident field with "almost arbitrary" time variation to determine the scattered field, find the Fourier transform of the scattered field, $E_{sm}(\omega)$, and then, by simple division, get the RCS:

$$\sigma(\omega, \theta_m, \phi_m, \theta_n, \phi_n) = 4\pi r^2 \frac{|E_{sm}(\omega)|^2}{|E_{in}(\omega)|^2} = 4\pi r^2 \frac{|\mathcal{F}\{E_{sm}(t)\}|^2}{|\mathcal{F}\{E_{in}(t)\}|^2} \tag{4.13}$$

It is clear from Eq. (4.13) that the RCS cannot be determined at frequencies that do not exist in the spectrum of the "almost arbitrary" incident field. This is why an incident field whose spectrum has no zeros, such as a *Gaussian pulse*, should be used.

There are several issues that need to be considered with regard to time-domain RCS calculation. First, time-domain calculations give (via the Fourier transform) the RCS at many discrete frequencies. The highest frequency at which the RCS will be available is inversely proportional to the time sampling interval Δt; the smaller Δt is, the higher this frequency will be. The difference between any two adjacent frequencies at which the RCS will be available is inversely proportional to the observation time window; the longer the observation time, the better the resolution. Therefore, wide-band, high-resolution RCS data require long observation times with very small time steps, which translate into very long computation times. This is a major tradeoff between single frequency RCS calculations in the frequency domain and the wideband RCS calculation in the time domain.

4.3 Finite Difference–Time-Domain Method

4.3.1 Spatial and Temporal Discretization

The calculation of RCS via the time domain implies the calculation of the scattered field $\vec{E}_s(\vec{r}, t)$ at a distant observation point $\vec{r}$ due to uniform plane wave illumination $\vec{E}_i(\vec{r}', t)$ of the scatterer. The vector $\vec{r}'$ denotes an observation point either on the scatterer surface or possibly within the scatterer (in the case of penetrable bodies). Note that the incident field is defined as the field that would have existed at $\vec{r}'$ without the scatterer present (i.e., in free space).

The first step in the FDTD procedure is spatial and temporal discretization. Just as with any computer solution of transient field problems, FDTD computer calculations require discretization of space and time; that is, *spatial* and *temporal* sampling. The incident and scattered fields are calculated at a large but finite number of points in space and at discrete time instants. The accuracy of the field representation depends on the size of the spatial and temporal discretization steps, $\Delta \ell$ and Δt, respectively. Note that we will assume a Cartesian coordinate system and a uniform spatial discretization with the spatial discretization step constant throughout the space and identical for all coordinate directions (axes). Thus, the spatial discretization elements will be $\Delta \ell$ line segments in one dimension, $\Delta \ell \times \Delta \ell$ squares in two dimensions, and $\Delta \ell \times \Delta \ell \times \Delta \ell$ cubes in three dimensions, as shown in Fig. 4.4.

Note that, in general, decreasing the spatial and temporal discretization steps $\Delta \ell$ and Δt improves the accuracy of the field representation, but one needs first to determine the upper and lower bounds (if any) for the two quantities. First, we examine the time step Δt. The maximum value of the

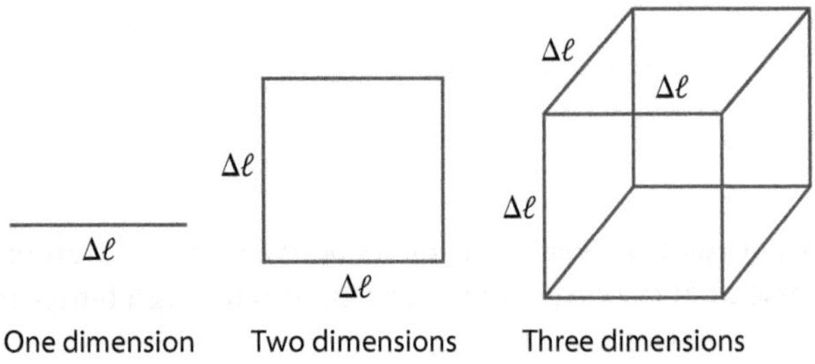

Fig. 4.4 Discretization cells in one, two, and three dimensions.

time step depends on the fastest temporal rate of change of the incident field and is given by the *Nyquist sampling theorem* [9]. It states that we need at least two samples per period of the highest frequency component in the spectrum of the signal being sampled:

$$\Delta t \leq \frac{T_{\min}}{2} = \frac{1}{2 f_{\max}} \tag{4.14}$$

Note that this is a *soft* requirement because the consequences of violating it are not catastrophic and manifest themselves as the so-called aliasing errors [9] that appear in the reconstruction of an undersampled waveform. Introducing the Nyquist sampling interval

$$\Delta t_{\text{Nyquist}} = \frac{1}{2 f_{\max}} \tag{4.15}$$

we can write

$$\Delta t \leq \Delta t_{\text{Nyquist}} \tag{4.16}$$

Note that, in practice, one typically *oversamples* a signal many times by taking more than just two samples per period of the highest frequency component of relevance. We can thus write

$$\Delta t \leq \frac{\Delta t_{\text{Nyquist}}}{K} \tag{4.17}$$

where K denotes the *oversampling factor* and $K \geq 1$.

One of the consequences of the Nyquist theorem is that it is possible to reconstruct an error-free waveform from its samples only if the waveform bandwidth B is finite. Because finite bandwidth implies infinite time duration [9], it follows that, for any signal of finite duration, error-free reconstruction of such a signal from its samples is not possible. However, this error can be made acceptably small by increasing the temporal sampling rate or, equivalently, reducing the time step Δt.

An issue related to the selection of the time step is the resolution of closely spaced scatterers or the resolution of different features of a scatterer of complex shape. So that the scattered fields of two closely spaced scatterers be easily resolvable in time (that is, nonoverlapping), we need an incident field waveform whose duration is small relative to the time that the electromagnetic wave needs to travel the distance between the scatterers. Short duration implies wide bandwidth which, in turn, implies high temporal sampling rates and small time steps.

Determining the limits for the spatial sampling rate or, equivalently, the minimum bound on $\Delta\ell$ is more complicated than determining the minimum temporal sampling rate. If the most common case of Cartesian coordinates and uniform spatial sampling in all coordinate directions is assumed, $\Delta\ell$ needs to satisfy two inequalities. The first one results from the stability requirement [7]:

$$\sqrt{\mathcal{D}}u_p\Delta t \leq \Delta\ell \tag{4.18}$$

where $\mathcal{D}$ denotes the number of spatial dimensions ($\mathcal{D} = 1, 2, 3$ for one-, two-, and three-dimensional problems, respectively), and u_p denotes the velocity of propagation (phase velocity). For the case of an inhomogeneous medium, the maximum velocity of propagation should be used, that is, the one for free space:

$$\sqrt{\mathcal{D}}\frac{1}{\sqrt{\mu_0\epsilon_0}}\Delta t \leq \Delta\ell \tag{4.19}$$

The rationale behind the lower bound for $\Delta\ell$ in Eq. (4.19) will be explained later for the one- and two-dimensional cases in their respective sections. Note that the lower bound for the spatial step $\Delta\ell$ depends on the time step Δt.

The upper bound for $\Delta\ell$ can be determined from the spatial analog to the Nyquist sampling theorem. Thus, we require that at least two spatial samples are taken per wavelength (which is the *spatial period*), corresponding to the highest frequency component in the incident field spectrum. This implies at least two samples per shortest wavelength. This condition must be satisfied regardless of the direction of propagation of the incident or scattered waves. As shown in Fig. 4.5, the "worst" case for a uniform two-dimensional grid is wave propagation along the diagonal of a square cell because the spatial samples are farthest apart from each other, $\sqrt{2}\Delta\ell = \sqrt{\mathcal{D}}\Delta\ell$. Conversely, the "best" case would be propagation along one of the axes of the grid where the spatial samples are closest, separated by $\Delta\ell$. Similarly, the worst-case propagation for a uniform three-dimensional grid is wave propagation along the diagonal of the cubic cell, because the spatial samples between those points are then the farthest apart from each other, a distance of $\sqrt{3}\Delta\ell = \sqrt{\mathcal{D}}\Delta\ell$. The corresponding best case is propagation along one of the grid axes where the spatial samples are closest, separated by $\Delta\ell$.

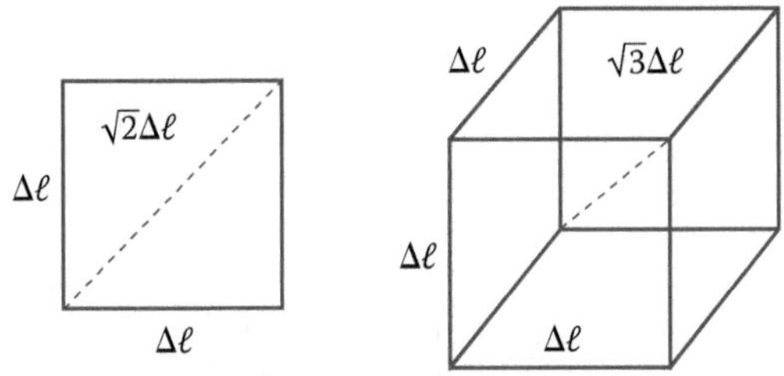

Fig. 4.5 Farthest node separations in two and three dimensions used to determine the spatial sampling interval $\Delta\ell$.

Requiring that we have at least two spatial samples per shortest wavelength for the worst-case propagation along a cell diagonal gives

$$\sqrt{D}\,\Delta\ell \leq \frac{\lambda_{\min}}{2} \tag{4.20}$$

Because the wavelength is the ratio of the velocity of propagation and frequency, the shortest wavelength corresponds to the highest frequency component in the medium with the lowest velocity of propagation (denoted $u_{\min}$):

$$\lambda_{\min} = \frac{u_{\min}}{f_{\max}} = \frac{1}{\sqrt{\mu\epsilon} f_{\max}} = \frac{1}{\sqrt{\mu_r\epsilon_r}} = \frac{1}{\sqrt{\mu_0\epsilon_0}\, f_{\max}} \tag{4.21}$$

where μ_r and ϵ_r denote, respectively, the relative permeability and the relative permittivity of the medium that has the lowest velocity of propagation within the problem domain. Substituting Eq. (4.20) in Eq. (4.21) yields

$$\sqrt{D}\Delta\ell \leq \frac{1}{\sqrt{\mu_r\epsilon_r}} \frac{1}{\sqrt{\mu_0\epsilon_0}} \frac{1}{2f_{\max}} \tag{4.22}$$

which gives the upper bound for the spatial step as

$$\Delta\ell \leq \frac{1}{\sqrt{D}} \frac{1}{\sqrt{\mu_r\epsilon_r}} \frac{1}{\sqrt{\mu_0\epsilon_0}} \frac{1}{2f_{\max}} \tag{4.23}$$

Thus, the spatial step must satisfy the following inequality:

$$\sqrt{D}\frac{1}{\sqrt{\mu_0\epsilon_0}}\Delta t \leq \Delta\ell \leq \frac{1}{\sqrt{D}} \frac{1}{\sqrt{\mu_r\epsilon_r}} \frac{1}{\sqrt{\mu_0\epsilon_0}} \frac{1}{2f_{\max}} \tag{4.24}$$

The lower bound is a *hard* requirement (in that its violation leads to divergent solutions), whereas the upper bound is a soft requirement (meaning that a gradual increase in error results as the upper bound is exceeded). The lower bound for $\Delta\ell$ should be lower or, at most, equal to the upper bound, which requires that

$$\sqrt{\mathcal{D}}\frac{1}{\sqrt{\mu_0\epsilon_0}}\Delta t \leq \frac{1}{\sqrt{\mathcal{D}}}\frac{1}{\sqrt{\mu_r\epsilon_r}}\frac{1}{\sqrt{\mu_0\epsilon_0}}\frac{1}{2f_{\text{max}}} \tag{4.25}$$

which can be simplified to

$$\Delta t \leq \frac{1}{\mathcal{D}}\frac{1}{\sqrt{\mu_r\epsilon_r}}\frac{1}{2f_{\text{max}}} \tag{4.26}$$

or, if the Nyquist sampling interval is used,

$$\Delta t \leq \frac{1}{\mathcal{D}}\frac{1}{\sqrt{\mu_r\epsilon_r}}\Delta t_{\text{Nyquist}} \tag{4.27}$$

Based on the preceding results, the following guidelines are suggested for selecting the temporal step Δt and the spatial step $\Delta\ell$. First, determine the highest relevant frequency f_{max} in the spectrum of the incident field. The temporal step is then selected such that the inequality

$$\Delta t \leq \frac{1}{\mathcal{D}}\frac{1}{\sqrt{\mu_r\epsilon_r}}\frac{1}{2f_{\text{max}}} \tag{4.28}$$

is satisfied, with the relative permeability and permittivity taken for the medium with the lowest velocity of propagation. Note that the smaller the time step, the more calculations must be performed within the observation time window, requiring longer computational times.

Next, select the spatial step such that the inequality in Eq. (4.24) is satisfied. Rearranging the inequality gives

$$\sqrt{\mathcal{D}}u_{\text{max}} \leq \frac{\Delta\ell}{\Delta t} \leq \frac{1}{\sqrt{\mathcal{D}}}u_{\text{min}}\frac{1}{2f_{\text{max}}\Delta t} \tag{4.29}$$

Introducing the *grid velocity*,

$$u_{\text{grid}} = \frac{\Delta\ell}{\Delta t} \tag{4.30}$$

which represents the rate at which the numerical solution "propagates" on an FDTD grid, we can write

$$\sqrt{\mathcal{D}}u_{\text{max}} \leq u_{\text{grid}} \leq \frac{1}{\sqrt{\mathcal{D}}}u_{\text{min}}\left(\frac{1}{\Delta t/\Delta t_{\text{Nyquist}}}\right) \tag{4.31}$$

or, using the oversampling factor,

$$\sqrt{\mathcal{D}}u_{\text{max}} \leq u_{\text{grid}} \leq \frac{K}{\sqrt{\mathcal{D}}}u_{\text{min}} \tag{4.32}$$

Equation (4.32) shows that the grid velocity needs to be at least $\sqrt{\mathcal{D}}$ times higher than the highest velocity of propagation but not higher than $K/\sqrt{\mathcal{D}}$ times the lowest velocity of propagation.

The oversampling factor must be chosen such that the upper bound exceeds the lower bound:

$$K \geq \mathcal{D}\left(\frac{u_{\max}}{u_{\min}}\right) \tag{4.33}$$

The oversampling factor is thus the product of the problem dimension and the ratio of the largest to the smallest propagation velocity. It is clear that a large velocity ratio requires a large oversampling factor, that is, a very small time step Δt. In case of homogeneous problems (involving only one medium), the maximum and the minimum velocities of propagation are the same and, thus,

$$K \geq \mathcal{D} \tag{4.34}$$

Equation (4.34) implies that, for a homogeneous medium, we need to sample at least twice the Nyquist sampling rate in two dimensions and at least three times the Nyquist sampling rate in three dimensions.

4.3.2 Waveform Selection

The ideal waveform for an FDTD calculation is one with finite duration in time and finite spectrum width. However, such a combination is not possible in a physical signal [9]. If we have an ideal limited-duration waveform, such as a rectangular pulse, its spectrum (a sinc function) will extend to infinity, eventually converging to zero at infinity. Conversely, if we want a spectrum whose components are zero above a certain maximum frequency, $f_{\max}$, we need a "pulse" in time domain that has infinite duration, eventually converging to zero at infinity. For example, a rectangular spectrum corresponds (by the symmetry property of Fourier transforms) to a sinc pulse in the time domain.

In general, there is an inverse relationship between a waveform's time duration and its frequency bandwidth: reducing the waveform duration in time increases its spectral width, and vice versa. This is expressed using the *time-bandwidth product* [10], which states that, for a waveform $f(t)$, the product of the effective duration and the effective bandwidth equals a constant:

$$T_{\text{eff}} B_{\text{eff}} = \text{constant} \tag{4.35}$$

The *effective duration* and the *effective bandwidth* for a time function $f(t)$ and its Fourier transform $F(\omega)$ are defined by [10]

$$T_{\text{eff}} = \sqrt{\frac{\int_{-\infty}^{\infty} t^2 |f(t)|^2 dt}{\int_{-\infty}^{\infty} |f(t)|^2 dt}} \tag{4.36}$$

and

$$B_{\text{eff}} = \sqrt{\frac{\int_{-\infty}^{\infty} \omega^2 |F(\omega)|^2 d\omega}{\int_{-\infty}^{\infty} |F(\omega)|^2 d\omega}} \tag{4.37}$$

The best waveform for an FDTD calculation is the one with the minimum time-bandwidth product. Such a waveform simultaneously minimizes the waveform duration and its bandwidth and, thus, minimizes aliasing errors and the required number of time steps. It can be shown [10] that the minimum theoretical value of the time bandwidth product is 1/2:

$$T_{\text{eff}} B_{\text{eff}} \geq 1/2 \tag{4.38}$$

The function $f(t)$ that gives the minimum time-bandwidth product of 1/2 is the *Gaussian pulse*:

$$f(t) = \exp\left[-t^2/(2\tau^2)\right] \tag{4.39}$$

The effective duration and the effective bandwidth for a Gaussian pulse are given by [10]:

$$T_{\text{eff}} = \frac{\tau}{\sqrt{2}} \quad \text{and} \quad B_{\text{eff}} = \frac{1}{\tau\sqrt{2}} \tag{4.40}$$

From Eq. (4.40), the parameter τ can be expressed in terms of either the effective duration or the effective bandwidth:

$$\tau = \sqrt{2} T_{\text{eff}} \quad \text{or} \quad \tau = \frac{1}{\sqrt{2} B_{\text{eff}}} \tag{4.41}$$

A Gaussian pulse can now be specified using either the effective duration or the effective bandwidth:

$$f(t) = \exp\left[-t^2/(4T_{\text{eff}}^2)\right]$$

or

$$f(t) = \exp\left[-B_{\text{eff}}^2 t^2\right] \tag{4.42}$$

The Gaussian pulse defined in Eq. (4.42) has infinite duration. We need to truncate it by multiplying the Gaussian pulse with a unit amplitude rectangular pulse $p(t/T_0)$ of duration T_0. The truncated Gaussian pulse will be denoted $g(t)$:

$$g(t) = f(t)\, p(t/T_0) \tag{4.43}$$

The truncation changes the effective duration and the effective bandwidth; it reduces T_{eff} but increases B_{eff}. However, we would typically truncate the Gaussian pulse such that these changes are very small. Therefore, we can usually approximate the effective duration and the effective bandwidth of the truncated Gaussian pulse by their values for the original Gaussian pulse.

The selection of the width of the truncated pulse T_0 can be based on the desired or achievable accuracy (precision). Suppose that $e_{\min}$ is the smallest number we have interest in (or perhaps confidence in, because of roundoff and other errors). There is no reason to consider the values of the Gaussian pulse that are less than $e_{\min}$; that is, we can truncate the Gaussian pulse at T_0 such that

$$g(T_0/2) = e_{\min} \tag{4.44}$$

The ratio of the peak value of the normalized Gaussian pulse (which will be 1) and the value of $e_{\min}$ can be expressed in decibels. This ratio is referred to as the *dynamic range*:

$$R_{\text{dyn}} = 20\log_{10}\left(\frac{1}{e_{\min}}\right) = -20\log_{10}(e_{\min}) \text{ (in dB)} \tag{4.45}$$

The duration of the truncated pulse can now be related to the desired dynamic range. First, we express $e_{\min}$ in terms of the dynamic range:

$$e_{\min} = 10^{(-R_{\text{dyn}}/20)} \tag{4.46}$$

and then solve

$$\exp[-(B_{\text{eff}}T_0/2)^2] = 10^{(-R_{\text{dyn}}/20)} \tag{4.47}$$

for T_0, which gives

$$T_0 = \frac{2}{B_{\text{eff}}}\sqrt{\frac{R_{\text{dyn}}}{20\log_{10}(e)}} \approx \frac{0.96}{B_{\text{eff}}}\sqrt{\frac{R_{\text{dyn}}}{2}} \approx \frac{1}{B_{\text{eff}}}\sqrt{\frac{R_{\text{dyn}}}{2}} = T_{\text{eff}}\sqrt{2R_{\text{dyn}}} \tag{4.48}$$

Equation (4.48) can also be written as

$$\frac{T_0}{T_{\text{eff}}} \approx \sqrt{2R_{\text{dyn}}} \tag{4.49}$$

where R_{dyn} is in decibels. A plot of the ratio T_0/T_{eff} vs R_{dyn} is shown in Fig. 4.6.

A dynamic range of 120 dB is likely to be sufficient in most cases. This range corresponds to $e_{\min} = 10^{-6}$, or six decimal places. From Fig. 4.6, this gives, approximately,

$$T_0 \geq 16T_{\text{eff}} \quad \text{or} \quad T_0 \geq \frac{8}{B_{\text{eff}}} \tag{4.50}$$

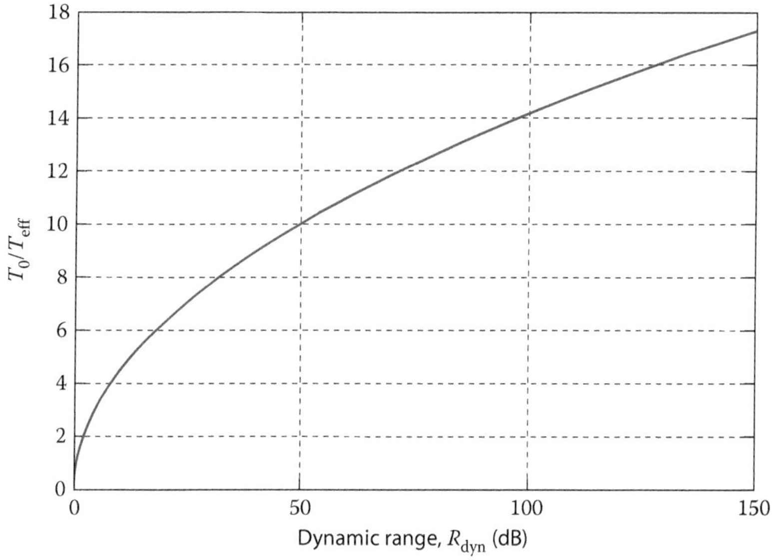

Fig. 4.6 Normalized T_0 vs dynamic range.

The truncated Gaussian pulse would normally be shifted by $T_0/2$ such that the pulse values are zero for $t < 0$. Thus, the expression for the truncated and shifted pulse is

$$g(t) = \exp[-B_{eff}^2(t - T_0/2)^2]p[(t - T_0/2)/T_0] \tag{4.51}$$

Selecting $T_0 = 8/B_{eff}$ for a dynamic range greater than 120 dB yields

$$g(t) = \exp[-(B_{eff}t - 4)^2]p[(t - 4/B_{eff})/(8/B_{eff})] \tag{4.52}$$

The truncated and shifted Gaussian pulse is centered at $4/B_{eff}$, and its duration is $8/B_{eff}$. Such a pulse is shown in Fig. 4.7 for $B_{eff} = 1$ GHz. Note that the time scale is in nanoseconds and the effective duration of the pulse is 0.5 ns. Thus, the truncated Gaussian pulse is completely specified by two parameters: the effective bandwidth B_{eff} (or, equivalently, T_{eff}) and the dynamic range R_{dyn} (or, equivalently, T_0).

4.3.3 Time and Spatial Step Selection

The selection of the time step depends on the selection of the waveform describing the time variation of the incident field. Having selected a truncated Gaussian pulse, we can now proceed to determine the time step Δt and the spatial step $\Delta \ell$. The selection of the time step is related to the highest component in the spectrum of the Gaussian pulse that we would like to sample properly. We denote the frequency of such a component as f_{max}. Note that all spectral components at frequencies higher than f_{max} will be undersampled, which will introduce errors. If the power contained in these frequency components is small, the errors due to undersampling will be small, and vice versa. Thus, we need to determine the power spectral density of the truncated Gaussian pulse. Because the truncated pulse is most likely to be truncated such that $T_0 \gg T_{eff}$, that is, at very small values, we can approximate the spectrum of the truncated pulse $G(\omega)$ with the spectrum of the original Gaussian pulse $F(\omega)$. This greatly simplifies the required calculations.

The spectrum of the Gaussian pulse is also a Gaussian pulse in the frequency domain [9]:

$$F(\omega) = \frac{\sqrt{\pi}}{B_{eff}} \exp\left\{-[\omega/(2B_{eff})]^2\right\} \tag{4.53}$$

The normalized magnitude of the Fourier transform of the Gaussian pulse with $B_{eff} = 1$ GHz is shown in Fig. 4.8. The *power spectral density* (PSD) is given by the magnitude squared of the Fourier transform:

$$S(\omega) = |G(\omega)|^2 \approx |F(\omega)|^2 = \frac{\pi}{B_{eff}^2} \exp\left\{-[\omega/(\sqrt{2}B_{eff})]^2\right\} \tag{4.54}$$

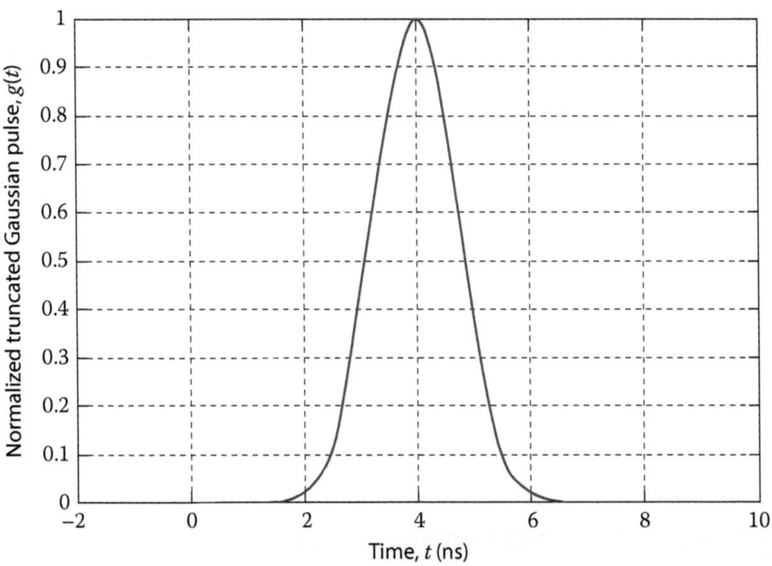

Fig. 4.7 Truncated and shifted Gaussian pulse.

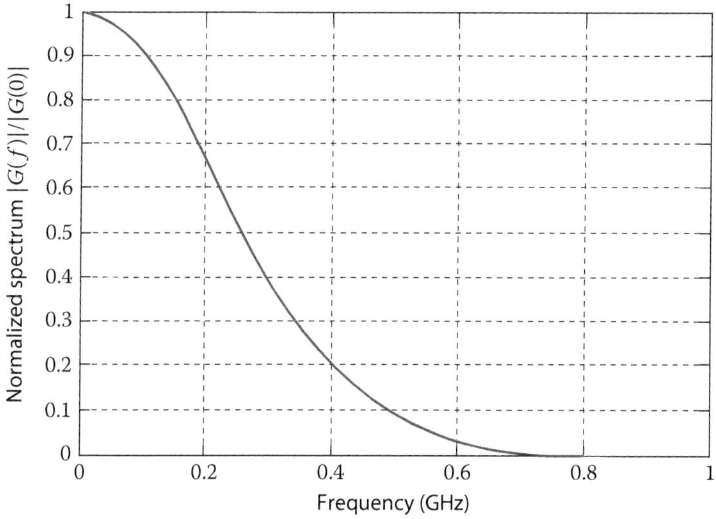

Fig. 4.8 Normalized spectrum of a Gaussian pulse.

The power between frequencies 0 and $f_{\max}$ is obtained by integrating the power spectral density between these limits:

$$P(f_{\max}) = \frac{1}{2\pi}\int_{-\omega_{\max}}^{\omega_{\max}} S(\omega)\mathrm{d}\omega = \frac{1}{B_{\mathrm{eff}}^2}\int_0^{\omega_{\max}} \exp\left\{-[\omega/(\sqrt{2}B_{\mathrm{eff}})]^2\right\}\mathrm{d}\omega \quad (4.55)$$

where $\omega_{\max} = 2\pi f_{\max}$. The fraction of the total power above $f_{\max}$ can be calculated by

$$\begin{aligned}\frac{\Delta P}{P_{\mathrm{total}}} &= \frac{P_{\mathrm{total}} - P(f_{\max})}{P_{\mathrm{total}}} = 1 - \frac{P(f_{\max})}{P_{\mathrm{total}}} \\ &= 1 - \frac{\int_0^{\omega_{\max}} \exp\left\{-[\omega/(\sqrt{2}B_{\mathrm{eff}})]^2\right\}\mathrm{d}\omega}{\int_0^{\infty} \exp\left\{-[\omega/(\sqrt{2}B_{\mathrm{eff}})]^2\right\}\mathrm{d}\omega} = 1 - \mathrm{erf}\left(\sqrt{2}\pi\frac{f_{\max}}{B_{\mathrm{eff}}}\right)\end{aligned} \quad (4.56)$$

where $\mathrm{erf}(x)$ is the so-called error function [11], defined by

$$\mathrm{erf}(x) = \frac{2}{\sqrt{\pi}}\int_0^x e^{-t^2}\,\mathrm{d}t \quad (4.57)$$

The percent of power above f_{max} is

$$100\left[\frac{P_{total} - P(f_{max})}{P_{total}}\right] = 100\left[1 - \text{erf}\left(\sqrt{2}\pi f_{max}/B_{eff}\right)\right]$$
$$= 100\,\text{erfc}\left(\sqrt{2}\pi f_{max}/B_{eff}\right) \tag{4.58}$$

where the complementary error function erfc(x) has been used [11]:

$$\text{erfc}(x) = 1 - \text{erf}(x) \tag{4.59}$$

The fraction of total power above f_{max} can be expressed in decibels by taking 10 $\log_{10}$ of Eq. (4.56)

$$10\,\log_{10}\left[\frac{\Delta P}{P_{total}}\right] = 10\,\log_{10}\left[1 - \text{erf}\left(\sqrt{2}\pi f_{max}/B_{eff}\right)\right] \tag{4.60}$$

Equation (4.60) is plotted versus the ratio $f_{norm} = f_{max}/B_{eff}$ (frequency normalized to the effective bandwidth) in Fig. 4.9.

For FDTD calculations using the truncated Gaussian pulse, one can consider the power contained in the spectral components above f_{max} as "noise" power. This is because the spectral components above f_{max} are not properly sampled and, thus, cause errors that have characteristics similar to noise on a signal. In this case, the power in the truncated Gaussian pulse is the "signal" power. Using this model the *signal-to-noise ratio* (S/N) in decibels corresponds exactly to the fraction of the power shown in Fig. 4.9. From the figure, we note that selecting $f_{max} = B_{eff}$ (or $f_{norm} = 1$) gives an S/N of about 90 dB. On the other hand, selecting $f_{max} = 1.2B_{eff}$

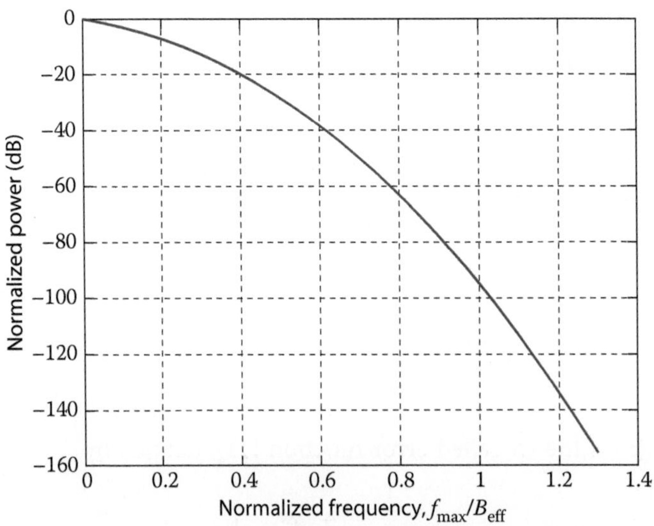

Fig. 4.9 Fraction of total power above f_{norm} in decibels.

gives an S/N of about 130 dB. Thus, the selection of f_{max} can be based on the desired S/N. In most cases, S/N of 120 dB is adequate, which gives the approximate condition:

$$f_{max} \geq 1.2B_{eff} \qquad \text{for} \qquad S/N \geq 120\,\text{dB} \tag{4.61}$$

At this point, it is possible to devise a procedure that provides us values of the parameters Δt and $\Delta \ell$ for an FDTD solution:

1. Select a unit amplitude Gaussian pulse as the waveform because it has the minimum time-bandwidth product.
2. Select the effective bandwidth B_{eff} based on the highest frequency needed to obtain sufficiently accurate RCS data (via the Fourier transform of the time-domain scattered field).
3. Determine the effective duration T_{eff} from the effective bandwidth B_{eff}.
4. Select the minimum significant amplitude, which sets the dynamic range R_{dyn}.
5. Determine the truncated pulse duration T_0 from the selected dynamic range.
6. Select the fraction of the total power of the Gaussian pulse that should be within the highest properly sampled frequency (the signal-to-noise ratio, S/N).
7. Determine the maximum frequency f_{max} such that the selected S/N is achieved.
8. Determine the Nyquist sampling interval $\Delta t_{Nyquist}$.
9. Determine the temporal oversampling factor K from the problem dimensionality $\mathcal{D}$ and the ratio of the highest to lowest propagation velocities u_{max}/u_{min}.
10. Select the actual calculation time sampling interval Δt based on the oversampling factor K and the Nyquist sampling interval $\Delta t_{Nyquist}$.
11. Determine the upper and the lower bounds for the spatial step $\Delta \ell$, based on the selected time step Δt.
12. Select the actual calculation spatial step Δt.
13. Select the desired frequency resolution Δf.
14. Determine the duration of the time observation window T_{max} from the selected Δf.
15. Compute the number of time steps $N = T_{max}/\Delta t$.

Example 4.1: Selection of the FDTD Calculation Parameters

Suppose that we need RCS data at frequencies up to 10 GHz with a resolution of 1 MHz. The FDTD calculation will be done in three dimensions, and the number of samples should be minimized. Using the 15 steps just listed, the FDTD calculation parameters can be determined. They are listed in Table 4.1.

(Continued)

Example 4.1: Selection of the FDTD Calculation Parameters *(Continued)*

Table 4.1 Finite Difference–Time-Domain Calculation Parameters for Example 4.1

Step	Parameter	Criterion
1	Gaussian pulse shape	$T_{\text{eff}} \cdot B_{\text{eff}}$ is minimum
2	$B_{\text{eff}} = 10$ GHz	Problem statement
3	$T_{\text{eff}} = 50$ ps	$T_{\text{eff}} = 1/(2B_{\text{eff}})$
4	$R_{\text{dyn}} = 120$ dB	Typical accuracy
5	$T_0 = 0.8$ ns	From Fig. 4.6
6	$S/N = 120$ dB	From R_{dyn}
7	$f_{\max} = 12$ GHz	From Fig. 4.9
8	$\Delta t_{\text{Nyquist}} = 41.67$ ps	$\Delta t_{\text{Nyquist}} = 1/(2f_{\max})$
9	$K = 3$	3-D ($\mathcal{D} = 3$) and minimize samples
10	$\Delta t = 10$ ps	$\Delta t \leq \Delta t_{\text{Nyquist}}/K$
11	$5.196 \text{ mm} \leq \Delta \ell \leq 7.217 \text{ mm}$	$\sqrt{3}c\Delta t \leq \Delta \ell \leq c\Delta t_{\text{Nyquist}}/\sqrt{3}$
12	$\Delta \ell = 6$ mm	Convenient value satisfying step 11
13	$\Delta f = 1$ MHz	Problem statement
14	$T_{\max} = 1\ \mu$s	$T_{\max} = 1/\Delta f_{\max}$
15	$N = 10^5$	$N = T_{\max}/\Delta t$

4.4 Finite Difference–Time-Domain Equations in One Dimension

4.4.1 Derivation of the Magnetic Field Update Equation

The incident and scattered electromagnetic fields and the media parameters in one-dimensional problems vary with only one spatial coordinate. We will choose this coordinate to be the z axis. The fields will also be functions of time t. The media will be assumed *stationary* (i.e., media parameters such as μ, ϵ, and σ are independent of t).* One-dimensional electromagnetic fields must be TEM, that is, the field vectors $\vec{E}$ and $\vec{H}$ are orthogonal and lie in a plane transverse to the direction of propagation. Because we have denoted the direction of propagation as z, the fields are TEM_z. The triplet of vectors $(\hat{z}, \vec{E}, \vec{H})$ are mutually orthogonal and form a right-handed system, as shown in Fig. 4.10. The coordinate axes are selected so that the electric field vector is x-directed and the magnetic field vector is y-directed.

$$\vec{E}(z,t) = E_x(z,t)\hat{x} \qquad \text{and} \qquad \vec{H}(z,t) = H_y(z,t)\hat{y} \tag{4.62}$$

*In this chapter σ represents conductivity, not the RCS, unless otherwise noted.

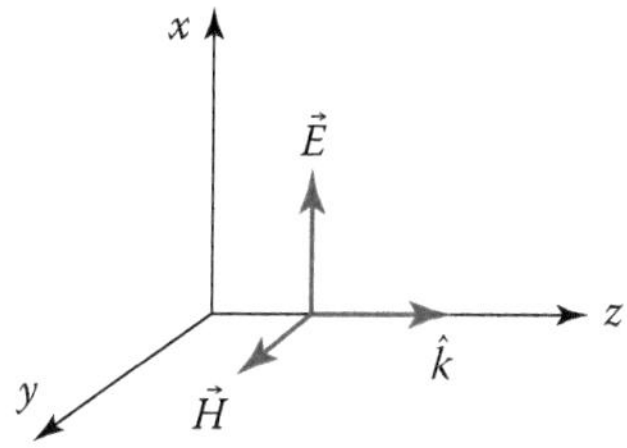

Fig. 4.10 TEM$_z$ field components in a one-dimensional medium.

The electric and magnetic fields satisfy Maxwell's equations, which express the electric and magnetic field coupling. They can be written in integral or differential form as described in Appendix A. We will use the integral forms because they are applicable to finite domains (lines, surfaces, or volumes), whereas the differential forms are applicable to infinitesimally small domains (points). The integral forms of Maxwell's first two equations for a TEM$_z$ field are:

$$\oint_{C_E} E_x(z,t)\hat{x} \cdot \vec{\mathrm{d}\ell}_E = -\frac{\partial}{\partial t}\left\{\iint_{S_E} \mu(z)H_y(z,t)\hat{y} \cdot \vec{\mathrm{d}s}_E\right\} \tag{4.63}$$

and

$$\oint_{C_H} H_y(z,t)\hat{y} \cdot \vec{\mathrm{d}\ell}_H = \frac{\partial}{\partial t}\left\{\iint_{S_H} \epsilon(z)E_x(z,t)\hat{x} \cdot \vec{\mathrm{d}s}_H\right\} + \iint_{S_H} \sigma(z)E_x(z,t)\hat{x} \cdot \vec{\mathrm{d}s}_H \tag{4.64}$$

Note that Eq. (4.64) does not have a source current term because it has been assumed that the domain of interest is a *source-free region*. The source for the scattered field calculations is a known incident field. The contours of integration for the electric and the magnetic field integrals [left-hand sides of Eqs. (4.63) and (4.64)] are, in general, different and are thus labeled C_E and C_H, respectively. Similarly, the surfaces associated with the contours are labeled S_E and S_H for the magnetic field surface integral and the electric field surface integral, respectively. A closed-path line integral is referred to as the *circulation*, and surface integrals, such as those occurring on the right-hand side, are *fluxes*.

The first step in discretizing Maxwell's equations is to select the contours for the electric and magnetic field circulations. Because the essence of Maxwell's equations is the coupling of the electric and magnetic fields, the contours will be selected such that this coupling is achieved in a straightforward manner, as shown in Fig. 4.11. The contours C_E and C_H are squares of dimensions $\Delta\ell \times \Delta\ell$ and are in orthogonal planes. The C_E contour is in the x–z plane, whereas the C_H contour is in the y–z plane. The contours can be compared to links of a chain, evoking the idea of electric and magnetic field linkage. This selection of paths leads to the so-called *dual discretization grid*, also referred to as the *Yee lattice* [1].

The discrete equivalent of the electric field circulation is determined as follows. A contour C_E *is* shown in Fig. 4.12, with the center of the surface

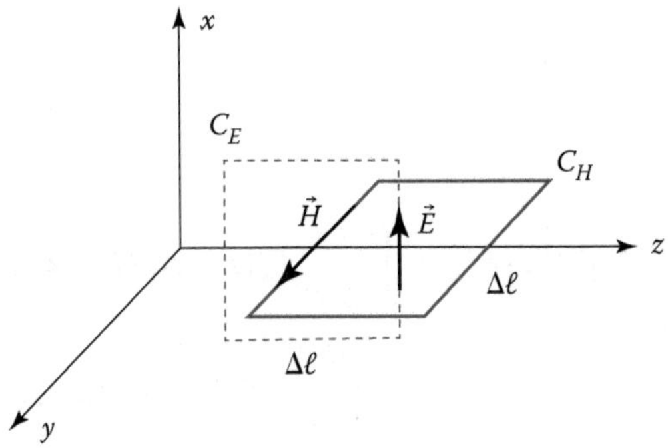

Fig. 4.11 Contours for electric and magnetic field circulations and fluxes.

S_E at the coordinates (x, z). A local coordinate system can be established, with its origin at the center of the contour. Any point (x', z') within or on the contour C_E (or, equivalently, any point on the surface S_E) can be specified by its local coordinates ξ and ζ, where

$$x' = x + \xi \qquad z' = z + \zeta \tag{4.65}$$

and

$$-\frac{\Delta\ell}{2} \leq \xi, \zeta \leq \frac{\Delta\ell}{2}$$

The local coordinates will be used in evaluation of the line and surface integrals that constitute the integral forms of Maxwell's equations. Note that the electric field E_x is not a function of the x coordinate. The electric field is zero along the top and the bottom sides of the contour because

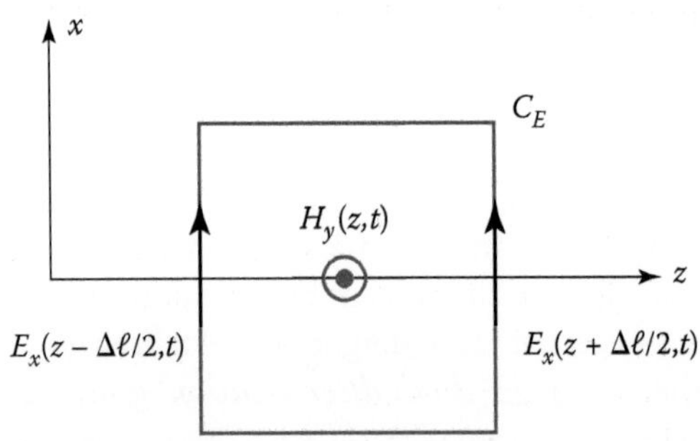

Fig. 4.12 Contour for electric circulation and magnetic flux.

there can be no z-directed field components for a TEM_z field. Now, define a counterclockwise reference direction such that the normal to the surface S_E is in the direction of the magnetic field $\vec{H}(z, t)$. The circulation of the electric field around the electric field contour C_E can be evaluated exactly:

$$\oint_{C_E} E_x(z+\zeta,t)\hat{x}\cdot \mathrm{d}\vec{\ell}_E = \int_{-\Delta\ell/2}^{\Delta\ell/2} E_x(z+\Delta\ell/2,t)\hat{x}\cdot \mathrm{d}\vec{\xi} - \int_{-\Delta\ell/2}^{\Delta\ell/2} E_x(z-\Delta\ell/2,t)\hat{x}\cdot \mathrm{d}\vec{\xi} \tag{4.66}$$

or

$$\oint_{C_E} E_x(z+\zeta,\, t)\hat{x}\cdot \mathrm{d}\vec{\ell}_E = [E_x(z+\Delta\ell/2,\, t) - E_x(z-\Delta\ell/2,\, t)]\Delta\ell \tag{4.67}$$

On the other hand, the rate of change of the magnetic flux through the surface S_E can be evaluated only approximately because the exact dependence of the magnetic flux density $B(z, t) = \mu(z)H(z, t)$ on the z coordinate is not known a priori. Thus we proceed to evaluate the magnetic flux approximately first,

$$\iint_{S_E} \mu\,(z+\zeta)H_y(z+\zeta,t)\hat{y}\cdot \mathrm{d}\vec{s}_E = \int_{-\Delta\ell/2}^{\Delta\ell/2}\int_{-\Delta\ell/2}^{\Delta\ell/2} \mu(z+\zeta)H_y(z+\zeta,t)\hat{y}\cdot\hat{y}\,\mathrm{d}\xi\mathrm{d}\zeta \tag{4.68}$$

However, there is no variation of the magnetic flux density with the x coordinate and, therefore, the surface integral reduces to a line integral:

$$\iint_{S_E} \mu(z+\zeta)H_y(z+\zeta,t)\hat{y}\cdot \mathrm{d}\vec{s}_E = \Delta\ell \int_{-\Delta\ell/2}^{\Delta\ell/2} \mu(z+\zeta)H_y(z+\zeta,t)\,\mathrm{d}\zeta \tag{4.69}$$

There are infinitely many ways to model the variation of the magnetic flux density within the contour as a function of the local coordinate ζ. Because the magnetic flux density $B_y(z+\zeta, t)$ is the product of the permeability $\mu(z+\zeta)$ and the magnetic field $H_y(z+\zeta, t)$, we first need to assume a certain variation of the magnetic field with z, which allows evaluation of the integral over S_E. The simplest approximation is that the contour width $\Delta\ell$ is small enough to allow the magnetic field $H_y(z+\zeta, t)$ within the contour to be assumed constant and equal to the value at the center $H_y(z, t)$. This is equivalent to using a piecewise constant (or pulse expansion) approximation of the actual magnetic field variation with the z coordinate. Note that the aforementioned assumption allows the magnetic field to change from contour to contour even though it is constant within a contour. This yields an approximate expression for the magnetic flux:

$$\iint_{S_E} \mu(z+\zeta)H_y(z+\zeta,t)\hat{y}\cdot \vec{\mathrm{d}s}_E \approx \Delta\ell H_y(z,t) - \int_{-\Delta\ell/2}^{\Delta\ell/2} \mu(z+\zeta)\mathrm{d}\zeta \tag{4.70}$$

The integral of the permeability $\mu(z+\zeta)$ can be rewritten in the following manner:

$$\int_{-\Delta\ell/2}^{\Delta\ell/2} \mu(z+\zeta)\mathrm{d}\zeta = \Delta\ell\left[\frac{1}{\Delta\ell}\int_{-\Delta\ell/2}^{\Delta\ell/2} \mu(z+\zeta)\mathrm{d}\zeta\right] \tag{4.71}$$

The term in the brackets is recognized as the average permeability within the contour C_E:

$$\mu_{\text{avg}}(z) = \frac{1}{\Delta\ell}\int_{-\Delta\ell/2}^{\Delta\ell/2} \mu(z+\zeta)\mathrm{d}\zeta \tag{4.72}$$

The approximate expression for the magnetic flux through S_E can now be written in terms of the average permeability:

$$\iint_{S_E} \mu(z+\zeta)H_y(z+\zeta,t)\hat{y}\cdot \vec{\mathrm{d}s}_E \approx (\Delta\ell)^2\mu_{\text{avg}}(z)H_y(z,t) \tag{4.73}$$

The approximation in Eq. (4.73) resulted from the piecewise constant approximation of the magnetic field with respect to the z coordinate. The main advantage of the piecewise constant expansion is its simplicity. Increased accuracy can be achieved by using more involved models for the field variation with z but at the expense of increasing the computational time.

Maxwell's first equation can now be replaced by the approximate form

$$[E_x(z+\Delta\ell/2,t) - E_x(z-\Delta\ell/2,t)]\Delta\ell \approx -\frac{\partial}{\partial t}\left[\mu_{\text{avg}}(z)H_y(z,t)(\Delta\ell)^2\right] \tag{4.74}$$

which, because of the stationary medium assumption, can be simplified to

$$E_x(z+\Delta\ell/2,t) - E_x(z-\Delta\ell/2,t) \approx -\frac{\partial}{\partial t}\{H_y(z,t)\}\mu_{\text{avg}}(z)\Delta\ell \tag{4.75}$$

Equation (4.75) can be rewritten as

$$\frac{\partial}{\partial t}\{H_y(z,t)\} \approx \frac{-1}{\mu_{\text{avg}}(z)\Delta\ell}[E_x(z+\Delta\ell/2,t) - E_x(z-\Delta\ell/2,t)] \tag{4.76}$$

At this point, the time derivative operator in Eq. (4.76) is typically implemented by a finite difference approximation [8]. However, an alternate approach is presented here, such that the approximation of the field time variation is shown to be analogous to the approximations already introduced for the field spatial variation. Equation (4.76) is integrated with respect to time to get an approximate equation for the magnetic field at time t:

$$H_y(z,t) \approx \frac{-1}{\mu_{\text{avg}}(z)\Delta\ell}\left[\int_0^t E_x(z+\Delta\ell/2,\tau)\mathrm{d}\tau - \int_0^t E_x(z-\Delta\ell/2,\tau)\mathrm{d}\tau\right] \quad (4.77)$$

A similar integral equation can be written for the magnetic field at the same spatial location but at an earlier time, t–Δt:

$$H_y(z,t-\Delta t) \approx \frac{-1}{\mu_{\text{avg}}(z)\Delta\ell}\left[\int_0^{t-\Delta t} E_x(z+\Delta\ell/2,\tau)\mathrm{d}\tau - \int_0^{t-\Delta t} E_x(z-\Delta\ell/2,\tau)\mathrm{d}\tau\right] \quad (4.78)$$

Subtracting Eqs. (4.77) and (4.78) gives

$$H_y(z,t) - H_y(z,t-\Delta t) \approx \frac{-1}{\mu_{\text{avg}}(z)\Delta\ell}\left[\int_{t-\Delta t}^{t} E_x(z+\Delta\ell/2,\tau)\mathrm{d}\tau - \int_{t-\Delta t}^{t} E_x(z-\Delta\ell/2,\tau)\mathrm{d}\tau\right] \quad (4.79)$$

When Eq. (4.79) is rearranged, it becomes the *magnetic field update equation*; it provides the field at the current location and time if it is assumed that the previous time value at the same location is known. Thus,

$$H_y(z,t) \approx H_y(z,t-\Delta t) - \frac{1}{\mu_{\text{avg}}(z)\Delta\ell}\left[\int_{t-\Delta t}^{t} E_x(z+\Delta\ell/2,\tau)\mathrm{d}\tau - \int_{t-\Delta t}^{t} E_x(z-\Delta\ell/2,\tau)\mathrm{d}\tau\right] \quad (4.80)$$

The integrals on the right-hand side of Eq. (4.80) cannot be evaluated exactly because the exact temporal variation of the electric fields within the Δt interval prior to t is generally not known. This is the same reason the spatial integral for the magnetic flux in Eq. (4.71) could not be evaluated exactly; the precise variation of the magnetic field within the contour was not known. However, the integrals can be evaluated approximately by assuming a certain variation of the electric field with the temporal variable t over the interval Δt. The simplest approach consistent with the assumptions made for the field spatial variation is to consider the interval Δt small enough so that the electric field is approximately constant within Δt and equal to the value at the center of the time interval $(t - \Delta t, t)$:

$$\begin{aligned} E_x(z+\Delta\ell/2, \tau) &\approx E_x(z+\Delta\ell/2, t-\Delta t/2) \\ E_x(z-\Delta\ell/2, \tau) &\approx E_x(z-\Delta\ell/2, t-\Delta t/2) \end{aligned} \tag{4.81}$$

for $t - \Delta t \le \tau \le t$. These are piecewise constant approximations of the electric field variation with respect to t.

The resulting approximate expression for the magnetic field at (z, t) obtained by using Eq. (4.81) in Eq. (4.80) is

$$\begin{aligned} H_y(z,t) \approx H_y(z,t-\Delta t) - \frac{\Delta t}{\Delta\ell\,\mu_{\text{avg}}(z)}[E_x(z+\Delta\ell/2, t-\Delta t/2) \\ -E_x(z-\Delta\ell/2, t-\Delta t/2)] \end{aligned} \tag{4.82}$$

Now, we make use of the identity

$$\frac{1}{\mu} = \frac{1}{\mu_0\mu_r} = \frac{1}{\sqrt{\mu_0\epsilon_0}}\frac{1}{\sqrt{\frac{\mu_0}{\epsilon_0}}}\frac{1}{\mu_r} = \frac{c}{\eta_0}\frac{1}{\mu_r} \tag{4.83}$$

where the free-space velocity of propagation is c, the intrinsic impedance of free space is η_0, and μ_r denotes relative permeability. Introducing the grid propagation velocity defined in Eq. (4.30), we can write the approximate update equation for the magnetic field as

$$\begin{aligned} H_y(z,t) \approx H_y(z,t-\Delta t) - \frac{1}{\eta_0}\frac{c}{u_{\text{grid}}}\frac{1}{\mu_{r\text{avg}}(z)}[E_x(z+\Delta\ell/2, t-\Delta t/2) \\ - E_x(z-\Delta\ell/2, t-\Delta t/2)] \end{aligned} \tag{4.84}$$

Equation (4.84) simplifies for the case of nonmagnetic media ($\mu_r = 1$):

$$\begin{aligned} H_y(z,t) \approx H_y(z,t-\Delta t) - \frac{1}{\eta_0}\frac{c}{u_{\text{grid}}}[E_x(z+\Delta\ell/2, t-\Delta t/2) \\ - E_x(z-\Delta\ell/2, t-\Delta t/2)] \end{aligned} \tag{4.85}$$

Equations (4.84) and (4.85) show that the magnetic field H_y and the electric field E_x are evaluated at points shifted in space by $\Delta\ell/2$ and at instants separated in time by $\Delta t/2$. The relationship between spatial and temporal samples of the electric and the magnetic fields is thus the same: The samples are shifted with respect to each other by one-half of the sampling interval. This is the essence of the Yee lattice, and it applies to two- and three-dimensional formulations as well.

4.4.2 Derivation of the Electric Field Update Equation

To derive the electric field update equation, we start with Maxwell's second equation for the circulation of the magnetic field:

$$\oint_{C_H} H_y(z,t)\hat{y}\cdot \vec{\mathrm{d}\ell}_H = \frac{\partial}{\partial t}\left\{\iint_{S_H} \epsilon(z)E_x(z,t)\hat{x}\cdot \vec{\mathrm{d}s}_H + \iint_{S_H} \sigma_c(z)E_x(z,t)\hat{x}\cdot \vec{\mathrm{d}s}_H\right\} \tag{4.86}$$

where σ_c is the conductivity of the medium.

The contour used to derive the discrete equivalent of the magnetic field is shown in Fig. 4.13. A local coordinate system (ψ, ζ) is established, with its origin at the center of the contour. Any point (y', z') within or on the contour C_H (or, equivalently, any point on the surface S_H) can be specified by its local coordinates, much as in Eq. (4.65):

$$y' = y + \psi, \qquad z' = z + \zeta \tag{4.87}$$

for $-\Delta\ell/2 \le \psi, \zeta \le \Delta\ell/2$. If a counterclockwise reference direction is assumed such that the normal to the surface S_H is in the direction of the electric field $\vec{E}(z,\ t)$, the circulation of the magnetic field around C_H is given exactly by the simple expression:

$$\oint_{C_H} H_y(z+\zeta,t)\hat{y}\cdot \vec{\mathrm{d}\ell}_H = \int_{-\Delta\ell/2}^{\Delta\ell/2} H_y(z-\Delta\ell/2,t)\hat{y}\cdot\hat{y}\mathrm{d}\psi - \int_{-\Delta\ell/2}^{\Delta\ell/2} H_y(z+\Delta\ell/2,t)\hat{y}\cdot\hat{y}\mathrm{d}\psi \tag{4.88}$$

or

$$\oint_{C_H} H_y(z+\zeta,t)\hat{y}\cdot \vec{\mathrm{d}\ell}_H = [H_y(z-\Delta\ell/2,t) - H_y(z+\Delta\ell/2,t)]\Delta\ell \tag{4.89}$$

The rate of change of the electric flux through the surface S_H can be evaluated only approximately because the exact dependence of the electric flux

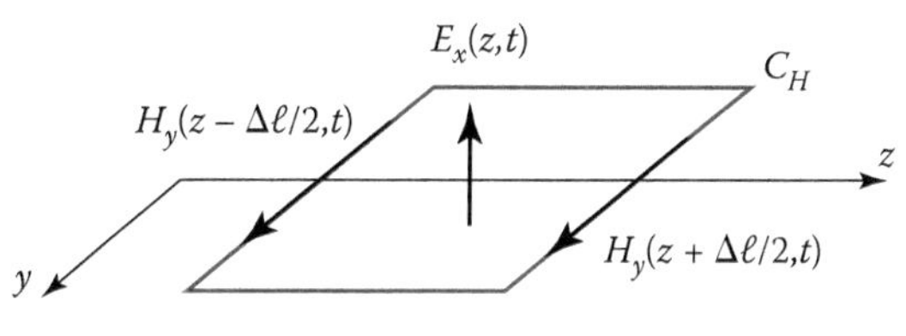

Fig. 4.13 Contour for magnetic circulation and electric flux.

density on the coordinate z is unknown a priori. First, write the flux integrals:

$$\iint_{S_H} \epsilon(z+\zeta)E_x(z+\zeta,t)\hat{x}\cdot\vec{\mathrm{d}s}_H = \int_{-\Delta\ell/2}^{\Delta\ell/2}\int_{-\Delta\ell/2}^{\Delta\ell/2} \epsilon(z+\zeta)E_x(z+\zeta,t)\hat{x}\cdot\hat{x}\mathrm{d}\psi\mathrm{d}\zeta \tag{4.90}$$

$$\iint_{S_H} \sigma_c(z+\zeta)E_x(z+\zeta,t)\hat{x}\cdot\vec{\mathrm{d}s}_H = \int_{-\Delta\ell/2}^{\Delta\ell/2}\int_{-\Delta\ell/2}^{\Delta\ell/2} \sigma_c(z+\zeta)E_x(z+\zeta,t)\hat{x}\cdot\hat{x}\mathrm{d}\psi\mathrm{d}\zeta \tag{4.91}$$

The integrands are not functions of the local y coordinate ψ and, thus, the surface integrals in Eqs. (4.90) and (4.91) reduce to line integrals:

$$\begin{aligned}\iint_{S_H} \epsilon(z+\zeta)E_x(z+\zeta,t)\hat{x}\cdot\vec{\mathrm{d}s}_H &= \Delta\ell\int_{-\Delta\ell/2}^{\Delta\ell/2} \epsilon(z+\zeta)E_x(z+\zeta,t)\mathrm{d}\zeta \\ \iint_{S_H} \sigma_c(z+\zeta)E_x(z+\zeta,t)\hat{x}\cdot\vec{\mathrm{d}s}_H &= \Delta\ell\int_{-\Delta\ell/2}^{\Delta\ell/2} \sigma_c(z+\zeta)E_x(z+\zeta,t)\mathrm{d}\zeta\end{aligned} \tag{4.92}$$

Now, let the contour width $\Delta\ell$ be small enough so that the electric field $E_x(z+\zeta, t)$ within the contour is approximately constant and equal to the value at the contour's center, $E_x(z, t)$. This assumption yields approximate expressions for the flux integrals:

$$\begin{aligned}\iint_{S_H} \epsilon(z+\zeta)E_x(z+\zeta,t)\hat{x}\cdot\vec{\mathrm{d}s}_H &= \Delta\ell E_x(z,t)\int_{-\Delta\ell/2}^{\Delta\ell/2} \epsilon(z+\zeta)\mathrm{d}\zeta \\ \iint_{S_H} \sigma_c(z+\zeta)E_x(z+\zeta,t)\hat{x}\cdot\vec{\mathrm{d}s}_H &= \Delta\ell E_x(z,t)\int_{-\Delta\ell/2}^{\Delta\ell/2} \sigma_c(z+\zeta)\mathrm{d}\zeta\end{aligned} \tag{4.93}$$

or, using the average permittivity and conductivity [defined similarly to μ_{avg} in Eq. (4.72)]:

$$\begin{aligned}\iint_{S_H} \epsilon(z+\zeta)E_x(z+\zeta,t)\hat{x}\cdot\vec{\mathrm{d}s}_H &= (\Delta\ell)^2\epsilon_{\text{avg}}(z)E_x(z,t) \\ \iint_{S_H} \sigma(z+\zeta)E_x(z+\zeta,t)\hat{x}\cdot\vec{\mathrm{d}s}_H &= (\Delta\ell)^2\sigma_{\text{avg}}(z)E_x(z,t)\end{aligned} \tag{4.94}$$

Maxwell's second equation can now be replaced by an approximate equation:

$$\begin{aligned}&[H_y(z-\Delta\ell/2,t) - H_y(z+\Delta\ell/2,t)]\Delta\ell \\ &\approx \frac{\partial}{\partial t}\{[\epsilon_{\text{avg}}(z)E_x(z,t)](\Delta\ell)^2\} + (\Delta\ell)^2\sigma_{\text{avg}}(z)E_x(z,t)\end{aligned} \tag{4.95}$$

and, if we take advantage of the stationary property of the medium,

$$H_y(z-\Delta\ell/2,t)-H_y(z+\Delta\ell/2,t) \approx \epsilon_{\text{avg}}(z)\Delta\ell\frac{\partial}{\partial t}\{E_x(z,t)\}+\Delta\ell\sigma_{\text{avg}}(z)E_x(z,t) \tag{4.96}$$

Equation (4.96) can be rewritten as

$$\frac{\partial}{\partial t}\{E_x(z,t)\} \approx \frac{1}{\epsilon_{\text{avg}}(z)\Delta\ell}[H_y(z-\Delta\ell/2,t)-H_y(z+\Delta\ell/2,t)] - \frac{\sigma_{\text{avg}}(z)}{\epsilon_{\text{avg}}(z)}E_x(z,t) \tag{4.97}$$

Equation (4.97) is integrated with respect to time to get an approximate equation for the electric field at time t:

$$E_x(z,t) \approx \frac{1}{\epsilon_{\text{avg}}(z)\Delta\ell}\left[\int_0^t H_y(z-\Delta\ell/2,\tau)\mathrm{d}\tau - \int_0^{t-\Delta t} H_y(z+\Delta\ell/2,\tau)\mathrm{d}\tau\right] - \frac{\sigma_{\text{avg}}(z)}{\epsilon_{\text{avg}}(z)}\int_0^t E_x(z,\tau)\mathrm{d}\tau \tag{4.98}$$

Equation (4.98) also holds at time $t-\Delta t$:

$$E_x(z,t-\Delta t) \approx \frac{1}{\epsilon_{\text{avg}}(z)\Delta\ell}\left[\int_0^{t-\Delta t} H_y(z-\Delta\ell/2,\tau)\mathrm{d}\tau. - \int_{t-\Delta t}^{t} H_y(z+\Delta\ell/2,\tau)\mathrm{d}\tau\right] - \frac{\sigma_{\text{avg}}(z)}{\epsilon_{\text{avg}}(z)}\int_0^{t-\Delta t} E_x(z,\tau)\mathrm{d}\tau \tag{4.99}$$

Subtracting Eq. (4.99) from Eq. (4.98) and rearranging slightly give

$$E_x(z,t) \approx E_x(z,t-\Delta t) + \frac{1}{\epsilon_{\text{avg}}(z)\Delta\ell}\left[\int_{t-\Delta t}^t H_y(z-\Delta\ell/2,\tau)\mathrm{d}\tau - \int_{t-\Delta t}^{t} H_y(z+\Delta\ell/2,\tau)\mathrm{d}\tau\right] - \frac{\sigma_{\text{avg}}(z)}{\epsilon_{\text{avg}}(z)}\int_{t-\Delta t}^{t} E_x(z,\tau)\mathrm{d}\tau \tag{4.100}$$

Now, if the interval Δt is small enough such that the magnetic field is approximately constant within Δt and equal to the value at the center of the time step, then,

$$\begin{aligned} H_y(z+\Delta\ell/2,\tau) &\approx H_y(z+\Delta\ell/2,t-\Delta t/2) \\ H_y(z-\Delta\ell/2,\tau) &\approx H_y(z-\Delta\ell/2,t-\Delta t/2) \end{aligned} \tag{4.101}$$

for $t - \Delta t/2 \leq \tau \leq t$. Thus, Eq. (4.100) becomes

$$\begin{aligned} E_x(z,t) \approx E_x(z,t-\Delta t) + \frac{\Delta t}{\epsilon_{\text{avg}}(z)\Delta\ell}[H_y(z-\Delta\ell/2, t-\Delta t/2) \\ - H_y(z+\Delta\ell/2, t-\Delta t/2)] - \frac{\sigma_{\text{avg}}(z)}{\epsilon_{\text{avg}}(z)}\int_{t-\Delta t}^{t} E_x(z,\tau)\mathrm{d}\tau \end{aligned} \tag{4.102}$$

The last term in Eq. (4.102) can be multiplied and divided by Δt and a new quantity defined that corresponds to the average electric field between times $t - \Delta t$ and t. The average value of E_x cannot be evaluated exactly because the variation of the electric field with time is not known. We will assume a linear variation within the interval Δt:

$$[E_x(z,t)]_{\text{avg}(\Delta t)} = \frac{1}{\Delta t}\int_{t-\Delta t}^{t} E_x(z,\tau)\mathrm{d}\tau \approx \frac{1}{2}[E_x(z,t-\Delta t) + E_x(z,t)] \tag{4.103}$$

When the fact that

$$\frac{1}{\epsilon} = \frac{1}{\epsilon_r}c\eta_0 \tag{4.104}$$

is used and the grid propagation velocity is introduced, Eq. (4.103) becomes

$$\begin{aligned} E_x(z,t) \approx E_x(z,t-\Delta t) + \frac{\eta_0 c}{u_{\text{grid}}\epsilon_{r\text{avg}}(z)}[H_y(z-\Delta\ell/2, t-\Delta t/2) \\ - H_y(z+\Delta\ell/2, t-\Delta t/2)] - \frac{\eta_0 c}{u_{\text{grid}}\epsilon_{r\text{avg}}(z)}\frac{\Delta\ell\sigma_{\text{avg}}(z)}{2} \\ \times [E_x(z,t-\Delta t) + E_x(z,t)] \end{aligned} \tag{4.105}$$

Finally, the *electric field update equation* can be written as

$$\begin{aligned} E_x(z,t) \approx E_x(z,t-\Delta t)\left[\frac{1 - \mathcal{A}\Delta\ell\sigma_{\text{avg}}(z)/2}{1 + \mathcal{A}\Delta\ell\sigma_{\text{avg}}(z)/2}\right] \\ + [H_y(z-\Delta\ell/2, t-\Delta t/2) \\ - H_y(z+\Delta\ell/2, t-\Delta t/2)]\left[\frac{\mathcal{A}}{1 + \mathcal{A}\Delta\ell\sigma_{\text{avg}}(z)/2}\right] \end{aligned} \tag{4.106}$$

where the coefficient $\mathcal{A}$ has been defined for convenience:

$$\mathcal{A} \equiv \frac{\eta_0 c}{u_{\text{grid}}\epsilon_{r\text{avg}}(z)} \tag{4.107}$$

Equation (4.106) holds for conducting media. If the media are nonconducting, the terms involving σ drop out. Thus,

$$E_x(z,t) \approx E_x(z,t-\Delta t) + \frac{\eta_0 c}{u_{\text{grid}}\epsilon_{\text{ravg}}(z)}[H_y(z-\Delta\ell/2, t-\Delta t/2) - H_y(z+\Delta\ell/2, t-\Delta t/2)] \tag{4.108}$$

Furthermore, if the medium is free space, then $\epsilon_r = 1$, and

$$E_x(z,t) \approx E_x(z,t-\Delta t) + \frac{\eta_0 c}{u_{\text{grid}}}[H_y(z-\Delta\ell/2, t-\Delta t/2) - H_y(z+\Delta\ell/2, t-\Delta t/2)] \tag{4.109}$$

As in the case of the magnetic field update equation [Eq. (4.84)], Eq. (4.109) indicates that the electric field and the magnetic field should be evaluated at points shifted spatially by $\Delta\ell/2$ and at instants separated in time by $\Delta t/2$.

4.4.3 One-Dimensional Grid

The electric and magnetic field update equations have been derived in local coordinate systems, with origins at the centers of the magnetic and electric contours, respectively. These equations now need to be converted to a global coordinate system, that is, to a common grid of equidistant sampling points along the z axis. Let us assume that our domain is a line segment of length L. The fields are sampled using a spatial step, $\Delta\ell = L/(N_z - 1)$. The locations of electric and magnetic field sampling points are interleaved, as shown in Fig. 4.14 for $N_z = 5$. The first and the last spatial sampling points form the grid edges. The spatial edge samples can be either electric field samples or magnetic field samples. Although the selection of the field for the grid edges makes no difference in principle, we will use electric field samples for the grid edges.

Note that, in general, we would evaluate the fields at the edges first because a known incident field enters the grid via edge nodes. This implies that we will start our updates in time for the same field (in our case, the electric field) that has been selected for the grid edges. In other words, the same field would be present at the spatial and the temporal edges.

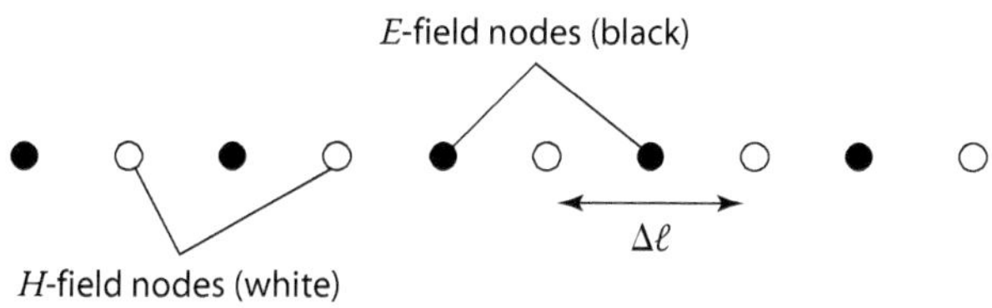

Fig. 4.14 E-field (black) and H-field (white) nodes forming a one-dimensional grid.

Electric field update equations can now be "converted" to electric field grid equations by replacing the variables z and t with the grid coordinates of the electric field spatial and temporal sampling points:

$$z \to m\Delta\ell, \qquad m = 1, 2, \ldots, N_z$$
$$t \to n\Delta t, \qquad n = 1, 2, \ldots, N_t$$

The electric field update equation for nonconductive media [Eq. (4.108)] becomes

$$\begin{aligned} E_x(m\Delta\ell,\, n\Delta t) \approx\; & E_x(m\Delta\ell,\, n\Delta t - \Delta t) \\ & + \frac{c\eta_0}{u_{\text{grid}}\epsilon_{r\text{avg}}(m\Delta\ell)}[H_y(m\Delta\ell - \Delta\ell/2,\, n\Delta t - \Delta t/2) \\ & - H_y(m\Delta\ell + \Delta\ell/2,\, n\Delta t - \Delta t/2)] \end{aligned} \tag{4.110}$$

Equation (4.110) can be written more concisely by making a change of notation and thereby omitting the common $\Delta\ell$ and Δt terms. A superscript on a field quantity will define the index of the temporal sampling point; the argument m denotes the grid index:

$$\begin{aligned} E_x^n(m) \approx\; & E_x^{n-1}(m) \\ & + \frac{c\eta_0}{u_{\text{grid}}\epsilon_{r\text{avg}}(n)}\left[H_y^{n-1/2}(m - 1/2) - H_y^{n-1/2}(m + 1/2)\right] \end{aligned} \tag{4.111}$$

The electric field grid equation for conductive media, using the same notation, is

$$\begin{aligned} E_x^n(m) \approx\; & \left[\frac{1 - \mathcal{A}\Delta\ell\sigma_{\text{avg}}(m)/2}{1 + \mathcal{A}\Delta\ell\sigma_{\text{avg}}(m)/2}\right]E_x^{n-1}(m) + \left[\frac{\mathcal{A}}{1 + \mathcal{A}\Delta\ell\sigma_{\text{avg}}(m)/2}\right] \\ & \times \left[H_y^{n-1/2}(m - 1/2) - H_y^{n-1/2}(m + 1/2)\right] \end{aligned} \tag{4.112}$$

where $\mathcal{A}$ is given by Eq. (4.107) with z replaced by m. Finally, the free-space electric field grid equation is

$$E_x^n(m) \approx E_x^{n-1}(m) + \frac{c\eta_0}{u_{\text{grid}}}[H_y^{n-1/2}(m - 1/2) - H_y^{n-1/2}(m + 1/2)] \tag{4.113}$$

Similarly, magnetic field update equations can be converted to magnetic field grid equations by replacing the variables z and t with the grid coordinates of the magnetic field spatial and temporal sampling points:

$$z \to (m + 1/2)\Delta\ell, \qquad m = 1, 2, \ldots, N_z - 1$$
$$t \to (n + 1/2)\Delta t, \qquad n = 1, 2, \ldots, N_t - 1$$

The magnetic field grid equation, using the same notation as for the electric field grid equation, is

$$H_y^{n+1/2}(m+1/2) \approx H_y^{n-1/2}(m+1/2) - \frac{c}{\eta_0 u_{\text{grid}} \mu_{r\text{avg}}(m+1/2)} \left[E_x^n(m+1) - E_x^n(m)\right] \tag{4.114}$$

or, for nonmagnetic media,

$$H_y^{n+1/2}(m+1/2) \approx H_y^{n-1/2}(m+1/2) - \frac{c}{\eta_0 u_{\text{grid}}} \left[E_x^n(m+1) - E_x^n(m)\right] \tag{4.115}$$

Note that the electric and magnetic field grid equations have the following general form:

$$E_x^{\text{new}} = C_{E_1} E_x^{\text{old}} + C_{E_2} \nabla H_y^{\text{old}} \tag{4.116}$$

$$H_y^{\text{new}} = C_{H_1} H_y^{\text{old}} + C_{H_2} \nabla E_x^{\text{old}} \tag{4.117}$$

where coefficients $C_{E,H}$ are real constants that depend on the media properties and the velocity ratio c/u_{grid}, and where the ∇ operator represents the spatial derivative (gradient). This general form of the grid equations can be interpreted as follows: the new value of E (or H) at a grid node is equal to a weighted sum of the old value of E (or H) at the same node and the spatial variation of the old H (or E) between the two nearest neighbor nodes.

4.4.4 Grid Termination

The grid equations derived in the preceding section describe how waves propagate along an infinite grid. In all practical applications, the grid will have to be terminated to a finite number of nodes. Therefore, a method of grid termination is required. Note that the update equations derived in the last section are valid for all nodes except those on the grid edges. The reason is that the grid *edge nodes* have only one neighbor instead of two like the *internal nodes.* Thus, the edge nodes with a single neighbor have update equations different from the equations derived for the internal grid nodes. The question is: How do we get these equations in a systematic manner?

The answer to this question depends on what we wish to accomplish at the grid edges. One objective is to absorb a wave incident from inside the grid onto a boundary edge such that there is no reflection back into the grid (i.e., no reflected wave from the grid edge). This would allow us to truncate the grid at some finite length yet yield a solution for an infinite grid. This is an *absorbing boundary condition* (ABC), as discussed in Chapter 3, and there are numerous references discussing various implementations and their limitations [12–14]. Here, however, we examine a different approach; one that is intuitive and can be implemented to any desired degree of accuracy.

We start by making only one demand on the grid termination condition: that it is "transparent" to the grid, meaning that, ideally, the solution within the grid should be the solution that one would obtain with an infinitely large grid. However, if the grid were infinitely large, the grid edges would be at infinity and the equations for the grid edge nodes would be irrelevant because the scattered field would take an infinitely long time to reach the grid edges; that is, the edge equations would never be needed. The *transparent grid termination* (TGT) requirement is conceptually very simple and, fortunately, straightforward to implement. To achieve a TGT, we will use the concept of a multiport (which, in one dimension, reduces to a two-port), introduced in Sec. 4.3.

Consider a node on the edge of the grid to be an output port of a linear two-port. It does not matter, in principle, whether the edge node is an E-field or H-field node. Consistent with our previous assumption, let the edge node for the two-port model be an E-field node. The input of this two-port will be the nearest node "of the same kind." Because we have assumed an E-field node for the output port, the input port will be the nearest E-field node inside the grid. Note that we could also have selected an E-field node as the output port and the nearest H-field node as the input port. However, the selection of the same kind of node for both input and output ports has an advantage in that the results obtained in this manner can be used to solve the wave equation (a second-order partial differential equation) that has only one field as the variable and the grid nodes of only one type. Figure 4.15 shows the two ports for the two grid edge nodes.

The fields at the output ports $E(0, t)$ and $E(L, t)$ where L is the spatial domain length can be expressed as convolutions of the fields at the input ports $E(\Delta\ell, t)$ and $E(L-\Delta\ell, t)$ with the two-port impulse response $h(t)$:

$$E_x(0,t) = E_x(\Delta\ell, t) * h(t) = \int_0^t E_x(\Delta\ell, \tau) h(t-\tau) \mathrm{d}\tau \tag{4.118}$$

$$E_x(L,t) = E_x(L-\Delta\ell, t) * h(t) = \int_0^t E_x(L-\Delta\ell, \tau) h(t-\tau) \mathrm{d}\tau \tag{4.119}$$

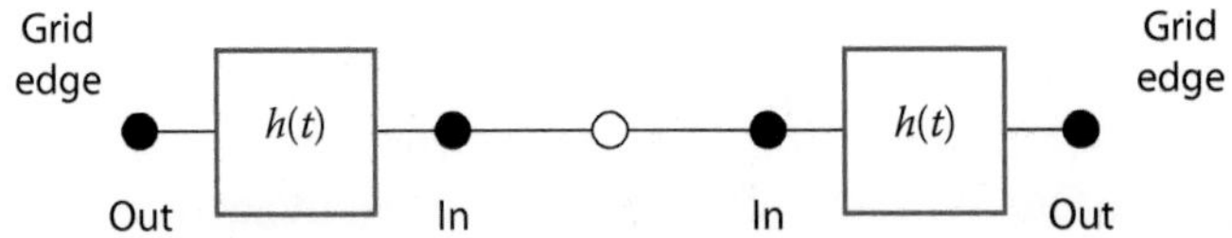

Fig. 4.15 One-dimensional grid terminations represented as two-ports.

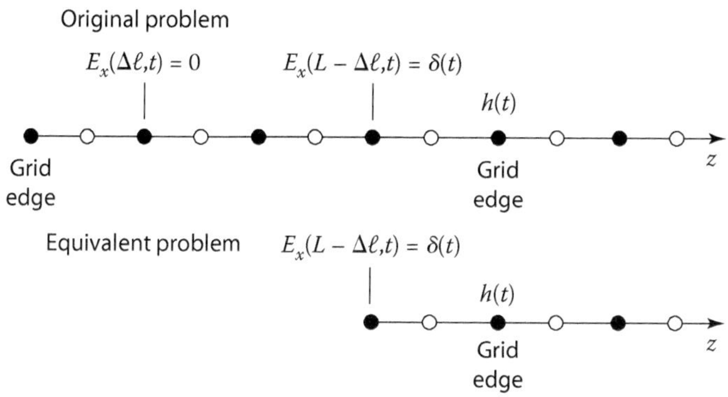

Fig. 4.16 Determining the grid impulse response: original and equivalent problems.

The impulse responses for both ports of the two ports are identical by symmetry. The discretized forms of these equations, involving samples of the fields at $t = n\Delta t$, are themselves discretized convolutions:

$$E_x^n(0) = \sum_{p=0}^{n} E_x^p(1)h^{n-p} \tag{4.120}$$

$$E_x^n(N_z) = \sum_{p=0}^{n} E_x^p(N_z - 1)h^{n-p} \tag{4.121}$$

The convolution equations express the fields at the edges as weighted sums of the time histories of the fields "just inside" the grid. In that respect, they are equations of the same type as the equations for the non-edge grid nodes [Eqs. (4.112) and (4.114)] except that they involve, in general, summations with more terms. However, it will be shown that the impulse response $h(t)$ is a function that rapidly converges to zero. This fact can be used to reduce the number of terms in the convolution sum. The impulse response $h(t)$, or more accurately its sampled form $h^n = h(n\Delta t)$, needs to be determined only once for a selected grid velocity $u_{\text{grid}} = \Delta\ell/\Delta t$.

The issue remains regarding how to determine the impulse response. The procedure is illustrated in Fig. 4.16. The discretized impulse response h^n will be determined using the discrete equivalent of the Dirac delta function that we will denote as δ^n. Because there are two grid edges and two input ports, the input at one port will be set to zero (grounded), and an excitation δ^n will be applied to the second port. Because, for one dimension, there is only one h^n to determine, we set $E_x(\Delta\ell,t)$ to zero and apply the

Dirac delta function as the field $E_x(L-\Delta\ell, t)$:

$$E_x^n(N_z - 1) = \delta^n \tag{4.122}$$

$$E_x^n(1) = 0 \tag{4.123}$$

Note that the impulse response h_n is observed at $z = L$ and that the grid extends, theoretically, past the observation point to infinity. Because $\delta^n = 0$ for $n > 0$, the grid to the left of the $L-\Delta\ell$ node is effectively isolated from the observation point at L. Thus, to find h^n, we need only consider the portion of the grid to the right of the $L-\Delta\ell$ node, as shown in Fig. 4.16.

The impulse response must be determined as if the grid were not terminated at $z = L$. If the grid were terminated, the TGT condition would be applied, but this is unknown and is exactly what we are trying to find. A grid extending to infinity does not require a termination and thus circumvents the termination problem. Although it is not possible in practice to extend the grid to infinity, the grid can be made large enough such that any "reflections" off the new grid edge would arrive after the impulse response has converged to a small selected value that we consider negligible. The time it takes the impulse to "propagate" to this new grid edge and back to the observation point is

$$T_d = \frac{2D}{u_{\text{grid}}} = \frac{2N_D\Delta\ell}{\Delta\ell/\Delta t} = 2N_D\Delta t \tag{4.124}$$

where D is the distance between the observation point and the new grid's edge and N_D is the number of nodes between the observation point and the new grid's edge. The duration of the impulse response h^n that will not be "corrupted" by the reflections will be $2N_D$.

Finally, it is assumed that the space outside the TGT boundary is free space and, consequently, that the grid equations for the electric and magnetic fields will be the free-space equations:

$$E_x^n(m) = E_x^{n-1}(m) + \frac{\eta_0 c}{u_{\text{grid}}}\left[H_y^{n-1/2}(m-1/2) - H_y^{n-1/2}(m+1/2)\right] \tag{4.125}$$

and

$$H_y^{n+1/2}(m+1/2) = H_y^{n-1/2}(m+1/2) + \frac{c}{\eta_0 u_{\text{grid}}}\left[E_x^n(m) - E_x^n(m+1)\right] \tag{4.126}$$

Note that the order of the terms in the brackets for the magnetic field equation has been reversed, so that both have the same sign (+) in front of the brackets. The preceding equations can also be written in a more computationally efficient form to reduce the number of multiplications that would need to be done at each time step. Multiplying Eq. (4.125) by η_0, we get

$$\eta_0 H_y^{n+1/2}(m+1/2) = \eta_0 H_y^{n-1/2}(m+1/2) + \frac{c}{u_{\text{grid}}}\left[E_x^n(m) - E_x^n(m+1)\right] \tag{4.127}$$

An auxiliary field quantity h_y may be introduced:

$$h_y \equiv \eta_0 H_y \tag{4.128}$$

and the electric and magnetic field grid equations rewritten as

$$E_x^n(m) = E_x^{n-1}(m) + \frac{c}{u_{\text{grid}}}\left[h_y^{n-1/2}(m-1/2) - h_y^{n-1/2}(m+1/2)\right] \tag{4.129}$$

$$h_y^{n+1/2}(m+1/2) = h_y^{n-1/2}(m+1/2) + \frac{c}{u_{\text{grid}}}\left[E_x^n(m) - E_x^n(m+1)\right] \tag{4.130}$$

The equations for the electric field (E_x) and for the magnetic field multiplied by η_0 ($\eta_0 H_y$) have identical forms, involving only one parameter: the velocity ratio c/u_{grid}. The boundary impulse response h_n, to be obtained using these equations, will thus be valid for a selected velocity ratio or, equivalently, for the selected grid velocity $u_{\text{grid}} = \Delta\ell/\Delta t$. The boundary impulse response h_n, obtained using Eqs. (4.129) and (4.130) for $c/u_{\text{grid}} = 1$, is shown in Fig. 4.17. Note that the amplitude is not a continuous curve; it has values only at the sample points $n\Delta t$.

The boundary impulse response for one dimension is simply

$$h^n = \delta^1 \tag{4.131}$$

The impulse response is nonzero only for $n = 1$, where it has the value of 1. This behavior is identical to a Dirac delta impulse delayed by Δt and may also be written as

$$h(t) = \delta(t - \Delta t) \tag{4.132}$$

We can now write the equations for the grid edges as follows:

$$E_x^n(0) = \sum_{p=0}^{n} E_x^p(1)h^{n-p} = E_x^{n-1}(1)h^1 = E_x^{n-1}(1) \tag{4.133}$$

$$E_x^n(N_z) = \sum_{p=0}^{n} E_x^p(N_z-1)h^{n-p} = E_x^{n-1}(N_z-1)h^1 = E_x^{n-1}(N_z-1) \tag{4.134}$$

The fields at the grid edges are updated by simply taking the previous values of their nearest neighbors inside the grid.

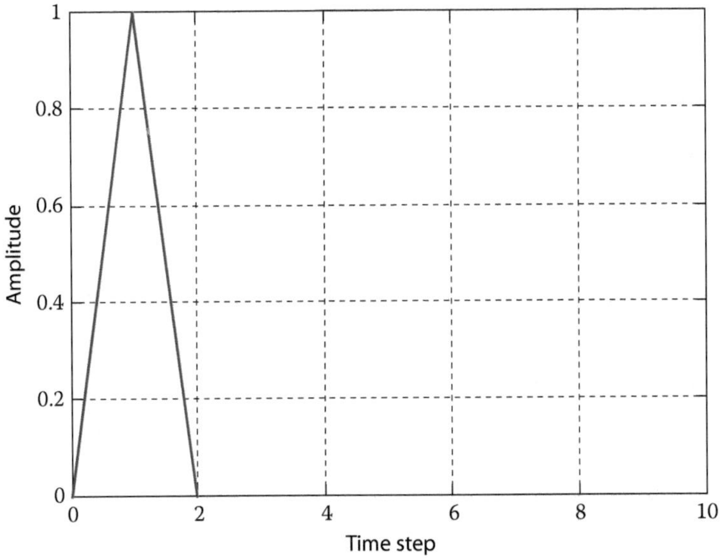

Fig. 4.17 Boundary impulse response.

4.4.5 Incident Field Injection into the Grid

The source for scattering problems is a known incident field. An incident field can be "injected" into the grid at the grid's edge nodes. Assuming that the grid's edge nodes are electric field nodes, we can consider the fields at the edges to be sums of the incident and the scattered electric fields:

$$\begin{aligned} E_x^n(0) &= E_{ix}^n(0) + E_{sx}^n(0) \\ E_x^n(N_z) &= E_{ix}^n(N_z) + E_{sx}^n(N_z) \end{aligned} \tag{4.135}$$

The "edge"-incident field is known, but the "edge"-scattered field is not. Note that an incident field propagating in the positive z direction is "injected" into the grid at the edge node at $z = 0$, whereas an incident field propagating in the negative z direction is "injected" into the grid at the edge node at $z = L$. The unknown scattered field at an "edge" node is, as shown in the preceding section, obtained from the scattered field at the node "just inside":

$$\begin{aligned} E_{sx}^n(0) &= E_{sx}^{n-1}(1) \\ E_{sx}^n(N_z) &= E_{sx}^{n-1}(N_{z-1}) \end{aligned} \tag{4.136}$$

However, the grid equations for the electric fields "just inside" give the total electric field, not just the scattered field:

$$E_x^n(1) = E_x^{n-1}(1) + \eta_0 \frac{c}{u_{\text{grid}}} [H_y^{n-1/2}(1/2) - H_y^{n-1/2}(3/2)] \tag{4.137}$$

$$E_x^n(N_z-1) = E_x^{n-1}(N_z-1) + \eta_0 \frac{c}{u_{\text{grid}}} [H_y^{n-1/2}(N_z-3/2) - H_y^{n-1/2}(N_z-1/2)] \tag{4.138}$$

The fields "just inside" can be written in terms of the incident and the scattered fields, much as with the "edge" fields:

$$\begin{aligned} E_x^n(1) &= E_{ix}^n(1) + E_{sx}^n(1) \\ E_x^n(N_z-1) &= E_{ix}^n(N_z-1) + E_{sx}^n(N_z-1) \end{aligned} \tag{4.139}$$

which means that we can find the "just inside" scattered fields by subtracting the incident fields from the total fields:

$$\begin{aligned} E_{sx}^n(1) &= E_x^n(1) - E_{ix}^n(1) \\ E_{sx}^n(N_z-1) &= E_x^n(N_z-1) - E_{ix}^n(N_z-1) \end{aligned} \tag{4.140}$$

Unlike the incident field at an edge node, the incident field at a "just inside" node is not known a priori. This is because the known incident field at an "edge" node has to "propagate" inward, through a single layer of the grid, to reach a "just inside" node. However, the incident field at a "just inside" node can be calculated using the reciprocity property of the two-ports we have used to "connect" an "edge" node to a node "just inside" (Fig. 4.16). The impulse responses $h(t)$ of the two-ports will not change if the input and output ports are reversed; that is, if we consider the "edge nodes" as the input ports and nodes "just inside" as the output ports. This means that, in one dimension, the incident field at a "just inside" node is simply a delayed (by Δt) replica of the known "edge" incident field:

$$\begin{aligned} E_{xi}^n(1) &= E_{xi}^{n-1}(0) \\ E_{xi}^n(N_z-1) &= E_{xi}^{n-1}(N_z) \end{aligned} \tag{4.141}$$

Combining the preceding information allows us to calculate the scattered fields at the "just inside" nodes:

$$\begin{aligned} E_{sx}^n(1) &= E_x^n(1) - E_{ix}^{n-1}(0) \\ E_{sx}^n(N_z-1) &= E_x^n(N_z-1) - E_{ix}^{n-1}(N_z) \end{aligned} \tag{4.142}$$

One might correctly conclude from Eqs. (4.135) through (4.142) that the order of operations is critical when "injecting" an incident field into the grid. The order is illustrated by the flowchart in Fig. 4.18. Note that each loop pass corresponds to one time step Δt.

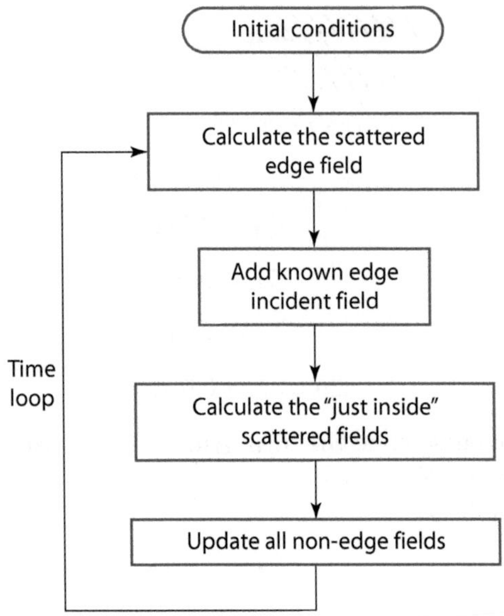

Fig. 4.18 Ordered procedure for injecting a signal into the grid.

Example 4.2: Scattering from a Conducting Half-Space

Given that this problem deals with only one dimension, the scatterers necessarily are infinitely large in extent, and the incident and the scattered fields experience no spreading as they propagate. The RCS in this case is RCS per unit area (m^2/m^2), a dimensionless ratio,

$$\sigma_{1D}(\omega) = \frac{|E_s(\omega)|^2}{|E_i(\omega)|^2} = \frac{|\mathcal{F}\{E_s(t)\}|^2}{|\mathcal{F}\{E_i(t)\}|^2} = |\Gamma(\omega)|^2 \tag{4.143}$$

It is simply the magnitude squared of the voltage reflection coefficient Γ, which we know has a value of 1 (=0 dB) for a PEC. The RCS per unit area, as given by Eq. (4.143) can be determined from the Fourier transforms of the time-domain incident and scattered electric fields.

The scatterer geometry is shown in Fig. 4.19. Note that the thickness is immaterial in the case of a PEC scatterer. To begin with, we choose a Gaussian pulse and assume that the following parameters are required:

$$B_{\text{eff}} = 1\,\text{GHz}$$
$$R_{\text{dyn}} = 120\,\text{dB}$$
$$S/N = 90\,\text{dB}$$
$$\Delta f = 1\,\text{MHz}$$

(Continued)

Example 4.2: Scattering from a Conducting Half-Space (Continued)

Proceeding as outlined in Fig. 4.18, the following values are calculated:

$$T_{\text{eff}} = 0.5\,\text{ns}$$

$$T_0 = 8\,\text{ns}$$

$$f_{\text{max}} = 1\,\text{GHz}$$

$$\Delta t_{\text{Nyquist}} = 0.5\,\text{ns}$$

The ratio of the phase velocities is $u_{\text{max}}/u_{\text{min}} = 1$ so that the oversampling factor should satisfy $K \geq 1$. The smallest allowable value gives $\Delta t = 0.5$ ns. Continuing with the parameter selection,

$$\Delta \ell = c\Delta t = 0.15\,\text{m}$$

$$T_{\text{max}} = 1.0\,\mu\text{s}$$

$$N = 2.0 \times 10^3$$

The last set of calculations dictate that 2000 time steps should be used, with a truncated Gaussian pulse of length 16 time steps. The pulse is shown in Fig. 4.20. To accelerate the calculation of the Fourier transform, the so-called fast Fourier transform algorithm (FFT) is used [9]. This algorithm requires that the number of time samples is an integer power of 2 and, therefore, we choose 2048 time steps, which is close to the desired 2000.

The incident Gaussian pulse is first injected into the grid free of any scatterers (i.e., a free-space grid). The scattered field for such a grid would

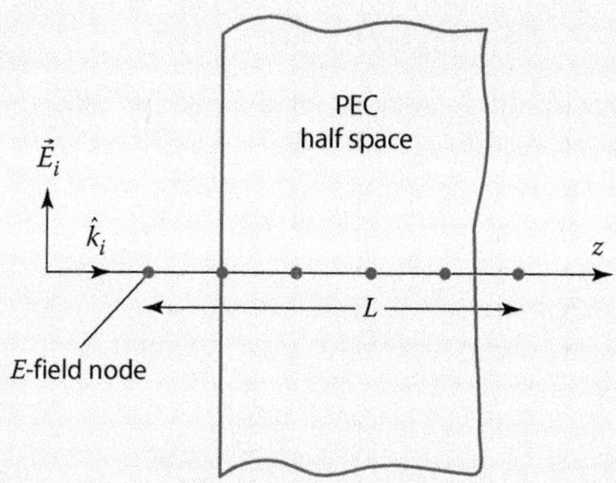

Fig. 4.19 One-dimensional semi-infinite PEC scatterer.

(Continued)

Example 4.2: Scattering from a Conducting Half-Space *(Continued)*

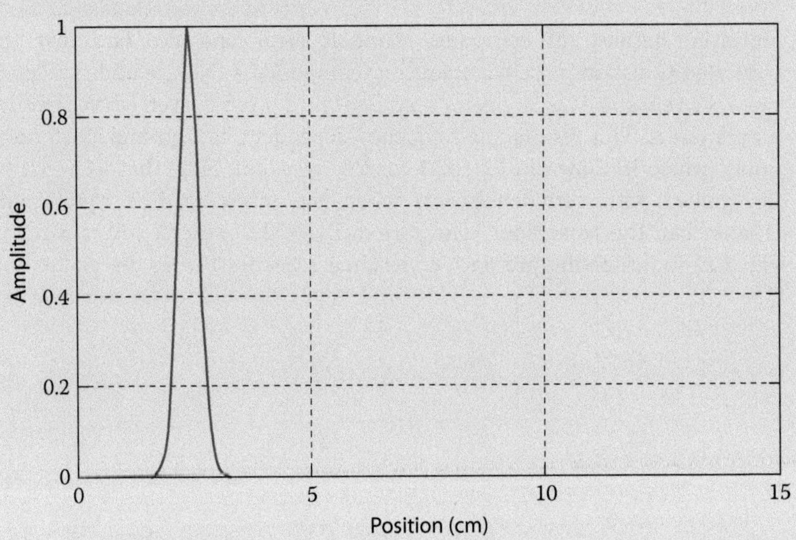

Fig. 4.20 Truncated and normalized Gaussian pulse for Example 4.2.

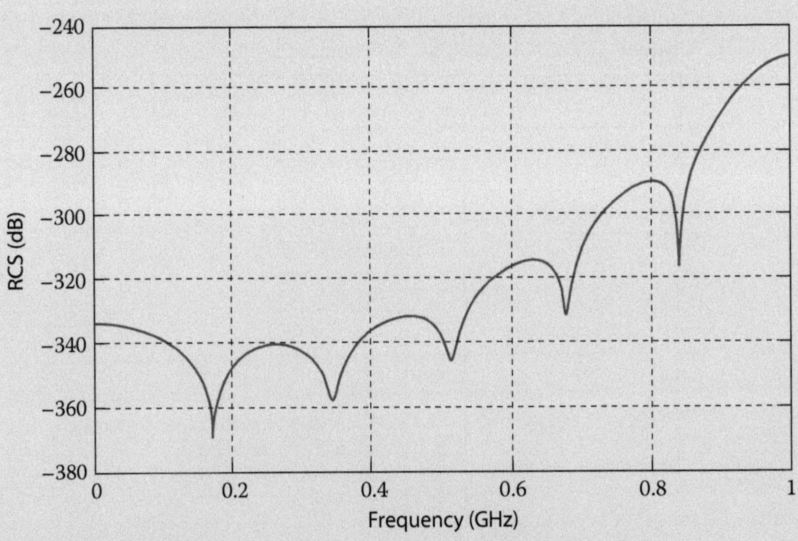

Fig. 4.21 RCS due to grid noise as a function of frequency.

(Continued)

Example 4.2: Scattering from a Conducting Half-Space (*Continued*)

ideally be zero and, consequently, the RCS per unit area would also be zero. However, because of computer roundoff error and the fact that the truncated Gaussian pulse has frequency components that are undersampled, there is a very small but nonzero "scattered" field present, even when there are no scatterers. This results in a frequency-dependent background "RCS noise floor," which is shown in Fig. 4.21 for this problem. Note that RCS can be determined with confidence only when the target RCS is significantly greater than the noise floor. The periodicity of the minima and maxima in Fig. 4.21 indicates the presence of residual reflections from the grid edges. The maxima occur at frequencies for which the reflections from the two edges add in phase, whereas the minima occur at the frequencies where the residual reflections cancel. The periods of cycles are functions of the size of the computational domain L. The larger the domain length, the more closely spaced are the minima and maxima, and vice versa.

Next, the RCS per unit area of the PEC target can be computed from Eq. (4.143). The RCS per unit area vs frequency is plotted in Fig. 4.22. The graph shows that the RCS is equal to 0 dB up to 1 GHz, with the RCS error extremely small and less than 10^{-10} dB. Note that the vertical scale is in decibels but multiplied by 10^{-11}.

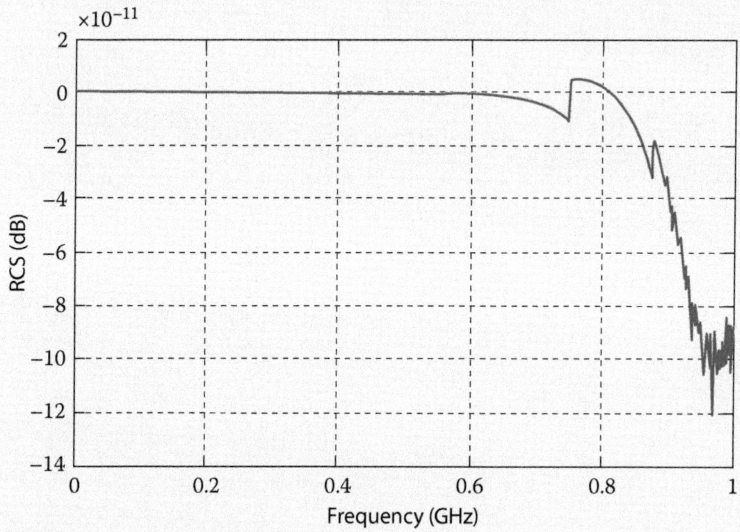

Fig. 4.22 RCS per unit area of the one-dimensional semi-infinite PEC target.

Example 4.3: Scattering from a Dielectric Half-Space

The PEC scatterer in Example 4.2 is replaced by a dielectric with relative permittivity $\epsilon_r = 4$, as shown in Fig. 4.23. The dielectric extends to $z = +\infty$, but the size of the domain is truncated to a length L. The length must be sufficient to keep the incident field from reaching $z = L$ until the observation time $T_{\max}$ has elapsed. This simulates the semi-infinite dielectric slab in a numerical sense because the eventual reflections from the grid edge at $z = L$ have been precluded.

The expected result for the RCS per unit area can be determined based on the intrinsic impedances of the dielectric and the free space. From Appendix A,

$$\sigma_{1D} = |\Gamma|^2 = \left(\frac{\sqrt{1/4} - 1}{\sqrt{1/4} + 1}\right)^2 = 1/9 \approx -9.5\,\text{dB}$$

Finite difference–time-domain computations were performed using two different values for the temporal oversampling factor K, with the ratio $u_{\text{grid}}/\text{c} = 1$ and the grid size kept the same in both media. (Note that reducing the spatial and temporal sampling intervals by the same factor does not alter u_{grid}.) The results are shown in Fig. 4.24 for the values of $K = 2$ and $K = 4$. Note that both K values give the exact RCS value as $f \to 0$ but give incorrect results as $f \to f_{\max}$.

The RCS error at higher frequencies can obviously be reduced by increasing the temporal oversampling factor, but at the expense of longer computation times. The presence of the dielectric accentuates the need for proper

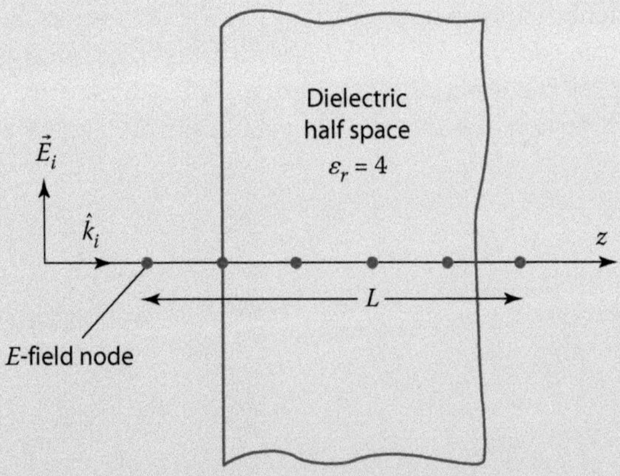

Fig. 4.23 One-dimensional semi-infinite dielectric scatterer.

(Continued)

Example 4.3: Scattering from a Dielectric Half-Space *(Continued)*

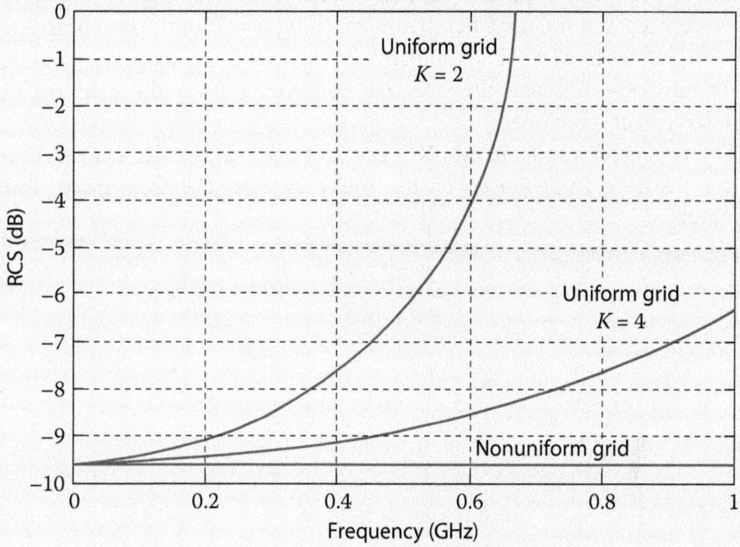

Fig. 4.24 RCS per unit area of the one-dimensional semi-infinite dielectric scatterer ($\epsilon_r = 4$).

sampling of higher frequency components in the spectrum of the incident wave, and one means of accomplishing this is to oversample. The undersampled frequency components contribute to the RCS errors in such a way that the RCS calculated via FDTD is larger than the true value at higher frequencies. Because the effect of the undersampled components is to *increase* the RCS, the behavior of the grid within the dielectric can be compared to that of a low-pass (rejection) filter. The bandwidth of this "grid filter" is controlled by the temporal sampling interval Δt.

A second means of achieving the proper sampling of higher frequency components is to use nonuniform grid spacing. The grid dimensions should be chosen so that $u_p/u_{\text{grid}} = 1$ in each medium. Thus, the grid spacing in a dielectric should be

$$\Delta \ell_d = \frac{\Delta \ell_0}{\sqrt{\epsilon_r}} \tag{4.144}$$

where $\Delta\, \ell_0$ is the free-space grid size. When nonuniform grid spacing is incorporated into the solution, the error in RCS is practically zero, as shown by the third curve in Fig. 4.24.

Example 4.4: Scattering from a Dielectric Slab of Finite Thickness

The geometry of Example 4.3 is modified so that the dielectric ($\epsilon_r = 4$) has finite thickness of 1 m as shown in Fig. 4.25. The domain length is $L = 15$ m and the dielectric slab's left edge is at $z = 6$ m. The RCS of the slab is plotted in Fig. 4.26. The minima occur at frequencies for which the slab thickness is an odd integer multiple of the quarter-wavelength within the slab, and the maxima occur at frequencies at which the slab thickness is an even integer multiple of $\lambda/2$.

Note that the envelope of the maxima has the same shape as the RCS curve for the semi-infinite slab. Theoretically, all maxima in Fig. 4.26a should have the same height. In fact, they are not equal because using a uniform grid spacing for the entire problem has resulted in undersampling errors. If a smaller grid size is used in the dielectric, $\Delta\ell_d$ given by Eq. (4.144), then the error can essentially be reduced to zero, as shown in Fig. 4.26b.

Snapshots of the reflected and transmitted pulses for the slab are shown in Fig. 4.27. The reflected pulse (A) propagates to the left (negative z direction), whereas the transmitted pulse (B) propagates to the right (positive z direction). The small pulse in the middle (C) is the pulse reflected off the right edge of the dielectric and is propagating in the negative z direction.

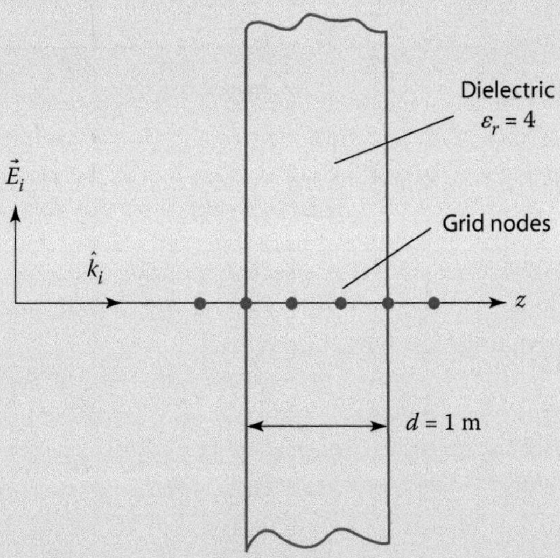

Fig. 4.25 One-dimensional dielectric slab of thickness *d*.

(Continued)

Example 4.4: Scattering from a Dielectric Slab of Finite Thickness *(Continued)*

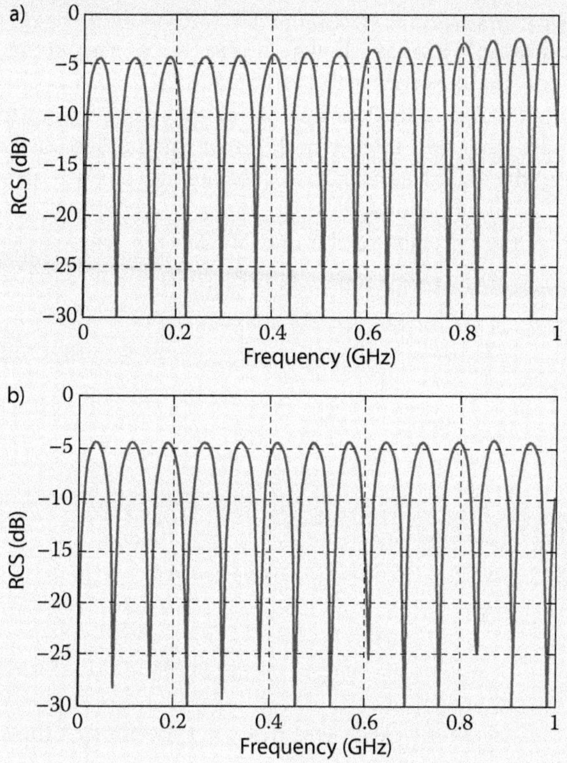

Fig. 4.26 RCS per unit area of the one-dimensional 1-m-thick dielectric slab ($\epsilon_r = 4$) for two grid sizes.

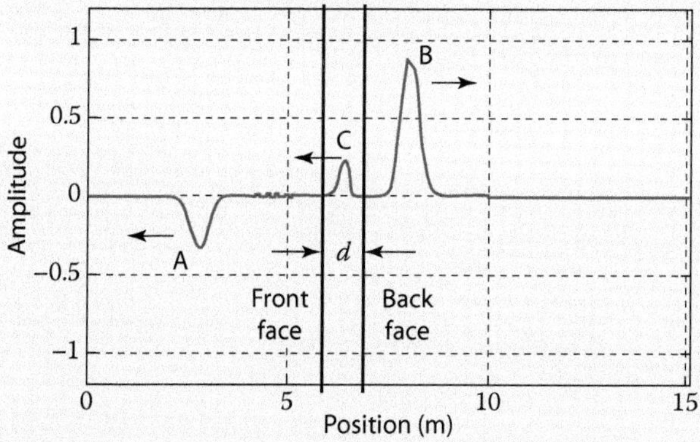

Fig. 4.27 Time snapshot of the fields for Example 4.4.

4.5 Finite Difference–Time-Domain Method in Two Dimensions

4.5.1 Introduction

The incident and scattered electromagnetic fields and the media parameters in two-dimensional problems are independent of one of the three spatial coordinates. There are cases where a two-dimensional problem may be solved in place of a related three-dimensional problem. For example, say that we are interested in finding the scattered field from a very long circular cylinder at an observation point that is close to the cylinder, as depicted in Fig. 4.28. In this case, the scattered field is determined primarily by the current on the cylinder surface in the vicinity of the observation point. Furthermore, if $d \ll L$, then the effects of the cylinder edges can be neglected, and the current distribution on the finite cylinder approaches that of an infinitely long cylinder. The three-dimensional RCS can be estimated from the two-dimensional RCS (often called the *echo width*) by [15]

$$\sigma_{3D} \approx \sigma_{2D} \frac{2L^2}{\lambda}. \tag{4.145}$$

Working in the Cartesian system, we will denote z as the independent axis. The field invariance with z can be expressed mathematically as

$$\frac{\partial \vec{E}}{\partial z} = 0, \qquad \frac{\partial \vec{H}}{\partial z} = 0 \tag{4.146}$$

The electric and magnetic fields will thus be functions of the spatial coordinates x and y and of time t. All media will be assumed stationary.

An arbitrary two-dimensional electromagnetic field can be expressed as a linear combination (superposition) of TE_z and TM_z components. The TE_z

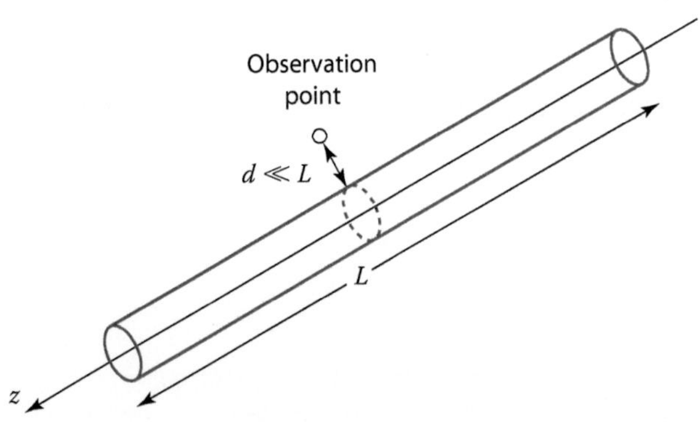

Fig. 4.28 Example of a three-dimensional problem that can be approximated by a two-dimensional problem.

field components are E_x, E_y, and H_z:

$$\vec{E}_{\mathrm{TE}}(x, y, t) = E_x(x, y, t)\hat{x} + E_y(x, y, t)\hat{y} \tag{4.147}$$

$$\vec{H}_{\mathrm{TE}}(x, y, t) = H_z(x, y, t)\hat{z} \tag{4.148}$$

The dual TM_z components are H_x, H_y, and E_z:

$$\vec{H}_{\mathrm{TM}}(x, y, t) = H_x(x, y, t)\hat{x} + H_y(x, y, t)\hat{y} \tag{4.149}$$

$$\vec{E}_{\mathrm{TM}}(x, y, t) = E_z(x, y, t)\hat{z} \tag{4.150}$$

To apply the FDTD method in two dimensions, discretized forms of Maxwell's equations are required. For TE_z fields,

$$\begin{aligned}&\oint_{C_E} \left[E_x(x, y, t)\hat{x} + E_y(x, y, t)\hat{y}\right] \cdot \mathrm{d}\vec{\ell}_E \\ &\quad = -\frac{\partial}{\partial t}\left\{\iint_{S_E} u(x, y) H_z(x, y, t)\hat{z} \cdot \mathrm{d}\vec{s}_E\right\}\end{aligned} \tag{4.151}$$

and

$$\begin{aligned}&\oint_{C_E} H_z(x, y, t)\hat{z} \cdot \mathrm{d}\vec{\ell}_H \\ &\quad = -\frac{\partial}{\partial t}\left\{\iint_{S_H} \epsilon(x, y)\left[E_x(x, y, t)\hat{x} + E_y(x, y, t)\hat{y}\right] \cdot \mathrm{d}\vec{s}_H\right\} \\ &\quad + \iint_{S_H} \sigma_c(x, y)\left[E_x(x, y, t)\hat{x} + E_y(x, y, t)\hat{y}\right] \cdot \mathrm{d}\vec{s}_H\end{aligned} \tag{4.152}$$

For TM_z fields, the dual equations are

$$\begin{aligned}&\oint_{C_E} E_z(x, y, t)\hat{z} \cdot \mathrm{d}\vec{\ell}_E \\ &\quad = -\frac{\partial}{\partial t}\left\{\iint_{S_E} \mu(x, y)[H_x(x, y, t)\hat{x} + H_y(x, y, t)\hat{y}] \cdot \mathrm{d}\vec{s}_E\right\}\end{aligned} \tag{4.153}$$

and

$$\begin{aligned}&\oint_{C_H} \left[H_x(x, y, t)\hat{x} + H_y(x, y, t)\hat{y}\right] \cdot \mathrm{d}\vec{\ell}_H \\ &\quad = \frac{\partial}{\partial t}\left\{\iint_{S_H} \epsilon(x, y) E_z(x, y, t)\hat{z} \cdot \mathrm{d}\vec{s}_H\right\} \\ &\quad + \iint_{S_H} \sigma_c(x, y) E_z(x, y, t)\hat{z} \cdot \mathrm{d}\vec{s}_H\end{aligned} \tag{4.154}$$

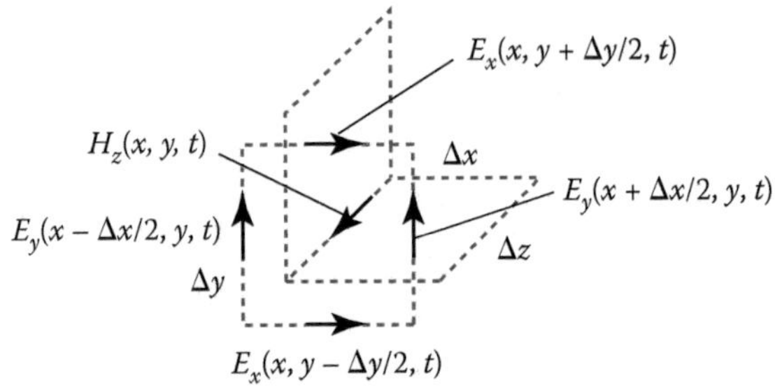

Fig. 4.29 Contours for TE_Z electric and magnetic field circulations and fluxes.

Note that Maxwell's second equations [Eqs. (4.152) and (4.154)] do not have the source current terms because it has been assumed that there are no source currents in the domain of interest. The source for scattered field calculations is a known incident field.

The contours of integration for the electric and magnetic field circulations are, in general, different and are denoted C_E for an electric field contour and C_H for a magnetic field contour. Similarly, the surfaces associated with the contours are labeled S_E for the magnetic flux and S_H for the electric flux. The electric and magnetic field contours for a TE_z field are shown in Fig. 4.29. The most basic forms are rectangles of dimensions $\Delta x \times \Delta y$ and $\Delta x \times \Delta z$ located in orthogonal planes. The C_E contours are in the x–y plane, whereas the C_H contours are in the y–z plane. They can be visualized as links in a chain fence, evoking the idea of electric and magnetic field linkage in two orthogonal directions. The dual C_E and C_H contours for a TM_z field are shown in Fig. 4.30.

4.5.2 Update Equations

Derivation of the TM_z and TE_z update equations for two dimensions proceeds along the same line as for the one-dimensional case described in

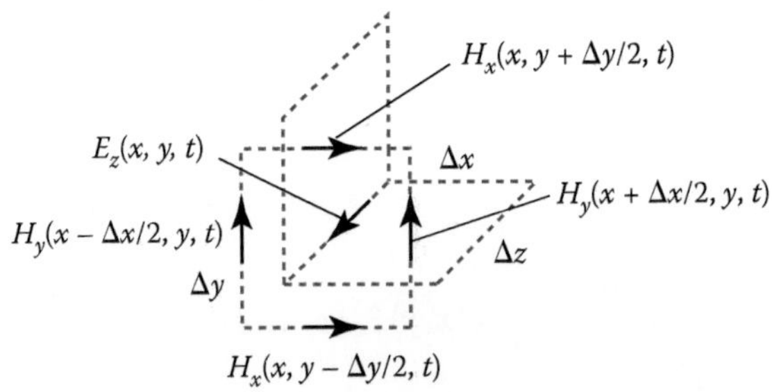

Fig. 4.30 Contours for TM_Z electric and magnetic field circulations and fluxes.

Sec. 4.4. Because of the similarities, the intermediate details of the derivations are skipped, and only the final form of the update equations is presented.

We will assume a uniform grid in both dimensions with $\Delta x = \Delta y = \Delta \ell$ as illustrated in Fig. 4.31. Furthermore, we restrict $\Delta \ell$ to be small enough so that the fields at all points inside of a cell are approximately constant and equal to the value at the center. Using the definition of average in Eq. (4.72) it is possible to obtain the average constitutive relations for each grid cell. For example, when averaging over both grid variables

$$\sigma_{\text{avg}} \equiv (\Delta \ell)^2 \int_{-\Delta \ell/2}^{\Delta \ell/2} \int_{-\Delta \ell/2}^{\Delta \ell/2} \sigma_c(x + \xi, y + \psi) \mathrm{d}\xi \mathrm{d}\psi \tag{4.155}$$

or, when averaging over only one grid variable

$$\sigma_{\text{avg}x} \equiv \Delta \ell \int_{-\Delta \ell/2}^{\Delta \ell/2} \sigma_c(x + \xi,\ y) \mathrm{d}\xi \tag{4.156}$$

The addition of x to the subscript on σ denotes the variable over which the parameter has been averaged.

Because we have either TE_z or TM_z two-dimensional electromagnetic fields, there will be TE_z and TM_z grids as well. Let us assume that our domain is an $L \times L$ square. We will assume a uniform grid, that is,

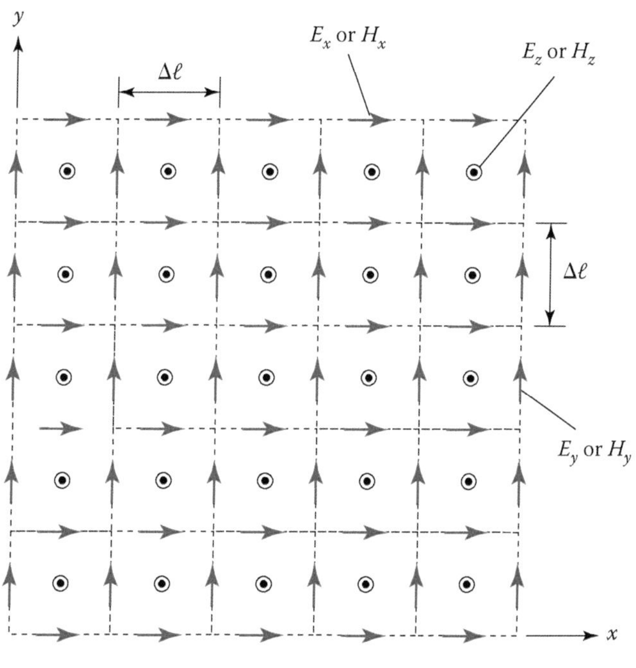

Fig. 4.31 Two-dimensional FDTD grid.

a discretization step $\Delta\ell$ that is not changing with position. The fields are sampled using a spatial step $\Delta\ell = L/N$. The electric and magnetic field sample locations are interleaved, as in Fig. 4.31 for $N = 5$. The spatial grid edge samples can be, in principle, either electric field samples or magnetic field samples. We will select electric fields for spatial edge samples for TE_z grids and magnetic fields for TM_z grids. The grid shown in Fig. 4.31 is thus applicable to either TE_z or TM_z fields.

An $N \times N$ TE_z grid has N^2 H_z magnetic field nodes, $N(N+1)$ E_x electric field nodes, and $N(N+1)$ E_y electric field nodes. Similarly, an $N \times N$ TE_z grid has N^2 E_z electric field nodes, $N(N+1)$ H_x magnetic field nodes, and $N(N+1)$ H_y magnetic field nodes. Note that the fields at the edges would be evaluated first because a known incident field enters the grid via edge nodes. This means that we will start the updates in time for the same field (in our case, the electric field) that has been selected for the grid edges, in other words, the same field that would be at the spatial and temporal edges. TE_z electric and magnetic field update equations can now be converted to electric and magnetic field grid update equations by replacing the variables x, y, and t with the grid coordinates of the electric field spatial and temporal sampling points:

$$
\begin{array}{lll}
x \rightarrow & i\Delta\ell, & i = 0, 1, \ldots, N \\
y \rightarrow & j\Delta\ell, & j = 0, 1, \ldots, N \\
t \rightarrow & n\Delta t, & n = 0, 1, \ldots, N_t
\end{array}
\tag{4.157}
$$

The notation can be simplified by omitting the common $\Delta\ell$ and Δt terms and using a superscript for the index of the temporal sampling point. The grid equations for conductive media are

$$
\begin{aligned}
E_x^n(i,j) \approx{}& \left[\frac{1 - A_y\Delta\ell\sigma_{\text{avgy}}(i,j)/2}{1 + A_y\Delta\ell\sigma_{\text{avgy}}(i,j)/2}\right] E_x^{n-1}(i,j) \\
&+ \left[\frac{A_y}{1 + A_y\Delta\ell\sigma_{\text{avgy}}(i,j)/2}\right] [H_z^{n-1/2}(i, j+1/2) \\
&- H_z^{n-1/2}(i, j-1/2)]
\end{aligned}
\tag{4.158}
$$

$$
\begin{aligned}
E_y^n(i,j) \approx{}& \left[\frac{1 - A_x\Delta\ell\sigma_{\text{avgx}}(i,j)/2}{1 + A_x\Delta\ell\sigma_{\text{avgx}}(i,j)/2}\right] E_y^{n-1}(i,j) \\
&+ \left[\frac{A_x}{1 + A_x\Delta\ell\sigma_{\text{avgx}}(i,j)/2}\right] [H_z^{n-1/2}(i-1/2, j) \\
&- H_z^{n-1/2}(i+1/2, j)]
\end{aligned}
\tag{4.159}
$$

$$H_z^n(i,j) \approx H_z^{n-1}(i,j) - \frac{c}{\eta_0 u_{\text{grid}} \mu_{\text{ravg}}(i,j)} \Big[E_x^{n-1/2}(i,j-1/2) - E_x^{n-1/2}(i,j+1/2) + E_y^{n-1/2}(i+1/2,j) - E_y^{n-1/2}(i-1/2,j) \Big] \tag{4.160}$$

The dual TM_z grid equations for conducting media are

$$H_x^n(i,j) \approx H_x^{n-1}(i,j) + \frac{c}{\eta_0 u_{\text{grid}} \mu_{\text{ravgy}}(i,j)} \Big[E_z^{n-1/2}(i+1/2,j) - E_z^{n-1/2}(i-1/2,j) \Big] \tag{4.161}$$

$$H_y^n(i,j) \approx H_y^{n-1}(i,j) + \frac{c}{\eta_0 u_{\text{grid}} \mu_{\text{ravgx}}(i,j)} \Big[E_z^{n-1/2}(i,j-1/2) - E_z^{n-1/2}(i,j+1/2) \Big] \tag{4.162}$$

$$E_z^n(i,j) \approx \left[\frac{1 - \mathcal{A}\Delta\ell\sigma_{\text{avgx}}(i,j)}{1 + \mathcal{A}\Delta\ell\sigma_{\text{avg}}(i,j)} \right] E_z^{n-1}(i,j) + \left[\frac{\mathcal{A}}{1 + \mathcal{A}\Delta\ell\sigma_{\text{avg}}(i,j)} \right] \Big[H_x^{n-1/2}(i,j-1/2) - H_x^{n-1/2}(i,j+1/2) + H_y^{n-1/2}(i+1/2,j) - H_y^{n-1/2}(i-1/2,j) \Big] \tag{4.163}$$

The quantities $\mathcal{A}$, $\mathcal{A}_x$, and $\mathcal{A}_y$, are functions of (x, y) and therefore dependent on the indices (i, j), although not specifically indicated. For example:

$$\begin{aligned} \mathcal{A}_y &\equiv \frac{\eta_0 c}{u_{\text{grid}} \epsilon_{\text{ravgy}}(x,y)} \\ \mathcal{A}_x &\equiv \frac{\eta_0 c}{u_{\text{grid}} \epsilon_{\text{ravgx}}(x,y)} \end{aligned} \tag{4.164}$$

The TE_z electric and magnetic field grid equations in two dimensions have the general form:

$$\begin{aligned} E_x^{\text{new}} &= C_{Ex1} E_x^{\text{old}} + C_{Ex2} \nabla H_z^{\text{old}} \\ E_y^{\text{new}} &= C_{Ey1} E_y^{\text{old}} + C_{Ey2} \nabla H_z^{\text{old}} \\ H_z^{\text{new}} &= C_{H1} H_z^{\text{old}} + C_{Hx2} \nabla E_x^{\text{old}} + C_{Hy2} \nabla E_y^{\text{old}} \end{aligned} \tag{4.165}$$

where the coefficients are real constants that depend on the media properties and the velocity ratio c/u_{grid} and ∇ represents the spatial derivative (gradient). Similarly, the TE_z electric and magnetic field equations can be written

$$\begin{aligned} H_x^{\text{new}} &= C_{Hx1} H_x^{\text{old}} + C_{Hx2} \nabla E_z^{\text{old}} \\ H_y^{\text{new}} &= C_{Hy1} H_y^{\text{old}} + C_{Hy2} \nabla E_z^{\text{old}} \\ E_z^{\text{new}} &= C_{E1} E_z^{\text{old}} + C_{Ex2} \nabla E_x^{\text{old}} + C_{Ey2} \nabla H_y^{\text{old}} \end{aligned} \tag{4.166}$$

4.5.3 Two-Dimensional Grid Termination

The grid equations derived in Sec. 4.5.2 are valid for all the nodes except those on the grid edges. The reason is that grid edge nodes for a field component in a transverse-to-z plane have only one neighbor for a z-directed field component instead of two. This is illustrated in Fig. 4.32 for an x-directed field component inside the grid and an x-directed field component on the grid's right-hand edge.

An equation for the circulation of the z-directed field component cannot be written for an edge node because the right-most z-directed field component shown in Fig. 4.32 cannot be updated using the grid equations discussed so far. To handle this situation, the concepts of TGT and the discrete boundary impulse response (DBIR) introduced in Sec. 4.3.3 are used. Update equations for edge nodes can be devised based on convolutions of the fields at one layer inside and the precalculated impulse responses from the inside layer. The edge field will be expressed as a superposition of responses, as depicted in Fig. 4.33 for the case of the top edge of the grid. Superposition applies to other edges as well.

The superposition can also be modeled using the concept of a multiport, as described in Sec. 4.2.2. The nodes on grid edges can be considered as output ports of a linear multiport. If it is assumed, for simplicity, that the grid is square, with $(N+1)$ edge field nodes on each side, the total number of output ports will be $4N$. The nodes on the layer just inside the grid boundary will be considered as input ports of the linear multiport. There will be a total of $4(N–2)$ input ports. Thus, the equivalent multiport is essentially the same as that shown in Fig. 4.3, except that N is replaced by $4(N–2)$ and M by $4N$.

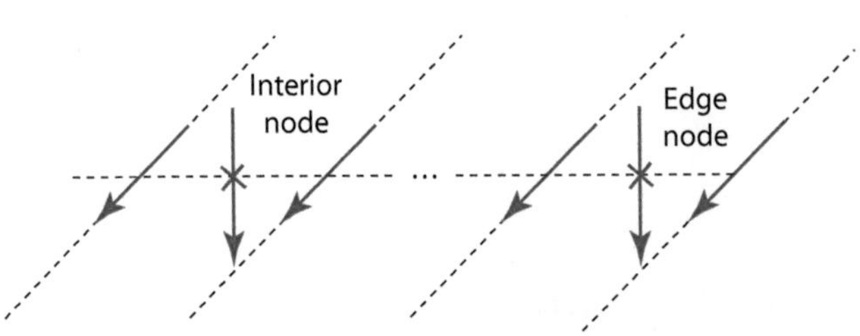

Fig. 4.32 Illustration of edge nodes versus inside nodes.

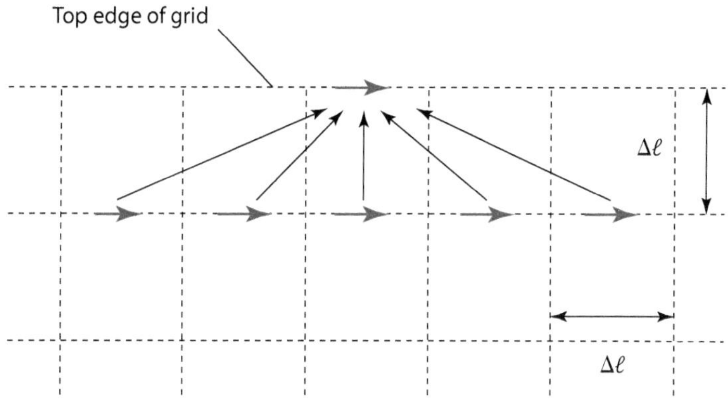

Fig. 4.33 Edge field determined as a superposition of internal node responses.

The input and output fields need not be of the same kind; that is, the input fields could be E fields and the output H fields, or vice versa. It is most convenient, however, to choose input and output fields of the same kind so that the impulse responses $h_{mn}(t)$ will be dimensionless. The input and output fields will be E fields for TE_z and H fields for TM_z fields. In this manner, the same impulse responses, $h_{mn}(t)$, can be used for both TE_z and TM_z grids. The following discussion is presented in terms of TE_z fields, that is, electric fields at the grid edges. Dual equations for the TM_z fields can be obtained simply by replacing E with H in the TE_z equations.

There are four sides of the square grid boundary. The top side has the electric field $E_y(0, y, t)$, the bottom side has the electric field $E_y(L, y, t)$, the left side has the electric field $E_x(x, 0, t)$, and the right side has the electric field $E_x(x, L, t)$. The layer just inside will have the corresponding fields as $E_y(\Delta\ell, y, t)$, $E_y(L- \Delta\ell, y, t)$, $E_x(x, \Delta\ell, t)$, and $E_x(x, L - \Delta\ell, t)$. The extreme values of x and y at the edges are 0 and L, whereas those for the layer just inside are $\Delta\ell$ and $L - \Delta\ell$. Using the grid coordinates (i, j) and starting at the upper left corner (0, 0) with a clockwise reference direction, the fields on the grid boundary will be

$$E_y^n(0,j) \rightarrow E_x^n(i,N) \rightarrow E_y^n(N,j) \rightarrow E_x^n(i,0) \tag{4.167}$$

These can be grouped into a vector of electric fields on the grid boundary:

$$\mathbf{E}_{\text{bd}}(t) = \begin{bmatrix} E_y^n(0,j,t) \\ E_x^n(i,N,t) \\ E_y^n(N,j,t) \\ E_x^n(i,0,t) \end{bmatrix} \tag{4.168}$$

where the subscript "bd" denotes "at the boundary."

In an analogous manner, an electric field vector can be formed for the layer just inside the grid boundary:

$$\mathbf{E}_{\text{in}}(t) = \begin{bmatrix} E_y^n(1,j,t) \\ E_x^n(i,N-1,t) \\ E_y^n(N-1,j,t) \\ E_x^n(i,1,t) \end{bmatrix} \tag{4.169}$$

In this case, the subscript "in" denotes a node just inside the boundary. The advantage of the vector arrangement is that we can now use a single index (m) to identify a field on the grid boundary and a second index (n) to identify an electric field on the layer just inside the grid boundary. The use of single indices m and n allows for a direct and compact reference to the multiport representation of the grid termination problem.

The electric field on the grid boundary $\mathbf{E}_{\text{bd}}(t)$ can be expressed as a convolution of the electric fields at the layer just inside the grid boundary $\mathbf{E}_{\text{in}}(t)$ and a matrix of impulse responses $\mathbf{h}(t)$:

$$\mathbf{E}_{\text{bd}}(t) = \mathbf{h}(t) * \mathbf{E}_{\text{in}}(t) \tag{4.170}$$

The $N_r = 4N \times N_c = 4(N-2)$ matrix of impulse responses can be written as

$$\mathbf{h}(t) = \begin{bmatrix} h_{11}(t) & h_{12}(t) & h_{13}(t) & \cdots & h_{1N_c}(t) \\ h_{21}(t) & \ddots & & & h_{2N_c}(t) \\ \vdots & & & \ddots & \vdots \\ h_{N_r1}(t) & h_{N_r1}(t) & h_{N_r3}(t) & \cdots & h_{N_rN_c}(t) \end{bmatrix} \tag{4.171}$$

The electric field at a boundary node identified by a particular value of the index m can be written as the product of the mth row of the impulse response matrix $\mathbf{h}(t)$ and the column vector of the electric fields just inside the boundary identified by index n:

$$\begin{aligned} E_{\text{bd}}(m,t) &= \sum_{n=1}^{N_c} h_{mn}(t) * E_{\text{in}}(n,t) \\ &= \sum_{n=1}^{N_c} \int_0^t E_{\text{in}}(n,\tau) h_{mn}(t-\tau) \mathrm{d}\tau \end{aligned} \tag{4.172}$$

Note the summation over all nodes on the layer just inside the boundary. The discretized forms of the preceding equations, involving samples of electric fields at $t = q\Delta t$, are the sums of discrete convolutions. Because a discrete convolution is itself a sum, the result for the boundary nodes will involve double summations. With superscripts used for time samples and convolution summation,

$$E_{\text{bd}}^{q}(m) = \sum_{n=1}^{N_c} \sum_{p=0}^{q} E_{\text{in}}^{p}(n) h_{mn}^{q-p} \tag{4.173}$$

The convolution equations express the fields at the edges as weighted sums of the time histories of the fields just inside the grid. The weighting coefficients are the values of the discrete boundary impulse responses h_{mn}. In this respect, they are equations of the same type as those for the non-edge grid nodes [Eqs. (4.157) and (4.158)], but they involve, in general, summations with more terms. Fortunately, the impulse responses $h_{mn}(t)$ converge to zero relatively fast, and this property limits the number of significant terms in the convolution sums. Thus, the convolution sums can be truncated such that only a certain number of the most recent values of the electric fields just inside the boundary are retained. The discrete boundary impulse responses $h_{mn}(q\Delta t)$ need to be determined only once for a selected grid velocity $u_{\text{grid}} = \Delta\ell/\Delta t$ and then stored in matrix form.

A typical matrix based on the grid arrangement shown in Fig. 4.34 would have $N_r = 4N$ rows (one row per boundary node) and $N_c = 4(N-2)N_h$ columns; N_h denotes the length of the truncated discrete boundary impulse responses. This form is suitable because the boundary electric fields can then be calculated as a product of this matrix and a column

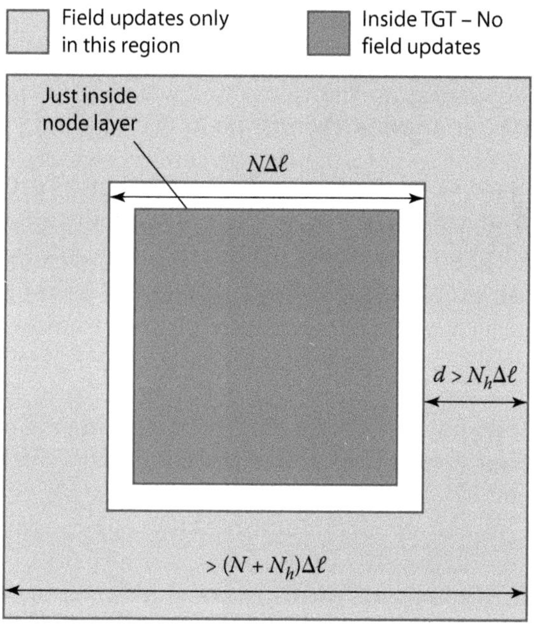

Fig. 4.34 Two-dimensional grid for calculation of TGT boundary impulse responses.

vector constructed of $4(N-2)$ subvectors of length N_h. The subvectors represent the N_h most recent values of the electric field at nodes just inside the boundary. The particular choice of N_h depends on the desired amplitude resolution, but values over 20 generally give good results for $u_{\text{grid}}/c = \sqrt{D}$. The subvectors are updated at each time step by shifting all of their values down by one (that is, further into the past), discarding the last value, and updating the first subvector element with the most recent field value calculated via standard grid update equations. The double summations are thus efficiently executed as matrix-vector multiplications.

The discretized impulse responses h^q_{mn} are determined in a manner analogous to the one-dimensional case, using the discrete equivalent of the Dirac delta function, which we will denote as δ^q. It is convenient to use the equivalent multiport to explain the procedure. To determine the discrete boundary impulse response for a particular input port n, all input ports except the nth are set to zero and the Dirac impulse is applied to the nth input port. Recording all the outputs from $t = 0$ to $t = N_h \Delta t$ provides us with $4N$ discrete boundary impulses of duration N_h:

$$E^q_{\text{in}}(n) = \delta^q \tag{4.174}$$

$$E^q_{\text{in}}(p) = 0 \tag{4.175}$$

for $p = 1, 2, \ldots, 4(N-2)$ and $p \neq n$. The process is repeated $4(N-2)$ times [because there are $4(N-2)$ input ports], and the results are stored in the $4N \times 4(N-2)N_h$ matrix. This matrix is then used to implement the transparent grid termination via the matrix multiplication by the $4(N-2)N_h$ column vector of the time histories of the fields just inside the boundary.

Note that, depending on the values of N_h and N, there may be input nodes sufficiently far from an output node that the time it takes for an input impulse to propagate to the output port exceeds N_h. In such a case, the contribution of the input node to the output node is known in advance to be zero. This may be used to reduce the number of operations in TGT implementation. The process of obtaining the two-dimensional DBIRs is carried out on an oversized grid whose size should be at least ($N + N_h$ by $N + N_h$) or, equivalently, the distance from the TGT boundary to the boundary of the oversized grid should be at least $N_h \Delta\ell/2$. The precise details of the termination of the oversized grid are inconsequential because the reflections off its boundary do not arrive until after the time stepping has been terminated at the output nodes. Thus, the reflections do not have an opportunity to corrupt the impulse responses. In a typical application, $N \gg N_h$, and the oversized grid would be only incrementally larger than the physical (target) grid.

When the DBIRs are calculated, the nodes inside of the TGT boundary need not be updated because the nodes on the layer just inside the TGT boundary (i.e., the input ports) are zero for $t > 0$. At $t = 0$, only one node on the layer just inside the TGT boundary has the value of 1. Thus, the DBIR calculations require updates only for nodes between the TGT boundary and the large grid boundary. This allows significant computational time saving if $N \gg N_h$. The update equations used for the region between the "inner" (TGT) and the outer boundaries are those for free space, given that this region is assumed to be free space for RCS applications.

The update equations for the TE_z fields can be simplified by scaling the magnetic field by the free-space intrinsic impedance. Writing the TE_z and TE_z update equations in this manner would clearly show that the only parameter affecting the equations is the velocity ratio c/u_{grid}. The DBIRs are thus specific for a selected velocity ratio and can be used only with that velocity ratio. The velocity ratio in the preceding equations is that for the medium assumed to exist between the TGT boundary and the large grid boundary which, in most cases, will be free space.

The most likely value for the velocity ratio would be $1/\sqrt{2}$ for two-dimensional fields. (Recall that the velocity ratio c/u_{grid} must be less than, or at most equal to, $1/\sqrt{\mathcal{D}}$ by the stability requirement.) This is equivalent to requiring that the numerical solution must not advance at a slower rate than that of the physical wave propagating through the same medium. Thus, the choice of $c/u_{\text{grid}} = 1/\sqrt{\mathcal{D}}$ matches the speed with which the numerical wave propagates through the grid to the speed with which the physical wave propagates in the physical medium (free space). The match in two dimensions occurs, unfortunately, only for directions of propagation that are 45 deg with respect to the grid axis. In one dimension, the match can be exact because there is only one allowed direction of propagation.

The shape of the discrete boundary impulse response $h_{mn}(t)$ determined by the procedure just described depends on the relative positions of the observation (output, index m) and the source (input, index n). In general, the greater the difference between the indices, the greater the following trends:

1. The impulse response starts later.
2. The maximum value of the impulse response is smaller.
3. The impulse response is more spread out in time.

To illustrate these effects, three sample impulse responses are shown in Fig. 4.35. The observation point (output port) is the same for all three responses and is located in the center of one side of the TGT boundary. The three DBIRs are for a delta function applied to the following circumstances.

1. Fig. 4.35a: Input node is immediately below the output node.
2. Fig. 4.35b: Input node is 5 nodes to the side of the node immediately below the output node.
3. Fig. 4.35c: Input node is 10 nodes to the side of the node immediately below the output node.

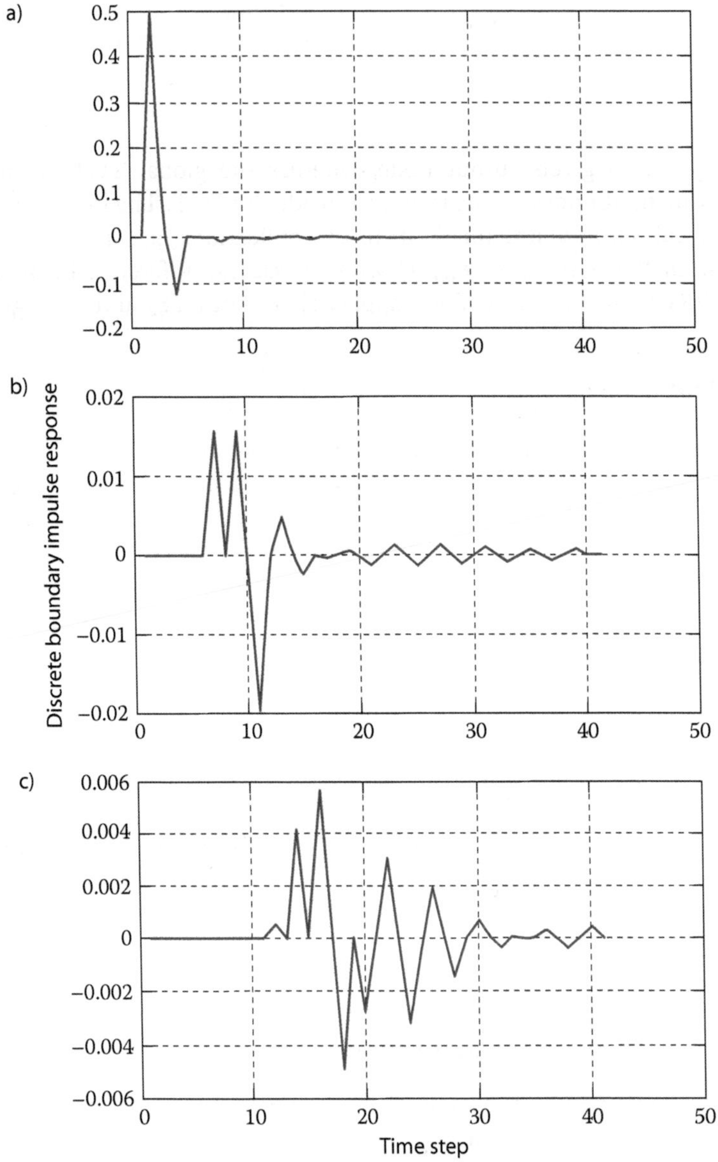

Fig. 4.35 Typical discrete boundary impulse responses.

The velocity ratio c/u_{grid} for the DBIR calculations is $1/\sqrt{2}$. Note that the plotting program interpolates linearly between the sampling points $q\,\Delta\,t$. The impulse responses were truncated to a duration of $N_h = 40$. Because of the large range of DBIR values, the impulse responses are shown individually. The largest value for a source node just below the input is $1/2$; for a source 5 nodes to the side, about 25 times smaller; and, for a source node 10 nodes to the side, about 1000 times smaller.

The rapid decrease in the DBIR with distance offers the possibility of implementing the TGT with local equations instead of global convolutions. The (approximate) local TGT implementation involves only a few input nodes nearest a given output node, whereas the global TGT calculation involves all input nodes on the layer just inside the TGT. The loss of accuracy due to localization is acceptable in most applications because of the rapid decrease in the maxima of $h_{mn}(t)$ with increasing difference in m and n. The DBIRs shown in Fig. 4.35 are applicable to both TE_z and TM_z grids.

4.5.4 Incident Field Injection into the Grid

The source of excitation in scattering problems is a known incident field that is injected into the grid via the edge nodes. We will assume that the incident field is a uniform plane wave. The time variation of the incident field (i.e., the waveform) is arbitrary, but we will restrict our analysis to Gaussian pulse shape $g(t)$. Furthermore, we will assume that an electric incident field is needed for the grid's edges (which corresponds to a TE_z grid, with TM_z grid equations deduced by duality).

In two dimensions there is no propagation in the z direction and, thus, any z variation is zero ($z = z_0$). Furthermore, the incident angle θ_i in the transverse plane (the x–y plane) is $\pi/2$. The expressions for a two-dimensional incident field thus simplify to

$$E_{ix}(x, y) = -E_{i\phi} \sin\phi_i \, g(t + R(x, y)/c) \tag{4.176}$$

$$E_{iy}(x, y) = E_{i\phi} \cos\phi_i \, g(t + R(x, y)/c) \tag{4.177}$$

$$E_{iz}(x, y) = -E_{i\theta} \, g(t + R(x, y)/c) \tag{4.178}$$

and

$$R(x, y) = \cos\phi_i(x - x_0) + \sin\phi_i(y - y_0) \tag{4.179}$$

is the distance between a point (x, y) where the field is calculated and a chosen reference point (x_0, y_0) outside the computational grid. The z-directed component of the electric field is zero for TE_z fields (edge nodes for TE_z fields are E_x or E_y electric field nodes), which implies that E_{iz} in Eq. (4.178) is zero.

Following the approach applied to one-dimensional problems, electric or magnetic fields at the grid's edge nodes are represented as sums of incident and scattered electric or magnetic fields. To simplify the notation, the electric field on the boundary will be denoted E_{bd}, with E_{bd} either E_x or E_y, depending on the boundary side. The total field is a sum of incident plus scattered terms:

$$E^q_{\text{bd}} = E^q_{i\text{bd}} + E^q_{s\text{bd}} \tag{4.180}$$

The unknown scattered field on the layer of nodes just inside the boundary is obtained by subtracting the known incident field from the total field which, in turn, is obtained from the update equations for the grid interior (boundary excluded). To simplify the notation, the total field for the layer of nodes just inside the boundary will be denoted E_{in}. The scattered field on the layer just inside the boundary is written

$$E_{\text{sin}} = E_{\text{in}} - E_{i\text{in}} \tag{4.181}$$

The unknown scattered field at an edge node can be obtained as described in Sec. 4.4.5 from the scattered field at the layer of nodes just inside the boundary:

$$E^q_{\text{sbd}}(m) = \sum_{n=1}^{4(N-2)} \sum_{p=0}^{q} E^p_{\text{sin}}(n) h^{q-p}_{mn} \tag{4.182}$$

The total field on the boundary is now updated by adding the calculated scattered field to the known incident field. Note that all boundary impulse responses have $h^0_{mn} = 0$, that is, the current value of the scattered field on the boundary depends only on the past values of the scattered field on the layer just inside the boundary. The order of operations in injecting the incident field into a two-dimensional grid is the same as that delineated in Fig. 4.18.

4.5.5 Near-to Far-Field Transformation

The calculation of RCS requires the far-zone scattered electric field in the frequency domain. The frequency-domain far-zone electric field can be determined using the electric vector potential $\vec{F}$ and magnetic vector potential $\vec{A}$, which were introduced in Sec. 2.5. In cylindrical coordinates and for surface currents, Eqs. (2.6) and (2.7) reduce to

$$\vec{A}(\rho, \phi) = \frac{e^{-jk\rho}}{\sqrt{8j\pi k\rho}} \iint_S \vec{J}_{\text{seq}}(\rho') e^{jk\rho' \cos(\phi-\phi')} ds' \tag{4.183}$$

$$\vec{F}(\rho, \phi) = \frac{e^{-jk\rho}}{\sqrt{8j\pi k\rho}} \iint_S \vec{J}_{m\text{seq}}(\rho') e^{jk\rho' \cos(\phi-\phi')} ds' \tag{4.184}$$

The sources are the electric and magnetic currents given by the equivalence principle:

$$\vec{J}_{\text{seq}} = \hat{n} \times \vec{H}_s \tag{4.185}$$

$$\vec{J}_{m\text{seq}} = -\hat{n} \times \vec{E}_s \tag{4.186}$$

where $\hat{n}$ denotes a position-dependent unit vector normal to the surface of the scatterer. The source point is in the direction (ρ', ϕ'), and the observation point direction is given by (ρ, ϕ). The far field is determined from Eqs. (2.15) and (2.16), with the far-field approximation applied:

$$E_\phi(\rho, \phi) = -j\omega\mu A_\phi - jkF_z \tag{4.187}$$

$$E_z(\rho, \phi) = -j\omega\mu A_z + jkF_\phi \tag{4.188}$$

The equivalent currents are calculated either over the surface of the scatterer or over a closed surface enclosing the scatterer. Because the far-zone scattered field is needed, the equivalent currents are not calculated from the total fields but only from the scattered fields. The scattered fields on the grid's boundary (or on the layer just inside the boundary) provide the information needed to determine the equivalent currents. The equivalent currents in the frequency domain are obtained by the Fourier transform of the recorded time-domain scattered fields on the grid's boundary. The RCS per unit length is obtained from

$$\sigma_{2D}(\phi) = \lim_{\rho\to\infty} 2\pi\rho\frac{|\vec{E}_s|^2}{|\vec{E}_i|^2} = 2\pi\rho\frac{|E_\phi|^2 + |E_z|^2}{|\vec{E}_i|^2} \tag{4.189}$$

The two-dimensional RCS has the unit meter and is frequently called the *echo width.*

The procedure for computing the frequency-domain RCS of a target is summarized by the following example.

Example 4.5: Scattering from an Infinite Cylinder with Square Cross Section

Consider the infinitely long PEC scatterer with square cross section as shown in Fig. 4.36. The cylinder axis is the z axis and each side has a width of $w = 1.697$ m partitioned into 16 cells ($\Delta\ell = 0.1061$ m). The incident waveform is a TM_z polarized Gaussian pulse (i.e., only an E_z component), and the *selected* calculation parameters are

$$B_{\text{eff}} = 1\,\text{GHz}$$
$$R_{\text{dyn}} = 120\,\text{dB}$$
$$S/N = 90\,\text{dB}$$
$$\Delta f = 10\,\text{MHz}$$

The effective pulse duration and the maximum frequency of interest are

$$T_{\text{eff}} = \frac{1}{2B_{\text{eff}}} = 0.5\,\text{ns} \tag{4.190}$$

$$f_{\max} = B_{\text{eff}} = 1\,\text{GHz} \tag{4.191}$$

(Continued)

Example 4.5: Scattering from an Infinite Cylinder with Square Cross Section *(Continued)*

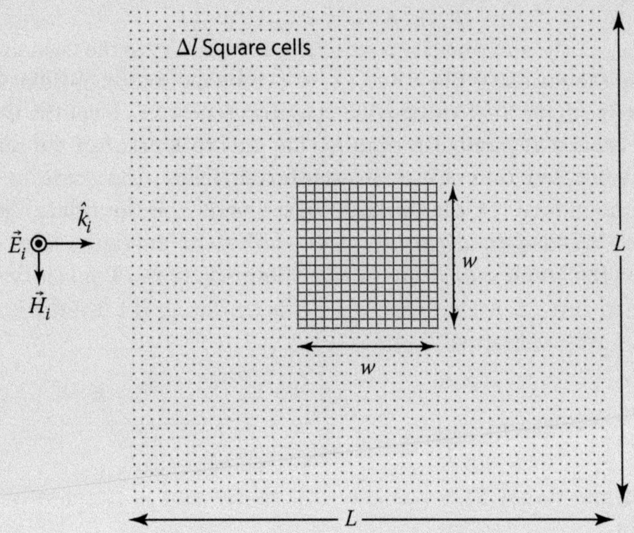

Fig. 4.36 Cross section of an infinitely long square cylinder with a TM_z incident field.

The Nyquist sampling rate is

$$\Delta t_{\text{Nyquist}} = \frac{1}{2f_{\max}} = 0.5\,\text{ns} \tag{4.192}$$

If a PEC scatterer in free space is assumed, the time oversampling factor K should be greater than or equal to two. For this example we choose $K = 2$. The spatial sampling interval and the grid velocity need to satisfy

$$c\Delta t \leq \Delta\ell \leq c\Delta t_{\text{Nyquist}}$$
$$c/u_{\text{grid}} \leq 1/\sqrt{2}$$

The desired resolution Δf of 10 MHz requires a time window duration of

$$T_{\max} = \frac{1}{\Delta f} = 0.1\,\mu\text{s}$$

Selecting $\Delta t = 0.5\Delta t_{\text{Nyquist}},$ the required number of steps is

$$N = \frac{T_{\max}}{\Delta t} = 400$$

Thus, we will need 400 time steps and a truncated Gaussian pulse with the length of 16 time steps. We will select 512 time steps (which is greater than 400, but an integer power of 2 as required by the FFT algorithm).

(Continued)

Example 4.5: Scattering from an Infinite Cylinder with Square Cross Section *(Continued)*

Figure 4.37 shows three sets of time snapshots: two each at *early, mid,* and *late* times. Early, mid, and late refer to the time relative to the Gaussian pulse's arrival at the scatterer. The x and y axes denote the spatial coordinates of the point at which the field is evaluated, and the vertical axis gives the magnitude of the field at that particular location. The square pedestal in the center represents the cylinder cross section, and thus no field exists inside.

At early times the Gaussian shape of the wave is clearly visible. The reflection from the front face of the cylinder is evident. At mid times, some edge and corner diffraction is noticeable. At late times, the scattering disturbances are

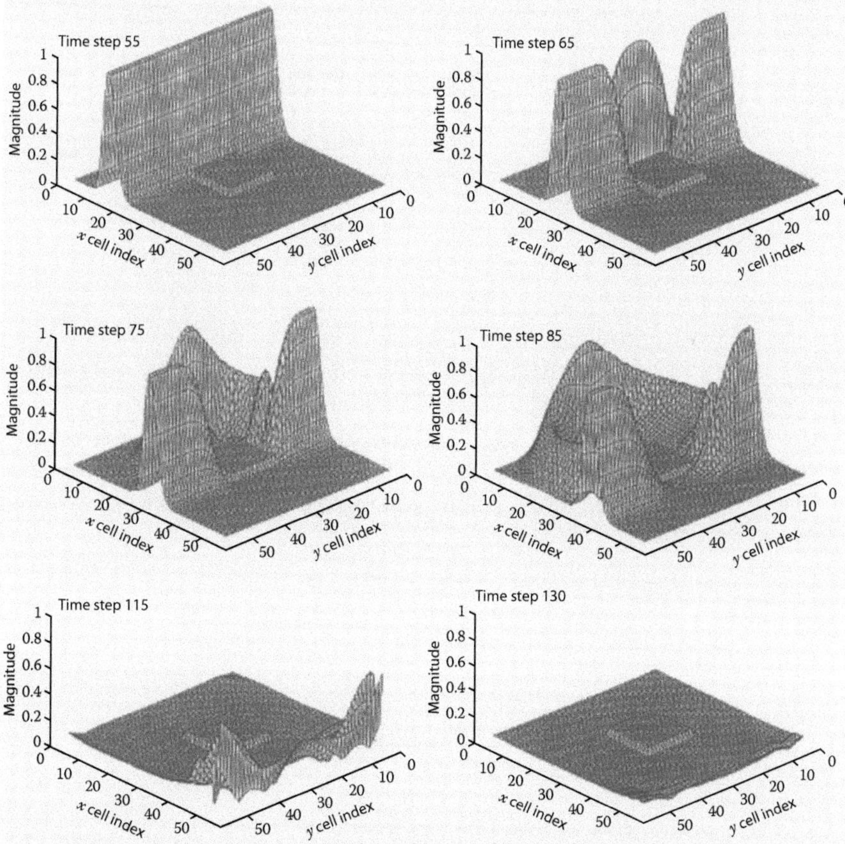

Fig. 4.37 Snapshots of the total field at early (top), middle (center), and late (bottom) times.

(Continued)

Example 4.5: Scattering from an Infinite Cylinder with Square Cross Section *(Continued)*

approaching the grid boundary and the shell of an outgoing cylindrical wave is barely visible.

The bistatic RCS of the cylinder obtained using the equivalent currents [Eqs. (4.185) and (4.186)] is shown in Fig 4.38 for frequencies of 150 and 300 MHz. In both cases the incidence angle is $\phi = 180$ deg (wave approaching from the $-x$ direction).

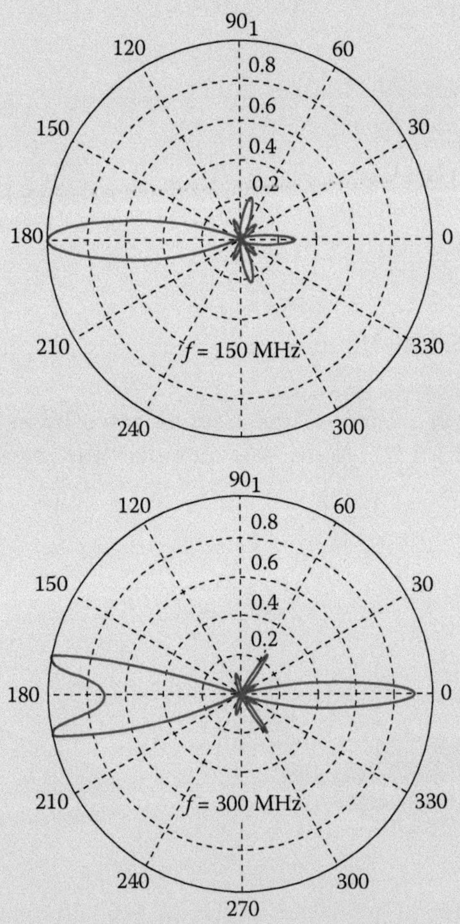

Fig. 4.38 Normalized bistatic RCS of the square cylinder in Example 4.5 ($L = 1.697$ m) at two frequencies.

4.6 Finite Difference–Time-Domain Method in Three Dimensions

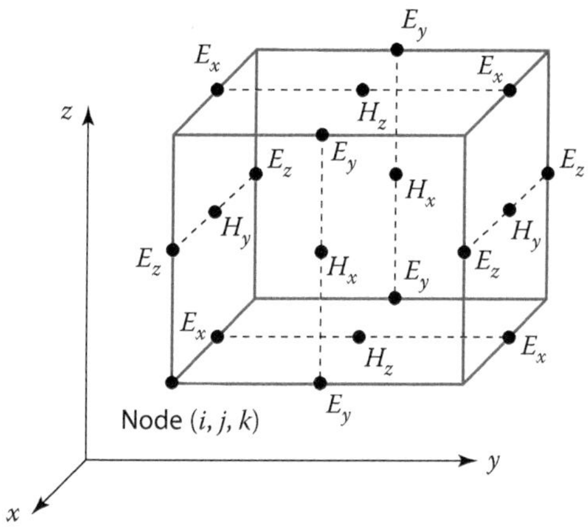

Fig. 4.39 Yee cell for three-dimensional implementation of the FDTD.

Clearly the most general application of the FDTD, and the one of most interest in the calculation of RCS for realistic targets, is the three-dimensional formulation. The development of the update equations in three dimensions follows the same procedure as for one and two dimensions. The details have been worked out in several references [2,7].

A three-dimensional *Yee cell* is shown in Fig. 4.39. As in the case for two-dimensions, we use the following shorthand notation:

Grid dimensions: $(\Delta x, \Delta y, \Delta z)$
Time step: Δt
Nodes indices: $(i, j, k) = (i\Delta x, j\Delta y, k\Delta z)$
Notation for field samples: $E_x(i\Delta x, j\Delta y, k\Delta z, n\Delta t) \equiv E_x^n(i, j, k,)$

The E components are defined in the middle of edges, whereas the H components are at the centers of the faces, denoted by integers plus one-half (for example, $i + 1/2$). With this notation the magnetic field update equations are [2]

$$
\begin{aligned}
H_x^{n+1/2}(i,j,k) &\approx H_x^{n-1/2}(i,j,k) + \frac{\Delta t}{\mu_x(i,j,k)} \\
&\times \left[\frac{E_y^n(i,j,k+1) - E_y^n(i,j,k)}{\Delta z} - \frac{E_z^n(i,j+1,k) - E_z^n(i,j,k)}{\Delta y}\right] \\
H_y^{n+1/2}(i,j,k) &\approx H_y^{n-1/2}(i,j,k) + \frac{\Delta t}{\mu_y(i,j,k)} \\
&\times \left[\frac{E_z^n(i+1,j,k) - E_z^n(i,j,k)}{\Delta x} - \frac{E_x^n(i,j,k+1) - E_x^n(i,j,k)}{\Delta z}\right] \\
H_z^{n+1/2}(i,j,k) &\approx H_z^{n-1/2}(i,j,k) + \frac{\Delta t}{\mu_z(i,j,k)} \\
&\times \left[\frac{E_x^n(i,j+1,k) - E_x^n(i,j,k)}{\Delta y} - \frac{E_y^n(i+1,j,k) - E_y^n(i,j,k)}{\Delta x}\right]
\end{aligned}
\tag{4.193}
$$

The electric field update equations are

$$
\begin{aligned}
E_x^{n+1}(i,j,k) \approx{} & \mathrm{K}_x(i,j,k)E_x^n(i,j,k) \\
& + \mathrm{P}_x(i,j,k)\left[\frac{H_z^{n+1/2}(i,j+1,k) - H_z^{n+1/2}(i,j,k)}{\Delta y}\right. \\
& \left. - \frac{H_y^{n+1/2}(i,j,k+1) - H_y^{n+1/2}(i,j,k)}{\Delta z}\right] \\
E_y^{n+1}(i,j,k) \approx{} & \mathrm{K}_y(i,j,k)E_y^n(i,j,k) \\
& + \mathrm{P}_y(i,j,k)\left[\frac{H_x^{n+1/2}(i,j,k+1) - H_x^{n+1/2}(i,j,k)}{\Delta z}\right. \\
& \left. - \frac{H_z^{n+1/2}(i+1,j,k) - H_z^{n+1/2}(i,j,k)}{\Delta x}\right] \\
E_z^{n+1}(i,j,k) \approx{} & \mathrm{K}_z(i,j,k)E_z^n(i,j,k) \\
& + \mathrm{P}_z(i,j,k)\left[\frac{H_y^{n+1/2}(i+1,j,k) - H_y^{n+1/2}(i,j,k)}{\Delta x}\right. \\
& \left. - \frac{H_x^{n+1/2}(i,j+1,k) - H_x^{n+1/2}(i,j,k)}{\Delta y}\right]
\end{aligned}
\tag{4.194}
$$

where

$$
\mathrm{K}_\alpha(i,j,k) = \frac{\epsilon_\alpha(i,j,k) - 0.5\Delta t\sigma_{c\alpha}(i,j,k)}{\epsilon_\alpha(i,j,k) + 0.5\Delta t\sigma_{c\alpha}(i,j,k)} \tag{4.195}
$$

$$
\mathrm{P}_\alpha(i,j,k) = \frac{\Delta t}{\epsilon_\alpha(i,j,k) + 0.5\Delta t\sigma_{c\alpha}(i,j,k)} \tag{4.196}
$$

with $\alpha = x$, y, or z. The three-dimensional formulation is the basis of several commercially available software packages.

4.7 Finite Integration Technique

The FIT is a time-domain numerical method based on the integral form of Maxwell's equations [16]. Because it involves integrals rather than differentials, the FIT has some desirable properties compared to the FDTD. Integral quantities are more stable when doing extensive numerical calculations, as the case for a large grid or many time steps. The calculations deal with physically measurable quantities such as voltage and current, and energy and charge conservation are easily checked, which can be used to evaluate convergence [17].

The detailed derivation of the discrete form of Maxwell's equations and their FIT application is described in Refs. 17 and 18. The purpose of this section is to highlight some of the important features of the FIT formulation and how it is implemented numerically. To begin, the target computational domain is discretized into subdomains, just as in the case of the FDTD method. Nonorthogonal and conformal grids are possible [18], but for illustrative purposes "bricks" are used as the cell geometry. A typical brick is shown in Fig. 4.40. The coordinates of node (x_i, y_j, z_k) are denoted by indices (i, j, k).

To illustrate the discretization process a couple of examples from Ref. [16] are repeated here. Maxwell's first equation (Faraday's law) is

$$\oint_L \vec{E}(\vec{r},t)\cdot \vec{d\ell} = -\iint_S \frac{\partial}{\partial t}\vec{B}(\vec{r},t)\cdot \vec{ds} \tag{4.197}$$

For the front face let

$$e_x(i,\ j,\ k) = \int_L \vec{E}(\vec{r},\ t)\cdot \vec{d\ell} \tag{4.198}$$

$$b_z(i,\ j,\ k) = \int_S \vec{B}(\vec{r},\ t)\cdot \vec{ds} \tag{4.199}$$

where L denotes the path around the perimeter of the face's surface S. Similar definitions follow for the other components of e and b. Now Faraday's law

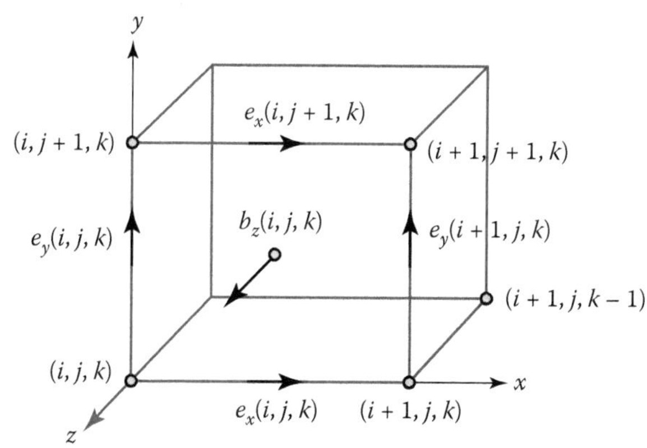

Fig. 4.40 Grid cell for implementing the FIT in three dimensions.

applied to the front face in Fig. 4.40 becomes

$$e_x(i,j,k) + e_y(i+1,j,k) - e_x(i,j+1,k) - e_y(i,j,k) = -\frac{\partial}{\partial t} b_z(i,j,k) \tag{4.200}$$

This representation can be extended to the remaining faces of the brick and then to all cells in the computational volume. The components are assembled into a column vector ordered by Cartesian coordinates, first in x, then y, and finally z, resulting in the vectors

$$\mathbf{e} = \left[\mathbf{e}_x | \mathbf{e}_y | \mathbf{e}_z\right] \tag{4.201}$$

$$\mathbf{b} = \left[\mathbf{b}_x | \mathbf{b}_y | \mathbf{b}_z\right] \tag{4.202}$$

These vectors are used to express the discretized Faraday's law for the entire computational volume

$$\underbrace{\begin{pmatrix} & & \cdots & & \\ 1 & \cdots & 1 \cdots & -1 \cdots & -1 \\ & & \cdots & & \end{pmatrix}}_{\mathbf{C}} \mathbf{e} = -\frac{\partial}{\partial t}\mathbf{b} \tag{4.203}$$

The matrix **C** represents the *discrete curl operator* with elements consisting of only three possible values:$\{-1, 0, 1\}$.

Similarly, a *discrete divergence operator* can be generated starting from Maxwell's fourth equation:

$$\iint_S \vec{B}(\vec{r}, t) \cdot \vec{ds} = 0 \tag{4.204}$$

That is, the sum of the magnetic flux out of a cell must be zero:

$$\begin{aligned} &- b_x(i,j,k) + b_x(i+1,j,k) - b_y(i,j,k) + b_y(i,j+1,k) \\ &- b_z(i,j,k) + b_z(i,j,k+1) = 0 \end{aligned} \tag{4.205}$$

Collecting these equations over the entire computational domain gives the discrete form of the divergence operator

$$\underbrace{\begin{pmatrix} & & & \cdots & & & \\ -1 & \cdots & 1 \cdots & -1 \cdots & 1 \cdots & -1 & \cdots & 1 \\ & & & \cdots & & & \end{pmatrix}}_{\mathbf{S}} \mathbf{b} = \mathbf{0} \tag{4.206}$$

The discretization of the remaining two Maxwell's equations (Ampere's law and Gauss's law) requires the introduction of a dual cell that is shifted 1/2 in all dimensions as shown in Fig. 4.41. The dual cell is analogous to the dual discretization grid used in the FDTD. Let $\tilde{G}$ denote the dual grid

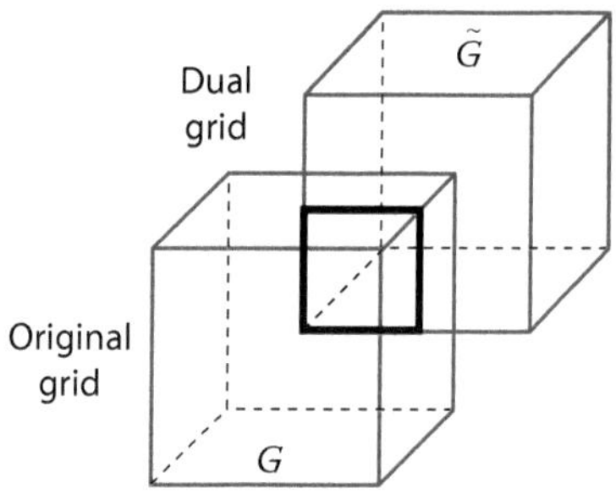

Fig. 4.41 Dual grid complex for development of the discretized equations for the FIT.

with surfaces $\tilde{S}$ and edges $\tilde{L}$. Thus Ampere's law is written as

$$\oint_{\tilde{L}} \vec{H}(\vec{r},t)\cdot \mathrm{d}\vec{\ell} = -\iint_{\tilde{S}} \left[\frac{\partial}{\partial t}\vec{D}(\vec{r},t)+\vec{J}(\vec{r},t)\right]\cdot \mathrm{d}\vec{s} \tag{4.207}$$

and Gauss's law is

$$\iint_{\tilde{S}} \vec{D}(\vec{r},t)\cdot \mathrm{d}\vec{s} = Q(\vec{r},t) \tag{4.208}$$

Collectively the discrete forms of Maxwell's equations can be written as

$$\begin{aligned} \boldsymbol{C}\,\boldsymbol{e} &= -\frac{\partial}{\partial t}\boldsymbol{b} \\ \tilde{\boldsymbol{C}}\,\boldsymbol{h} &= -\frac{\partial}{\partial t}\boldsymbol{d}+\boldsymbol{j} \\ \boldsymbol{S}\,\boldsymbol{b} &= \boldsymbol{0} \\ \tilde{\boldsymbol{S}}\,\boldsymbol{d} &= \boldsymbol{q} \end{aligned} \tag{4.209}$$

The vectors **h**, **d**, **j**, and **q** are the discrete representation of magnetic field intensity, electric flux density, current density, and charge density. They are defined in a manner analogous to **b** and **e** in Eqs. (4.201) and (4.202). $\tilde{\mathbf{C}}$ and $\tilde{\mathbf{S}}$ are the discrete curl and divergence operators for the dual grid.

The discretization procedure can be continued with the constitutive relationships, potentials, and continuity equation. The resulting set of equations uniquely and completely describes the fields and currents in a cell. Equations (4.209) are continuous functions of time, but they are easily discretized (i.e., applied to "marching in time") [19]. As with the case for any numerical technique with computational boundaries, absorbing or radiation boundary conditions must be applied at the edges.

Example 4.6: Bistatic RCS of an Aircraft Using *Microwave Studio*

FIT is the computational approach used in a commercially available software package *Microwave Studio* by CST.† Figure 4.42 shows two aircraft models

(Continued)

†Information on *Microwave Studio* by Computer Simulation Technology is available at www.cst.com.

Example 4.6: Bistatic RCS of an Aircraft Using *Microwave Studio* (Continued)

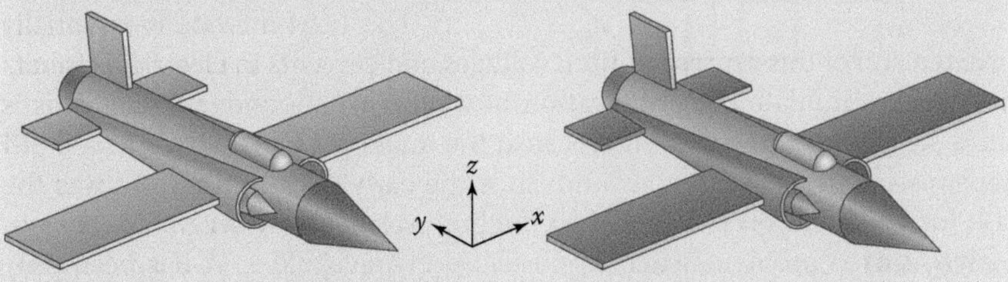

Fig. 4.42 Two aircraft models built in *Microwave Studio*: conventional (left) and "stealthy" (right).

assembled using the geometry builder in *Microwave Studio*. The model on the right has a few rudimentary stealth features incorporated (rounded leading edges and composite materials). The bistatic RCS of the two models are shown in Fig. 4.43, indicating a significant reduction in RCS for the "stealthy" version. The wingspan and length of the aircraft are approximately 8 m. The frequency is 300 MHz, the polarization is horizontal, and $\phi_i = 0$ deg.

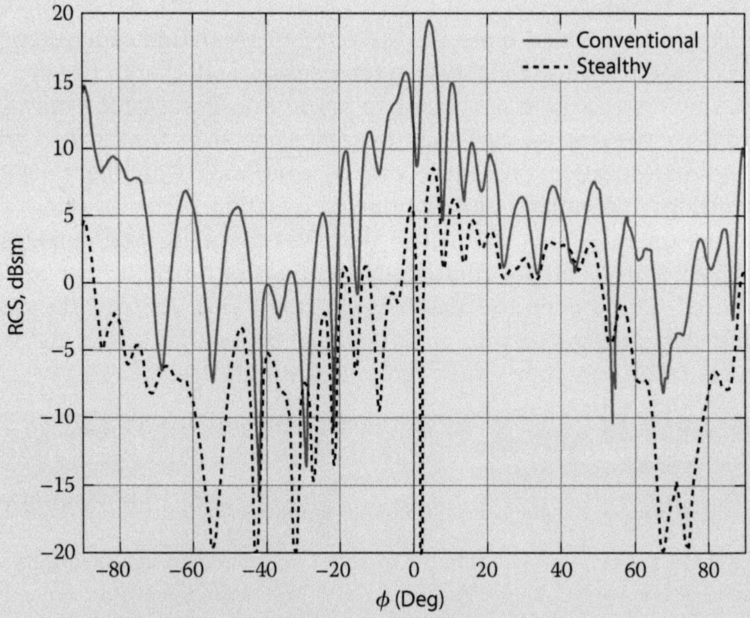

Fig. 4.43 Azimuth bistatic RCS of the two aircraft in Example 4.6.

4.8 Transmission Line Matrix Method

As a conclusion to this chapter on time domain numerical methods, we briefly examine the *transmission line matrix (TLM) method* [20, 21]. In earlier chapters, equivalent transmission line circuits have been used to represent the behavior of electromagnetic fields. The TLM method is essentially an extension of this representation: voltages and currents in electrical circuits are used to simulate the propagation of waves in unbounded space. This is not a surprising development, because the analogy of free space as a TEM transmission line has been around since the early 20th century. As was the case for FEM and FDTD, the TLM method was used for several decades to analyze closed problems such as obstacles in waveguides. It has been only since the development and refinement of absorbing and transparent boundary conditions that the method has been extended to open problems (i.e., antenna radiation and scattering) [22].

Similar to FEM and FDTD, the computational domain must be discretized. The discretization points in space, or nodes, become junctions of transmission lines. The basic building block is a differential length (Δx, Δy, Δz) of lumped element transmission lines in the x, y, and z directions. As illustrated in Fig. 4.44 for two dimensions, there are two types of nodes: *shunt nodes* used for TM_z modes and *series nodes* for TE_z modes. (There are several variations in the capacitance and inductance parameters, depending on how the particular formulation relates the voltage and current to fields.) The voltages at the series nodes represent the electric field components, and the currents of the shunt nodes represent the magnetic field components. The inductance

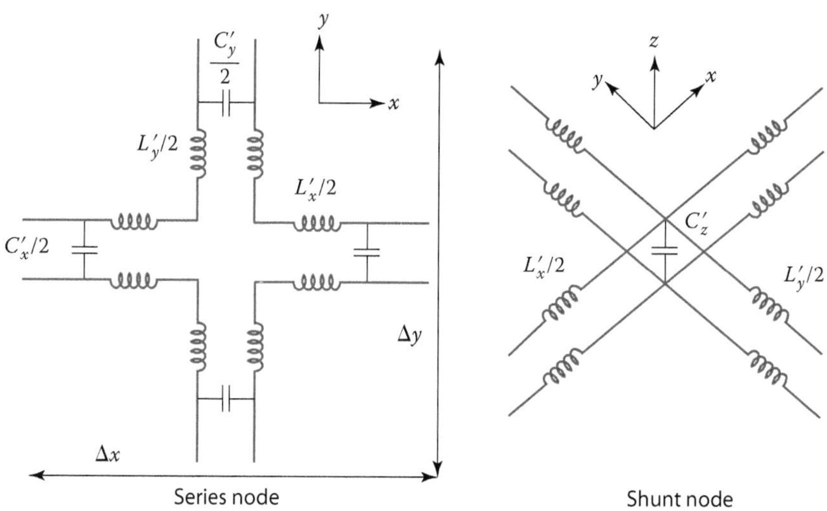

Fig. 4.44 Series and shunt nodes for modeling two-dimensional problems.

and capacitance per unit length (primed quantities) are determined by the physical and electrical properties of the cell. The electrical characteristics of a transmission line cell can be set based on the type of material that fills each particular cell. In general, a boundary can be represented by a complex impedance, or an open or short circuit stub. In its simplest form, an absorbing termination is represented by an impedance matched to the characteristic impedance of the medium, with some adjustment to account for the angle of incidence. Radiating boundaries require the application of absorbing boundary conditions.

The early formulations of TLM used *expanded nodes* where the series and shunt nodes were displaced one-half step. This arrangement arises naturally for the same reason that the contours for electric and magnetic field circulations and fluxes were separated in the FDTD. More recently the *symmetric condensed node* (SCN) formulation was developed. It combines the shunt and series nodes in multiple coordinate planes into a single point. It is a variation of TLM that allows for the currents and voltages to be computed at the same node points, rather than separating them by half a cell.

In three dimensions the representation of series and shunt nodes becomes quite a bit more complicated. A cubic cell has equal lengths, $\Delta x = \Delta y = \Delta z \equiv \Delta \ell$. A set of input/output planes is required for each coordinate axis, and each one carries two polarizations. Therefore a three-dimensional node consists of 12 ports, as depicted in Fig. 4.45 (each termination plane serves as two ports). The scattering properties of the nodes are obtained from basic circuit and conservation laws: conservation of charge and flux, and continuity of the electric and magnetic fields. Precisely which ports are allowed to couple is based on the field coupling in Maxwell's equations. Based on these considerations, a 12 by 12 scattering matrix can be defined for each node. A uniformly discretized computational

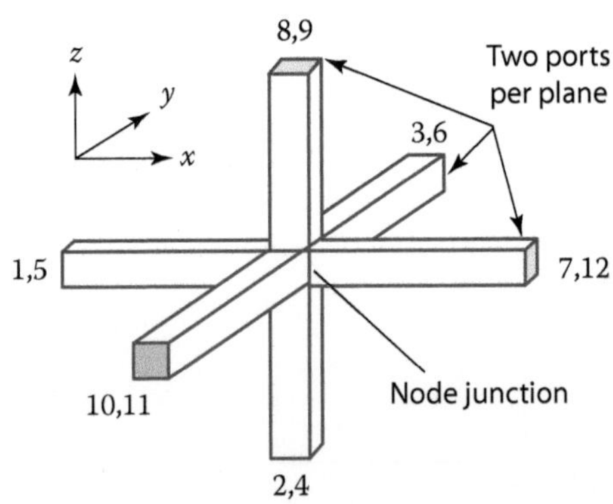

Fig. 4.45 Simplified representation of a symmetrical condensed node in three dimensions.

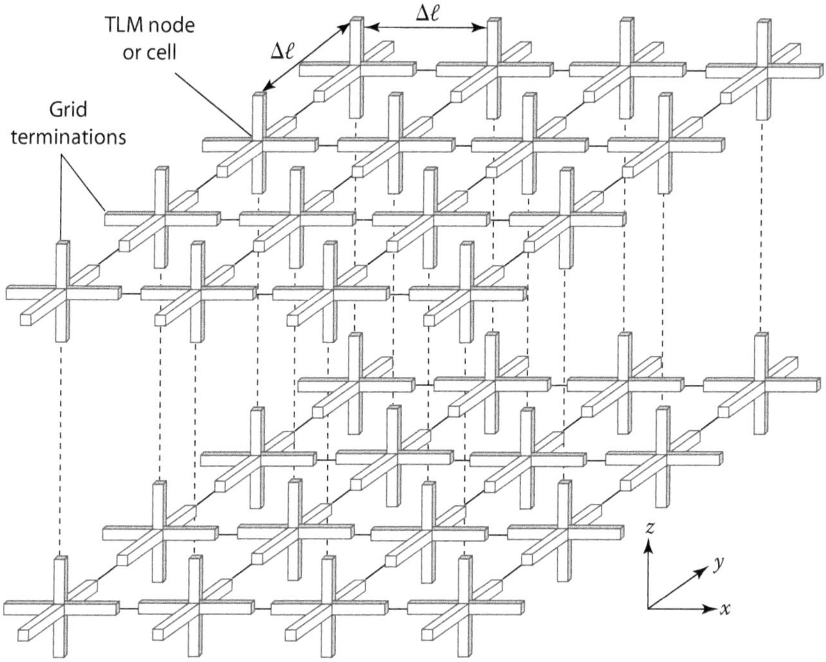

Fig. 4.46 TLM representation of a three-dimensional computational domain.

domain in Cartesian coordinates is illustrated in Fig. 4.46. A scattering matrix for the entire mesh (a *supermatrix*) comprises all of the individual node matrices. A frequency domain solution deals with solving the supermatrix. Obviously the complete matrix can be extremely large for an electrically large target, although it is a sparse matrix for which efficient and fast solution algorithms exist.

In the time domain a matrix solution is not required because all nodes are not excited simultaneously. For a time-domain solution, a voltage pulse is introduced at the nodes corresponding to the incidence direction of the plane wave. A marching-in-time process is initiated, allowing the wave to propagate throughout the grid. The velocity of propagation in each cell is determined by the cell's characteristic impedance. The time step is restricted by $\Delta\ell$ and the velocity of propagation in the cell, analogous to Eq. (4.19) for the FDTD. The time domain fields at the computational boundaries are saved and then Fourier transformed to obtain the frequency domain fields. Finally, the far-scattered fields are computed using a near-to-far transformation based on the radiation integral [23].

Many enhancements and variations of the TLM method have been developed. They include variable mesh sizes to improve convergence and reduce dispersion, extension to anisotropic materials (represented by stubs with different admittances in the proper shunt or series nodes), and time-varying media (modeled by time-varying cell parameters). Because mechanical and thermal processes have electrical circuit equivalents, it is possible to include the effects of their energies on the electromagnetic performance of the model.

From this brief description it is clear that there are many similarities between TLM and FDTD. It has been shown that under certain circumstances the two methods are exactly equivalent and give identical outputs [24]. The TLM method is the basis of the computational engine for the commercially available software package MEFiSTO. The TLM method has demonstrated sufficient accuracy for calculating the RCS of low observable targets [25].

References

[1] Yee, K. S., "Numerical Solution of Initial Boundary Value Problems Involving Maxwell's Equations in Isotropic Media," *IEEE Transactions on Antennas and Propagation*, Vol. AP-14, No. 3, May 1966, pp. 302–307.

[2] Taflove, A., and Brodwin, M. E., "Numerical Solution of Steady-State Electromagnetic Scattering Problems Using the Time-Dependent Maxwell's Equations," *IEEE Transactions on Microwave Theory and Techniques*, Vol. MTT-23, No. 8, Aug. 1975, pp. 623–630.

[3] Sun, E.-Y., and Rusch, W., "Time-Domain Physical Optics," *IEEE Transactions on Antennas and Propagation*, Vol. 42, No. 1, Jan. 1994, pp. 9–15.

[4] Miller, E. K., Poggio, A. J., and Burke, G. J., "An Integro-Differential Equation Technique for the Time-Domain Analysis of Thin Wire Structures: I. The Numerical Method" *Journal of Computational Physics*, Vol. 12, 1973, pp. 24–48.

[5] Miller, E. K., Poggio, A. J., and Burke, G. J., "An Integro-Differential Equation Technique for the Time-Domain Analysis of Thin Wire Structures: II. Numerical Results," *Journal of Computational Physics*, Vol. 12, 1973, pp. 210–233.

[6] Liu, T. K., and Mei, K. K., "A Time-Domain Integral-Equation Solution for Linear Antennas and Scatterers," *Radio Science*, Vol. 8, Nos. 8, 9, Aug.–Sept. 1973, pp. 797–804.

[7] Kunz, K. S., and Luebbers, R. J., *The Finite Difference Time Domain Method for Electromagnetics*, CRC Press, Boca Raton, FL, 1993.

[8] Sadiku, M. N. O., "Numerical Techniques in Electromagnetics," CRC Press, Boca Raton, FL, 1992.

[9] Brigham, E. O., *The Fast Fourier Transform and Its Applications*, Prentice-Hall, Englewood Cliffs, NJ, 1988.

[10] Stremler, F. G., *Introduction to Communications Systems*, 3rd ed., Addison-Wesley, Reading, MA, 1990.

[11] Spiegel, M. R., *Mathematical Handbook of Formulas and Tables*, Schaum's Outline Series in Mathematics, McGraw-Hill, New York, 1968.

[12] Mur, G., "Absorbing Boundary Conditions for the Finite-Difference Approximation of the Time-Domain Electromagnetic Field Equations," *IEEE Transactions on Electromagnetic Compatibility*, Vol. EMC-23, No. 4, April 1981, pp. 377–382.

[13] Blaschak, J., and Kriegsmann, G., "A Comparative Study of Absorbing Boundary Conditions," *Journal of Computational Physics*, Vol. 77, No. 1, July 1988, pp. 109–139.

[14] Engquist, B., and Majda, A., "Absorbing Boundary Conditions for the Numerical Simulation of Waves," *Mathematics of Computation*, Vol. 31, No. 139, July 1977, pp. 629–651.

[15] Balanis, C. A., *Advanced Engineering Electromagnetics*, Chap. 11, Wiley, New York, 1989.
[16] Tonti, E., "Finite Formulation of the Electromagnetic Field," *Progress in Electromagnetic Research (PIER) Monograph Series*, PIER 32, EMW Publishing, Cambridge, MA, 2001, pp. 1–44.
[17] Clemens, M., and Weiland, T., "Discrete Electromagnetism with the Finite Integration Technique," *Progress in Electromagnetic Research (PIER) Monograph Series*, PIER 32, EMW Publishing, Cambridge, MA, 2001, pp. 65–87.
[18] Schuhmann, R., and Weiland, T., "Stability of the FDTD Algorithm on Nonorthogonal Grids Related to the Spatial Interpolation Scheme," *IEEE Transactions on Magnetics*, Vol. 34, No. 5, May 1998, pp. 2751–2754.
[19] Mattiussi, G., "The Geometry of Time-Stepping," *Progress in Electromagnetic Research (PIER) Monograph Series*, PIER 32, EMW Publishing, Cambridge, MA, 2001, pp. 123–149.
[20] Christopoulos, C., *The Transmission-Line Modeling Method (TLM)*, IEEE Press, Piscataway, NJ, 1995.
[21] Hoefer, W. J. R., "The Transmission Line Matrix Method–Theory and Applications," *IEEE Transactions on Microwave Theory and Techniques*, Vol. 33, No. 10, Oct. 1985, pp. 882–893.
[22] Chen, Z., Ney, M., and Hoefer, W. J. R., "Absorbing and Connecting Boundary Conditions for the TLM Method," *IEEE Transactions on Microwave Theory and Techniques*, Vol. 41, No. 11, Nov. 1993, pp. 2016–2024.
[23] Khalladi, M., Morente, J., Porti, J., and Gimenez, G., "Two Near- to Far-Zone Approaches for Scattering Problems Using the TLM Method," *IEEE Transactions on Antennas and Propagation*, Vol. 41, No. 4, April 1993, pp. 502–505.
[24] Johns, P., "On the Relationship Between TLM and Finite-Difference Methods for Maxwell's Equations," *IEEE Transactions on Microwave Theory and Techniques*, Vol. 35, No. 1, Jan. 1987, pp. 60–61.
[25] Porti, J., Morente, J., Magan, H., and Ruiz, D., "RCS of Low Observable Targets with the TLM Method," *IEEE Transactions on Antennas and Propagation*, Vol. 46, No. 5, May 1998, pp. 741–743.

Problems

4.1 Calculate the effective duration and effective bandwidth (T_{eff} and B_{eff}) of a Gaussian pulse with $f(t)$ given by Eq.(4.39).

4.2 The physical optics approximation can be applied to the time domain as well as the frequency domain [3]. The radiation integral with the frequency dependence explicitly written is

$$\vec{E}(\omega) = \frac{-j\omega\mu_0}{4\pi}\int_V \{\vec{J}(\omega) - [\vec{J}(\omega)\cdot\hat{r}]\hat{r}\}\frac{e^{-jkr}}{r}\mathrm{d}v'$$

For $r/2 \gg 1/2\pi$.

a) Use the Fourier transform pair

$$\vec{E}(\omega) = \frac{1}{2\pi}\int_{-\infty}^{\infty}\vec{E}(t)e^{-j\omega t}\mathrm{d}t$$

$$\vec{E}(t) = \frac{1}{2\pi}\int_{-\infty}^{\infty}\vec{E}(\omega)e^{+j\omega t}\mathrm{d}t$$

to derive the time-domain form of the radiation integral

$$\vec{E}(t) = \int_{-\infty}^{\infty} \left\{ \frac{-j\omega\mu_0}{4\pi} \int_V (\vec{J}(\omega) - [\vec{J}(\omega) \cdot \hat{r}]\hat{r}) \frac{e^{-jkr}}{r} dv' \right\} e^{j\omega t} d\omega$$

b) Use the fact that, for time-harmonic fields,

$$\frac{\partial}{\partial t}[\vec{J}(\omega)e^{j\omega t}] = j\omega\vec{J}(\omega)e^{j\omega t}$$

and

$$I(t - r/c) \equiv \int_{-\infty}^{\infty} \vec{J}(\omega)e^{j\omega(t-r/c)} d\omega$$

with $k = \omega/c$ to obtain

$$\vec{E}(t) = \frac{\mu_0}{4\pi} \frac{\partial}{\partial t} \int_V \{I(t - r/c) - [I(t - r/c) \cdot \hat{r}]\hat{r}\} \frac{dv'}{r} \qquad \text{(P4.1)}$$

Note that $\vec{E}(t)$ depends on the derivative of the current and that the $\hat{r}$ current component does not contribute to the radiated field, just as in the harmonic case.

4.3 In this problem, we use Eq. (P4.1), derived in Problem 4.2, to determine the approximate time-domain scattering from a flat conducting plate. Assume the following:

1. Plate of dimensions $L_x \times L_y$ lies in the $x - y$ plane;
2. Wave incident in the $y - z$ plane; and
3. Observation point in the far field (i.e., parallel rays from the plate surface to the observation point).

The incident field is given by

$$\vec{H}_i = -H_0\hat{x} \sin\left(\frac{2\pi t}{T}\right) U(t)$$

where $T = 1/f$ is the carrier period and $U(t)$ the unit step function.

a) What is the current induced on the plate by the incident wave according to the physical optics approximation?

b) Show that the scattered field due to the current in part a is

$$E(t) = -\frac{\mu_0}{4\pi} \frac{4\pi H_0}{T} \int_{S_{\text{illum}}} \cos\left(\frac{2\pi t}{T}\right) \frac{ds}{r}$$

where S_{illum} denotes the surface illuminated by the incident wave front at time t.

c) Break the plate into N strips (parallel to x), and assume that, if the incident wave front touches any portion of a strip, the entire strip is illuminated. (Alternately, we could adjust the time step for the given incident angle so that the wave front traverses one entire strip each time step.) Compute the scattered field for $L_y = 1$ m, $f = 10$ GHz, $\theta_i = 45$ deg, $\theta_s = 135$ deg, $r = 50$ m, and $N = 60$. For a simple model, consider only the current induced by the incident wave front in computing the scattered field. (In other words, neglect the scattering from one strip that impinges on another strip and thus changes the total current on that strip.)

4.4 Consider the infinite dielectric slab of Example 4.4.

a) Determine the grid size $\Delta\ell'$ inside the dielectric so that the numerical solution "propagates" at the same velocity as in free space for the grid spacing $\Delta\ell$.

b) Write a computer program to compute the one-dimensional RCS, and compare the error to that for the uniform grid size used in the example.

4.5 Consider the cylinder with square cross section of Example 4.5.

a) Use the method of moments to compute the three-dimensional bistatic RCS of the cylinder if $L = \lambda/2$ at 1 GHz. Assume that the length of the cylinder in the z dimension is $L_z = 2\lambda$. The incidence direction is $\theta_i = 90$ deg and $\phi_i = 0$ deg. Plot the RCS as a function of ϕ for $\theta = 90$ deg.

b) Write a computer program to compute the two-dimensional RCS using the FDTD method, and compare it to the result obtained in part a. Use the approximation in Eq. (4.145):

$$\sigma_{3D} \approx \sigma_{2D} \frac{2L_z^2}{\lambda}$$

to convert RCS data from two to three dimensions.

4.6 Derive the TE$_z$ update Eqs. (4.158–4.160) starting with Eqs. (4.151) and (4.152) using the contours of Fig. 4.29.

4.7 Derive the TM$_z$ update Eqs. (4.161–4.163) starting with Eqs. (4.153) and (4.154) using the contours of Fig. 4.30.

4.8 Consider a one-dimensional grid with absorbing boundary conditions at the edges. Write a program that implements a unit impulse (delta-function) source at a node close to either edge of the grid. If the grid were ideal, the responses at any of the nodes within the grid would be just delayed replicas of the source impulse. Show that this indeed is the case when the velocity ratio is selected as one. Does this also hold for other velocity ratios that satisfy the stability condition?

4.9 Consider a square grid with N nodes on the side. Assume that the two-dimensional field is TM_z and that the edge nodes are electric field (E_z) nodes. The top of the grid is illuminated with a Gaussian plane wave as depicted in Fig. P4.9. Assume that the plane wavefront is parallel to the top edge of the grid.

a) Write a two-dimensional FDTD program implementing a Neumann-like boundary condition for the grid sides, that is, update the left and right side edge nodes such that E_z of a side-edge node equals the value of its nearest neighbor just inside the grid. Use homogeneous Dirichlet boundary condition ($E_z = 0$) for the bottom side of the grid. Select the velocity ratio as $1/\sqrt{2}$. Plot the incident field at the node in the center of the grid, before the reflection off the bottom side reaches the center of the grid.

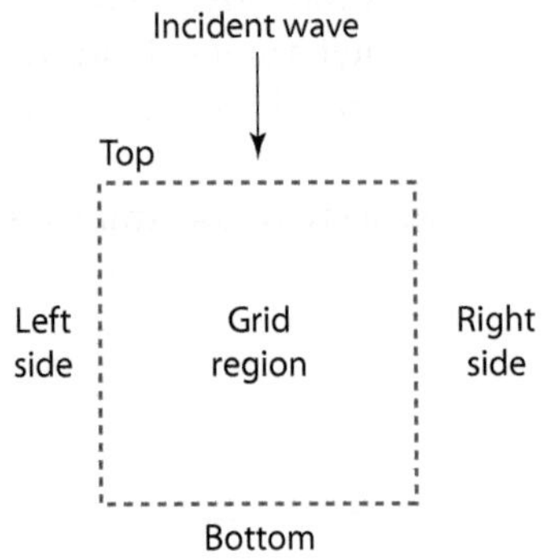

Fig P4.9 Two-dimensional grid box with a wave incident from the top.

b) Compare the incident field waveform at the center of the grid to the incident field at the top edge of the grid. Note any distortion caused by the grid "dispersion".

c) Modify the program developed in part a such that the velocity ratio is one. Although this theoretically violates the stability requirement, the waveform observed at the grid center (again before the reflection off the bottom reaches the center) will actually be closer to the incident field waveform. Verify this by modifying your program accordingly. Explain why the velocity ratio of 1 works under these circumstances.

d) Devise a simple boundary condition for the bottom of the grid such that no reflections occur there. Implement this boundary condition in your program and verify that it works.

4.10 Consider a two-dimensional grid with a point source (E_z) in the center. The source produces a unit impulse (a delta pulse) at $t = 0$. If the grid were ideal, the response at any other E_z node within the grid would be just a delayed and attenuated replica of the source pulse. Select the grid velocity ratio as $1/\sqrt{2}$ and plot the impulse responses at several nodes, at various distances from the source node. What do you conclude about the "distortion" of the pulse as it travels through the grid? How does this compare with the one-dimensional case with the velocity ratio of 1?

4.11 A single cycle of a sinusoid is often used as an ultra-wideband waveform. Plot the Fourier transform (positive frequencies) if the sinusoid has a frequency of 1 GHz.

4.12 Run the MATLAB® program for Example 4.4 (*fdtdpanel.m*) for the default values. Note the amplitudes of the Gaussian wave inside of the panel compared to the outside. Explain the relative amplitudes.

4.13 Reflection and transmission by a plane wave incident on an infinite planar interface (e.g., a panel) can be computed using a finite sample with the appropriate PMC and PEC boundary conditions, as shown in Fig. P4.13. The cell can be any convenient size. Waveguide ports 1 and 2 are assigned at the front and back faces. Using commercial software such as CST Microwave Studio® or Altair FEKO®, calculate S_{11} and S_{21} for a dielectric panel (with dielectric constant and thickness of your choosing) using the boundary conditions in the figure.

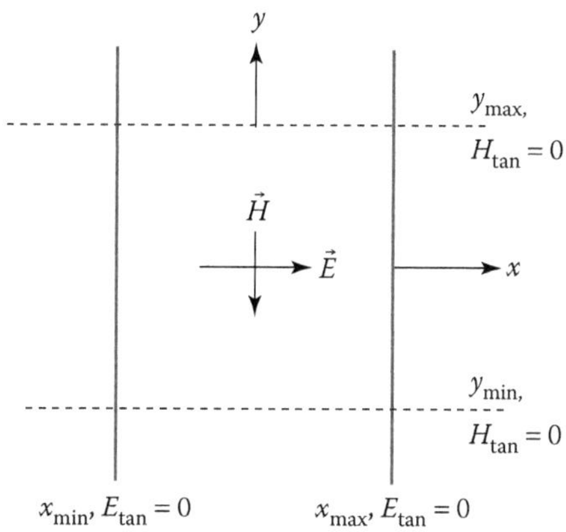

Fig P4.13 Equivalent boundary conditions for a plane wave normally incident on an infinite structure.

4.14 Consider a simple target such as a plate, cylinder, or sphere of moderate size (several wavelengths). Conduct a set of simulations using a frequency domain solver (MM or FEM) versus a time domain solver (FDTD or FIT). The objective is to observe some of the fundamental tradeoffs in methods such as computer memory and CPU requirements, and run time. Comment on how this comparison might change if the frequency band of interest is increased.

Chapter 5 Microwave Optics

5.1 Introduction

Microwave optics refers to a collection of ray-tracing methods. In previous chapters, the fundamental mechanism of specular reflection (i.e., a reflection that obeys Snell's law) has been used to justify concepts such as image theory. Ray tracing has an advantage over the numerical methods of Chapters 3 and 4 in that the propagation of rays gives physical insight into scattering behavior. For example, when a plane wave is reflected from an interface, it is easy to see that the maximum reflection will occur when the observation direction satisfies Snell's law.

Geometrical optics (GO) [1–3] is a version of ray tracing that has been used for centuries at optical frequencies to design systems of lenses (telescopes, microscopes, etc.). It works well at optical frequencies because the lens dimensions are much larger than a wavelength and, thus, the interaction of the wave with the lens becomes a "localized" phenomenon. Under these circumstances, the reflected ray appears to originate from a single point on the surface called a *specular point.*

There are several problems with geometrical optics, and they are more pronounced as the scattering body becomes smaller. For example, a reflection from an infinite PEC surface would result in a reflected field observed only when $\theta_r = \theta_i$ and in a zero field elsewhere. This agrees with the physical optics result for the same situation; an infinite surface gives a scatter pattern that has zero beamwidth. For a finite surface, PO yields scattering patterns with finite beamwidths, whereas GO still gives only nonzero fields in the specular direction. A second shortcoming of GO is the discontinuity of the field at shadow boundaries. GO predicts zero fields in shadowed regions. In Sec. 2.11, it was shown, using Huygen's principle, that the field at the shadow boundary is -6 dB relative to the field on the illuminated side; it decays to zero in the shadow region at a rate that increases with frequency. Most optics problems are scalar in nature and polarization is ignored.

At microwave frequencies, a useful ray-tracing theory must include the following scattering mechanisms: 1) specular contributions from large surface areas, 2) edge scattering, and 3) interactions between and among surfaces and edges. The inclusion of polarization is also important at microwave frequencies.

Geometrical and physical optics are capable of satisfying condition 1 and, to some extent, condition 3. The edge-related effects require a mathematical model for *edge diffraction.* The edge diffraction model akin to GO is the *geometrical theory of diffraction* (GTD)[4]. The corresponding model used in

Table 5.1 Summary of Two Approaches

Mechanism	Current Based	Ray Optics Based
Surface reflection	*GO surface current*—physical optics (PO)	*Specular rays*—geometrical optics (GO)
Edge scattering	*Fringe currents*—physical theory of diffraction (PTD)	*Diffracted rays*—geometrical theory of diffraction (GTD)
Surface waves	Interaction between fringe currents	Multiple diffractions

conjunction with PO is the *physical theory of diffraction* (PTD)[5]. Although the PTD is not a ray-tracing technique, it will be discussed in this chapter because it is a *high-frequency technique* for predicting edge diffraction.

In the discussions that follow, we consider ray optics to include GO and the GTD. As first developed, the GTD had several shortcomings with regard to singularities in the field. Several modified versions of the GTD, such as the *uniform theory of diffraction* (UTD)[6] and the *asymptotic theory of diffraction* (ATD)[7], have eliminated the problems by introducing correction factors. In the following discussions, the GTD will refer to the collection of these diffraction theories.

For the GTD, rays are hypothesized that obey diffraction laws similar to reflection laws. A diffraction coefficient is defined for an edge that depends on the edge geometry and polarization of the incident wave. Upon diffraction, the scattered field is given by the incident field times the diffraction coefficient. Furthermore, the diffracted wave follows prescribed straight-line paths in free space. The total field at an observation point is the vector sum of all the reflected and diffracted fields arriving at that point.

The edge correction for PO is provided by *fringe currents* that flow along edges. The magnitude, phase, and location of these currents are given by the PTD. To obtain the edge-scattered field at a point in space, it is necessary to use the fringe currents in the radiation integrals. The total field is the sum of the standard PO scattered field plus the edge-scattered field. The two approaches are summarized in Table 5.1.

Each approach has its advantages and disadvantages. As already mentioned, ray optics can be tied to the physics of the scattering problem. Another advantage of the GTD is that it provides a nonzero field in shadow regions, but the inclusion of surface waves is often difficult. This is especially true for curved surfaces with edges. For complex targets, many multiple reflections and diffractions may be needed to predict the RCS accurately. This can result in long run times for computer codes, even though no complicated integrations or matrix operations are involved.

Whereas the PTD is simple in principle, it can be inconvenient in practice. The fringe currents can extend an appreciable distance away from the edge, and, therefore, a two-dimensional integration must be performed.

This disadvantage has been eliminated for the most part by the development of *equivalent edge currents*. These are filaments of current that flow at the edge, as opposed to the PTD surface currents that flow on "ribbons" along the edge. These currents are a function of incidence angle, however, and so, for monostatic calculations, the integrations must be performed with each angle change. Another disadvantage is that a comprehensive theory for fringe current interaction has not yet been developed. On the plus side, the edge-scattered field is simply added to the PO field. This is the major attraction of the PTD because PO is so convenient and widely used.

5.2 Geometrical Optics*

5.2.1 Postulates

Geometrical optics is the most basic theory that describes wave behavior upon reflection or refraction at an interface between two materials. Geometrical optics is based on the following six assumptions or postulates:

1. ***Waves are everywhere locally plane.*** This TEM approximation implies that reflection surfaces are far from sources and that the radii of curvature of scattering surfaces are much greater than the wavelengths.
2. ***The wave direction is specified by the normal to the equiphase surface.*** Equiphase surfaces are referred to as *eikonal surfaces*. As shown in Fig. 5.1, these are spherical surfaces for isotropic point sources and planes for plane waves.

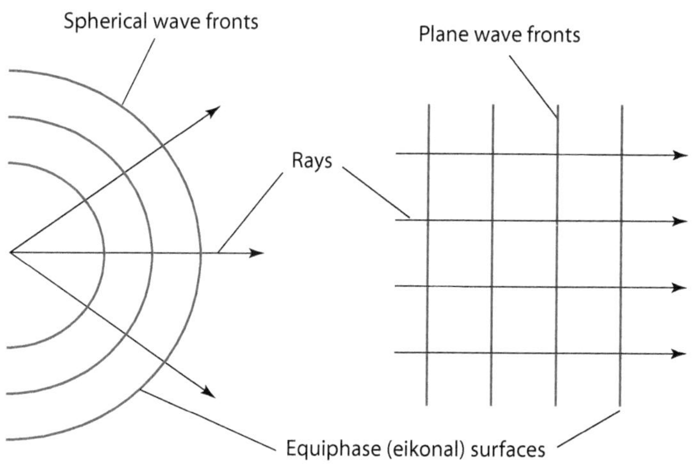

Fig. 5.1 Eikonal surfaces for spherical and planar waves.

*Much of the material in Secs. 5.2 and 5.3 is adapted from unpublished notes by W. V. T. Rusch.

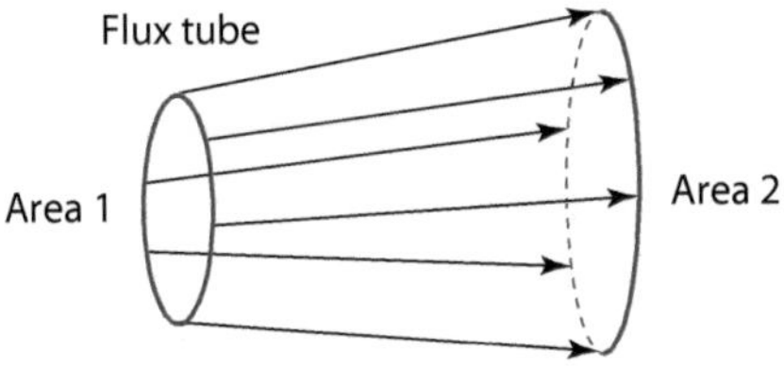

Fig. 5.2 Flux tube.

3. ***Rays travel in straight lines.*** In a homogeneous medium, rays travel in straight lines.
4. ***Polarization is constant along a ray.*** Polarization can change on reflection or refraction. After reflection, however, the wave polarization will not change if the medium is isotropic.
5. ***Power in a flux tube is conserved.*** Power is viewed as flowing in *flux tubes*, which are essentially "bundles" of rays as shown in Fig. 5.2. A flux tube has the following properties:
   a) Eikonal surfaces (wave fronts) are perpendicular to all rays in the tube.
   b) There is no energy flux through the sides of the tube.
   c) The cross section of the tube is characterized by the radii of curvature in the two principal planes.

Figure 5.3 shows an *astigmatic* wave front, that is, one with different radii of curvature in two principal planes. How the two principal planes are defined will be discussed later. The power density at points along the tube depends on the cross-sectional area because power is conserved. In the notation of Fig. 5.3,

$$|\vec{W}_0|\mathrm{d}s_0 = |\vec{W}|\mathrm{d}s \tag{5.1}$$

or, for the field,

$$|\vec{E}| = |\vec{E}_0|\sqrt{\frac{\mathrm{d}s_0}{\mathrm{d}s}} \tag{5.2}$$

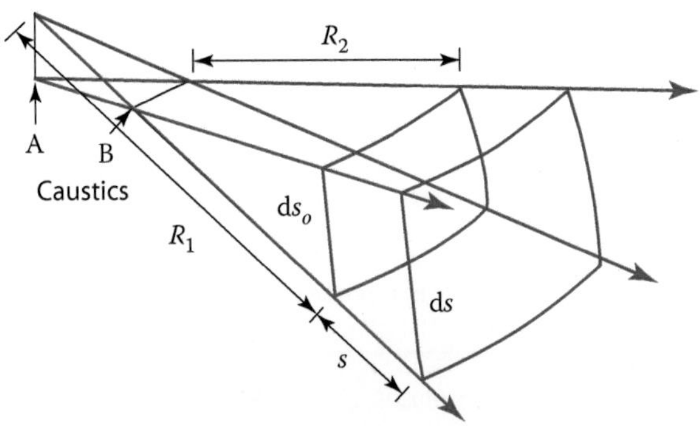

Fig. 5.3 Astigmatic flux tube.

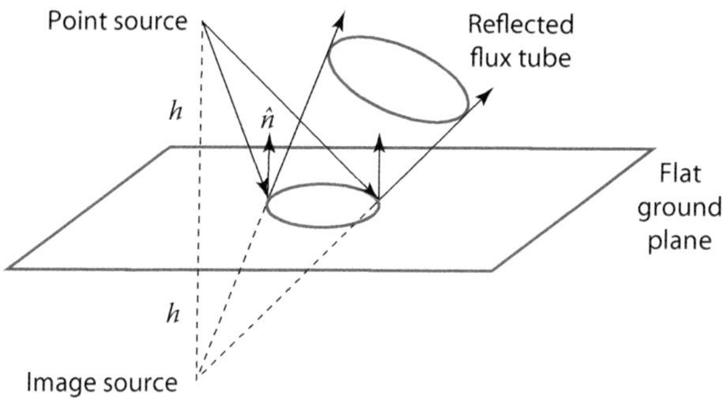

Fig. 5.4 Reflection of a ray tube at a planar interface.

The square root term is the *spread factor*. Including the phase delay for the distance s gives

$$\vec{E} = \vec{E}_0 \sqrt{\frac{R_1 R_2}{(R_1 + s)(R_2 + s)}} e^{-jks} = \vec{E}_0 A(s) e^{-jks} \tag{5.3}$$

There are three primary cases of interest:

a) Spherical wave, $R_1 = R_2 = R$ and $s \gg R$:

$$\frac{R}{R+s} e^{-jks} \rightarrow \frac{e^{-jks}}{s} \tag{5.4}$$

b) Plane wave, $R_1 = R_2 = \infty$: The spread and phase factors become e^{-jks}.

c) Cylindrical wave, $R_1 = \infty$, $R_2 = \rho$ and $s \gg \rho$:

$$\sqrt{\frac{\rho}{\rho + s}} e^{-jks} \rightarrow \frac{e^{-jks}}{\sqrt{s}} \tag{5.5}$$

6. ***The reflection of rays obeys Snell's law.*** Each ray in a tube satisfies Snell's law as depicted in Fig. 5.4 for a planar interface. Snell's law is a consequence of *Fermat's principle*, which states: *The ray (or rays) from a source to an observer is a curve with a length that is stationary with respect to infinitesimal variations in path.*

Assume that the source and observation points are at (x', y', z') and (x, y, z) and the reflection points at (x_s, y_s, z_s). Then, the distances from the reflection point on the surface to the source and observation points are

$$\begin{aligned} r_1 &= \sqrt{(x_s - x)^2 + (y_s - y)^2 + (z_s - z)^2} \\ r_2 &= \sqrt{(x_s - x')^2 + (y_s - y')^2 + (z_s - z')^2} \end{aligned} \tag{5.6}$$

Fermat's principle requires that

$$\frac{\partial}{\partial x_s}(r_1 + r_2) = 0 \quad \text{and} \quad \frac{\partial}{\partial y_s}(r_1 + r_2) = 0 \tag{5.7}$$

or, in vector form, $\hat{k}_i \times \hat{n} = \hat{k}_r \times \hat{n}$.

Example 5.1: Vector Form of Snell's Law

Snell's law requires that the incident and reflected rays satisfy $\theta_r = \theta_i$. If the incident and reflected ray vectors are given by $\hat{k}_i$ and $\hat{k}_r$, respectively then,

$$\begin{aligned} \hat{k}_r &= \hat{k}_i - 2(\hat{n} \cdot \hat{k}_i)\hat{n} \\ \vec{E}_r &= -\vec{E}_i + 2(\hat{n} \cdot \vec{E}_i)\hat{n} \\ \vec{H}_r &= \vec{H}_i - 2(\hat{n} \cdot \vec{H}_i)\hat{n} \end{aligned} \tag{5.8}$$

The cross section of a flux tube is related to the field intensity in the region. Equation (5.2) presents a problem if the cross section is zero, as at locations A and B in Fig. 5.3. Energy must be conserved, but the tube cross section is zero. These are *caustics*, that is, points at which the energy density is infinite. Depending on the wave-front curvature, caustics can be points, lines, or surfaces. GO fails at caustics, and other methods must be used to obtain the field. However, there are no general techniques that are valid for determining the *caustic correction factors* for arbitrary geometries.

An example of a point caustic is shown in Fig. 5.5. A reflecting spherical shell is illuminated by the point source at the origin. All the incident rays are normal to the spherical surface and, therefore, all reflected rays converge back at the origin.

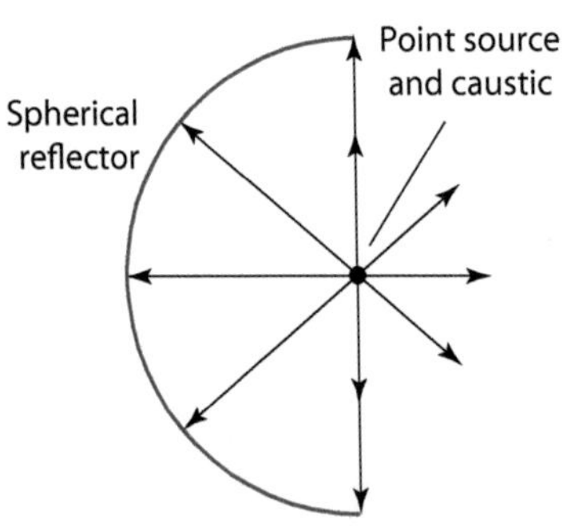

Fig. 5.5 Caustic at the center of a reflecting sphere.

5.2.2 Doubly Curved Surfaces

If the reflecting surface is curved rather than flat, a divergence factor must be included. The curvature causes the flux tube on reflection to spread at a different rate than the incident flux tube. Figure 5.6 illustrates the effect for a spherical surface. In most cases, the surface curvature is more complicated, as shown in Fig. 5.7. Just as in the case of the astigmatic flux tube, we define two principal planes with R_1^s and R_2^s the radii of curvature in these two planes.

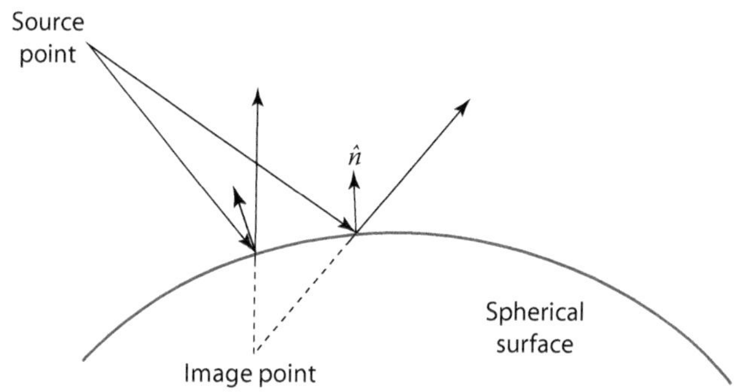

Fig. 5.6 Ray tube divergence at a spherical reflecting surface.

The two principal planes are normally determined by approximating the surface curvature by circular arcs lying in orthogonal planes intersecting at the reflection point as depicted in Fig. 5.7.

Unit vectors tangent to the surface that lie in the principal planes are denoted $\hat{u}_1$ and $\hat{u}_2$. The coordinate system has arbitrarily been ordered so that $\hat{u}_1 \times \hat{u}_2 = \hat{n}$, where $\hat{n}$ is the unit surface normal at the reflection point Q.

The angles the incident ray makes with the tangent vectors and normal are

$$\theta_1 = \cos^{-1}(\hat{u}_1 \cdot \hat{k}_i)$$

$$\theta_2 = \cos^{-1}(\hat{u}_2 \cdot \hat{k}_i)$$

$$\theta_i = \cos^{-1}(\hat{n} \cdot \hat{k}_i)$$

Note that the ray may not necessarily be incident in a principal plane. The distance along the incident ray from the source to Q is denoted s'. Similarly, the distance along the reflected ray from Q to the observation point P is s.

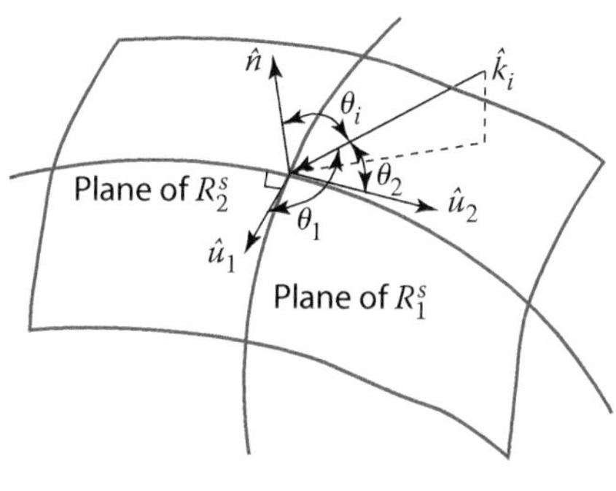

Fig. 5.7 Geometry for reflection from an arbitrary doubly curved surface.

If the principal planes of the incident wavefront and surface can be aligned then the *focal lengths* of the surface are given by

$$\frac{1}{f_{1,2}} = \frac{1}{\cos\theta_i}\left(\frac{\sin^2\theta_2}{R_1^s} + \frac{\sin^2\theta_1}{R_2^s}\right) \pm \left[\frac{1}{\cos^2\theta_i}\left(\frac{\sin^2\theta_2}{R_1^s} + \frac{\sin^2\theta_1}{R_2^s}\right)^2 - \frac{4}{R_1^s R_2^s}\right]^{\frac{1}{2}} \tag{5.9}$$

More general formulas are available in Ref. [5].

Now, the radii of curvature for the reflected wave front can be determined from the radii of curvature of the incident wave front and the focal lengths of the surface:

$$\begin{aligned} \frac{1}{R_1^r} &= \frac{1}{2}\left[\frac{1}{R_1^i} + \frac{1}{R_2^i}\right] + \frac{1}{f_1} \\ \frac{1}{R_2^r} &= \frac{1}{2}\left[\frac{1}{R_1^i} + \frac{1}{R_2^i}\right] + \frac{1}{f_2} \end{aligned} \tag{5.10}$$

Two important incident wave shapes are spherical ($R_1^i = R_2^i = s'$) and planar ($R_1^i = R_2^i = \infty$).

5.2.3 Matrix Form for the Reflected Field

The reflected field at observation point P can be written in matrix form in terms of the quantities just defined:

$$\mathbf{E}_r(s) = \mathbf{R} \cdot \mathbf{E}_i(s')A(s)e^{-jks}e^{j\phi_c} \tag{5.11}$$

The incident and reflected field vectors are defined as

$$\boldsymbol{E}_r(s) = \begin{bmatrix} E_{r\perp}(s) \\ E_{r\parallel}(s) \end{bmatrix} \tag{5.12}$$

and

$$\boldsymbol{E}_i(s') = \begin{bmatrix} E_{i\perp}(s') \\ E_{i\parallel}(s') \end{bmatrix} \tag{5.13}$$

$A(s)$ is the divergence and spreading factor:

$$A(s) = \left[\frac{R_1^r R_2^r}{(R_1^r + s)(R_2^r + s)}\right]^{\frac{1}{2}} \tag{5.14}$$

The phase term ϕ_C accounts for any phase changes that occur when the path traverses a caustic:

$$\phi_C = \begin{cases} 0, & \text{if neither center of curvature lies between } Q \text{ and } P \\ \pi/2, & \text{if one center of curvature lies between } Q \text{ and } P \\ \pi, & \text{if both centers of curvature lie between } Q \text{ and } P \end{cases}$$

The reflection coefficient matrix is

$$\boldsymbol{R} = \begin{bmatrix} \Gamma_{\perp\perp} & \Gamma_{\perp\parallel} \\ \Gamma_{\parallel\perp} & \Gamma_{\parallel\parallel} \end{bmatrix} \tag{5.15}$$

The first of each pair of subscripts on Γ refers to the polarization of the reflected wave and the second to the polarization of the incident wave. For a PEC, the reflection coefficient matrix simplifies to

$$\boldsymbol{R} = \begin{bmatrix} \Gamma_{\perp} & 0 \\ 0 & \Gamma_{\parallel} \end{bmatrix} = \begin{bmatrix} -1 & 0 \\ 0 & -1 \end{bmatrix} \tag{5.16}$$

The next few examples illustrate how these formulas are applied for plane wave scattering calculations.

Example 5.2: Monostatic Scattering by a Sphere

A plane wave is incident on a sphere of radius a located at the origin. The polarization does not matter for this problem because of the symmetry of the sphere. The observation point is a distance s from the reflection point. The principal radii of curvature of the wave front are $R_1^i = R_2^i = \infty$. The focal lengths in the two principal planes, which are arbitrary in the case of the sphere, are determined from Eq. (5.9) with $\theta_1 = \theta_2 = 90$ deg:

$$f_1 = f_2 = \sqrt{\frac{R_1^s R_2^s}{4}}$$

where $R_1^s = R_2^s = a$. Thus,

$$R_1^r = R_2^r = \sqrt{\frac{R_1^s R_2^s}{4}} \quad \Rightarrow \quad R_1^r R_2^r = \frac{R_1^s R_2^s}{4} \tag{5.17}$$

The divergence and spreading factor is

$$A(s) = \sqrt{\frac{(a/2)^2}{[(a/2)+s]^2}} \approx \frac{a}{2s} \tag{5.18}$$

where the fact that $s \gg a/2$ has been used.

The magnitude of the reflected field is

$$E_r(s) = E_i(Q)\frac{a}{2s}e^{-jks} \tag{5.19}$$

(Continued)

Example 5.2: Monostatic Scattering by a Sphere (Continued)

The RCS is computed from

$$\sigma = \lim_{s\to\infty} 4\pi s^2 \frac{|E_i(Q)(a/2s)e^{-jks}|^2}{|E_i(Q)|^2} \tag{5.20}$$

or

$$\sigma = 4\pi\left[\frac{a}{2}\right]^2 = \pi a^2 \tag{5.21}$$

This agrees with the data of Fig. 1.6 for the optical region (large a/λ).

Intuitively, it may not seem reasonable that the RCS is independent of frequency. For high frequencies, the surface of the sphere is "flat" in terms of wavelength. The scattered field can be considered to originate from a small disk centered at the specular point. As the frequency is increased, the area of this disk decreases, but the beamwidth of its scattered field also decreases because λ is smaller. Less energy is scattered, but it is concentrated in a narrower beam. These two effects compensate for each other, resulting in an RCS that is independent of frequency.

Example 5.3: Monostatic Scattering by a Paraboloid

Figure 5.8 shows a paraboloid obtained by rotating a parabolic curve about the z axis. The focal length is f and the diameter D. The angle ψ is measured from the negative z axis to a point on the reflector. The value of ψ at the edge is

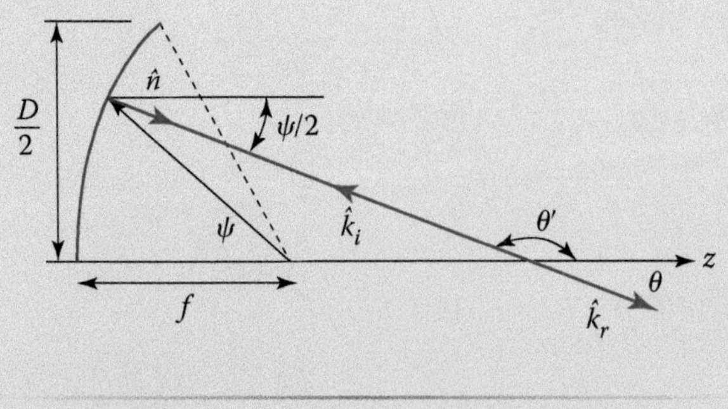

Fig. 5.8 Monostatic scattering by a paraboloid.

(Continued)

Example 5.3: Monostatic Scattering by a Paraboloid (Continued)

denoted ψ_e and given by

$$\psi_e = 2\tan^{-1}\left[\frac{D}{4f}\right] \tag{5.22}$$

For a monostatic RCS, according to GO, there is only a reflected field in the direction of P if there is a specular point on the paraboloid surface; that is, the surface normal must point at P. In terms of ψ, the surface normal is

$$\hat{n} = -\hat{r}\cos(\psi/2) - \hat{\theta}\sin(\psi/2) \tag{5.23}$$

Therefore, if $\theta > (\psi_e/2)$ (or $\theta' > \pi - \psi_e/2$), then, $E_s = 0$.

For plane wave incidence,

$$R_1^r R_2^r = f_1 f_2 = \frac{R_1^s R_2^s}{4} \tag{5.24}$$

and, from formulas for *conic mirrors* [8],

$$R_1^s = R_2^s = \frac{2f}{\cos^2(\psi/2)} \tag{5.25}$$

Using Eqs. (5.24) and (5.25) in Eq. (5.14) yields

$$A(s) = \frac{f}{s\cos^2(\psi/2)} = \frac{2f}{s(1+\cos\psi)} \tag{5.26}$$

The RCS of the paraboloid is

$$\sigma = \lim_{s\to\infty} 4\pi s^2 \frac{|[2f/s(1+\cos\psi)]E_i(Q)e^{-jks}|^2}{|E_i(Q)|^2} \tag{5.27}$$

Finally, for $\theta' > \pi - \psi_e$,

$$\sigma = 4\pi f^2\left[\frac{2}{1+\cos(2\theta)}\right]^2 \tag{5.28}$$

The RCS of a 5λ paraboloid is shown in Fig. 5.9. The focal length-to-diameter ratio (f/D) is 0.4, which results in an edge angle ψ_e of 64 deg.

Results for PO and the MM are included in Fig. 5.9 for comparison [9]. The PO and MM curves essentially follow the GO curve but have about 3 dB of ripple. The GO field drops to zero at the *reflection boundary* ($\psi_e/2$). As the diameter of the reflector is increased, the rate of decay of the field beyond the reflection boundary increases. For very large diameters ($D \approx 1000\lambda$), the decay is so rapid that the field appears almost discontinuous, as in the GO case. The difference between the MM and PO at wide angles is due to a traveling wave, which is not included in the PO calculation.

(Continued)

Example 5.3: Monostatic Scattering by a Paraboloid (*Continued*)

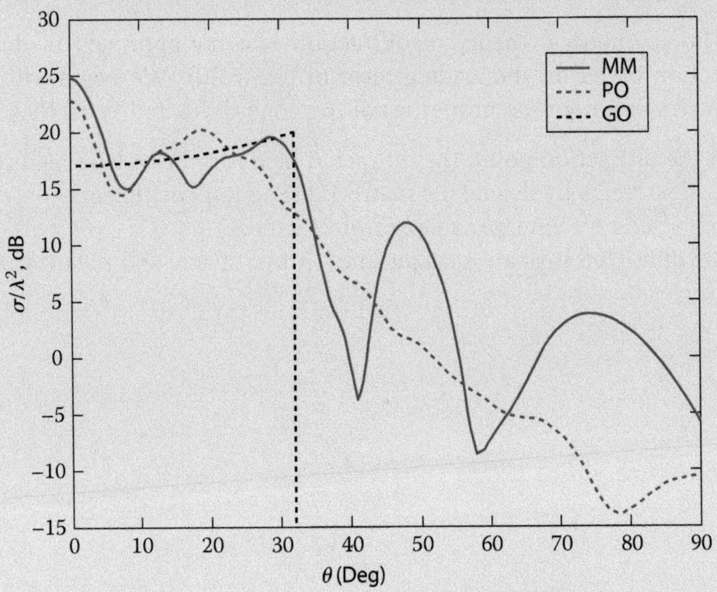

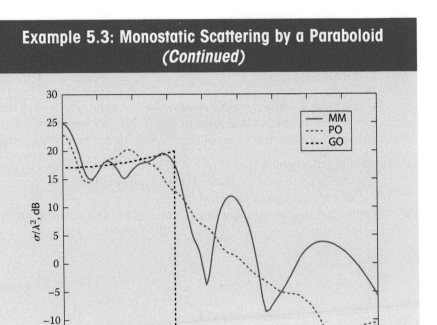

Fig. 5.9 Radar cross section of a paraboloid.

5.3 Geometrical Theory of Diffraction

GO has several shortcomings that present serious problems. First, the source and observation points must satisfy Snell's law or the reflected field is zero. GO also yields discontinuous fields at reflection and shadow boundaries and zero fields in shadowed regions. Thus, GO is too imprecise for most engineering applications dealing with targets that are small or only moderately large in terms of the wavelength of the incident radar wave.

The GTD has been developed to supplement GO and eliminate many of the problems just mentioned. The total field at an observation point is decomposed into GO and diffracted components:

$$\vec{E}_r(\vec{r}) \equiv \vec{E}_g(\vec{r}) + \vec{E}_d(\vec{r}) \tag{5.29}$$

It is assumed that the diffracted field can be expressed as an asymptotic series in inverse powers of k:

$$\vec{E}_d(\vec{r}) = e^{-jks(\vec{r})} \frac{1}{\sqrt{k}} \sum_{m=0}^{\infty} (-jk)^{-m} \vec{E}_m^d(\vec{r}) \tag{5.30}$$

as $k \to \infty$. This is a *Luneberg–Klein* series, and the leading term is of order $1/\sqrt{k}$. (The GO field is of order k^0.)

5.3.1 Postulates

The geometrical theory of diffraction is a ray approach to describing diffraction based on the leading term of Eq. (5.30). We begin with several postulates and note the properties of the edge-diffracted fields that result:

1. At the diffraction point, the diffracted fields are linearly related to the incident fields by dyadic (or matrix) diffraction coefficients.
2. Diffracted rays emerge radially from the edge.
3. The diffracted rays are straight lines in free space, and polarization remains constant along a ray.
4. The diffracted fields are TEM.
5. The field strength is inversely proportional to the square root of the cross-sectional area of the flux tube.
6. At a caustic, there is a phase advance of 90 deg; an edge is always a caustic.
7. At a focus, there is a phase advance of 180 deg.

The general edge diffraction geometry is shown in Fig. 5.10. The edge is formed by the joining of two semi-infinite planes that define a *wedge* with internal angle $(2-n)\pi$. When $n = 2$, the wedge degenerates to a single semi-infinite plane, and the resulting edge is referred to as a *knife edge*. In general, n does not have to be an integer.

In Fig. 5.10, a ray is shown incident on the edge at an angle β'. (In this chapter, the symbol β refers to an angle, not a phase constant.) A cone of

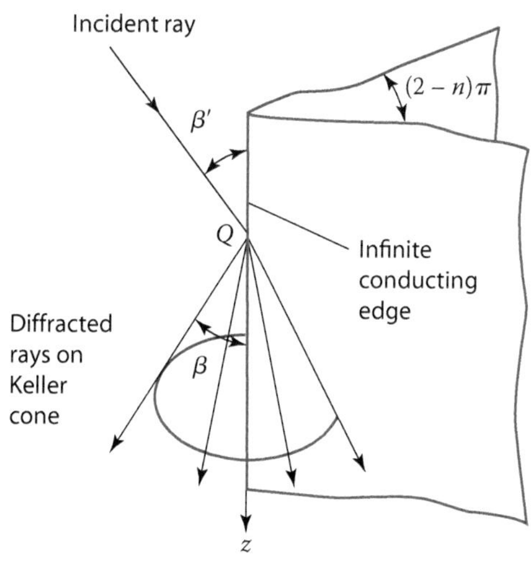

Fig. 5.10 Diffracted ray geometry and the Keller cone.

diffracted rays emerges from the diffraction point at an angle β. If the edge lies along the z axis and $x_s = y_s = 0$, the path length between the source and observation points is

$$d = \sqrt{x^2 + y^2 + (z_s - z)^2} + \sqrt{x'^2 + y'^2 + (z_s - z')^2} \tag{5.31}$$

Fermat's principle requires that

$$\frac{\partial d}{\partial z_s} = 0 \tag{5.32}$$

which is equivalent to $\beta = \beta'$. Thus, for Q to be a diffraction point, the observation point must lie on the surface of a cone with its vertex at Q, and the cone half-angle is β. The cone of diffracted rays is called the *Keller cone* after J. B. Keller [1,4]. Note that, when $\beta' = 90$ deg, the cone degenerates into a disk.

5.3.2 Edge-Fixed and Ray-Fixed Coordinate Systems

For an *edge-fixed* coordinate system, it is natural to decompose the polarization of the incident wave into components parallel and perpendicular to the reference plane defined by the incident ray and the edge. The diffraction coefficients are denoted D and, in general, are different for these two field components.
Thus,

1. Parallel (soft) polarization (D_s or $D_{\|}$) refers to the component of $\vec{E}$ parallel to the plane. The boundary condition for this component is $E_z = 0$.
2. Perpendicular (hard) polarization (D_h or $D_{\|}$) refers to the component of $\vec{E}$ perpendicular to the plane. The boundary condition for this component is $H_z = 0$.

For convenience, most diffraction calculations are performed in the *ray-fixed* coordinate system (β, ϕ) defined in Fig. 5.11. In this system,

$$\vec{E}_\beta = \vec{E}_{\|} \text{ and } \vec{E}_\phi = \vec{E}_{\perp} \tag{5.33}$$

In terms of the *ray-fixed* coordinate system, the GTD field can be expressed as

$$\vec{E}_d(P) = \left[\mathbf{D} \cdot \vec{E}_i(Q) + \mathbf{D}' \cdot \frac{\partial \vec{E}_i}{\partial n}\Big|_Q\right] A(s, s')e^{-jks} \tag{5.34}$$

where $\vec{E}_i(Q)$ is the incident field at Q and $\partial \vec{E}_i / \partial n|_Q$ denotes the derivative (slope) of the incident field at Q with respect to the normal to the edge. $\mathbf{D}$ is a matrix of diffraction coefficients and $\mathbf{D}'$ a matrix of *slope diffraction coefficients.*

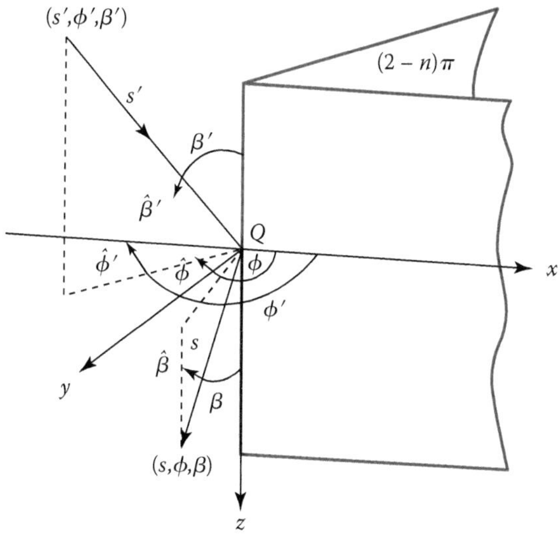

Fig. 5.11 Ray-fixed coordinate system.

The spreading and divergence factor is more complicated than in the GO case. It depends on the type of wave incident, its radii of curvature and angle of incidence, and the curvature of the edge in the vicinity of the diffraction point. In general, it is given by

$$\sqrt{\frac{\rho_c}{s(\rho_c + s)}}\left(\approx \frac{\sqrt{\rho_c}}{s} \text{ if } s \gg \rho_c\right) \tag{5.35}$$

where ρ_c is the caustic distance from the edge.

Expanding Eq. (5.34) in the ray-fixed system,

$$\begin{bmatrix} E_{d\beta} \\ E_{d\phi} \end{bmatrix} = \left\{ \begin{bmatrix} -D_s & 0 \\ 0 & -D_h \end{bmatrix} \begin{bmatrix} E_{i\beta'} \\ E_{i\phi'} \end{bmatrix} + \frac{1}{jk \sin\beta} \begin{bmatrix} -\dfrac{\partial D_s}{\partial \phi'} & 0 \\ 0 & -\dfrac{\partial D_h}{\partial \phi'} \end{bmatrix} \begin{bmatrix} \dfrac{\partial E_{i\beta'}}{\partial n} \\ \dfrac{\partial E_{i\phi'}}{\partial n} \end{bmatrix} \right\} A(s, s')e^{-jks} \tag{5.36}$$

This equation defines the diffracted field components at the observation point once the diffraction coefficients are known. The origin of the diffraction coefficients is discussed in the following section.

5.4 Diffraction Coefficients

The diffraction coefficients are derived from *canonical problems.* They are the simplest boundary-value problems that have the same fundamental local geometry as the problem of interest. For instance, if diffraction from a thin rectangular plate is being analyzed, an appropriate canonical problem is an infinite knife edge. Because diffraction is a local phenomenon, these two problems are essentially the same as long as the diffraction point on the plate is not near a corner. (Corners have their own diffraction coefficients.)

5.4.1 Half-Plane Diffraction

Scattering from an infinite half-plane was solved by Sommerfeld in the late nineteenth century. Only a brief outline of the procedure for obtaining the coefficients is given here. More detailed discussion is provided in Ref. [10]. For a given incidence direction, reflection and shadow boundaries can be defined as shown in Fig. 5.12. In region I, all three field components are present: incident, reflected, and diffracted. In region II, there are only direct and diffracted components, whereas region III has only a diffracted field. According to Eq. (5.29), the total field in each region is decomposed into GO and diffracted components. The GO terms are

$$E_g(k\rho) = \begin{cases} e^{jk\rho\cos(\phi-\phi')} \pm e^{jk\rho\cos(\phi+\phi')}, & \text{in region I} \\ e^{jk\rho\cos(\phi-\phi')}, & \text{in region II} \\ 0, & \text{in region III} \end{cases} \tag{5.37}$$

Here $\vec{r}$ has been replaced by ρ because the edge is infinite and, therefore, the scattered wave is cylindrical.

The diffracted field exists in all three regions. It is broken into two terms: one associated with the incident field and the other with the reflected field:

$$E_d(k\rho) \equiv V = V_i(\rho, \phi - \phi', n) \pm V_r(\rho, \phi + \phi', n) \tag{5.38}$$

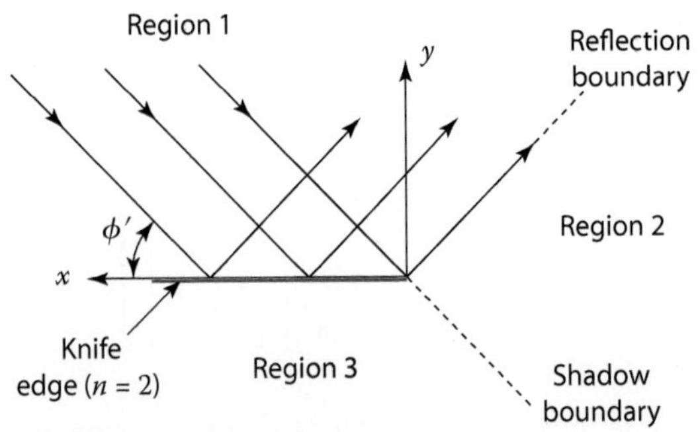

Fig. 5.12 Geometry for calculating diffraction from a knife edge.

Expressions for the diffracted field result from subtracting Eq. (5.34) from Sommerfeld's solution. If the incident ray is normal to the edge ($\beta' = \pi/2$), then,

$$V_{h,s} = \frac{e^{-j\pi/4}}{\sqrt{2\pi k}} \frac{\sin(\pi/n)}{n} \frac{e^{-jk\rho}}{\sqrt{\rho}} \times \left[\frac{1}{\cos(\pi/n) - \cos[(\phi - \phi')/n]} \pm \frac{1}{\cos(\pi/n) - \cos[(\phi + \phi')/n]} \right] \tag{5.39}$$

From Eq. (5.34) with no slope diffraction, the form of the diffracted field is given by

$$E_{dp} = E_{ip}(Q) D_p \frac{e^{-jk\rho}}{\sqrt{\rho}} \tag{5.40}$$

where $p = s, h$ for soft and hard polarizations. Thus, the diffraction coefficients are

$$D_{h,s} = \frac{e^{-j\pi/4}}{\sqrt{2\pi k}} \frac{\sin(\pi/n)}{n} \times \left[\frac{1}{\cos(\pi/n) - \cos[(\phi - \phi')/n]} \pm \frac{1}{\cos(\pi/n) - \cos[(\phi + \phi')/n]} \right] \tag{5.41}$$

Sommerfeld's solution is based on the asymptotic evaluation of an integral that is not valid in the vicinity of the shadow and reflection boundaries. For example, $n = 2$ for a knife edge, and the diffraction coefficients blow up when $\phi = \phi' + \pi$, which is a shadow boundary. This is a shortcoming of Keller's original GTD.

5.4.2 Definition of the Diffraction Coefficients

Diffraction coefficients that are valid near shadow boundaries have been obtained by using transition functions that cancel out the singularities. The most popular of these is the UTD [5], which uses a Fresnel integral as the transition function:

$$F(x) = 2j|\sqrt{x}|e^{jx} \int_{|\sqrt{x}|}^{\infty} e^{-j\tau^2} \, d\tau \tag{5.42}$$

The corrected coefficients are determined from

$$D_{s,h}(\ell, \phi, \phi', \beta, \beta') = \frac{e^{-j\pi/4}}{2\sqrt{2\pi k}\sin\beta} \left[\frac{F(k\ell a(\phi^-))}{\cos(\phi^-/2)} \pm \frac{F(k\ell a(\phi^+))}{\cos(\phi^+/2)} \right] \tag{5.43}$$

where $\phi^{\pm} = \phi \pm \phi'$, $a(\phi^{\pm}) = 2\cos[2]\ (\phi^{\pm}/2)$, and

$$A(s, s') = \begin{cases} \sqrt{\dfrac{s'}{s(s'+s)}}, & \text{spherical wave incidence} \\ \dfrac{1}{\sqrt{s}} & \text{cylindrical } (s \rightarrow \rho = s \sin\beta) \text{ and plane wave incidence} \end{cases} \tag{5.44}$$

$$\ell = \begin{cases} \dfrac{s's}{s'+s}\sin^2\beta, & \text{spherical} \\ \dfrac{\rho'\rho}{\rho'+\rho}, & \text{cylindrical} \\ s\sin^2\beta', & \text{plane waves} \end{cases} \tag{5.45}$$

Example 5.4: Bistatic Scattering from a Knife Edge (After Ref. [11])

A simple example of the application of the GTD is illustrated in Fig. 5.13. A plane wave is incident from the direction ϕ'. This is a two-dimensional problem in that the field distribution will be independent of z. However, the characteristics of the edge-scattered field are essentially the same as those of a long finite edge when viewed in a plane perpendicular to the edge.

When cylindrical coordinates are used, the source is at a fixed location (ρ', ϕ'), and the observation point is at (ρ, ϕ). If perpendicular polarization is assumed, the incident, reflected, and diffracted field components are given by

$$E_{i\perp} = e^{-j\vec{k}_i \cdot \vec{r}} = e^{jk\rho\cos(\phi-\phi')} \tag{5.46}$$

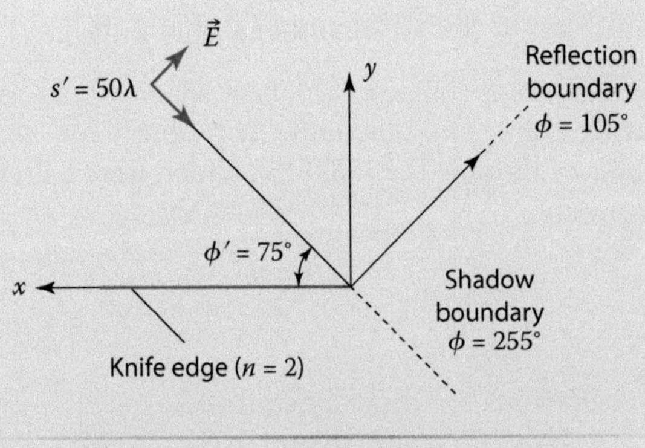

Fig. 5.13 Bistatic scattering from a knife edge.

(Continued)

Example 5.4: Bistatic Scattering from a Knife Edge (After Ref. [11]) *(Continued)*

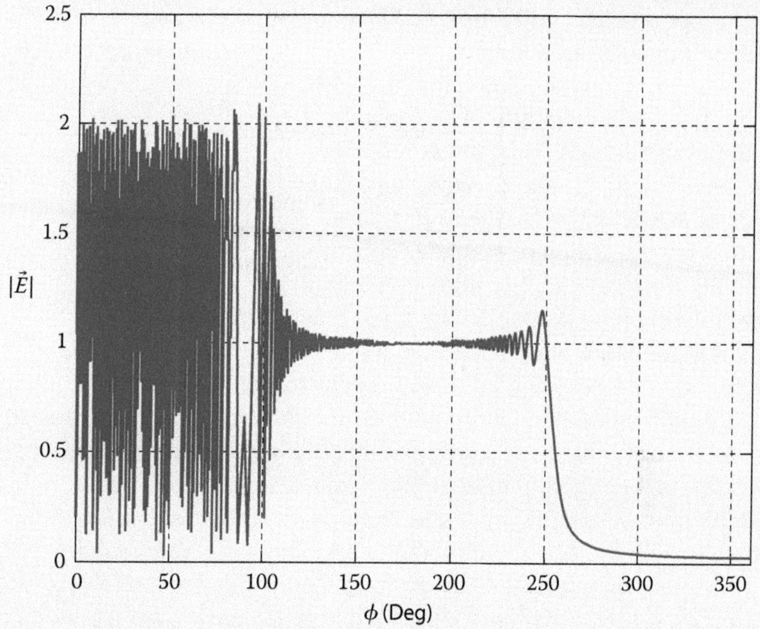

Fig. 5.14 Total field scattered from a knife edge (plane wave incidence).

$$E_{r\perp} = e^{-j\vec{k}_r \cdot \vec{r}} = e^{jk\rho\cos(\phi+\phi')} \tag{5.47}$$

$$E_{d\perp} = E_{i\perp}(Q) D_\perp \frac{e^{-jk\rho}}{\sqrt{\rho}} \tag{5.48}$$

The total field at the observation point is

$$E_\perp = E_{i\perp} + E_{r\perp} + E_{d\perp}$$

To evaluate D, specific values of s, s', β, ϕ, and ϕ' are needed. For $\phi' = 75$ deg, $\beta = 90$ deg, and $s' = 50\lambda$, the total field is plotted in Fig. 5.14.

From Fig. 5.14, it is evident that the structure of the scattered field is distinctly different in the three regions. In region I (0 deg $\leq \phi \leq$ 105 deg), the oscillations are due to the incident and reflected field adding and canceling. These two components are much greater than the diffracted field in this region.

In region II (105 deg $\leq \phi \leq$ 225 deg), the field is determined essentially by the incident (or direct) value. The oscillations increase near the reflection and shadow boundaries, where the strength of the diffracted field is greatest (because the diffraction coefficients are larger in these directions).

Finally, in region III (225 deg $\leq \phi \leq$ 360 deg), only the diffracted field is present. Note that the field strength at the shadow boundary is one-half of the direct value, which is −6 dB, as predicted by Huygen's principle.

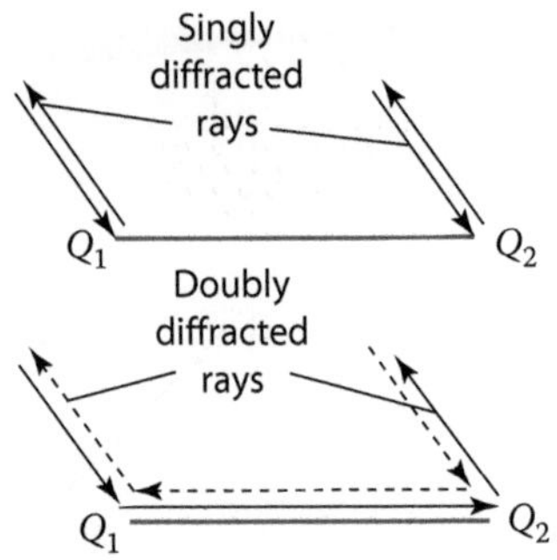

Fig. 5.15 Singly and doubly diffracted field components from parallel edges.

Consider the simplest example with two edges: the infinite strip of width w, a cross section of which is shown in Fig. 5.15. The incident plane wave will hit both edges, and a diffracted ray from each edge will return to the observation point. These are *singly diffracted rays* (first-order diffraction). One of the diffracted rays from each edge will travel along the surface and hit the opposite edge, as illustrated in Fig. 5.15. These *surface rays* are diffracted again and, as in the case of the original ray, one returns to the observation point and another travels back to the opposite edge. These are *doubly diffracted rays* (second-order diffraction). The process continues indefinitely to include third-order, fourth-order, etc., diffracted rays. In most cases, it is not necessary to include higher than second-order diffractions. The following example illustrates the significance of multiple diffracted rays.

Example 5.5: Monostatic Scattering Components from an Infinite Strip

An infinite thin strip of width w is shown in Fig. 5.16. The monostatic RCS is to be computed using the GTD. The plane wave is incident normal to the edge ($\beta = \pi/2$) and is perpendicularly polarized (TM_y). With the quantities defined in the figure,

$$s_1' = s' + \frac{w}{2}\cos\phi$$

$$s_2' = s' - \frac{w}{2}\cos\phi$$

$$\phi_1 = \phi$$

$$\phi_2 = \pi - \phi$$

(Continued)

Example 5.5: Monostatic Scattering Components from an Infinite Strip *(Continued)*

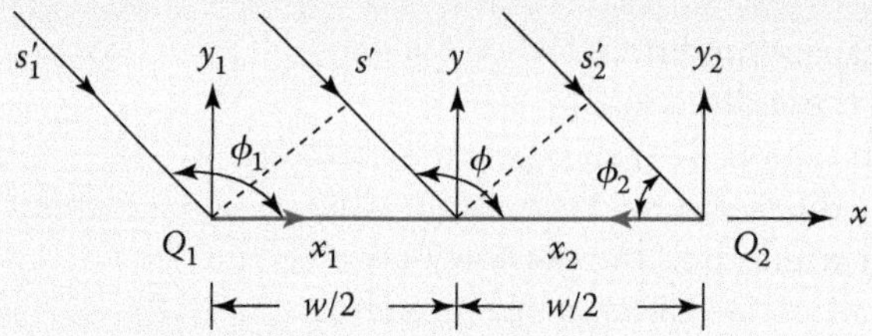

Fig. 5.16 Infinitely long strip of width *w*.

The incident field at the diffraction points is given by

$$E_i(Q_m) = e^{-jks_m}$$

where $m = 1, 2$. The subscript $\perp$ has been omitted, but it is understood that all field components are perpendicularly polarized.

For the singly diffracted rays, the incident wave is planar, and the observation point is in the same direction as the source, so that $s_m = s_m'$ and $\phi_m = \phi_m'$.

Thus,

$$\ell_m = s_m$$

$$A_m = \frac{1}{\sqrt{s_m}}$$

The total singly diffracted field is

$$E_{d1}(P) = E_i(Q_1)D_\perp\left(s_1, \phi_1, \phi_1, \frac{\pi}{2}, \frac{\pi}{2}\right)\frac{e^{-jks_1}}{\sqrt{s_1}} + E_i(Q_2)D_\perp\left(s_2, \phi_2, \phi_2, \frac{\pi}{2}, \frac{\pi}{2}\right)\frac{e^{-jks_2}}{\sqrt{s_2}} \tag{5.49}$$

Because $\beta = \pi/2$, the Keller cone is a disk, and one diffracted ray at Q_1 travels along the surface to Q_2, at which point it is diffracted again. Similarly, a surface-diffracted ray from Q_2 hits Q_1, as illustrated in Fig. 5.15. The singly diffracted rays are diffracted a second time at the opposite edge. The total doubly diffracted field returning to the observation point is

$$E_{d2} = E_i(Q_1)D_\perp\left(s_1, 0, \phi_1, \frac{\pi}{2}, \frac{\pi}{2}\right)\frac{e^{-jkw}}{\sqrt{w}}\left(\frac{1}{2}\right) \times D_\perp\left(\frac{s_1 w}{s_1 + w}, \phi_2, 0, \frac{\pi}{2}, \frac{\pi}{2}\right)\frac{e^{-jks_2}}{\sqrt{s_2}} + E_i(Q_2)D_\perp\left(s_2, 0, \phi_2, \frac{\pi}{2}, \frac{\pi}{2}\right)\frac{e^{-jkw}}{\sqrt{w}}\left(\frac{1}{2}\right) \times D_\perp\left(\frac{s_2 w}{s_2 + w}, \phi_1, 0, \frac{\pi}{2}, \frac{\pi}{2}\right)\frac{e^{-jks_1}}{\sqrt{s_1}} \tag{5.50}$$

(Continued)

Example 5.5: Monostatic Scattering Components from an Infinite Strip *(Continued)*

The first term corresponds to the dashed line in Fig. 5.15 and the second term to the solid line. This process can be continued indefinitely by adding higher-order diffraction terms (E_{d3}, E_{d4}, ..., etc.). However, terms greater than second-order do not contribute significantly in this case because the observation point is in the principal plane of the strip. The factor of 1/2 is added for surface rays because, at grazing ($\phi = 0$ or $m\pi$), the field at a point is composed of both incident and reflected components. Only the incident part should be used in Eq. (5.50).

Figure 5.17 shows the monostatic field for a strip when only the singly diffracted rays are used compared to both singly and doubly diffracted components. Note that, without the second-order diffraction term, the traveling wave lobe is absent. For a parallel-polarized incident field, the second-order diffraction term is zero. Further refinement of the solution is obtained using the slope diffraction term.

For complex structures, the ray tracing can become overwhelming. Even for the polygonal plate in Fig. 5.18, the diffracted ray geometry is complicated.

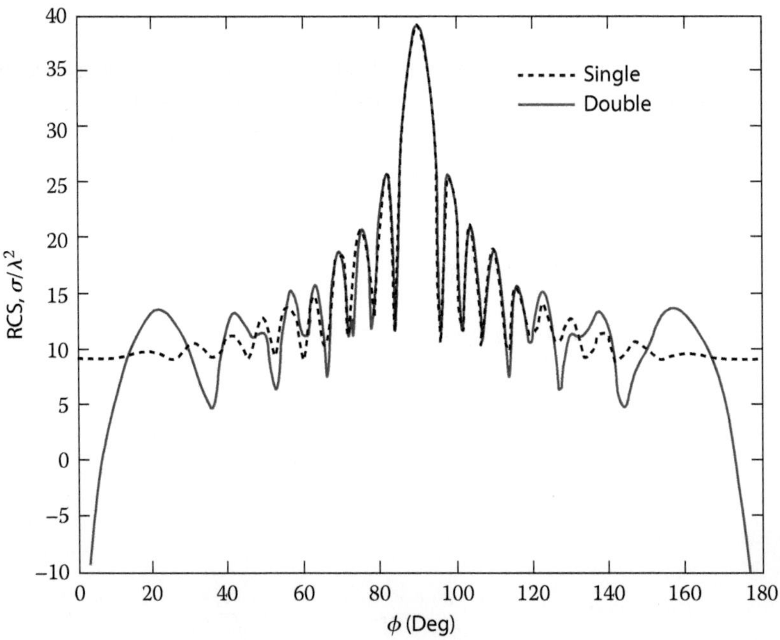

Fig. 5.17 Monostatic scattering from an infinitely long strip (TM$_y$, $w = 5\lambda$).

At each edge, the Keller cone is used to find the rays impinging on other edges. If multiple surfaces are present, reflected–reflected, reflected–diffracted, and diffracted–reflected terms may also be significant. Finally, the field polarization and phase relative to a specified reference must be tracked along the ray's path to the observation point. The ray tracing is generally done on a computer using vector relationships between ray paths and surface and edge normals. Often, such ray-tracing methods are just a brute-force search for reflection and diffraction points, in which case the computer run time can become extremely long.

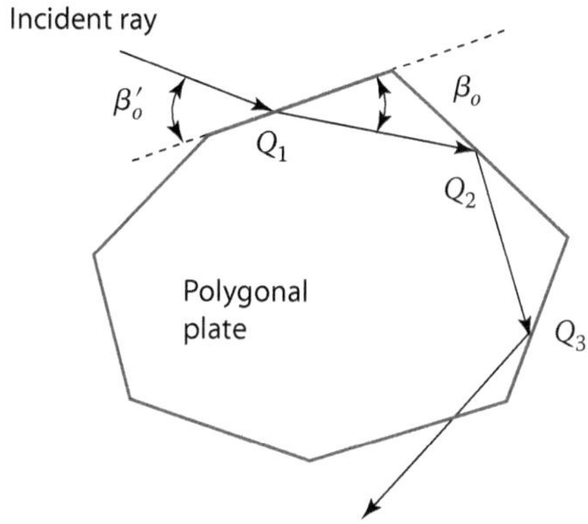

Fig. 5.18 Multiple diffraction for a polygonal plate.

5.5 Geometrical Theory of Diffraction Equivalent Currents

Figure 5.18 illustrates one of the problems with the GTD: at some observation points, a diffracted or reflected ray may never arrive, even after many diffractions occur. Thus, according to GO, there is no scattered field at that observation point. It is possible that a ray will arrive at an observation point slightly displaced from the first ray, giving the appearance of a sharp discontinuity in the field strength. In practice, this does not occur; the field strength can change rapidly, but it is still continuous. The discontinuity occurs because surface reflection and edge diffraction do not completely define scattering from a complex body. Other sources of scattering are corners and tips. Diffraction coefficients for these structures are available but are empirically derived and relatively difficult to implement [10].

One way around this problem is to define GTD-based equivalent currents as illustrated in Fig. 5.19. This is referred to as the *method of edge currents* (MEC). Fictitious electric and magnetic current filaments are considered to flow along the edges of the plate. The magnitude and phase of the currents depend

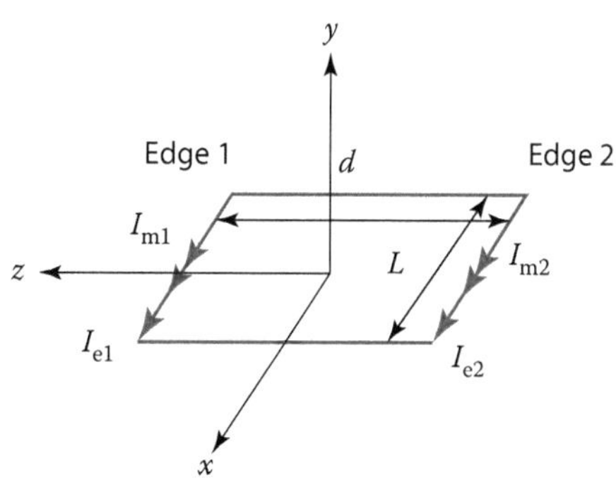

Fig. 5.19 GTD-based equivalent edge currents for a rectangular plate.

on the diffraction coefficient at the edge:

$$
\begin{aligned}
I_e &= \frac{-2j}{\eta k} E_i|_{\tan} D_{\parallel}(\ell, \phi, \phi', \beta)\sqrt{2\pi k} e^{j(\pi/4)} \exp(-jkz' \cos\beta') \\
I_m &= \frac{-2j\eta}{k} H_i|_{\tan} D_{\perp}(\ell, \phi, \phi', \beta)\sqrt{2\pi k} e^{j(\pi/4)} \exp(-jkz' \cos\beta')
\end{aligned}
\quad (5.51)
$$

The currents are now used in the radiation integral to compute the far-scattered fields. If the filaments are aligned with the z axis, then,

$$
\begin{aligned}
E_\phi &= \frac{-jk\sin\theta}{4\pi r} e^{-jkr} \int_{-(L/2)}^{L/2} I_m(z') \exp(jkz' \cos\theta) dz' \\
E_\theta &= \frac{jk\eta\sin\theta}{4\pi r} e^{-jkr} \int_{-(L/2)}^{L/2} I_e(z') \exp(jkz' \cos\theta) dz'
\end{aligned}
\quad (5.52)
$$

These currents are considered to radiate in free space (in the absence of the plate). Note that the observation point does not have to lie on the Keller cone to have a nonzero value for the scattered field. The current is usually assumed to be constant at all points along the edge. Strictly speaking, this is not true near bends or corners. If the current is constant, the integrals in Eqs. (5.52) reduce to the form

$$
L\mathrm{sinc}\left[\frac{kL}{2}(\cos\theta - \cos\beta')\right] \quad (5.53)
$$

which has a maximum when $\theta = \beta'$(the specular direction).

Example 5.6: Scattering from a Plate Using GTD-Based Equivalent Currents

A square plate with edge length L is located at the origin of a spherical coordinate system, with the y axis normal to its face, as shown in Fig. 5.19. A TM_y wave (polarized perpendicular to the edge) is incident from an angle ϕ in the x−y plane ($\theta = 90$ deg). If multiple diffractions are neglected, the equivalent electric current is given by Eqs. (5.52) with the diffraction coefficients,

Edge 1: $D_\perp(s, \phi, \phi, \pi/2, \pi/2)$
Edge 2: $D_\perp(s, \pi - \phi, \pi - \phi, \pi/2, \pi/2)$

Note that, for perpendicular polarization, $I_e(z') = 0$. The principal plane RCS of a 5λ square plate will be the same as the dashed pattern in Fig. 5.17. To obtain the accuracy of the solid line in Fig. 5.17, multiple diffractions can be included in the calculation of $I_m(z')$.

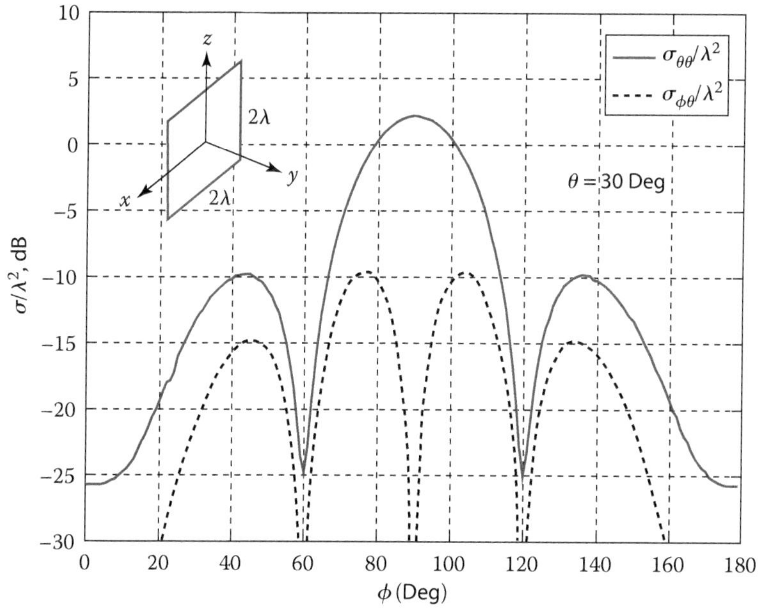

Fig. 5.20 Radar cross section of a square plate using equivalent currents.

The equivalent current approach to edge scattering has the advantage of giving a nonzero scattered field for observation points off the GTD diffraction cone. Figure 5.20 shows the RCS of a 2λ square plate for a cut off of the principal plane. The computed data are in close agreement with published measured data [10].

5.6 Physical Theory of Diffraction

The PTD, also known as the *method of edge waves* (MEW), is a current-based approach to obtaining the edge-scattered field [12]. The total field on a conductor with an edge is considered to be composed of the physical optics component and a nonuniform (or fringe) component. The situation is illustrated in Fig. 5.21. If the nonuniform current is known, the edge-scattered field can be computed by using this current in the radiation integral. The total field at an observation point will be the sum of the PO and edge-scattered fields.

The exact nonuniform current near edges on complex scatterers can be determined only by solving a boundary-value problem, and this is exactly what we are trying to avoid. However, as in the case of the GTD, edge scattering can be considered a local phenomenon. Therefore, the scattered field from a discontinuity can be approximated by the corresponding canonical scattering problem. If the total current can be determined from a

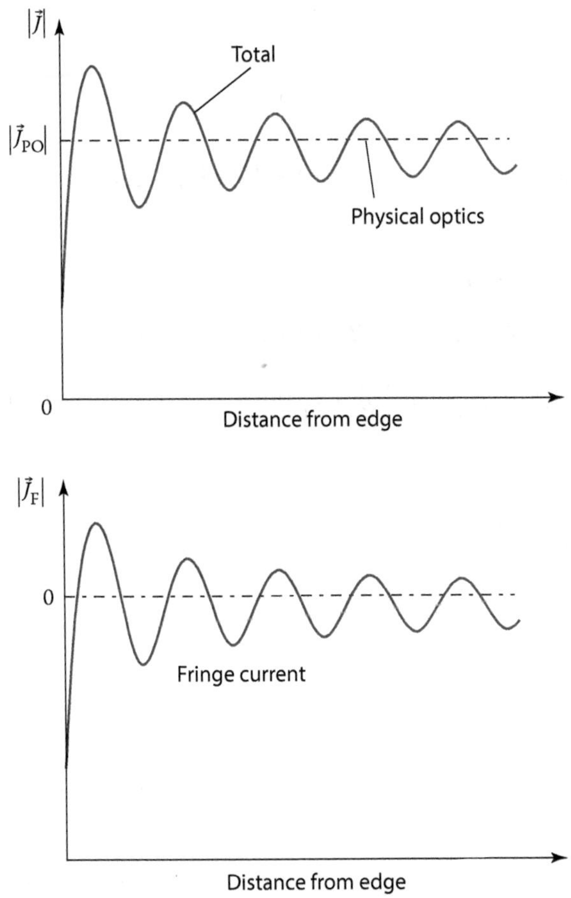

Fig. 5.21 Total current at a knife edge as a sum of physical optics and nonuniform (fringe) current components.

canonical problem, for example, an infinite knife edge, the PO component can be subtracted out and the result used as an estimate for the fringe current on a finite edge.

This approach has several attractive features, but they come at a price. The edge-scattered fields remain finite at shadow and reflection boundaries, unlike the GTD coefficients. Furthermore, the observation point is not restricted to the diffraction cone. However, the fringe currents extend a significant distance from the edge (usually $\approx 0.5\lambda$), so that two-dimensional integrals must be evaluated. Also, interactions between edges are more difficult to model because canonical problems with multiple edges are limited, if they exist at all.

Michaeli [13, 14] has reduced the PTD currents to filaments flowing along the body edge, which are similar to the GTD-based equivalent currents.

The electric and magnetic currents for a *thin screen* (knife edge) are given in Eqs. (24) and (25) of Ref. [14] and are repeated here for convenience. The electric current is

$$
\begin{aligned}
I_e = & \frac{-2j}{\eta k \sin^2\beta'} \frac{\sqrt{2}\sin(\phi'/2)}{\cos\phi' + \mu}\left\{\sqrt{1-\mu} - \sqrt{2}\cos(\phi'/2)\right\} E_i|_{\tan} \\
& + \left\{\frac{2j}{k\sin\beta'(\cos\phi' + \mu)}\right\} \\
& \times \left[\cot\beta'\cos\phi' + \cot\beta\cos\phi + \sqrt{2}\cos(\phi'/2)\right. \\
& \times \left.(\mu\cot\beta' - \cot\beta\cos\phi)/\sqrt{1-\mu}\right] H_i|_{\tan}
\end{aligned} \tag{5.54}
$$

and the magnetic current is

$$
I_m = \frac{2j\eta\sin\phi}{k\sin\beta'\sin\beta(\cos\phi' + \mu)}\left(1 - \frac{\sqrt{2}\cos(\phi'/2)}{\sqrt{1-\mu}}\right) H_i\Big|_{\tan} \tag{5.55}
$$

In the preceding formulas,

$$
\mu = \frac{\cos\gamma - \cos^2\beta'}{\sin^2\beta'} \tag{5.56}
$$

and

$$
\cos\gamma = \cos\beta\cos\beta' + \sin\beta\sin\beta'\cos\phi \tag{5.57}
$$

The only singularity occurs for $\beta = \beta'$ when $\phi = 0$ and $\phi' = \pi$ simultaneously. (This is the forward scatter case.) The singularity can be integrated and, therefore, the edge-scattered fields are finite.

Example 5.7: Scattering from a Plate Using PTD Currents

As in Example 5.6, a square plate with edge length L is located at the origin with the y axis normal to its face. A TM_y wave (polarized perpendicular to the edge) is incident from an angle ϕ in the $x-y$ plane ($\theta = 90$ deg). To compute the backscattered field (monostatic RCS) in the principal plane, $\phi = \phi'$ and $\beta = \beta' = 90$ deg. In this case,

$$
\mu = \cos\gamma = \cos\phi \,\text{and}\, \cot\beta = \cot\beta' = 0
$$

(Continued)

Example 5.7: Scattering from a Plate Using PTD Currents *(Continued)*

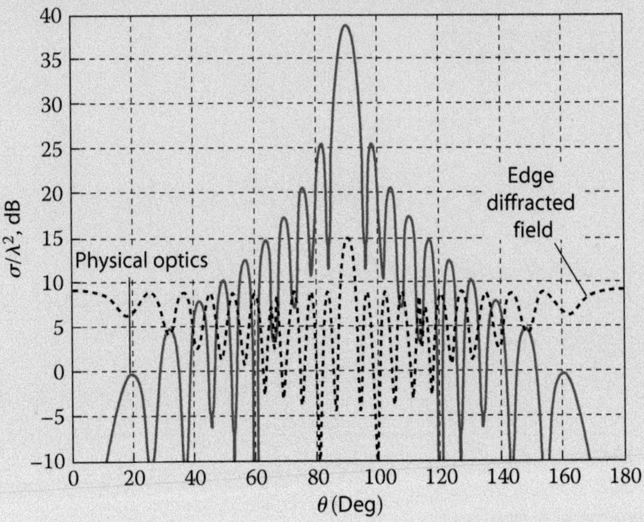

Fig. 5.22 PTD-edge-scattered field and PO component for a 5λ square plate.

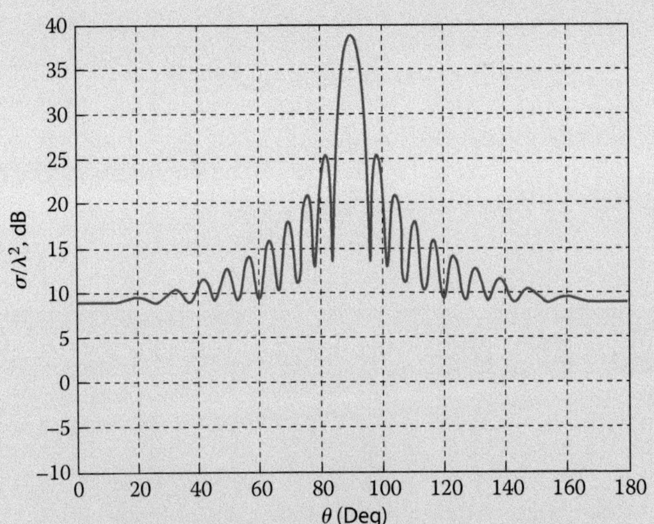

Fig. 5.23 Total scattered field for a 5λ square plate computed using PO and the PTD.

(Continued)

Example 5.7: Scattering from a Plate Using PTD Currents (Continued)

resulting in the following currents:

$$I_e = \frac{-j}{k\eta}\frac{1-\cos\phi-\sin\phi}{\cos\phi}H_i\bigg|_{\tan} \tag{5.58}$$

$$I_m = \frac{j\eta}{k}\tan\phi\left[1-\frac{\sqrt{2}\cos(\phi'/2)}{\cos\phi}\right]H_i\bigg|_{\tan} \tag{5.59}$$

These currents are used in the radiation integrals of form similar to Eqs. (5.52) to obtain the edge-scattered fields. Figure 5.22 shows the edge-scattered field from a square 5λ plate along with the PO field. Figure 5.23 shows the total field (PO plus edge-scattered fields). Note that the traveling wave lobe is not present because the interaction of the edge currents has not been included. Modification of the currents to include second-order diffraction is described in Ref. [14].

5.7 Incremental Length Diffraction Coefficients

Both the GTD and PTD are based on the postulate that high-frequency scattering is a local phenomenon. Consequently, diffraction from a finite edge or other scattering source can be derived from the appropriate canonical problem. For the case of an edge, the methods work well as long as the edge has no discontinuities or sharp bends and the incident field does not vary appreciably in amplitude. When these conditions are not satisfied, the diffracted fields are not well approximated by the diffraction coefficients of the corresponding two-dimensional canonical problem.

To alleviate this problem, Mitzner [15] has defined *incremental length diffraction coefficients* (ILDC) based on the PTD nonuniform current. These coefficients vary along the edge of the scatterer and are multiplied by the incident field at each point to obtain the diffracted fields of the nonuniform current. They can be applied to a wide range of geometries and hold for arbitrary angles of incidence and scattering.

Consider the infinite strip of width w located along the z axis as in Fig. 5.24. The edge-scattered field can be written

$$\vec{E}_s = \int_L \mathbf{dE}_s \cdot \mathbf{E}_i \tag{5.60}$$

where L denotes the contour of the edge and $\mathbf{dE}_s$ is a matrix of ILDCs,

$$\mathbf{dE}_s = d\ell\begin{bmatrix} F_{\theta\theta} & F_{\theta\phi} \\ F_{\phi\theta} & F_{\phi\phi}\end{bmatrix} \tag{5.61}$$

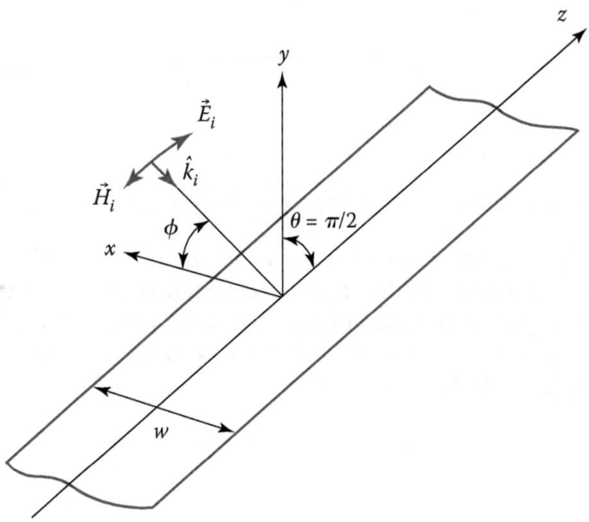

Fig. 5.24 Geometry for calculating ILDCs on an infinitely long strip of width *w*.

and

$$\mathbf{E}_i = \begin{bmatrix} E_{i\theta} \\ E_{i\phi} \end{bmatrix} \tag{5.62}$$

The elements of Eq. (5.61) depend on the wedge geometry and s, ϕ, ϕ', β, and β'.

Formulas have been derived for the ILDCs in terms of the canonical scattered fields[16]; that is, the fringe currents have been integrated and the edge-scattered field has been obtained as a function of the canonical scattered fields. For a knife edge, it is given by

$$\mathbf{dH}_s(r) \underset{r\to\infty}{=} \mathrm{d}z' \frac{C}{C_0}\left\{ \hat{\theta} \sin\phi \frac{H_{s\theta}(\cos^{-1}t)}{\sin(\cos^{-1}t)} + \left[\frac{H_{s\theta}(\cos^{-1}t)}{\sin(\cos^{-1}t)} \cos\phi \right.\right.$$
$$\left.\left. \times \left(\cos\theta + \frac{\sin^2\theta\cos\theta_0}{\sin^2\theta_0} \right) + H_{s\phi}(\cos^{-1}t) \frac{\sin\theta}{\sin\theta_0} \right] \hat{\phi} \right\} \tag{5.63}$$

where

$$\frac{C}{C_0} = \frac{e^{-jkr}}{4\pi r}\sqrt{8\pi k\rho\sin\theta_0}\,\exp\left[j(k\rho\sin\theta_0 - kz\cos\theta_0 + \pi/4)\right] \tag{5.64}$$

and

$$t = \frac{\sin\theta\cos\phi}{\sin\theta_0} \tag{5.65}$$

The incidence direction is given by (θ_0, ϕ_0). Finally,

$$d\mathbf{E}_s \underset{r\to\infty}{=} -\eta\hat{r} \times d\mathbf{H}_s \tag{5.66}$$

Example 5.8: Scattering from a Strip Using ILDCs

An infinitely long strip of width w is parallel to the z axis and a TM_y wave (polarized perpendicular to the edge) is incident from a direction $\beta' = \pi - \theta_0$ and $\phi' = \pi + \phi_0$. For backscattering, $\beta = \beta' = \theta_0 = \pi$ and $\phi = \phi' = \pi + \phi_0$. In this case, $\cos^{-1} t = \phi$, and the ILDC matrix (5.61) has only one nonzero element $(F_{\theta\theta})$. The ILDC is given by

$$d\mathbf{H}_s(r) \underset{r\to\infty}{=} dz' \frac{C}{C_0} H_{s\theta}(\phi)\hat{\theta}$$

where

$$\frac{C}{C_0} = \frac{e^{-jkr}}{4\pi r}\sqrt{8\pi k\rho}\exp\left[-j(k\rho + \pi/4)\right]$$

The electric field intensity is found from the TEM relationship,

$$d\mathbf{E}_s \underset{r\to\infty}{=} -\eta\hat{\rho} \times dz'\left[\frac{C}{C_0} H_{s\theta}(\phi)\hat{\theta}\right] = -\eta\frac{C}{C_0} dz' H_{s\theta}(\phi)\hat{\phi}$$

Expressions for the scattered field from the strip $H_{s\theta}$ are available in the literature (see, for example, Ref. [18]). Thus, the ILDCs are completely determined, and the line integral along the edge in Eq. (5.58) can be evaluated.

5.8 Summary

In this chapter, we have examined geometrical optics and several high-frequency techniques for computing edge-diffracted fields. All the edge-diffraction models are based on the postulate that edge scattering is a local phenomenon and, therefore, approximations to the current or scattered field can be obtained from the appropriate two-dimensional canonical problem.

The advantages and disadvantages of each approach have been pointed out along the way. Table 5.2 is a brief summary.

In spite of the apparent diversity in these approaches, it can be shown that they are all equivalent [17]. The particular method employed to solve a problem is primarily a matter of personal taste and convenience. Because of the emphasis on RCS in the 1970s and 1980s, GTD codes are more abundant than those based on the other methods. The geometrical theory of diffraction has the benefit of exposure in the open technical literature, whereas the PTD was developed in the Soviet Union and has not received extensive attention by the international community until recently.

Table 5.2 Advantages and Disadvantages of Techniques for Computing Edge-Diffracted Fields

Approach	Advantages	Disadvantages
GTD	Uniform theories are continuous across shadow boundaries Physical insight	Multiple diffractions are cumbersome Caustics are a problem Observation point must lie on Keller cone
PTD	Continuous everywhere Arbitrary observation point location	Edge interaction difficult to model Two-dimensional integral must be evaluated
MEC	Observation points can be off Keller cone	GTD diffraction coefficients must be known Only rigorous in the specular direction
MEW	Only line integral required Observation points can be off Keller cone	Multiple diffractions difficult

Probably the most significant disadvantage of all these methods is difficulty in *accurately* modeling the *interaction* between edges of a *complex target.* They do well for simple arrangements such as parallel edges (i.e., the strip or plate) but are not easily applied to more complex edge orientations. As we have seen in earlier chapters, edge interaction can be the dominant contributor to RCS in some directions. Consequently, it is important to be able to predict edge-scattering levels accurately when developing low-observable platforms.

References

[1] Keller, J. B., "Determination of Reflected and Transmitted Fields by Geometrical Optics," *Journal of the Optical Society of America,* Vol. 40, No. 1, Jan. 1950, p. 48.

[2] Deschamps, G. A., "Ray Techniques in Electromagnetics," *Proceedings of the IEEE,* Vol. 60, No. 9, Sept. 1972, p. 1022.

[3] Maurer, S. J., and Felsen, L. B., "Ray-Optical Techniques for Guided Waves," *Proceedings of the IEEE,* Vol. 55, No. 10, Oct. 1967, p. 1718.

[4] Keller, J. B., "Geometrical Theory of Diffraction," *Journal of the Optical Society of America,* Vol. 52, No. 2, Feb. 1962, p. 116.

[5] Kouyoumjian, R. G., and Pathak, P. H., "A Uniform Geometrical Theory of Diffraction for an Edge in a Perfectly Conducting Surface," *Proceedings of the IEEE,* Vol. 62, No. 11, Nov. 1974, p. 1448.

[6] Boersma, J., and Rahmat-Samii, Y. Y., "Comparison of Two Leading Theories of Edge Diffraction with the Exact Uniform Asymptotic Solution," *Radio Science,* Vol. 15, No. 6, Nov.–Dec. 1980, p. 1179.

[7] Lee, S. W., "Uniform Asymptotic Theory of Electromagnetic Edge Diffraction: A Review," *Electromagnetic Scattering,* edited by P. L. E. Uslenghi, Academic Press, New York, 1968, p. 67.

[8] Brueggemann, H. P., *Conic Mirrors,* Focal Press, New York, 1968.

[9] Rusch, W. V. T., "A Comparison of Geometrical and Integral Fields from High-Frequency Reflectors," *Proceedings of the IEEE,* Vol. 62, Nov. 1974, p. 1603.

[10] Sikta, F. A., et al., "First-Order Equivalent Current and Comer Diffraction Scattering from Flat Plate Structures," *IEEE Transactions on Antennas and Propagation,* Vol. AP-31, No. 4, July 1983, p. 584.

[11] Stutzman, W., and Thiele, G., *Antenna Theory and Design,* Wiley, New York, 1981, p. 496.
[12] Ufimtsev, P. Y., *Method of Edge Waves in the Physical Theory of Diffraction,* translated from Russian by the USAF Systems Command, Foreign Technology Division.
[13] Michaeli, A., "Equivalent Edge Currents for Arbitrary Aspects of Observation," *IEEE Transactions on Antennas and Propagation,* Vol. AP-32, No. 3, March 1984, p. 252 (also see correction in Vol. AP-33, No. 2, Feb. 1985).
[14] Michaeli, A., "Elimination of Infinities in Equivalent Edge Currents, Part I: Fringe Current Components," *IEEE Transactions on Antennas and Propagation,* Vol. AP-34, No. 7, July 1986, p. 912.
[15] Mitzner, K. M., "Incremental Length Diffraction Coefficients," Tech. Rept. AFALTR-73-296, April 1974 (NTIS document AD918861).
[16] Michaeli, A., "Equivalent Currents for Second-Order Diffraction by the Edges of Perfectly Conducting Polygonal Surface," *IEEE Transactions on Antennas and Propagation,* Vol. AP-35, No. 5, Feb. 1987, p. 42.
[17] Shore, R. A., and Yaghjian, A. D., "Incremental Diffraction Coefficients for Planar Surfaces," *IEEE Transactions on Antennas and Propagation,* Vol. AP-36, No. 1, Jan. 1988, p. 55.
[18] Bowman, J., et al., *Electromagnetic and Acoustic Scattering by Simple Shapes,* Hemisphere Publishing, New York, 1987.
[19] Ruck, G. T., et al., *Radar Cross Section Handbook,* Plenum Press, 1970.

Problems

5.1 A cylindrical surface is infinitely long in the z direction and finite in x and y such that

$$(x - x_0)^2 + (y - y_0)^2 = r_0^2, \quad \text{for} \quad y \geq y_0$$

Source and observation points are located at $O(x_1, y_1)$ and $P(x_2, y_2)$. Let $d = d_1 + d_2$ be the total path length between the two points as shown in Fig. P5.1.

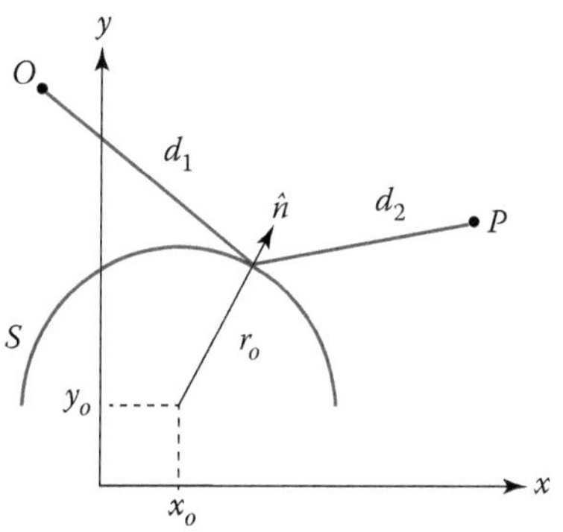

Fig. P5.1

a) Assume that $P(1, 2)$, $O(-1, 2)$, $x_0 = y_0 = 0$, and $r_0 = 1$, and evaluate

$$\mathcal{D} = \hat{n} \times (\nabla d \times \hat{n}) = \nabla d - (\nabla d \cdot \hat{n})\hat{n}$$

at both edges of S. $\mathcal{D}$ is the projection of ∇d on a plane tangent to the surface. If a reflection point exists, it will be directed away from S.

b) Find the reflection point using bisections.

5.2 A doubly curved surface has principal radii of curvature R_1^s and R_2^s. Define the *diagonalized curvature matrix* of S at R by[5]

$$\mathbf{C}_0^s = \begin{bmatrix} \frac{1}{R_1^s} & 0 \\ 0 & \frac{1}{R_2^s} \end{bmatrix}$$

Similarly, for the incident wave,

$$\mathbf{C}_0^i = \begin{bmatrix} \frac{1}{R_1^i} & 0 \\ 0 & \frac{1}{R_2^i} \end{bmatrix}$$

The principal directions of the incident wave are defined by $\hat{u}_{i1}$ and $\hat{u}_{i2}$, where (arbitrarily) $\hat{u}_{i1} \times \hat{u}_{i2} = \hat{k}_i$. As shown in Fig. 5.7, $\hat{u}_1$ and $\hat{u}_2$ are the principal direction unit vectors of the surface. Define a projection matrix $\mathbf{P}$, where the elements are given by $P_{pq} = \hat{u}_{ip} \cdot \hat{u}_q$, with p, $q = 1, 2$. The *undiagonalized curvature matrix* of the reflected wave is

$$\mathbf{C}^r = \mathbf{C}_0^i + 2\{\mathbf{P}^{-1}\}^T \mathbf{C}_0^s \mathbf{P}^{-1} \cos\theta_i$$

Diagonalizing this matrix yields the principal radii of curvature of the reflected wave. Apply this procedure to a spherical wave normally incident on a sphere of radius a. Verify the result using Eq. (5.9). *Hint:* Assume that the principal planes of the incident wave and surface are the same.

5.3 A "flying wing" appears on the horizon as it approaches a radar (i.e., the wing is viewed directly edge-on). Each half of the leading edge has a length d and makes an angle α with respect to the centerline (Fig. P5.3).

a) Estimate the monostatic RCS due to one-half of the leading edge for a horizontally (soft) polarized incident wave using a GTD equivalent edge current. Assume a knife edge, and specify the parameters of the diffraction coefficient $D_{\|}(\ell, \phi, \phi', \beta, \beta')$ in terms of the ray-fixed coordinate system. Express the RCS as a function of α.

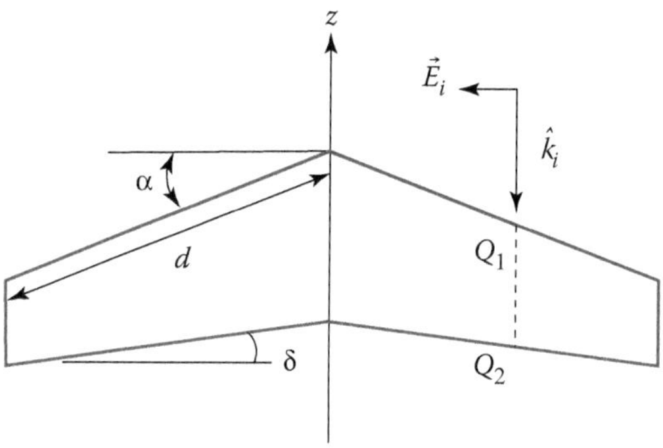

Fig. P5.3

b) At what angle δ will the diffracted ray from the leading edge to the trailing edge (Q_1 to Q_2) return in the backscatter direction?

5.4 Multiple diffractions are to be considered in calculating the RCS of the block shown in Fig. P5.4. When many multiple diffractions must be included, a matrix method is more efficient than simply adding more terms to Eqs. (5.49) and (5.50). This approach is called the *self-consistent* method.[18] Let C_1 represent the sum of all diffracted rays traveling from Q_1 to Q_2, and let C_2 represent the sum of all diffracted rays traveling from Q_2 to Q_1. Denote the direct ray contributions at the two points as V_1 and V_2. Then,

$$C_1 = C_2 R_{12} + V_1$$

$$C_2 = C_1 R_{21} + V_2$$

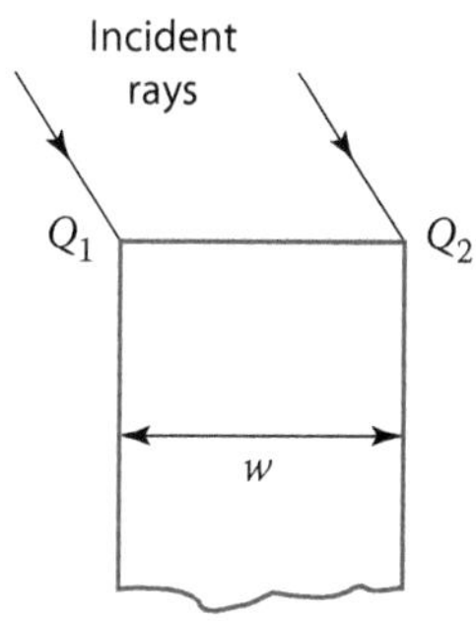

Fig. P5.4

where R_{pq} (p, $q = 1, 2$) are equivalent to reflection coefficients for a diffraction at point Q_q of a ray approaching from point Q_p. In matrix form,

$$\begin{bmatrix} 1 & -R_{12} \\ -R_{21} & 1 \end{bmatrix} \begin{bmatrix} C_1 \\ C_2 \end{bmatrix} = \begin{bmatrix} V_1 \\ V_2 \end{bmatrix}$$

a) Find expressions for V_p and R_{pq} in terms of the GTD diffraction coefficients.

b) Solve the equations directly for C_1 and C_2.

5.5 A corner reflector consists of two plates joined at a right angle as shown in Fig. P5.5. A source is located at O, and the observation point is denoted P. Possible paths between O and P include:

a) Direct

b) Singly reflected

c) Singly diffracted

d) Diffracted – reflected

e) Reflected – diffracted

f) Doubly reflected

g) Doubly diffracted

For the dimensions shown, determine which of the listed paths are possible. Give the locations of reflection points if applicable, and specify the arguments of the diffraction coefficients ℓ, β, β', ϕ and ϕ'. Consider

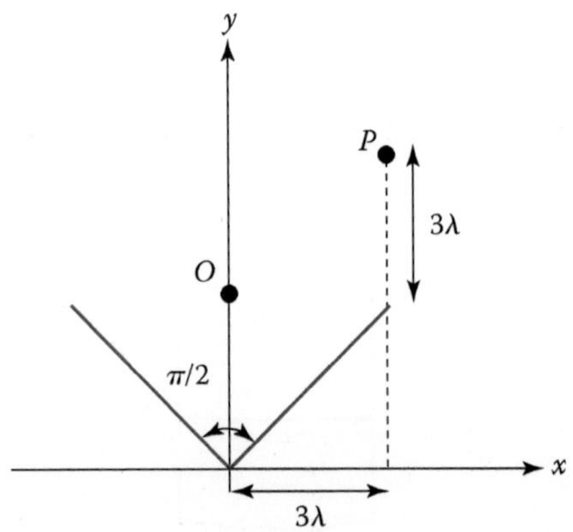

Fig. P5.5

that this is a two-dimensional problem and that the source radiates a TM_z-polarized cylindrical wave.

5.6 A spherical wave is normally incident on a surface with radii of curvature R_1^s and R_2^s. Find an expression for the radii of curvature of the reflected wave front R_1^r and R_2^r. The spherical wave source is at distance s' from the reflection point on the surface.

5.7 A plane wave is incident on a surface with radii of curvature R_1^s and R_2^s. Show that, if $s \gg R_1^s, R_2^s$ the RCS is

$$\sigma \approx \pi R_1^s R_2^s \tag{P5.1}$$

5.8 Scattering from the dielectric-coated ground plane in Problem 2.15 is to be computed by tracing rays. Show that the reflection coefficient at the air/dielectric interface is given by

$$\Gamma(z = t) = \frac{\Gamma_d - P_d P_0}{1 - P_d^2 \Gamma_d P_0}$$

where $P_d = e^{-\gamma \ell}$, $\ell = t/\cos\ \theta_t$, $P_0 = e^{-jk\Delta}$, $\Delta = 2kt\ \tan\ \theta_t\ \sin\ \theta$, $Z_0 = \eta_0\ \cos\ \theta$, and

$$\Gamma_d = \frac{Z_L - Z_0}{Z_L + Z_0}$$

where

$$Z_L = \begin{cases} \eta\ \cos\theta_t & \text{For TM} \\ \eta/\cos\theta_t & \text{For TE} \end{cases}$$

5.9 Use the formula derived in Problem 5.7 to calculate the RCS of the paraboloid shown in Fig. 5.8.

5.10 Sommerfeld found that the current on a knife edge is $\vec{J}_s = J_{sx}\hat{x} + J_{sz}\hat{z}$, where, for perpendicular polarization the components are [19]

$$J_{sz}(x, z) = \frac{4e^{-j\pi/4}}{\eta\sqrt{\pi}} \sin\theta_i \sin\phi_i \exp[-jk(x\sin\theta_i\cos\phi_i + z\cos\theta_i)]$$
$$\times \left\{ \sqrt{\frac{\pi}{2}}\{C(\alpha) + jS(\alpha)\} + \frac{j\exp\left[2jkx\sin\theta_i\cos^2(\phi_i/2)\right]\sin(\phi_i/2)}{4\sin\phi_i\sqrt{(kx\sin\theta_i)/2}} \right\}$$
$$J_{sx}(x, z) = 0$$

and for parallel polarization the components are

$$J_{sx}(x, z) = \frac{4e^{-j\pi/4}}{\eta\sqrt{\pi}} \sin^2\theta_i \exp[-jk(x \sin\theta_i \cos\phi_i + z\cos\theta_i)]$$

$$\times\sqrt{\frac{\pi}{2}}\{C(\alpha) + jS(\alpha)\}$$

$$J_{sz}(x, z) = \frac{4e^{-j\pi/4}}{\eta\sqrt{\pi}} \sin\theta_i \cos\phi_i \exp[-jk(x \sin\theta_i \cos\phi_i + z\cos\theta_i)]$$

$$\times\left\{\sqrt{\frac{\pi}{2}}\{C(\alpha) + jS(\alpha)\} - \frac{j\exp\left[2jkx \sin\theta_i \cos^2(\phi_i/2)\right]\cos^2(\phi_i/2)}{4\cos\phi_i\sqrt{(kx\sin\theta_i)/2}}\right\}$$

for $\alpha = 2\sqrt{\frac{kx\sin\theta_i}{\pi}}\cos(\phi_i/2)$. The Fresnel sine and cosine integrals are defined by

$$S(\alpha) = \int_0^\alpha \sin(\pi\, u^2/2)\, du$$

$$C(\alpha) = \int_0^\alpha \cos(\pi\, u^2/2)\, du$$

Plot the current on the knife edge as a function of distance from the edge, x for $\theta_i = \phi_i = 90$ deg and $\lambda = 1$ m. Compare the actual current far from the edge to that given by the physical optics approximation.

Note: Evaluating J requires calculation of the Fresnel sine and cosine integrals. Some mathematical packages contain these functions, but most do not. It is a simple matter to evaluate the integrals numerically using the rectangular rule. The integrand is approximated by a series of steps of width Δ. The integral of a function f is approximately

$$\int_{\alpha_1}^{\alpha_2} f(u)\, du \approx \Delta \sum_{n=1}^{N} f(u_n)$$

where $\Delta = (\alpha_2 - \alpha_1)\, N$ and N is the number of rectangles used to approximate f between α_1 and α_2, and u_n is the center point of rectangle n.

5.11 A unit amplitude spherical wave source S is located at (0,0,5) above an infinite flat PEC plane at $z = 0$ in Fig. P5.11. The coordinates are in wavelengths and the polarization is horizontal. An observation point P is located at (10,0,10).

a) Find the reflection point.

b) Using the formulas of GO, find the reflected field at P.

c) Find the reflected field at P using the method of images.

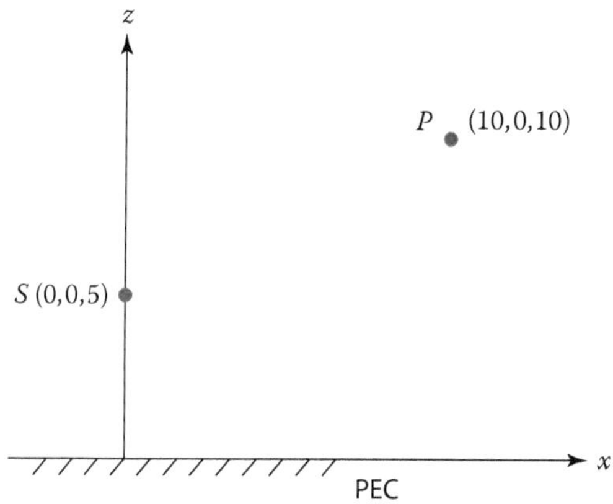

Fig. P5.11

5.12 A spherical wave source is located at S in Fig. P5.12, and an observation point at P. Two infinitely long perpendicular walls have a gap near the corner as shown and P and S have the same z coordinate. You can use geometry (it comes down to applying the properties of similar triangles) or you can use images.

a) Draw the singly reflected rays arriving at P and calculate the locations of reflection points.

b) Draw the doubly reflected rays arriving at P and calculate the locations of reflection points.

c) Draw the singly diffracted rays arriving at P.

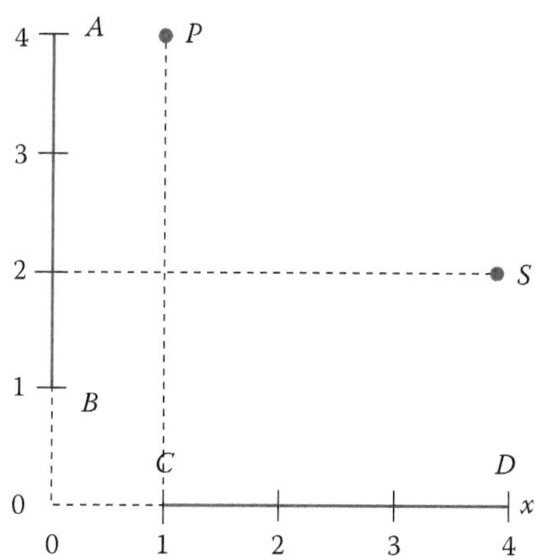

Fig. P5.12

5.13 A point source is located at a height $a = 1.5\lambda$ above a PEC sphere of radius $b = 4\lambda$ as shown in Fig. P5.13. Find the amplitude of the reflected ray emerging from the source at 30 deg at an observation point that is 10λ from the reflection point along the reflected ray path. Use the method described in Problem 5.2.

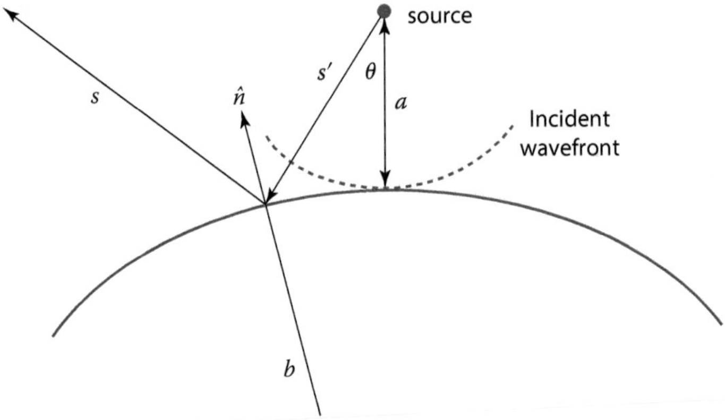

Fig. P5.13

5.14 A circular ogive of revolution is generated by taking an arc of radius a and rotating it about the x axis as shown in Fig. P5.14. The half angle subtended by the circular arc is $\beta = 30$ deg at the edge and $a = 1$ m.

a) Using the formulas from geometrical optics, find the broadside RCS. You can check your result using the formula given in Problem 5.7.

b) Repeat the calculation for incidence midway between the center and edge, $\beta = 15$ deg.

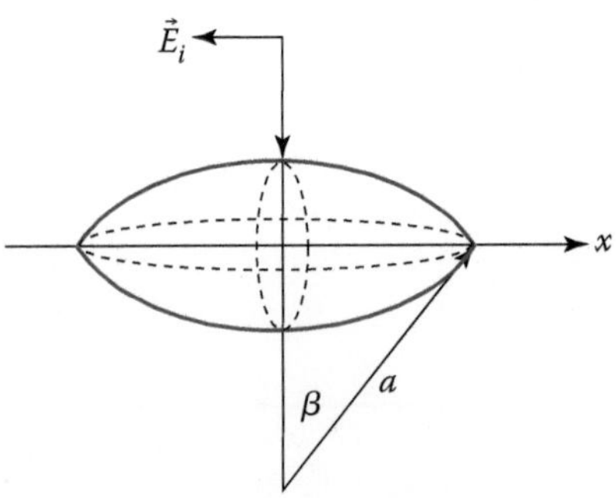

Fig. P5.14

Chapter 6 Complex Targets

6.1 Introduction

In previous chapters, various analytical and numerical tools have been presented and applied to simple targets, such as plates and spheres. Most radar targets, with few exceptions (such as weather balloons), are complex geometries with many scattering sources that interact. The interaction between multiple scatterers can cause rapid variations in the RCS as a function of aspect angle. For electrically large targets, the sources can be many wavelengths apart and, therefore, can add constructively and destructively just a fraction of a degree apart.

For example, an aircraft is constantly buffeted by turbulence, which causes its aspect angle, as viewed from the radar, to change. Furthermore, the stressed aircraft frame will deform and flex, changing the separation between scattering sources on the body. This, in turn, causes the aircraft's RCS to fluctuate rapidly as the scattering sources add and cancel as a function of time as well as angle. This apparently random fluctuation in the peaks and nulls of the RCS pattern gives rapid signal changes at the radar's receiving antenna (in both magnitude and phase). A phenomenon referred to as *glint* can result, yielding large tracking errors.

From a systems point of view, a target can be classified as one of four *Swerling* types,* depending on its scattering statistics [1]:

S1. The target's return appears to be composed of many independent scattering sources of approximately equal strength. These sources interact at a rate such that the total return is approximately constant for all pulses during a scan of the target but the return is independent from scan to scan.

S2. This type of target has the same scattering characteristics as S1. However, the fluctuations are more rapid and appear to be random from pulse to pulse rather than scan to scan.

S3. The return is characteristic of a large scattering source that dominates the RCS, along with many small, apparently random scatterers. The return varies at a rate such that it appears independent from scan to scan as in S1.

S4. As is the case for S3, one large reflector dominates the RCS but, in this case, the fluctuation appears to be random from pulse to pulse as in S2.

The fluctuation of the target RCS is a function of many variables in the radar system and the environment. In addition to change in aspect and target shape, there are multipath and atmospheric inhomogeneities. It will be

*Sometimes nonfluctuating targets are designated as type S0.

seen that imperfections and errors in the assembly and manufacturing processes will also contribute to spatial random scattering. System errors that contribute to the fluctuation would be a variation in transmitted power from pulse to pulse. These types of errors are not significant contributors in modern high-performance radars.

In this chapter, the methods of RCS prediction developed earlier are applied to complex targets. *Complex* refers not only to the geometric shapes involved but also to the presence of joints, multiple materials, and protruding objects, such as sensors and military ordnance.

6.2 Geometrical Components Method

For electrically large targets, the geometrical components method is one of the most common approaches to estimating the RCS. A complex target can usually be decomposed into basic geometric shapes, such as spheres, cones, plates, and ellipsoids, as illustrated for the aircraft in Fig. 6.1. A suitable "first-cut" estimate of the total target RCS is to sum the contributions from each of these *primatives*, neglecting the interactions between them. This approach is the *geometrical components method*, and it does a good job of predicting the peak RCS levels and locations versus frequency.

When the contribution from each of the target components is added, the summation can be performed *coherently* or *noncoherently*. In the first case, the relative phase of each term is included in the sum, whereas in the second case, only the magnitudes of each term are summed. For N scatterers, the coherent scattered field is

$$\vec{E}_s = \sum_{n=1}^{N} (\vec{E}_s)_n \tag{6.1}$$

We can define a complex *voltage* RCS, denoted as M, that is the square root of the scattered power pattern with the defining RCS constants included:

$$\sigma_{pq} \equiv M_{pq} M_{pq}^* = |M_{pq}|^2 = \left| \sum_{n=1}^{N} M_{npq} \right|^2 \sim \left| \sum_{n=1}^{N} (E_s)_{npq} \right|^2 \tag{6.2}$$

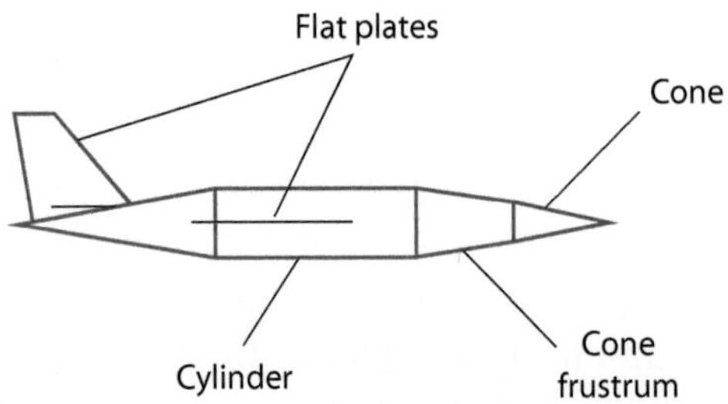

Fig. 6.1 Geometrical components method for an aircraft.

In Eq. (6.2), p and q represent either θ or ϕ. The subscript q denotes the polarization of the incident wave, which induces the current on scatterer n. This current, in turn, gives rise to a scattered field $\vec{E}_s$, which can have both θ and ϕ components. The subscript p denotes the particular scattered field component at the receiving antenna.

6.2.1 Expected Value of the Radar Cross Section

For the present discussion, we have dropped the polarization subscripts p and q, with the realization that the formulas developed can be applied to any component of the target scattering matrix. We can write the scattered field of the nth source as

$$M_n = \sqrt{\sigma_n} e^{j\psi_n} \tag{6.3}$$

Consider only two sources, in which case the voltage RCS becomes

$$M = \sqrt{\sigma_1} e^{j\psi_1} + \sqrt{\sigma_2} e^{j\psi_2} \tag{6.4}$$

If all phases are equally likely (i.e., their phases are random and uniformly distributed between 0 and 2π), the expected value of the RCS is

$$\langle MM^* \rangle = \left(\frac{1}{2\pi}\right)^2 \int_0^{2\pi} \int_0^{2\pi} \left(\sqrt{\sigma_1} e^{j\psi_1} + \sqrt{\sigma_2} e^{j\psi_2}\right) \times \left(\sqrt{\sigma_1} e^{j\psi_1} + \sqrt{\sigma_2} e^{j\psi_2}\right)^* \mathrm{d}\psi_1 \, \mathrm{d}\psi_2 \tag{6.5}$$

where $\langle\ \rangle$ denotes expected value. Evaluating the integrals yields

$$\langle MM^* \rangle = \sigma_1 + \sigma_2$$

In general, for N sources,

$$\langle MM^* \rangle = \sum_{n=1}^{N} \sigma_n \tag{6.6}$$

Note that this is equivalent to adding the magnitudes of radar cross sections of the N sources.

There are two cases for which the square of the sum in Eq. (6.2) can be approximated by a sum of squares:

1. If only one scattering source dominates at any given angle (say σ_i, where $\sigma_i \gg \sigma_n$, $n \neq i$), it is possible to add the powers noncoherently:

$$|\vec{E}_s|^2 \approx \sum_{n=1}^{N} |(\vec{E}_s)_n|^2 \rightarrow \sigma \approx \sum_{n=1}^{N} \sigma_n \approx \sigma_i \tag{6.7}$$

2. If the target consists of a large number of "randomlike" scattering sources of approximately equal magnitude that can be described on the basis of

equally likely relative phases, the noncoherent approach is justified:

$$\sigma \approx \sum_{n=1}^{N} \sigma_n = \sigma_{\text{ave}} \tag{6.8}$$

The variance of the RCS is

$$\text{var}\{\sigma\} = \left(\sum_{n=1}^{N} \sigma_n \right)^2 - \sum_{n=1}^{N} \sigma_n^2 \tag{6.9}$$

The situation described by Eq. (6.12) represents Swerling types 1 and 2 targets. The shape is essentially a $\cos\theta$ and is characteristic of a *diffuse* scatterer. It is commonly encountered when waves are scattered from very rough random surfaces.

As the number of random sources increases, the nulls fill and the sidelobe peaks disappear. Eventually, a broad isotropic pattern dominates, the level of which increases with the number of random sources.

Example 6.1: Target with Deterministic and Random Scattering Components

To illustrate the effects of random phase errors on the RCS, it is possible to construct a fictitious target consisting of a strip of width b along y and length L along x that is broken into N equal length a $(=L/N)$ segments. The reflection coefficient on segment n is $\Gamma_n = \Gamma_0 e^{j\psi_n}$ where the phases of the reflection coefficient are a random variable uniformly distributed over $[0, 2\pi]$. The random scattered field in the x-z plane varies as

$$E_\theta \sim \Gamma_0 ab \cos\theta \sum_{n=1}^{N} e^{j\psi_n} \text{sinc}(ka \sin\theta) \exp(j2kx_n \sin\theta) \tag{6.10}$$

The locations of the centers of the patches are given by $x_n = (L/N)(n - 1/2)$. For large N (small a) $\text{sinc}(ka \sin\theta) \approx 1$ and

$$|\vec{E}_\theta|^2 \sim \left| \Gamma_0 ab \cos\theta \sum_{n=1}^{N} e^{j\psi_n} \exp(j2kx_n \sin\theta) \right|^2 \tag{6.11}$$

or, if the noncoherent sum is used

$$|\vec{E}_{s\theta}|^2 \sim (\Gamma_0 ab \cos\theta)^2 \Big\{ \left| e^{j\psi_1} \exp(j2kx_1 \sin\theta) \right|^2 + \left| e^{j\psi_2} \exp(j2kx_2 \sin\theta) \right|^2 + \cdots + \left| e^{j\psi_N} \exp(j2kx_N \sin\theta) \right|^2 \Big\} \tag{6.12}$$

(Continued)

Example 6.1: Target with Deterministic and Random Scattering Components *(Continued)*

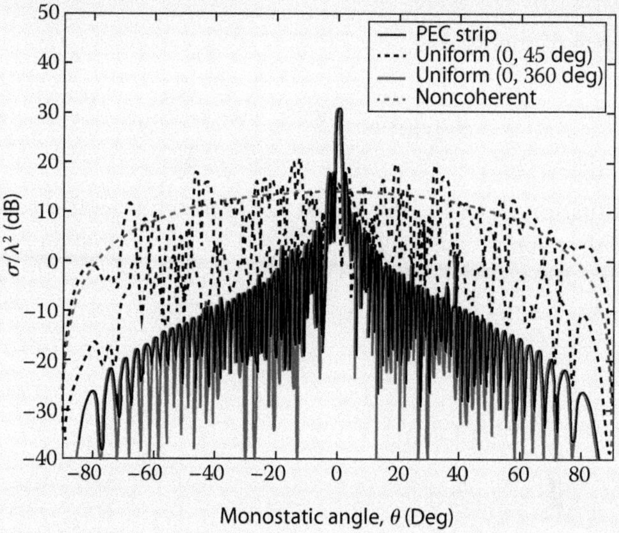

Fig. 6.2 Comparison of coherent and noncoherent RCS of an array of scatterers.

Several cases are compared in Fig. 6.2 for $\Gamma_0 = 1$, $N = 50$, $a = 0.5\lambda$, and $L = 20\lambda$. The coherent result is computed for random phases distributed uniformly over $[0, \pi/4]$ and $[0, 2\pi]$. From the figure it is evident that the noncoherent result is accurate for a large number of uncorrelated scatterers. The advantage of Eq. (6.12) is that the phases of the individual contributors do not have to be tracked.

Example 6.2: Corner Reflector

A corner reflector is formed by two plates joined at an angle $2\theta_o$, as shown in Fig. 6.3. Let each plate be a rectangle of dimension a by b, with the centers located at

$$\text{Plate 1: } (x_o, 0, z_o)$$

$$\text{Plate 2: } (-x_o, 0, z_o)$$

Where

$$x_o = \frac{a}{2}\sin\theta_o$$

$$z_o = \frac{a}{2}\cos\theta_o$$

(Continued)

Example 6.2: Corner Reflector *(Continued)*

A TM_z wave is incident in the $x-z$ plane. The RCS will be compared for the coherent and noncoherent cases. The angles measured from the normals to the plates are

$$\theta_2 = \theta - (\pi/2 - \theta_o)$$
$$\theta_1 = \theta + (\pi/2 - \theta_o)$$

For a single plate the scattered field in the $\phi = 0$ deg plane is

$$E_{s\theta} = \frac{-jk}{2\pi r} e^{-jkr} E_{o\theta} ab \cos\theta \,\mathrm{sinc}(kau) \tag{6.13}$$

The total field is the combination of two terms of the form of Eq. (6.11), one for each plate. With the origin used as the phase reference,

$$E_{s\theta} = \frac{-jk}{2\pi r} e^{-jkr} A E_{o\theta} \{ \cos\theta_1 \,\mathrm{sinc}(kau_1) \exp\left[j2k(ux_o + wz_o)\right] + \cos\theta_2 \,\mathrm{sinc}(kau_2) \exp\left[j2k(-ux_o + wz_o)\right]\} \tag{6.14}$$

Using Eq. (6.11), the coherent RCS is

$$\sigma_{\theta\theta} = \frac{4\pi A^2}{\lambda^2} \left| \cos\theta_1 \,\mathrm{sinc}(kau_1) \exp\left[jka\cos(\theta - \theta_o)\right] + \cos\theta_2 \,\mathrm{sinc}(kau_2) \exp\left[jka\cos(\theta + \theta_o)\right]\right|^2 \tag{6.15}$$

whereas the noncoherent sum (6.12) gives

$$\sigma_{\theta\theta} \approx \frac{4\pi A^2}{\lambda^2} \left(|\cos\theta_1 \,\mathrm{sinc}(kau_1)|^2 + |\cos\theta_2 \,\mathrm{sinc}(kau_2)|^2 \right) \tag{6.16}$$

A sample result is shown in Fig. 6.4 for a corner angle of 90 deg. In this case the difference between the two is not significant because, at any given angle, the scattering from one plate dominates. Note that neither result agrees

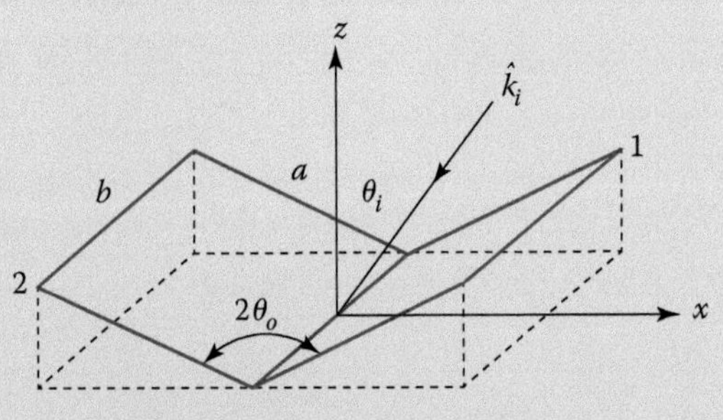

Fig. 6.3 Corner reflector geometry.

(Continued)

Example 6.2: Corner Reflector *(Continued)*

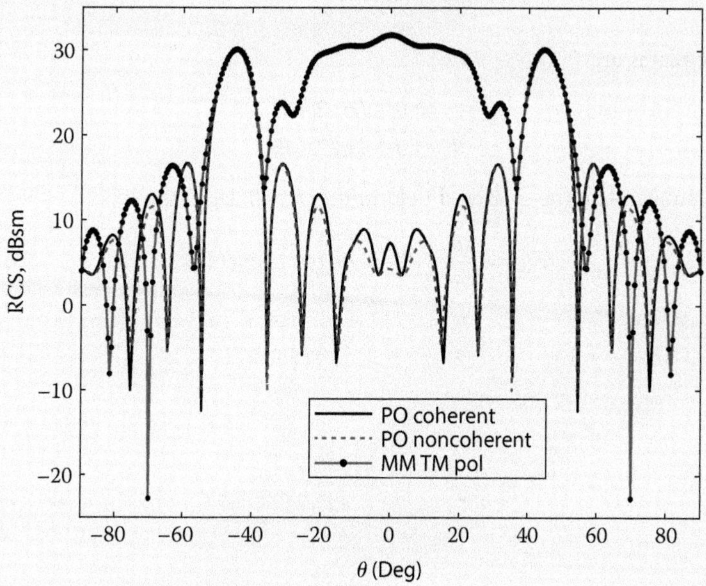

Fig. 6.4 Corner reflector scattering using MM and PO (coherent and noncoherent). The plate dimensions are 3λ by 0.5λ.

very well with the actual RCS (as given by the MM curve) in the region between the plates. This is because the multiple reflections are not included in the PO calculation.

6.2.2 Patch Models and Physical Optics

Geometrical components can be used to represent crude approximations to almost any complex target. The next step in improving the accuracy of a target model is a patch representation similar to that used in a MM solution. Triangular patches are the most flexible for approximating arbitrary doubly curved surfaces with edges. Figure 6.5 shows how aircraft and ships can be represented by triangular patches. Physical optics can be used on each patch, and the total field is obtained by superimposing the scattering from all illuminated and viewed patches. Physical theory of diffraction edge currents can be included for triangle edges that coincide with physical edges. Potentially, a large number of triangles must be used so that the surface curvature is accurately reproduced. θ is chosen so that the maxima from plates *CDBI* and *DFGJ* are viewed.

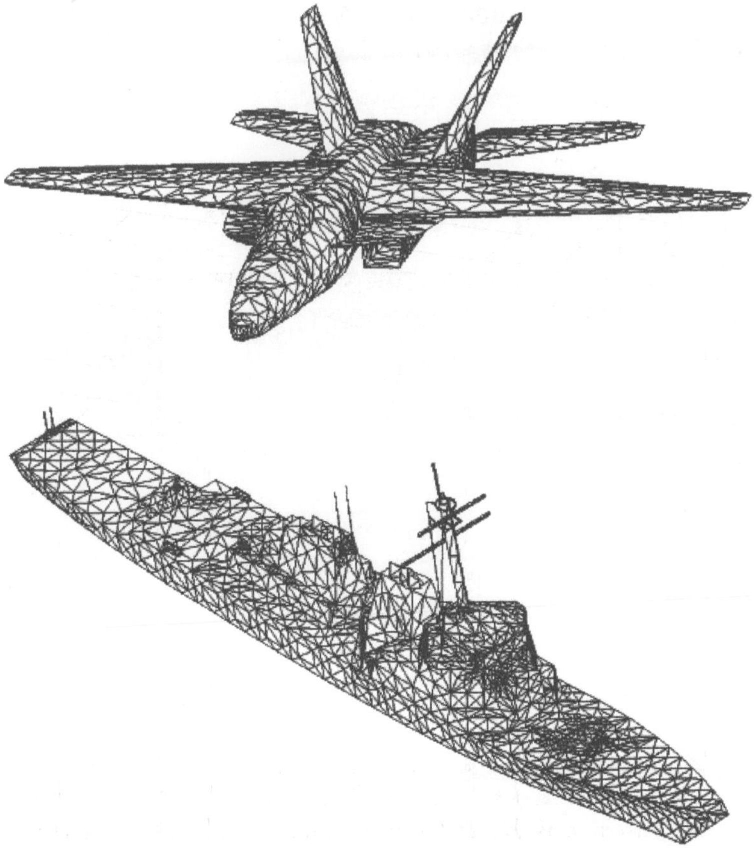

Fig. 6.5 Patch models of an aircraft and ship.

Example 6.3: Radar Cross Section Calculation for a Multiplate Structure

Figure 6.6 shows a multiplate structure constructed of trapezoidal plates. The plates are defined by the corner coordinates in Table 6.1. If the corners of each plate are ordered in a right-handed direction (i.e., counterclockwise), the outward normal is easily determined, as described in Problem 2.10. For a closed body such as this, the scalar (dot) product of the outward normal and the incident wave propagation vector must be negative. (This assumes that the coordinate system origin is located inside the structure.) Thus, for plate m, the scattering is included only if $\hat{n}_m \cdot \hat{k}_i < 0$. Shown in Fig. 6.7

(Continued)

Example 6.3: Radar Cross Section Calculation for a Multiplate Structure *(Continued)*

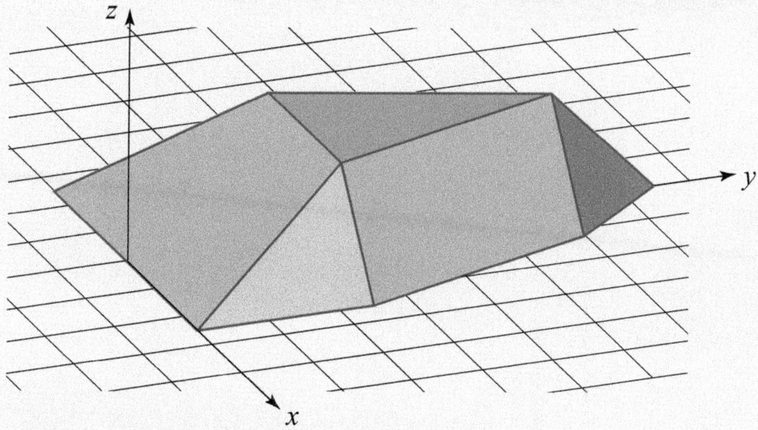

Fig. 6.6 Multiplate structure constructed of trapezoidal plates.

and Fig. 6.8 are the RCS patterns for this structure at 300 MHz. These are azimuth cuts (ϕ varies for a constant θ). The elevation angle θ is chosen so that the pattern cuts are through the peaks of the beams of the large side plates.

Table 6.1 Vertex Coordinates (in Meters) for the Multiplate Structure Shown in Fig. 6.6

Point	(x, y, z)
A	(5, 0, 0)
B	(5, 5, 0)
C	(2.5, 5, 3)
D	(0, 12, 3)
E	(0, 15, 0)
F	(−2.5, 5, 3)
G	(−5, 5, 0)
H	(−5, 0, 0)
I	(2.5, 12, 0)
J	(−2.5, 12, 0)

(Continued)

Example 6.3: Radar Cross Section Calculation for a Multiplate Structure *(Continued)*

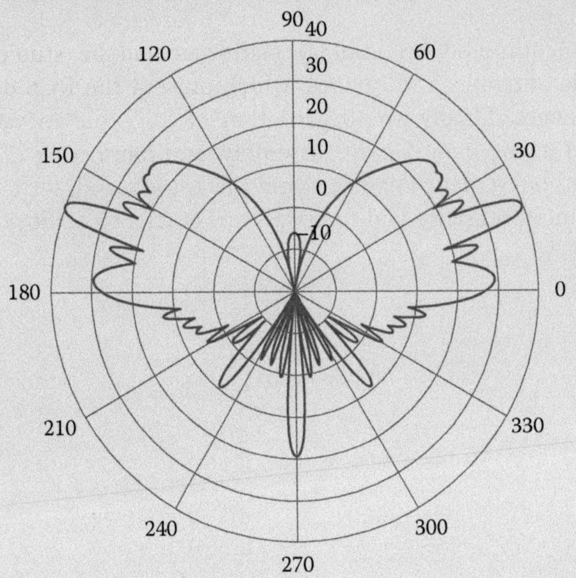

Fig. 6.7 TM RCS ($\sigma_{\phi\phi}$) of the multiplate structure for $\theta = 53$ deg.

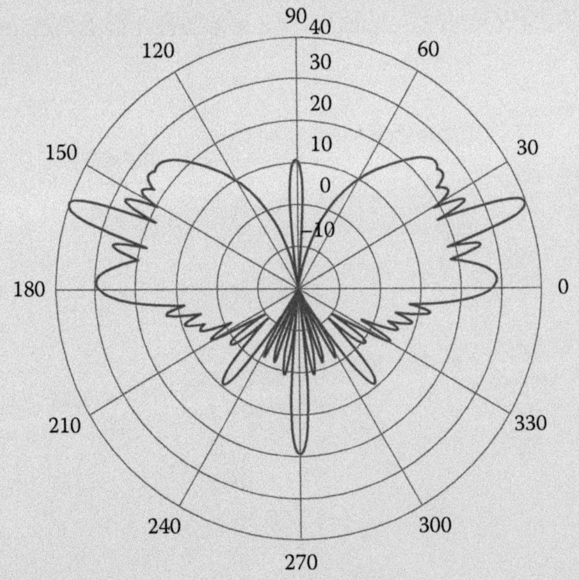

Fig. 6.8 TM RCS ($\sigma_{\theta\theta}$) of the multiplate structure for $\theta = 53$ deg.

All of the analysis methods discussed thus far can be used to predict the RCS of complex targets. The approach for large targets is essentially the same as for simple structures that consist of only a few plates; it is the bookkeeping that becomes challenging.

In the remainder of this chapter, special scattering structures that are commonly encountered on military platforms will be studied in detail. These include antennas and cavities, which present the RCS designer with unique challenges. Finally, the degradation of RCS due to variations and imperfections in the manufacture, assembly, and fabrication of materials is examined. Although many engineers find this topic bland, setting and maintaining tolerances on parts and materials are crucial to achieving low radar cross section.

6.3 Antenna Scattering Characteristics

6.3.1 General Comments

Almost all military vehicles use communication links, and most large combat vehicles have some type of radar or direction-finding equipment. These systems require an antenna to radiate or receive signals. Antennas are mounted either external to the platform surface (as in the case of ships) or integrated into the surface. The latter antenna type is referred to as a *conformal antenna* because it conforms to the platform structure. These approaches are depicted in Fig. 6.9 along with a third type in which the antenna is hidden from the threat radar's view. The screening surface can be a conductor that is closed when the antenna is not in use (i.e., a

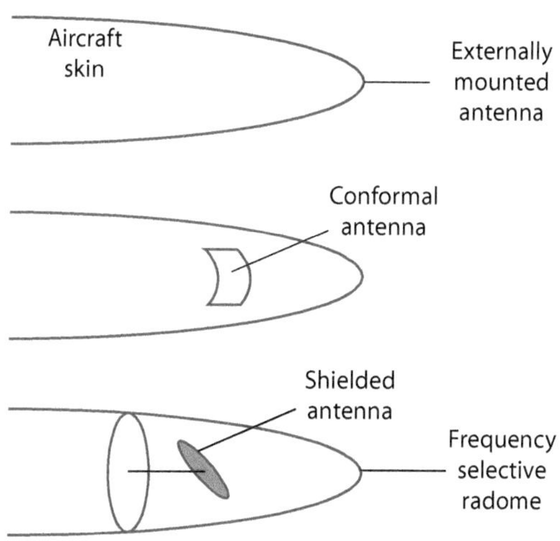

Fig. 6.9 Some examples of antenna mounting methods.

shutter) or a material that is transparent in the antenna's operating frequency band and opaque outside this band. In the latter case, the surface behaves like a filter and is called a *frequency selective surface* (FSS).

The antenna's impact on RCS is twofold [2,3]. First, the antenna frequently disturbs the continuity of the surface. For example, mounting an array on the side of an aircraft requires cutting a hole in the skin. This introduces more edges, with a potential for increasing wide-angle scattering. Furthermore, if the antenna is well designed, it absorbs almost all the energy incident in its operating band. Thus, a plate with an antenna installed has scattering characteristics like a plate with a hole in it, which gives rise to wide-angle scattering, as illustrated in Problem 2.3.

A second aspect of antenna scattering is that the threat radar's signal can penetrate into the feed and be reflected at internal mismatches and junctions. Even though these mismatches are small in the operating band (for a well-designed antenna), there can be a large number of scattering sources that add constructively under some conditions. This scattering is dependent on the type of feed and the devices incorporated therein.

6.3.2 Frequency Regions for Antenna Scattering

Generally, antenna scattering characteristics can be associated with three frequency regions, as indicated in Fig. 6.10.[†] In the operating band, signals enter the feed and are reflected at internal mismatches and then reradiated, as just discussed. Although the individual reflected signal levels are small, they can add coherently at some frequencies and angles, yielding a large RCS.

At low out-of-band frequencies, the radiating elements are not resonant and, therefore, little or no energy enters the feed. If $\lambda \gg \ell$, where ℓ is the element length, scattering from the element itself is negligible. The incident wave passes by the element relatively undisturbed and is reflected by surfaces behind the aperture (usually a metallic ground plane).

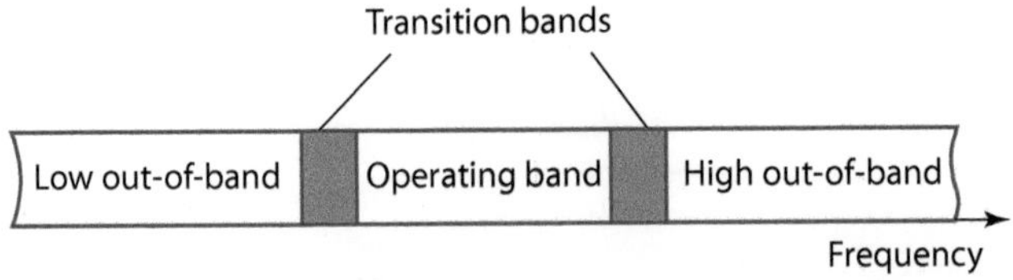

Fig. 6.10 Antenna frequency regions.

[†]The following discussion is presented in terms of phased arrays, but the principles can be applied to any antenna type.

At high out-of-band frequencies, scattering becomes complicated and hard to predict. At most frequencies, the elements are not matched, so that reflections inside the feed are generally not significant. (Exception: When the threat frequency is a harmonic of the antenna operating frequency, the elements may be resonant.) In this frequency region, the elements can be large in terms of wavelength and, consequently, present a significant RCS in themselves. The scattering characteristics of these three regions are summarized in the following list:

Low out-of-band region

1. Elements not matched (feed decoupled).
2. Closely packed elements (no Bragg diffraction).
3. Most scattering from ground plane surface.

In-band region

1. Elements well matched (feed contributes).
2. Internal reflections small but can add coherently.
3. RCS depends on feed network and devices.

High out-of-band region

1. Elements not matched (except at harmonics).
2. Element scattering significant.
3. Elements far apart (Bragg diffraction occurs).

The transition bands between the three regions (shown shaded in Fig. 6.10) are rarely sharp. They can extend up to an octave for standard wideband radar antennas, and an accurate RCS prediction requires that the entire antenna be modeled precisely.

6.4 Basic Equation of Antenna Scattering

As the preceding section indicates, the prediction of antenna RCS is difficult because of the many scattering sources and mechanisms. Traditionally, the antenna scattered field has been decomposed into two components called the *structural* and *antenna (radiation)* modes [4]. The antenna mode is determined completely by the radiation characteristics of the antenna and vanishes when the antenna is *conjugate-matched* to its radiation impedance. The structural mode contribution arises from currents induced on the antenna surfaces. For most antennas, the two terms do not have any practical significance, particularly when the antenna is installed on a platform. Surfaces that would contribute to the structural mode for a free-space antenna are not necessarily exposed when installed.

The basic equation of antenna scattering has been presented by several authors [4–7]. It expresses the total scattered field for a linearly polarized antenna when the antenna port is terminated with a load Z_L:

$$\vec{E}_s(Z_L) = \vec{E}_s(Z_a^*) + \left[\frac{j\eta}{4\lambda R_a}\vec{h}(\vec{h}\cdot\vec{E}_i)\frac{e^{-jkr}}{r}\right]\Gamma_0 \tag{6.17}$$

where

$Z_a = R_a + jX_a =$ the antenna impedance
$R_a = R_r + R_d =$ the antenna resistance
$R_r =$ the radiation resistance
$R_d =$ the ohmic resistance
$\vec{h} =$ effective height (see Appendix C)

and where

$$\Gamma_0 = \frac{Z_L - Z_a^*}{Z_L + Z_a^*} \tag{6.18}$$

When the load impedance is the complex conjugate of the radiation impedance, Z_L is called a conjugate-matched load. In Eq. (6.17), $\vec{E}_s(Z_a^*)$ is the structural mode; the second term on the right-hand side is the antenna mode.

Example 6.4: Radiation Mode Scattering from a Planar Array

The following assumptions are made:

1. The elements are identical, lie in the $z = 0$ plane, and are uniformly spaced, with dimensions d_x and d_y.
2. There are M elements in x and N elements in y.
3. The elements are linearly polarized in the x direction.
4. The incident field is TM_z-polarized, so that there is only a θ component of $\vec{E}$.
5. Neglect ohmic losses ($R_d = 0$), and assume resonance ($X_a = 0$).

For a single element (m, n) the radiation mode is given by

$$(\vec{E}_s)_{mn} = \left[\frac{j\eta}{4\lambda R_r}\vec{h}(\vec{h}\cdot\vec{E}_i)\frac{e^{-jkR}}{R}\right]\Gamma_0 \tag{6.19}$$

The effective height can be related to the maximum effective area of a single element by [8]

$$|\vec{h}| = 2\sqrt{\frac{A_{em}R_r}{\eta}} \tag{6.20}$$

(Continued)

Example 6.4: Radiation Mode Scattering from a Planar Array *(Continued)*

and, for a unit magnitude plane wave, its dot product with the incident field by

$$\vec{h} \cdot \vec{E}_i = (\hat{x} \cdot \hat{\theta})|\vec{h}|e^{-j\vec{k} \cdot \vec{r}'_{mn}}$$

where $\vec{r}'_{mn}$ is a position vector to element (m, n). Thus,

$$(E_s)_{mn} = \frac{j}{\lambda} A_e e^{-j\vec{k} \cdot \vec{r}'_{mn}} \frac{e^{-jkr}}{r} \Gamma_0 \tag{6.21}$$

where $A_e = \hat{x} \cdot \hat{\theta} A_{em}$ is the element's effective area presented to the incident wave. For a large array with a rectangular grid of elements and no Bragg lobes, $A_{em} \approx d_x\, d_y$. The total scattered field is the sum over all elements

$$\vec{E}_s = \sum_{m=1}^{M} \sum_{n=1}^{N} (\vec{E}_s)_{mn}$$

With the usual far-field approximations for the scattered wave,

$$E_{s\theta}(r, \theta, \phi) = \left(\frac{j\Gamma_0 A_e e^{-jkr}}{\lambda r} \right) \times \sum_{m=1}^{M} \sum_{n=1}^{N} \exp\{j2k[(m-1)d_x u + (n-1)d_y v]\} \tag{6.22}$$

Finally, the summations can be reduced to closed-form expressions:

$$E_{s\theta}(r, \theta, \phi) = \frac{j\Gamma_0 A_e e^{-jkr}}{\lambda r} \left[\frac{\sin(M\alpha)}{\sin \alpha} \right] \left[\frac{\sin(N\beta)}{\sin \beta} \right]$$

where $\alpha = kd_x\, u$ and $\beta = kd_y\, v$. From this, we obtain the RCS

$$\sigma_{\theta\theta} = \frac{4\pi A_e^2}{\lambda^2} |\Gamma_0|^2 \left[\frac{\sin(M\alpha)}{\sin \alpha} \right]^2 \left[\frac{\sin(N\beta)}{\sin \beta} \right]^2 \tag{6.23}$$

Several important observations can be made with respect to Eq. (6.23):

1. For $\theta = \phi = 0$, $\alpha = \beta = 0$, and the maximum values of the sine functions in parentheses are M and N. Thus, the maximum RCS is

$$\sigma_{\max} = \frac{4\pi (MNA_e)^2}{\lambda^2} |\Gamma_0|^2$$

However, with no Bragg diffraction, $MN\, A_e = MN\, A_{em} \cos \theta \approx MN\, d_x d_y \cos \theta$. The total array aperture area is $A \approx MN\, d_x\, d_y$, so that

$$\sigma_{\max} = \frac{4\pi A^2}{\lambda^2} |\Gamma_0|^2 \cos^2 \theta$$

Thus, the array RCS is reduced by a factor $|\Gamma_0|^2$ relative to a perfectly conducting plate with the same area.

(Continued)

Example 6.4: Radiation Mode Scattering from a Planar Array *(Continued)*

2. A relationship between the monostatic RCS of an array and the radiation pattern can be deduced by comparing the following pattern functions:

$$\text{RCS} \sim \left(\frac{\sin(N\beta)}{\sin\beta}\right)^2 \quad \text{and} \quad \text{GAIN} \sim \left(\frac{\sin(N\beta/2)}{\sin(\beta/2)}\right)^2$$

The scattering *pattern* from an array can be viewed as equivalent to the radiation pattern from the same array at twice the frequency. Alternately, the RCS pattern is equivalent to radiation pattern at the same frequency but from an array with the element spacings doubled.

3. In light of item 1, if the antenna is perfectly matched, then $|\Gamma_0| = 0$ and its RCS is zero. In practice, this is not achievable simultaneously at all angles θ because the conjugate-matched load required to make $\Gamma_0 = 0$ varies with the angle. Furthermore, Eq. (6.23) assumes that the reflected wave arises from a single scattering source at the antenna terminals; it neglects reflections that would occur at other points in the feed network.

It is a straightforward matter to derive a more general form of Eq. (6.23) that applies to arrays with arbitrary element locations and reflection coefficients (see Problem 6.18). In this case,

$$\sigma_{\theta\theta}(\theta, \phi) = \frac{4\pi A_e^2}{\lambda^2}\left|\sum_{m=1}^{M}\sum_{n=1}^{N}\Gamma_{mn}(\theta, \phi)e^{j\vec{k}\cdot\vec{d}_{mn}}\right|^2 \tag{6.24}$$

where $\Gamma_{mn}(\theta, \phi)$ represents the total reflected signal for element (m, n) and $\vec{d}_{mn}$ is the element's position vector.

6.5 Conjugate-Matched Antennas

The conjugate-matched antenna has received much attention in the past because it provides maximum efficiency [9]. Most studies have been based on the dipole because of the relatively simple equations that occur. Figure 6.11 shows a loaded dipole and its equivalent circuit. We desire to maximize the power absorbed by the antenna, and minimize the scattered power.

From circuit theory, it is known that the maximum power P_L absorbed by a complex load Z_L occurs when [10]

$$Z_L = Z_a^*$$

The power absorbed is a maximum for this condition, but it can be shown [8] that for a lossless dipole it is also equal to the power scattered, P_s

$$P_L = P_s = \frac{v^2}{8R_a} \tag{6.25}$$

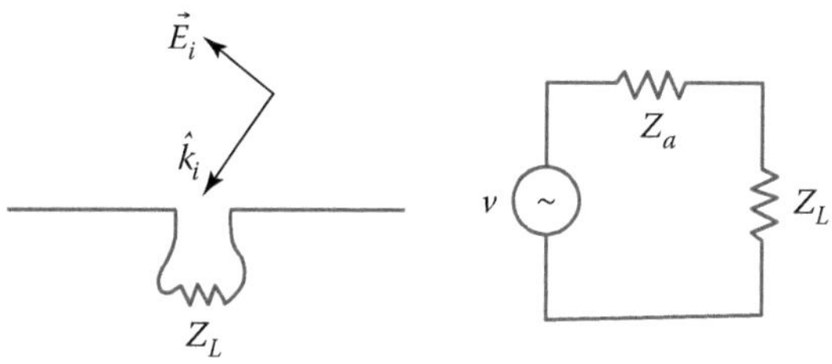

Fig. 6.11 Scattering from a dipole and its equivalent circuit.

which suggests that $P_L \leq P_s$ for a conjugate-matched antenna. In other words, to reduce scattering, the load must be mismatched so that more power is reflected so as to cancel the scattered field.

The reason for this apparent contradiction is that conjugate matching eliminates only the antenna mode term; the structural mode remains. By slightly mismatching the terminal impedance, just the right amount of antenna mode scattering can be introduced to cancel the structural mode component. The two contributions do not have the same angle or frequency characteristics and, therefore, the load mismatch must be readjusted if one of these two parameters changes. Figure 6.12 shows backscattering from cylindrical antennas for various terminal loads. In some cases, shorted terminals yield a lower RCS than the conjugate-matched condition.

It would be unfortunate if it were necessary to suffer a loss in gain to reduce the antenna's RCS, as Eq. (6.25) seems to imply. The reason for the large loss is that the forward scatter and backscatter are equal for a thin dipole. For a moderate- to high-gain antenna, this is not the case. As depicted in Fig. 6.13, the backscatter is much higher than the forward scatter. The high

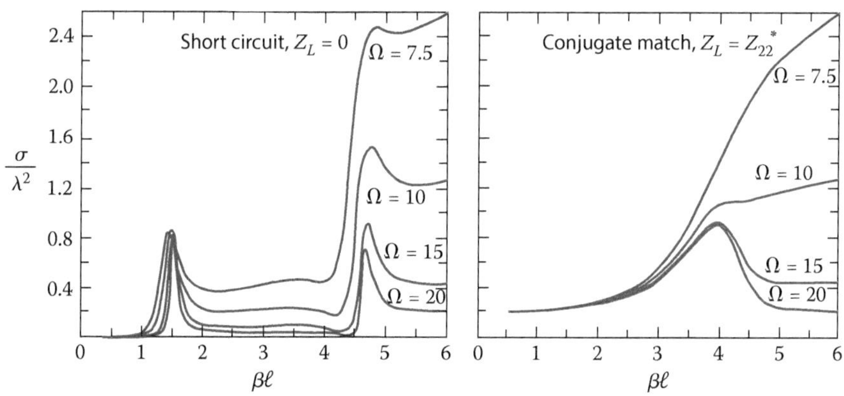

Fig. 6.12 RCS of cylindrical antennas for several load conditions [radius, a; half-length, ℓ; $\Omega = 2\ ln(2\ell/a)$ (from Ref. [7], courtesy IEEE)].

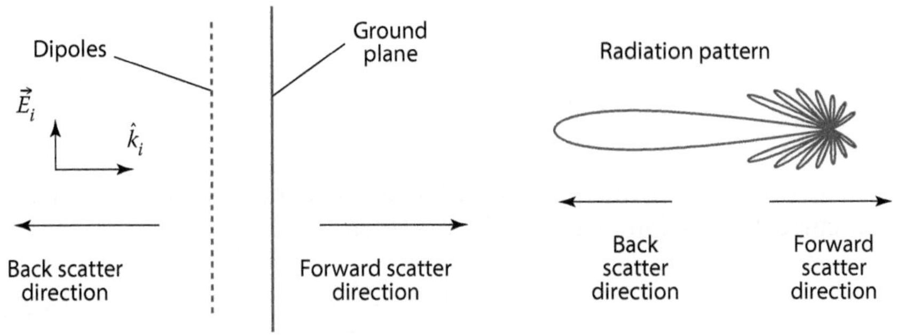

Fig. 6.13 Scattering from a high-gain array of dipoles above a ground plane.

gain is achieved by adding the ground plane, which essentially doubles the number of elements by introducing images. The image dipoles also scatter fields composed of antenna and structural modes. It is possible to adjust the image distance (i.e., ground plane location) to force the image and dipole structural modes to cancel. This leaves only the radiation modes, which can be eliminated (in principle) by conjugate matching.

In general, *if a lossless, linearly polarized antenna absorbs more power than it scatters under matched conditions, its gain in the back direction must exceed its gain in the forward direction* [9].

6.6 Rigorous Solutions for Antenna Radar Cross Section

Rigorous solutions for antenna radar cross sections are based on the MM or the finite difference (FD) method. The formulation of any given problem can be exact but, in practice, the resulting integral or differential equations must be solved numerically, as described in Chapters 3 and 4. In this section, three specific examples are discussed: 1) scattering from an array of dipoles above a finite ground plane, 2) scattering from a cavity-backed dipole, and 3) scattering from a parabolic reflector with a cavity-backed dipole feed.

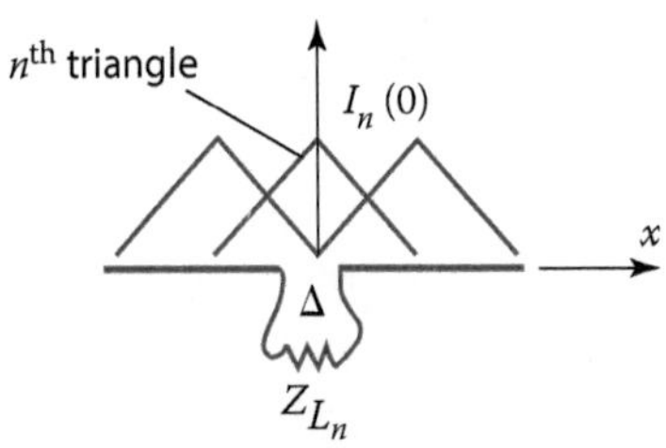

Fig. 6.14 MM expansion functions on a dipole in the vicinity of the feed point.

The discussion in the preceding section emphasized the importance of the terminal load impedance on the RCS. Thus, an accurate model for antenna RCS must include the terminal load in the MM or FD formulation. Figure 6.14 shows a triangular basis function spanning the segment of a dipole that contains the feed point. To compute the transmit pattern, the current on the dipole is determined by exciting the gap with a voltage V_0. When receiving, there is electric field in the gap denoted by E_0. If the gap length is Δ, the gap voltage and field are related by $V_0 = E_0\,\Delta$. The boundary condition that must be satisfied in the gap is

$$E_s + E_0 = \frac{Z_L I_n(0)}{\Delta} \tag{6.26}$$

where all field components are assumed to be tangential and $I_n(0)$ is the feed-point current. Equation (6.26) can be rewritten

$$E_0 \equiv E_i = -E_s + \frac{Z_L I_n(0)}{\Delta} \tag{6.27}$$

Equation (6.27) is essentially the EFIE of Eq. (3.97) with the load term $Z_L\, I_n(0)/\Delta$. A MM testing procedure would result in the following matrix equation:

$$\mathbf{V} = (\mathbf{Z} + \mathbf{Z}_L)\mathbf{I} \tag{6.28}$$

The elements of the load matrix are given by

$$[\mathbf{Z}_L]_{mn} = \delta_{mn} Z_L \tag{6.29}$$

where δ_{mn} is the Kronecker delta. In words, if the nth triangle spans a feed point with a load Z_L attached, the value of the load is simply added to the self-term of the MM impedance matrix.

Example 6.5: Radar Cross Section of an Array of Dipoles Above a Finite Ground Plane

Figure 6.15 shows an array of dipoles arranged in a rectangular grid above a finite ground plane. The RCS can be computed rigorously using the MM if basis functions are defined on all the surfaces [11]. Rooftop functions are used on the ground plane (as illustrated in Fig. 6.11), and overlapping triangles are used on the wires. The thin-wire approximation is assumed to hold, so that the circumferential currents on the dipoles can be neglected.

The RCS of a 4 ($=N_x$) by 5 ($=N_y$) array is shown in Fig. 6.16 ($d_x = 0.5\lambda$, $d_y = 0.45\lambda$). Each element of the array is conjugate-matched to its radiation impedance for normal incidence ($\theta = 0$ deg). The ground plane is $1.6\lambda \times 2.1\lambda$, and the dipoles are 0.38λ long and located 0.18λ above the ground plane.

Figure 6.16 also shows the array RCS for shorted dipoles, along with the scattering from the ground plane alone. Note that the conjugate-matched case still yields significant scattering at $\theta = 0$ deg. The scattering in this direction is due entirely to the structural mode. The primary contributor is edge diffraction from the ground plane. The RCS in this direction could be reduced somewhat by mismatching the load slightly. However, it would cause the RCS at other angles to increase.

(Continued)

Example 6.5: Radar Cross Section of an Array of Dipoles Above a Finite Ground Plane *(Continued)*

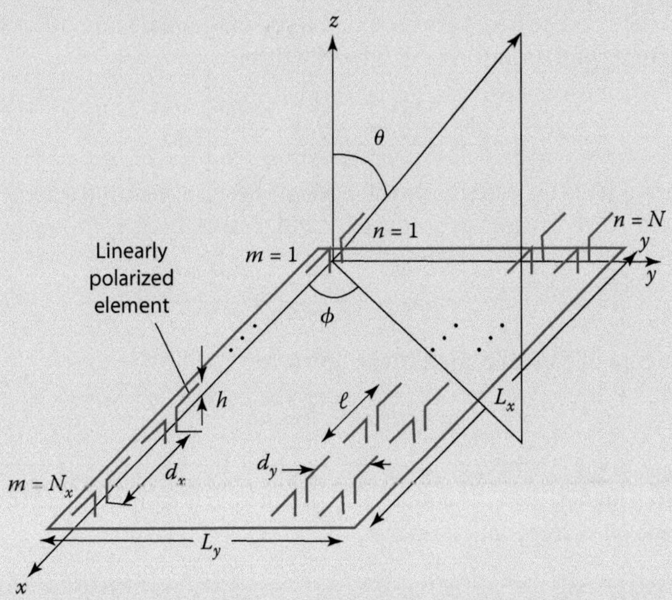

Fig. 6.15 Array of terminated dipoles above a finite ground plane.

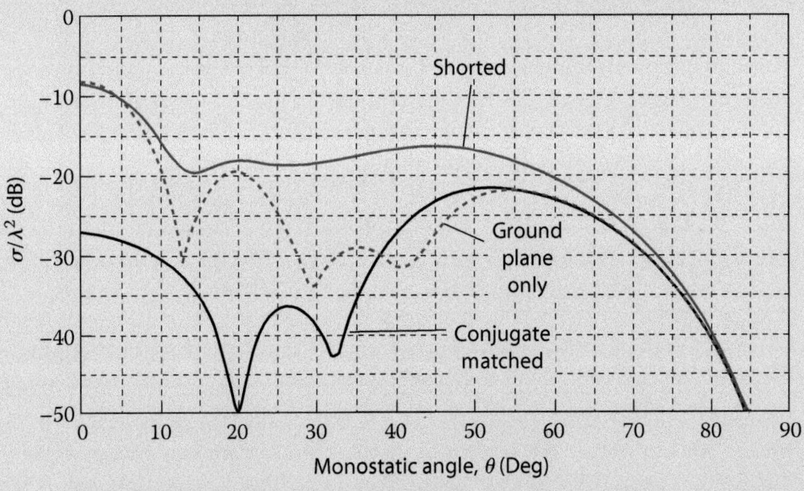

Fig. 6.16 Radar cross section of an array of terminated dipoles above a finite ground plane. The reference level is 30 dB.

Example 6.6: Radar Cross Section of a Cavity-Backed Dipole Antenna

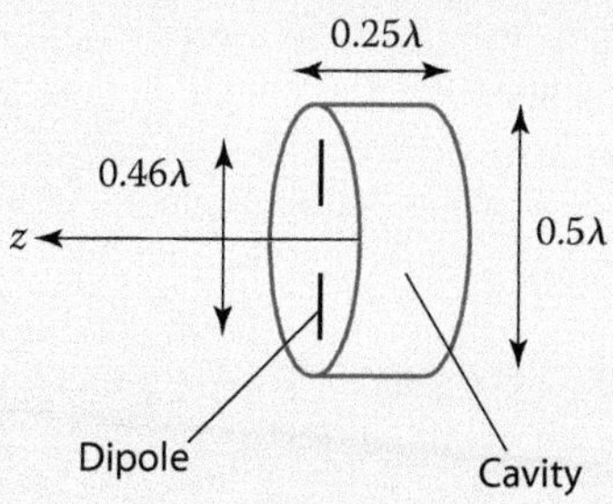

Fig. 6.17 Cavity-backed dipole antenna.

The RCS of the cavity-backed dipole, shown in Fig. 6.17, can also be computed using the MM [12]. The BOR basis functions described in Sec. 3.6.3 are used on the cavity surface. Overlapping triangles are used on the wire, and the thin-wire approximation is assumed. The RCS for several load conditions is given in Fig. 6.18.

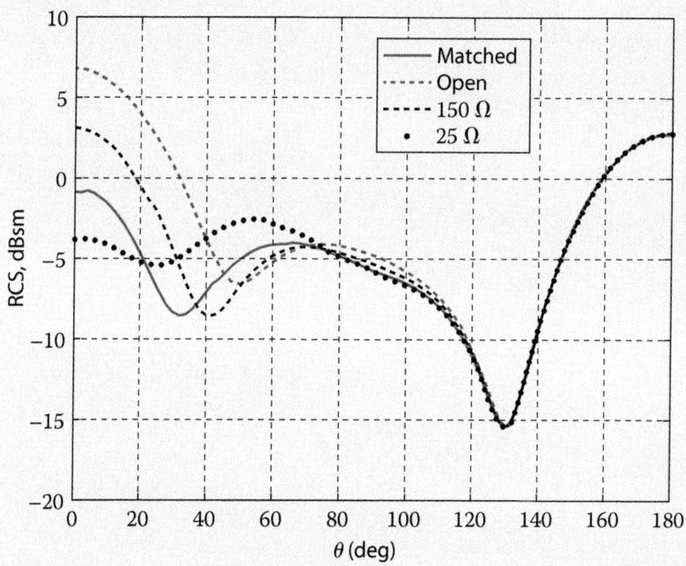

Fig. 6.18 Radar cross section of a cavity-backed dipole antenna (from Ref. [12]).

(Continued)

Example 6.6: Radar Cross Section of a Cavity-Backed Dipole Antenna *(Continued)*

The dimensions used are typical of an in-band frequency. $Z_L = 117 - j\,24\ \Omega$ is the conjugate match to the radiation impedance. Again, it is evident from the patterns that the lowest RCS requires mismatching the load so that additional scattering is provided to cancel the structural mode.

Example 6.7: Radar Cross Section of a Reflector with a Cavity-Backed Dipole Feed

The feed of the last example is used in conjunction with a paraboloid, as illustrated in Fig. 6.19. The current on the surfaces of revolution is computed using BOR basis functions (Sec. 3.6.3), and triangles are used on the dipole. The RCS at a low out-of-band frequency is shown in Fig. 6.20 for several f/D ratios. The dimensions shown are typical for an antenna with a 10λ main reflector diameter at its operating frequency. The f/D ratio affects the RCS level in the vicinity of $\theta = 0$ deg (broadside) because it determines the relative phase of the cavity scattered field with respect to the paraboloid scattered field. For $f/D = 0.4$, the round-trip difference is roughly an integer multiple of λ whereas, for $f/D = 0.49$, it is approximately an odd multiple of a half-wavelength.

Figure 6.21 shows the in-band RCS of the reflector and feed for several dipole load conditions. (The feed dimensions are twice the values shown in Fig. 6.17.)

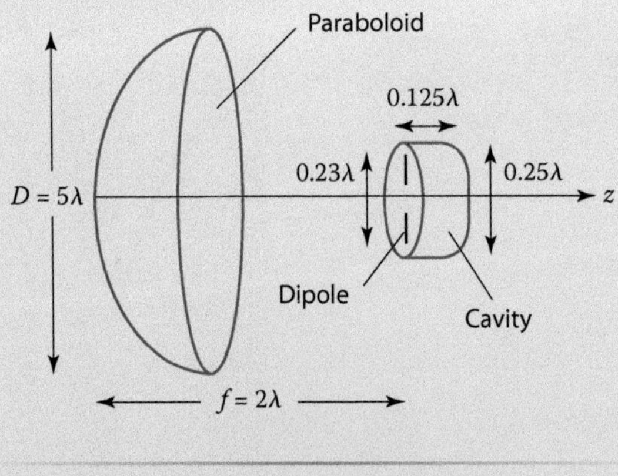

Fig. 6.19 Parabolic reflector antenna with a cavity-backed dipole feed. Dimensions are for a frequency of one-half of the operating frequency.

(Continued)

Example 6.7: Radar Cross Section of a Reflector with a Cavity-Backed Dipole Feed *(Continued)*

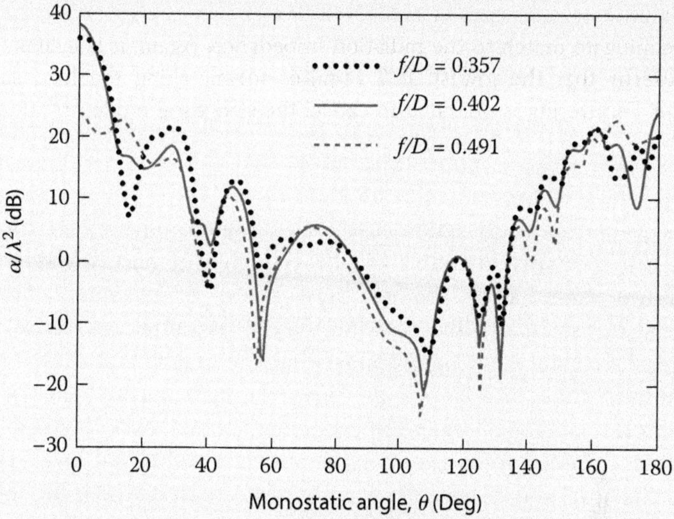

Fig. 6.20 Radar cross section of a parabolic reflector with a cavity-backed dipole feed at a low out-of-band frequency (from Ref. [12]).

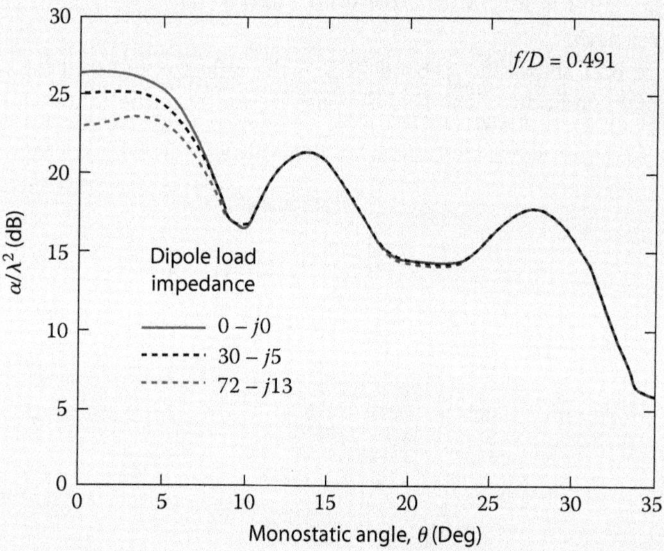

Fig. 6.21 Radar cross section of a parabolic reflector with a cavity-backed dipole feed at an in-band frequency (from Ref. [12]).

A comparison of the two reflector cross sections shows that, at low out-of-band frequencies, the dipole is not a significant contributor to the total RCS. To reduce computer run time, the dipole could be neglected entirely.

6.7 Scattering from Antenna Feeds

In the case of an array antenna, the terminal loads in the preceding examples represent the input impedances that the wave encounters propagating into the feed network. The incident threat signal power that was delivered to the load actually enters the feed network. The signal transmitted through the aperture elements will be reflected at points inside the feed circuit. For a phased array, the feed network can be extremely complicated, with every transmission line discontinuity a potential source of reflected signals. At first glance, it might appear that the majority of these reflections are random in phase and amplitude. However, high-performance arrays (i.e., arrays with low sidelobes) maintain very close tolerances on phase and amplitude errors. Consequently, the many small reflected signal levels inside the beam former can add coherently for the same reason that a highly focused radiation beam is obtained: path lengths to the aperture are designed to provide a linear phase.

6.7.1 Tracing Signals Through a Feed

The reflection sources for a typical phased array with a *parallel* feed [13] are shown in Fig. 6.22. The first scattering source encountered by the incident wave is the array element. If the threat frequency is in the operating band of the array and the wave is arriving at near-normal incidence, the reflection

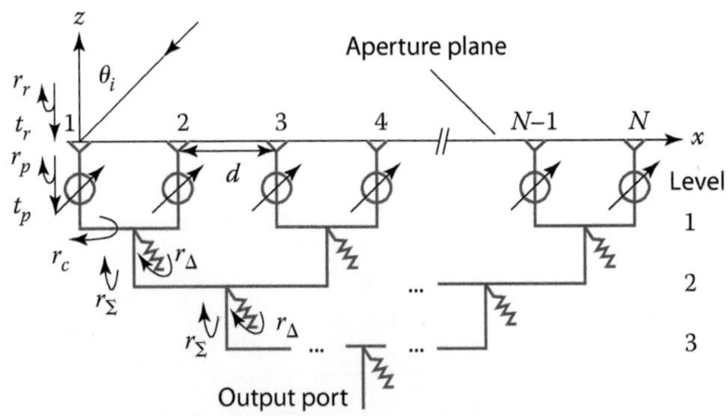

Fig. 6.22 Scattering sources for a phased array with a parallel feed.

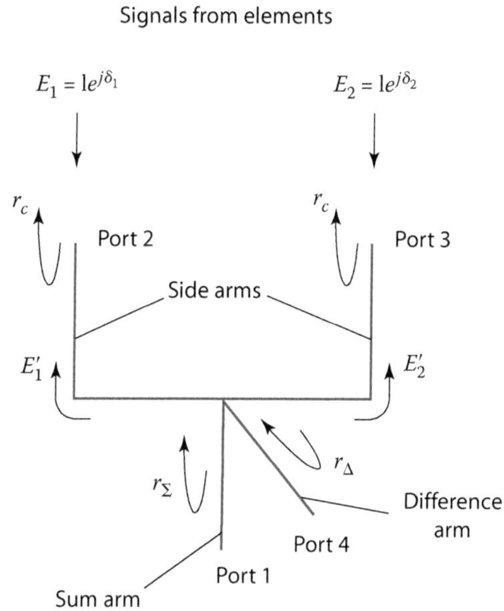

Fig. 6.23 Reflections at the ports of a magic tee.

coefficient r_r should be small if the antenna is well designed.[‡] The portion of the signal not reflected (t_r) is transmitted into the feed, where it proceeds to the phase-shifter input port. If the phase shifter is not exactly matched to the transmission line, a small reflection results. This reflected signal (determined by the reflection coefficient r_p) propagates back to the aperture. A portion is again reflected by the radiating element (because it is assumed to be reciprocal), and the remainder is reradiated.

The fraction of the signal that enters the phase shifter will have its phase shifted, depending on the device's state. This phase shift is included in the transmission coefficient if it is allowed to be complex:

$$t_p = |t_p|e^{j\chi}$$

The phase-shifted threat signal travels through a transmission line and enters the side arm of the first level of couplers. A general magic tee is shown in Fig. 6.23, with the fourth port loaded. This configuration can be used to represent any three-port power divider as well. Reflections can arise from mismatches at the inputs of both side arms (r_c at 2 and 3) as well as the sum and difference output ports (r_Σ at 1 and r_Δ at 4). The angles δ_1 and δ_2 represent the total signal phase at the coupler inputs relative to a phase reference. If this coupler is used in an array feed network, δ_1 and δ_2 include all

[‡]In this discussion the reflection and transmission coefficients of network devices will be denoted by the symbols r and t, respectively.

the insertion phases of the devices between the couplers and the aperture in addition to any space path delay relative to the origin.

6.7.2 Scattering at the Junction of a Magic Tee

The scattering matrix of a perfectly matched magic tee ($r_c = 0$) is

$$\mathbf{C} = \frac{\sqrt{2}}{2}\begin{bmatrix} 0 & 1 & 1 & 0 \\ 1 & 0 & 0 & 1 \\ 1 & 0 & 0 & -1 \\ 0 & 1 & -1 & 0 \end{bmatrix} \tag{6.30}$$

Figure 6.23 shows waves incident on the side arms of the first coupler in the array. They are unit amplitude but differ in phase. The subscripts 1 and 2 on E refer to the element numbers to which this tee's side arms are connected. The combined signal in the sum arm is

$$\begin{aligned} E_\Sigma = E_1 + E_2 &= \frac{\sqrt{2}}{2}e^{j\delta_1} + \frac{\sqrt{2}}{2}e^{j\delta_2} \\ &= \sqrt{2}\exp\left[j(\delta_1+\delta_2)/2\right]\cos\left(\frac{\delta_1+\delta_2}{2}\right) \end{aligned} \tag{6.31}$$

Similarly, for the difference arm,

$$\begin{aligned} E_\Delta = E_1 - E_2 &= \frac{\sqrt{2}}{2}e^{j\delta_1} - \frac{\sqrt{2}}{2}e^{j\delta_2} \\ &= j\sqrt{2}\exp\left[j(\delta_1+\delta_2)/2\right]\sin\left(\frac{\delta_1+\delta_2}{2}\right) \end{aligned} \tag{6.32}$$

Reflections inside the magic tee can arise from several sources. The net effect of internal scattering can be modeled by two reflection coefficients. One is for the "sum type" of scattering sources (r_Σ) and the other for the "difference type" of scattering sources (r_Δ). The reflected signals emerging from the sum and difference arms are denoted by primes and have the form

$$\begin{aligned} E'_\Sigma &= E_\Sigma r_\Sigma \\ E'_\Delta &= E_\Delta r_\Delta \end{aligned} \tag{6.33}$$

Thus, the total reflected signal arriving back at the side arm inputs is

$$\begin{aligned} E'_1 = E'_\Sigma + E'_\Delta = \Bigg[& r_\Sigma\cos\left(\frac{\delta_1+\delta_2}{2}\right) \\ & + jr_\Delta\sin\left(\frac{\delta_1+\delta_2}{2}\right)\exp\Bigg]\left[j(\delta_1+\delta_2)/2\right] \end{aligned} \tag{6.34}$$

and

$$E_2' = E_\Sigma' - E_\Delta' = \left[r_\Sigma \cos\left(\frac{\delta_1 + \delta_2}{2}\right) - jr_\Delta \sin\left(\frac{\delta_1 + \delta_2}{2}\right) \right] \exp[j(\delta_1 + \delta_2)/2] \qquad (6.35)$$

The significance of the exponentials in Eqs. (6.34) and (6.35) will become apparent later when the couplers are assembled in an array. These equations indicate that the RCS is dependent not only on the reflection coefficients but also on the relative phases of the signals entering the side arms. The portion of the signal not reflected by the first level of couplers is transmitted down to the next coupler, where the entire process is repeated.

6.7.3 Assumptions for an Approximate Solution

The RCS is computed by summing the scattering from all the signals that return to the aperture. It is evident that tracing the signal flow inside a feed network can be very tedious. To be precise, multiple reflections should also be included. In most practical cases, the higher order reflections can be neglected because they vary as r^2, r^3, and so forth, where in-band $r \ll 1$. For cases in which multiple reflections are important, a more rigorous approach based on a numerical method combined with scattering matrices is more appropriate. This method is described in the following section.

For a simplified "first-cut" analysis of the feed scattering, the following approximations can be made:

1. All the devices of the same type are assumed to have identical electrical characteristics. For example, all the radiating elements have the same reflection and transmission coefficients, r_r and t_r. This is not the same as saying that the elements are *ideal*, in which case $r_r = 0$. (Note that the array edge effects are being neglected.)
2. All the couplers will be represented by magic tees, which implies equal power to all elements (i.e., this is not a low sidelobe feed).
3. At frequencies in the operating band, all the feed devices are well matched so that $r \ll 1$ and, therefore, higher-order reflections can be neglected.
4. Lossless devices will be assumed: $|r|^2 + |t|^2 = 1$.

For a linear array along the x axis with the above-mentioned limitations, the fraction of the threat signal entering the radiating elements that is reflected and then returns to the aperture is given by a set of equations with the following form:

$$\Gamma_n \approx r_r e^{j(n-1)\alpha} + t_r^2 r_p e^{j(n-1)\alpha} + t_r^2 t_p^2 r_c e^{j2\chi_n} e^{j(n-1)\alpha} + t_r^2 t_p^2 t_c^2 e^{j\chi_n} E_n' + \cdots \quad n = 1, 2, \ldots, N \qquad (6.36)$$

where $\alpha = kdu$ is the incident wave interelement space delay. It does not explicitly occur in the last term; it is embedded in the E'_n factor. The variable subscript n is the element index. The phase-shifter phase has been written explicitly in the argument of the exponential; thus, t_p represents the magnitude of its transmission coefficient $|t_p|$. The quantity E'_n has the form of Eqs. (6.34) and (6.35), with the angles δ_1 and δ_2 replaced with δ_n and δ_m. For the first level of couplers in the feed network, the index m is that of the element that shares the coupler with element n. The terms occurring after the ellipses (...) include reflection and transmission coefficients for higher levels of couplers (*higher* referring to deeper into the feed or farther from the aperture).

The RCS resulting from the signals reflected by the feed and aperture is derived from the reflection coefficients in Eq. (6.36). For an array of isotropic elements spaced d along the x axis, the scattered field becomes

$$E_s \sim \frac{e^{-jkr}}{r}\sum_{n=1}^{N}\Gamma_n e^{jkd(n-1)u} = \frac{e^{-jkr}}{r}\sum_{n=1}^{N}\Gamma_n e^{j\alpha(n-1)} \tag{6.37}$$

The leading constants associated with the electric field have been neglected. Using Eq. (6.36) in Eq. (6.37) gives

$$E_s \sim \frac{e^{-jkr}}{r}\sum_{n=1}^{N}\Big[(r_r + t_r^2 r_p + t_r^2 t_p^2 e^{j2\chi_n} r_c)e^{j2\alpha(n-1)} + (t_r^2 t_p^2 e^{j\chi_n} e^{j\alpha(n-1)} E'_n) + \cdots\Big] \tag{6.38}$$

Now rearrange Eq. (6.38) so that scattering from each type of device is combined under a single summation operator:

$$E_s \sim \frac{e^{-jkr}}{r}\left(\sum_{n=1}^{N} r_r e^{j2\alpha(n-1)} + \sum_{n=1}^{N} t_r^2 r_p e^{j2\alpha(n-1)} + \sum_{n=1}^{N} t_r^2 t_p^2 e^{j2\chi_n} r_c e^{j2\alpha(n-1)} + \sum_{n=1}^{N} t_r^2 t_p^2 t_c^2 e^{j\chi_n} E'_n e^{j\alpha(n-1)} \ldots\right) \tag{6.39}$$

The first two sums can easily be reduced to closed form using the technique described in Sec. 2.12. For example,

$$r_r \sum_{n=1}^{N} e^{j2\alpha(n-1)} = r_r e^{j(N-1)\alpha}\frac{\sin(N\alpha)}{\sin(\alpha)}$$

The last two terms contain the factor $e^{j2\chi_n}$, and further reduction requires an assumption as to the form of χ_n. It is the phase-shifter setting for scanning the antenna beam to a direction θ_s. For a linear phase across the aperture,

$$\chi_n = (n-1)\chi_0 = (n-1)kd\sin\theta_s \tag{6.40}$$

Now, the third term shown in Eq. (6.39) takes the form

$$\sum_{n=1}^{N} t_r^2 t_p^2 e^{j2\chi_n} r_c e^{j2\alpha(n-1)} = t_r^2 t_p^2 r_c e^{j(N-1)\zeta} \frac{\sin(N\zeta)}{\sin(\zeta)}$$

where $\zeta = \alpha + \chi_0$ has been defined for convenience.

Reducing the remaining terms is more difficult because of the phase factors δ_n in E_n'. For an incident plane wave, the δ_n are determined by the total phase of the signal at the coupler inputs relative to a specified phase reference. This term also includes the phase progression between elements due to the angle of arrival of the wave front. As illustrated in Fig. 6.22, upon arriving at the coupler inputs, the signal has been delayed in space an amount $(n-1)kd\sin\theta$ and has acquired the transmission phase modifications χ_n due to the phase shifter. Therefore,

$$\delta_n = (n-1)(\alpha + \chi_0) \equiv (n-1)\zeta \tag{6.41}$$

Using Eq. (6.41) allows detailed evaluation of the E_n':

$$E_1' = \left[r_\Sigma \cos\left(\frac{\zeta}{2}\right) + jr_\Delta \sin\left(\frac{\zeta}{2}\right)\right] e^{j(\zeta/2)}$$

$$E_2' = \left[r_\Sigma \cos\left(\frac{\zeta}{2}\right) - jr_\Delta \sin\left(\frac{\zeta}{2}\right)\right] e^{j(\zeta/2)}$$

$$E_3' = \left[r_\Sigma \cos\left(\frac{\zeta}{2}\right) + jr_\Delta \sin\left(\frac{\zeta}{2}\right)\right] e^{j(5\zeta/2)}$$

$$E_4' = \left[r_\Sigma \cos\left(\frac{\zeta}{2}\right) - jr_\Delta \sin\left(\frac{\zeta}{2}\right)\right] e^{j(5\zeta/2)}$$

$$\vdots$$

By induction, it is evident that, for $n = 1, 3, 5, \ldots, N/2 - 1$,

$$\begin{aligned} E_n' &= \left[r_\Sigma \cos\left(\frac{\zeta}{2}\right) + jr_\Delta \sin\left(\frac{\zeta}{2}\right)\right] \exp[j(2n-1)\zeta/2] \\ E_{n+1}' &= \left[r_\Sigma \cos\left(\frac{\zeta}{2}\right) - jr_\Delta \sin\left(\frac{\zeta}{2}\right)\right] \exp[j(2n-1)\zeta/2] \end{aligned} \tag{6.42}$$

The fourth sum in Eq. (6.39) can be reduced to closed form after some rearranging. First, replace χ_n with its equivalent given in Eq. (6.40), and then substitute in the detailed expressions for E'_n and E'_{n+1}. The result is

$$\sum_{n=1}^{N} \exp[j(n-1)\chi_0]E'_n \exp[j\alpha(n-1)]$$
$$= \left[r_\Sigma \cos\left(\frac{\zeta}{2}\right) + jr_\Delta \sin\left(\frac{\zeta}{2}\right)\right] \sum_{n=1,3,\ldots}^{N} \exp[j(4n-3)\zeta/2]$$
$$+ \left[r_\Sigma \cos\left(\frac{\zeta}{2}\right) - jr_\Delta \sin\left(\frac{\zeta}{2}\right)\right] \sum_{n=2,4,\ldots}^{N} \exp[j(4n-5)\zeta/2] \tag{6.43}$$

With a change of index, the *normalized* sums reduce to

$$\left[r_\Sigma \cos^2\left(\frac{\zeta}{2}\right) \frac{\sin(N\zeta)}{(N/2)\sin(2\zeta)} - r_\Delta \sin^2\left(\frac{\zeta}{2}\right) \frac{\sin(N\zeta)}{(N/2)\sin(2\zeta)}\right] e^{j(N-1)\zeta} \tag{6.44}$$

Thus, Eq. (6.39) has been reduced to a sum of trigonometric functions:

$$E_s \sim \frac{e^{-jkr}}{r}\left\{\left[r_r \frac{\sin(N\alpha)}{N\sin(\alpha)} + r_p t_r^2 \frac{\sin(N\alpha)}{N\sin(\alpha)}\right] e^{j(N-1)\alpha}\right.$$
$$+ t_r^2 t_p^2 r_c \frac{\sin(N\zeta)}{N\sin(\zeta)} e^{j(N-1)\zeta} + t_c^2 t_r^2 t_p^2 \left[r_\Sigma \cos^2\left(\frac{\zeta}{2}\right) - jr_\Delta \sin^2\left(\frac{\zeta}{2}\right)\right]$$
$$\left. \times \left[\frac{\sin(N\zeta)}{(N/2)\sin(2\zeta)}\right] e^{j(N-1)\zeta} + \cdots \right\} \tag{6.45}$$

Up to this point, we have been dealing only with a linear array, and the properties of the radiating elements in the y direction have been ignored. An element factor must be included to account for the pattern characteristics in the $y-z$ plane. If it is assumed that the element is essentially an isotropic scatterer but has a short length ℓ in the y direction, the total *projected* area of a linear array can be approximated by

$$A_p = A\cos\theta \approx Nd\ell\cos\theta$$

Now, Eq. (6.45) can be cast into the form of Eq. (6.4) if each of the terms in the brackets is defined as a complex voltage RCS:

$$M_r = \sqrt{\sigma_r} e^{j\psi_r} = \sqrt{\frac{4\pi A_p^2}{\lambda^2}} r_r \frac{\sin(N\alpha)}{N\sin(\alpha)} e^{j(N-1)\alpha}$$

$$M_p = \sqrt{\sigma_p} e^{j\psi_p} = \sqrt{\frac{4\pi A_p^2}{\lambda^2}} r_p t_r^2 \frac{\sin(N\alpha)}{N\sin(\alpha)} e^{j(N-1)\alpha}$$

$$M_c = \sqrt{\sigma_c} e^{j\psi_c} = \sqrt{\frac{4\pi A_p^2}{\lambda^2} t_r^2 t_p^2 r_c} \frac{\sin(N\zeta)}{N\sin(\zeta)} e^{j(N-1)\zeta}$$

$$M_\Sigma = \sqrt{\sigma_\Sigma} e^{j\psi_\Sigma} = \sqrt{\frac{4\pi A_p^2}{\lambda^2} t_r^2 t_p^2 t_c^2 r_\Sigma} \frac{\cos^2(\zeta/2)\sin(N\zeta)}{(N/2)\sin(2\zeta)} e^{j(N-1)\zeta} \tag{6.46}$$

$$M_\Delta = \sqrt{\sigma_\Delta} e^{j\psi_\Delta} = \sqrt{\frac{4\pi A_p^2}{\lambda^2} t_r^2 t_p^2 t_c^2 r_\Delta} \frac{j\sin^2(\zeta/2)\sin(N\zeta)}{(N/2)\sin(2\zeta)} e^{j(N-1)\zeta}$$

The complex RCS is now given by

$$M = M_r + M_p + M_c + M_\Sigma + M_\Delta + \cdots \tag{6.47}$$

Finally, the total RCS is

$$\sigma = |M_r + M_p + M_c + M_\Sigma + M_\Delta + \cdots|^2 \tag{6.48}$$

If the array size is very large in terms of the threat radar wavelength, then Nd is large and the pattern functions in Eqs. (6.46) have very high, narrow peaks. Thus, it may be sufficient to approximate the magnitude squared of the sum in Eq. (6.48) by the sum of the magnitudes squared:

$$\sigma \approx |M_r|^2 + |M_p|^2 + |M_c|^2 + |M_\Sigma|^2 + |M_\Delta|^2 + \cdots \tag{6.49}$$

Substituting in the explicit expressions for the complex voltage RCS yields

$$\begin{aligned}\sigma \approx \frac{4\pi A^2 \cos^2\theta}{\lambda^2} \Bigg[& r_r^2 \left(\frac{\sin(N\alpha)}{N\sin(\alpha)}\right)^2 + r_p^2 t_r^4 \left(\frac{\sin(N\alpha)}{N\sin(\alpha)}\right)^2 \\ & + t_r^4 t_p^4 r_c^2 \left(\frac{\sin(N\zeta)}{N\sin(\zeta)}\right)^2 + t_r^4 t_p^4 t_c^4 r_\Sigma^2 \cos^4(\zeta/2) \left(\frac{\sin(N\zeta)}{(N/2)\sin(2\zeta)}\right)^2 \\ & + t_r^4 t_p^4 t_c^4 r_\Delta^2 \sin^4(\zeta/2) \left(\frac{\sin(N\zeta)}{(N/2)\sin(2\zeta)}\right)^2 \cdots \Bigg]\end{aligned} \tag{6.50}$$

Note that the relative phases of the scattered field components are neglected in this final form.

Example 6.8: Radar Cross Section of a Large Linear Array

Consider a phased array antenna with the following parameters:

1. $N = 50$
2. $d = 0.5\lambda_0$, $\ell = \lambda_0$ (λ_0 is the array operating wavelength)

(Continued)

Example 6.8: Radar Cross Section of a Large Linear Array *(Continued)*

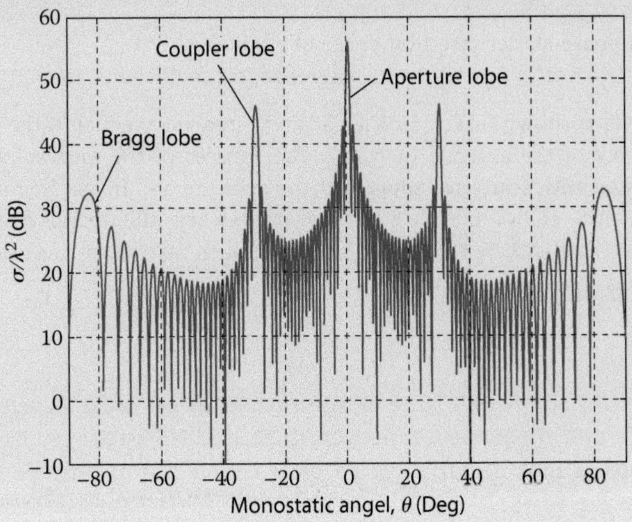

Fig. 6.24 Radar cross section of a linear array with a parallel feed for a threat frequency equal to the operating frequency (from Ref. [13]).

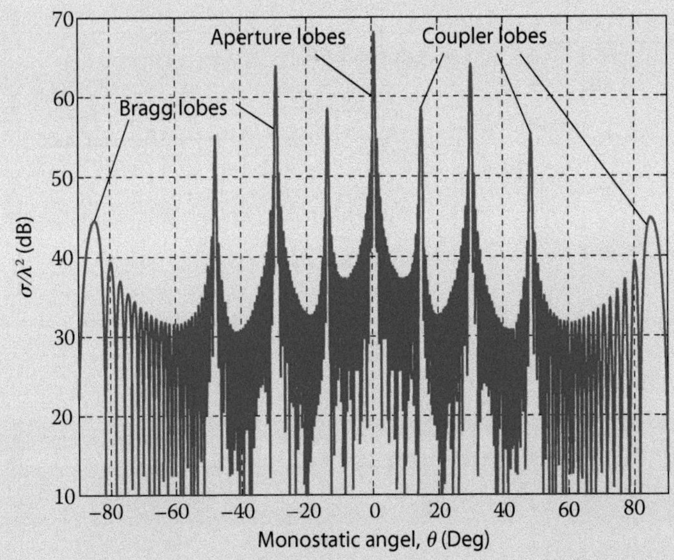

Fig. 6.25 Radar cross section of a linear array with a parallel feed for a threat frequency equal to two times the operating frequency (from Ref. [13]).

(Continued)

Example 6.8: Radar Cross Section of a Large Linear Array *(Continued)*

3. $\chi_0 = \psi_0 = 0$ (equal path lengths between the couplers and phase shifters, and phase-shifter insertion phase of zero)
4. $r_r = r_p = r_c = r_\Sigma = r_\Delta = 0.2$ (all reflection coefficients are 0.2)

The RCS is shown in Fig. 6.24 for a threat frequency equal to the operating frequency of the antenna ($\lambda = \lambda_0$). The sources of the high spikes can be associated with scattering sources at the aperture and in the feed as labeled. Figure 6.25 shows the RCS of the array when the threat frequency is doubled; that is, $\lambda = 2\lambda_0$, which is equivalent to using $d = \lambda_0$ in the pattern formulas. All other parameters are assumed to remain unchanged.

Example 6.9: Radar Cross Section of a Large Scanning Array

The array antenna of Example 6.8 is scanned to an angle of 45 deg. Therefore, $\chi_0 = kd\ \sin(\pi/4)$, with all the remaining parameters the same as in Example 6.8. The RCS for a threat frequency equal to the operating frequency is shown in Fig. 6.26.

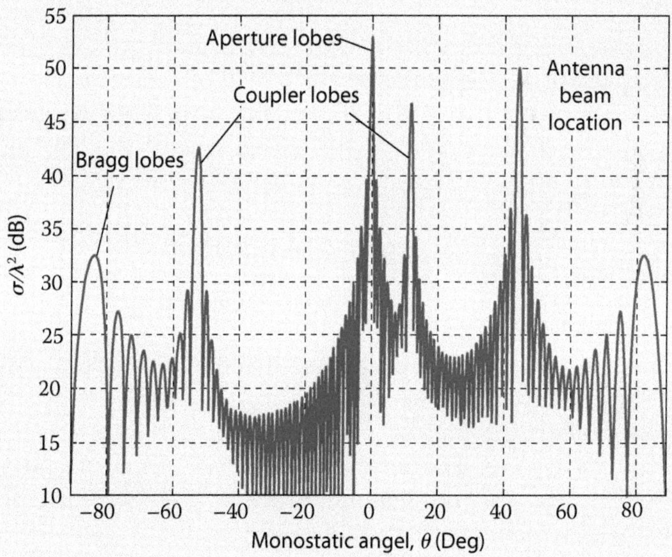

Fig. 6.26 Radar cross section of a scanned linear array ($\theta_S = 45$ deg) with a parallel feed for a threat frequency equal to the operating frequency (from Ref. [13]).

An examination of the RCS obtained in the last two examples yields several important generalizations:

1. The level of the lobes increases with the magnitudes of the reflection coefficients and the area A. The area depends on the number of elements and their spacing.
2. The lobe spacing in the RCS pattern for the first level of couplers is determined by the physical spacing of the couplers, $2d$. This holds for other levels of couplers as well. For instance, the second level of couplers is spaced a distance $4d$ and, therefore, a new set of RCS lobes would appear between the lobes generated by the first level of couplers.
3. Lobes associated with reflections occurring behind the phase shifters contain the factor χ_0 and, thus, their positions are determined by the phase-shifter settings. Note that the large lobe at $\theta = 45$ deg in Fig. 6.26 corresponds to the antenna beam pointing direction. This characteristic of a *reciprocal* antenna can be a problem for a low-observable platform if the antenna is used in a radar. When the radiation beam is trained on the target, the antenna presents a rather large RCS. In general, a target scattering source that collects an incident signal and returns it in the direction of the source is called a *retroreflector*. The associated lobe in the RCS pattern is referred to as a *retroreflection.*
4. As the frequency of the threat radar is increased relative to the antenna's operating frequency, more lobes appear in the RCS pattern.

6.7.4 Extension to Two-Dimensional Arrays

If the array is two-dimensional with a rectangular grid, each of the RCS terms in Eq. (6.50) will be multiplied by a pattern factor for the $y-z$ plane. The arguments of the functions will contain the y direction cosine v. For instance, an array of $N_x \times N_y$ elements spaced $d_x \times d_y$ will have an aperture reflection contribution given by

$$\sigma_r \approx \frac{4\pi A^2}{\lambda^2} r_r \left[\frac{\sin(N_x k d_x u)}{N_x \sin(k d_x u)} \frac{\sin(N_y k d_y v)}{N_y \sin(k d_y v)} \cos\theta \right]^2 \tag{6.51}$$

This is the same pattern factor encountered in the study of Bragg diffraction, and the RCS characteristics are similar.

If phase shifters are located behind each radiating element in a two-dimensional array, lobes due to reflection sources behind the phase shifters can move together with the antenna beam in both scanning planes. For example, the coupler sum arm term becomes

$$\sigma_\Sigma \approx \frac{4\pi A^2}{\lambda^2} t_r^4 t_p^4 t_c^4 r_\Sigma^2 \cos^4(\zeta_x/2) \left[\frac{\sin(N_x \zeta_x)}{(N_x/2)\sin(2\zeta_x)} \frac{\sin(N_y \zeta_y)}{N_y \sin(\zeta_y)} \cos\theta \right]^2 \tag{6.52}$$

where

$$\zeta_x = kd_x u + \chi_{0x}$$
$$\zeta_y = kd_y v + \chi_{0y} \tag{6.53}$$

Hence, $\chi_{0x} = k\,d_x \sin\theta_s \cos\phi_s$ and $\chi_{0y} = k\,d_y \sin\theta_s \sin\phi_s$ are the interelement phases to scan the antenna beam to (θ_s, ϕ_s). The preceding formula assumes that the first level of signal coupling is done along the x direction and that reflections from the remaining levels of couplers are neglected.

Figure 6.27 illustrates the characteristics of the high-RCS regions typically present for two-dimensional arrays. The contours are plotted in direction cosine space and denote spatial regions of RCS above a specified level. High RCS is present along the principal planes of the array, but the levels drop off quickly away from these planes because of the separable product in Eq. (6.52). The RCS of a scanned array is shown in Fig. 6.28. As in the linear array case, the lobes originating from reflections behind the phase shifters scan along with the antenna beam. The RCS terms are separable in

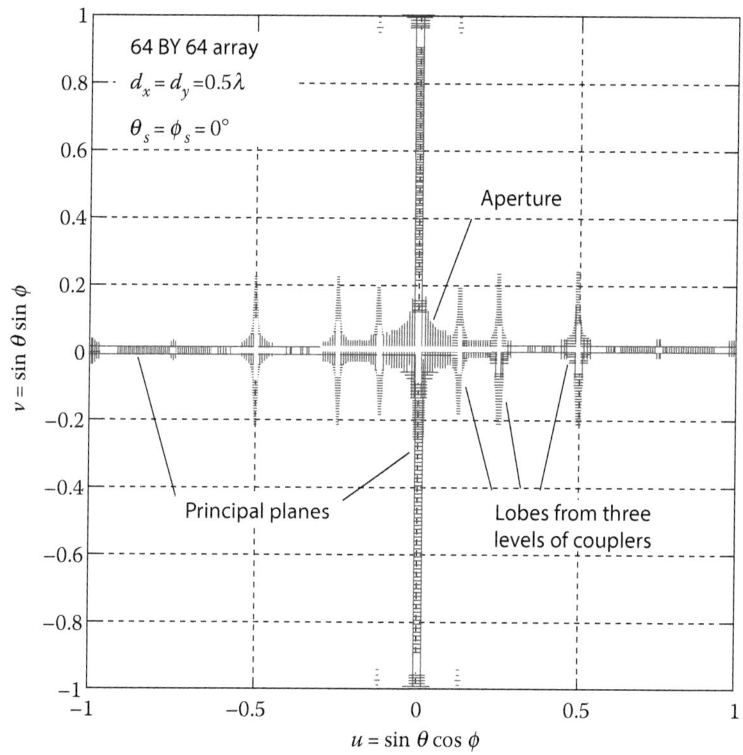

Fig. 6.27 Radar cross section of a two-dimensional array. Contours denote $\sigma/\lambda^2 > 20$ dB and there are three levels of couplers along the x direction.

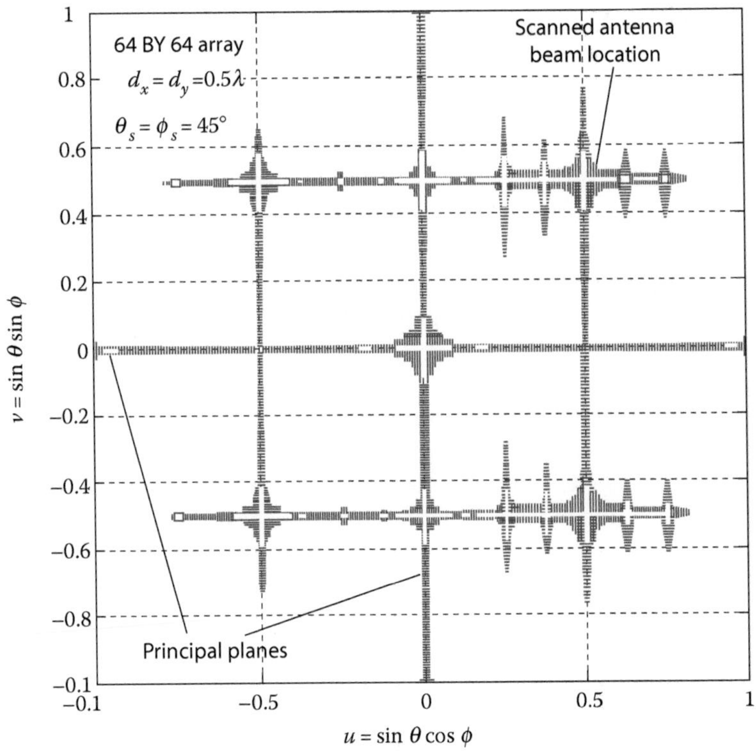

Fig. 6.28 Radar cross section of a scanned two-dimensional array. Contours denote $\sigma/\lambda^2 >$ 20 dB and there are three levels of couplers along the x direction.

direction cosine space and, therefore, the contours are straight lines (i.e., u constant and v constant). In visible space, the lobes trace out cones about the x and y axes.

In a large phased array, there can be many levels of couplers with physical spacings that are several wavelengths. Thus, even though the angular width of each spike is very small, the spikes are so numerous and closely spaced that the RCS can be considered, for all practical purposes, to have a continuous large value in the array principal planes. This point is evident in the contours in Fig. 6.27 and Fig. 6.28.

The preceding analysis is a good starting point for more detailed analyses. The formulas derived are conservative (i.e., yield higher than actual RCS levels) for several reasons. First, a lossless feed network was assumed. All practical antennas have some loss, and feeds with solid-state devices can have significant losses. It is not unusual to encounter phase shifters with 2 dB of loss and, because the scattered signals travel through the device twice, the net reduction on RCS can be significant (note the t_p^4 dependence for reflections that occur behind the phase shifters). Second, equal power splitters have been assumed. For a low sidelobe antenna, the feed is designed to provide a tapered amplitude distribution. The required coupler distribution will also yield some reduction in RCS sidelobes relative to that obtained using equal power splitters. Finally, all reflection coefficients of similar devices will not be the same as postulated in Sec. 6.7.3. They have random amplitudes and phases characterized by a mean and variance. The randomness does not permit all the scattering sources to add constructively at one angle.

6.8 Feed Scattering Characteristics

6.8.1 Feed Types

In the preceding section, formulas have been developed to predict the RCS of an array with a *parallel feed* (also known as *corporate feeds*). It was found that the number of lobes and their locations depend on the number of couplers in the feed. The lobe levels are a function of the reflection coefficients of the devices used in the feed. In addition to the parallel feed, there are two other commonly used feed types: *space feeds* and *series feeds*. These networks have distinctly different signal paths and, therefore, the RCS of each will also have unique features.

6.8.2 Series Feeds

A series feed is one in which the power to each element is tapped off a main line sequentially, as illustrated in Fig. 6.29. The coupling values are chosen to provide the desired amplitude distribution. The most common method of terminating the main line is matched loading. Another option is to use a short. The latter version is called a *standing wave* feed. For the loaded case, some power is always reflected from the load at the end of the

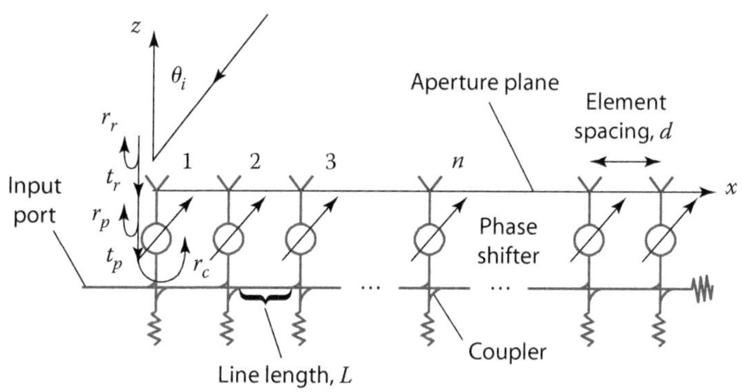

Fig. 6.29 Scattering sources for an array with a series feed.

main line. This gives rise to a reflection lobe in the radiation pattern that is the mirror image of the mainbeam about the broadside axis. It is reduced in magnitude relative to the main beam by an amount determined by the load reflection coefficient.

Figure 6.29 also identifies some potential scattering sources in the feed [14]. Tracing signals is a complicated task because of the many paths by which reflections can return to the aperture. As in the case of the parallel feed, there are reflections at the aperture, at the phase shifter, and at the coupler inputs, and their RCS contributions are given by the same expressions:

$$\sigma_r \approx \frac{4\pi A^2 \cos^2\theta}{\lambda^2} r_r^2 \left[\frac{\sin(N\alpha)}{N\sin(\alpha)}\right]^2 \tag{6.54}$$

$$\sigma_p \approx \frac{4\pi A^2 \cos^2\theta}{\lambda^2} r_p^2 t_r^4 \left[\frac{\sin(N\alpha)}{N\sin(\alpha)}\right]^2 \tag{6.55}$$

$$\sigma_c \approx \frac{4\pi A^2 \cos^2\theta}{\lambda^2} t_r^4 t_p^4 r_c^2 \left[\frac{\sin(N\zeta)}{N\sin(\zeta)}\right]^2 \tag{6.56}$$

Reflections originating at the coupler ports move in space as determined by the phase-shifter setting.

Figure 6.30 illustrates the signal distribution for a four-port coupler. If the inputs are matched ($r_c = 0$), the scattering matrix is given by

$$\mathbf{C} = \begin{bmatrix} 0 & t & jc & 0 \\ t & 0 & 0 & -jc \\ jc & 0 & 0 & t \\ 0 & -jc & t & 0 \end{bmatrix} \tag{6.57}$$

Analogous to the reflection beam mentioned earlier for the radiation case, the received incident threat signal can be decomposed into forward and backward traveling waves on the main line. Figure 6.31 illustrates that a signal

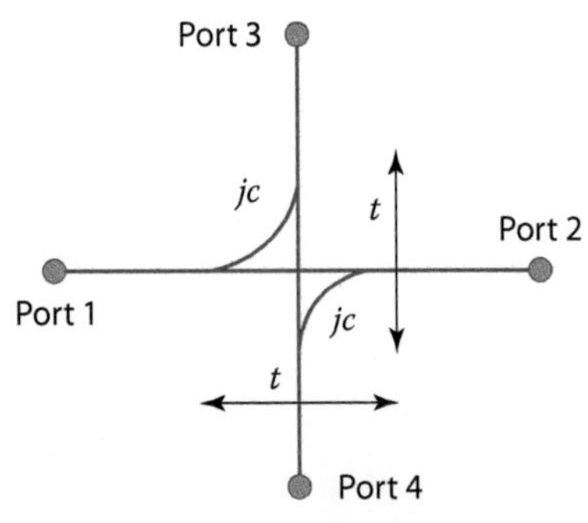

Fig. 6.30 Coupling and transmission paths for a four-port coupler.

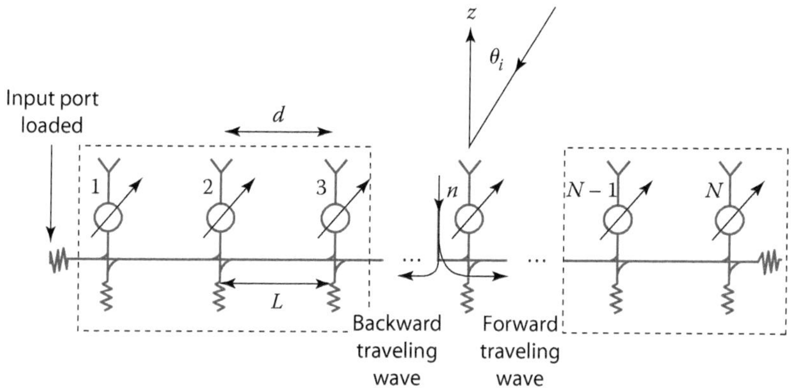

Fig. 6.31 Forward and backward traveling waves in a series feed network excited by a plane wave.

received by element n appears as an excitation for the $N - n$ elements toward the load and the $n - 1$ elements toward the input. A third scattering term not included in the forward or backward beams is the "self-beam" that is due to reflection from the load terminating element n. Thus, for each element in the array, three beams are reradiated, and their beamwidths and peak levels depend on the element's location in the array. The total scattered field is obtained by the superposition of the forward, backward, and self-waves from all elements.

An approximate formula for the RCS of a series network can be obtained by tracing signals through the feed (see Problem 6.8). If r_L is the reflection coefficient of the loads, the forward, self, and backward traveling fields can be expressed as

$$E_f \sim -r_L \sum_{n=1}^{N-1} c_n e^{j(n-1)\zeta} \left[\sum_{m=n+1}^{N} c_m e^{j(m-1)\zeta} \left(\prod_{i=1}^{m-1} t_i e^{j\psi_0} \right) \right] \tag{6.58}$$

$$E_b \sim -r_L \sum_{n=2}^{N-1} c_n e^{j(n-1)\zeta} \left[\sum_{m=1}^{n-1} c_m e^{j(m-1)\zeta} \left(\prod_{i=m}^{n-1} t_i e^{j\psi_0} \right) \right] \tag{6.59}$$

and

$$E_{\text{self}} \sim r_L t_n^2 e^{j2(n-1)\zeta} \tag{6.60}$$

Now, $\zeta = \alpha + \chi_0$, and $\psi_0 = kL$ is the main-line electrical length between couplers. The minus signs occur because the reflected beams pass through two coupled arms, yielding a j^2 factor. The total RCS due to the series feed beams is

$$|E_s|^2 \approx |E_f|^2 + |E_b|^2 + |E_{\text{self}}|^2 \tag{6.61}$$

The correct normalization can be obtained by comparing the total scattered field in Eq. (6.61) to the maximum that would be obtained by an array with shorted coupler inputs. Thus,

$$\sigma_s \approx \frac{4\pi A^2 \cos^2\theta}{\lambda^2} \left| \frac{E_s}{N} \right|^2 t_r^4 t_p^4 \tag{6.62}$$

Example 6.10: Radar Cross Section of an Array with a Series Feed

Equations (6.54–6.56) and (6.62) were used to compute the RCS of a 50-element array with a series feed network having interelement line lengths $\psi_0 = kL = 0.5\lambda = \pi$. Figure 6.32 and Fig. 6.33 show the RCS for the array beam at broadside and 45 deg, respectively. All reflection coefficients are 0.2, and the element spacing is 0.4λ at the threat frequency. The effective scattering height of the elements in the plane transverse to the array axis is

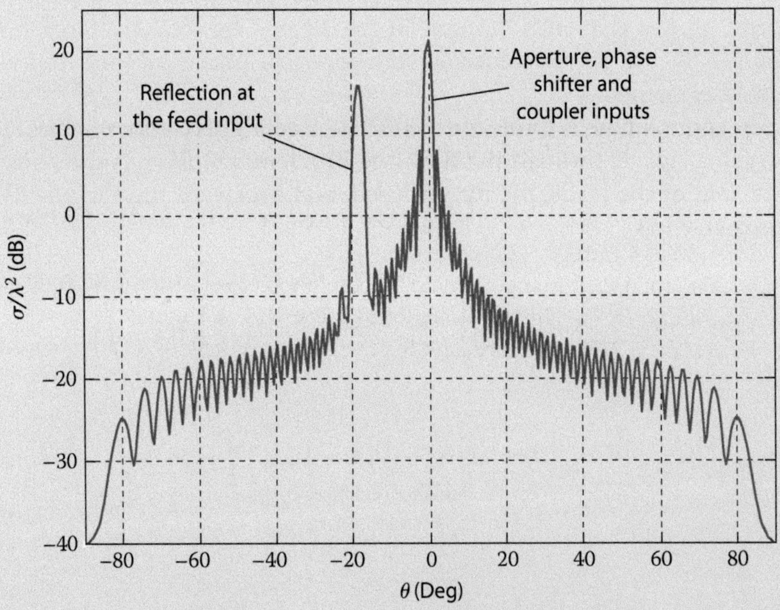

Fig. 6.32 Radar cross section of a series-fed array for a threat frequency equal to the operating frequency ($kL = \pi/4$, $\theta_s = 0$ deg).

(Continued)

Example 6.10: Radar Cross Section of an Array with a Series Feed *(Continued)*

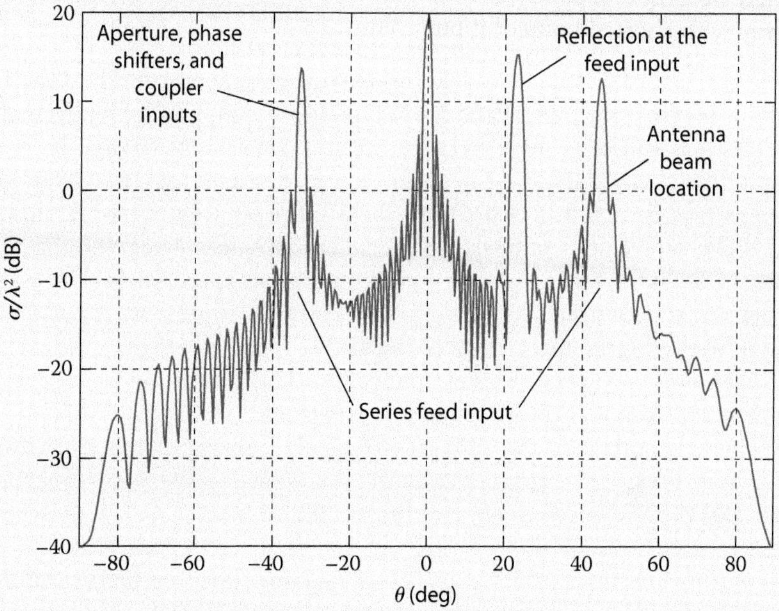

Fig. 6.33 Radar cross section of a series-fed array for a threat frequency equal to the operating frequency ($kL = \pi/4$, $\theta_S = 45$ deg).

assumed to be 0.5λ. Thus, the area of the array aperture is $A \approx (50)(0.4\lambda)(0.5\lambda)$.

The total RCS is approximated by the noncoherent sum of the individual contributions. In Fig. 6.32, all terms have a maximum at broadside. In Fig. 6.33, the self, backward, and forward terms scan along with the radiation beam, whereas the aperture and phase-shifter reflection terms do not.

The forward and backward terms in Eqs. (6.58) and (6.59) are usually much smaller than the self-term because of the product factors. Furthermore, it is common practice to design the line length between couplers to be an odd multiple of a quarter-wavelength ($\lambda/4$, $3\lambda/4, \ldots$) so that reflections between adjacent couplers will cancel. This tends to diminish the importance of the forward and backward terms even more. The disadvantage of the series feed is the broad lobe that scans with the radiation beam. It is not a sharp spike but is spread out because of the variation in coupling coefficients along the main line. Equation (6.61) neglects any reflection of the backward waves from the load at the input port, which can be a significant RCS contribution, as demonstrated in Problem 6.18.

6.8.3 Space Feeds

The third type of beam-forming network is the space feed, which includes a wide variety of antennas. The basic principle of operation is illustrated in Fig. 6.34. A horn or small array illuminates an array aperture. The array functions as a lens; it introduces the proper phase and amplitude modifications to the incident spherical wave to convert it to a plane wave having the desired amplitude characteristics. These feeds can be used in two- and three-dimensional forms. In a two-dimensional space feed, the network shown in Fig. 6.34 would be sandwiched between parallel plates. Bootlace and Rotman lenses are examples of two-dimensional space feeds. For a three-dimensional space feed, the diagram in Fig. 6.34 represents a cut through one of the two principal planes of the array. Three-dimensional space feeds are used only for ground-based applications because of the attendant large volume and weight, at least at radar frequencies.

The lens portion of the space feed typically consists of two back-to-back array apertures. The region between the two apertures contains transmission lines and other devices to control the amplitude and phase. When the feed radiates, the incident wave at the internal face of the lens (referred to as the *pickup aperture*) is spherical. Transmission lines or phase shifters in the lens will delay the signal near the center relative to the edges, so that a linear phase front is formed at the external face. Sources of reflections

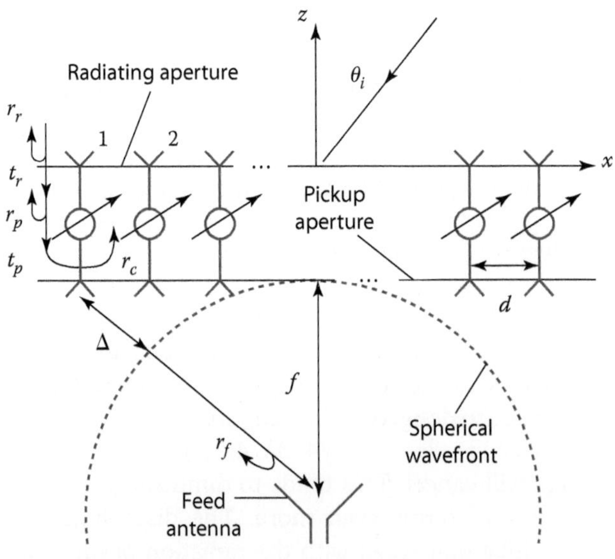

Fig. 6.34 Scattering sources for a space-fed array.

include the external and internal faces of the lens and phase-shifter inputs. A reflection from the feed element is possible only when the threat radar incident wave is focused; in other words, the antenna beam must be pointed in the direction of the threat radar.

As in the case of the parallel and series feeds, the reflections from the aperture and phase-shifter inputs exist and have the same pattern shape as the RCS from a series or parallel feed. Thus, Eqs. (6.54) and (6.55) still hold. For reflections at the pickup aperture, the phase of the signals returned to the external face is not linear. If the pickup elements have the same reflection coefficient as the aperture (r_r), the reflected signal is

$$E_a \sim r_r \sum_{n=1}^{N} \exp\{j2[\zeta(n-1) + \psi_n]\} \tag{6.63}$$

where ψ_n is the phase introduced by the transmission lines between the lens faces. Frequently, the spherical wave correction phase is added with fixed-length cables, but beam scanning is accomplished using the phase shifters. Cables provide true time delay, whereas the phase-shifter insertion phases have a more complicated frequency dependence. True time delay yields a wider *instantaneous* antenna bandwidth.

In Fig. 6.34, the path difference between center and edge rays is

$$\Delta = \sqrt{(D/2)^2 + f^2} - f \tag{6.64}$$

where D is the total array length ($D \approx N\,d$). If $\Delta \ll r$ (which implies a large f/D ratio), the error as a function of distance from the center of the array x is

$$\Delta \approx \frac{x^2}{2f}$$

Therefore, the correction term is approximately a quadratic function of aperture position x. Quadratic error distributions occur in the design of horn antennas, and their effects include loss in gain, broadening of the mainbeam, and increase in sidelobe levels. The same characteristics are observed in the scattered field E_a in Eq. (6.63), and the effects are even more pronounced because of the factor of 2 in the exponent.

If the antenna beam is scanned in the direction of the threat radar, the incident plane wave will focus at the feed. The reflected wave propagates back through the lens just as if a transmit signal had been injected into the feed. Referring to Example 6.4 (and Problem 6.12), we find that the RCS in the $x-z$ plane due to feed scattering is approximately

$$\sigma_f \approx t_r^8 t_p^4 r_f^2 \frac{4\pi A^2}{\lambda^2} \left[\frac{A_f^2}{\mathrm{FOD}_x \mathrm{FOD}_y}\right] |A_N(\alpha)|^4 \tag{6.65}$$

The feed gain is $G_f = (4\pi A_f)/\lambda^2$, and its reflection coefficient is r_f. The lens aperture area is $A = D_x D_y$. $A_N(\alpha)$ is the normalized array factor for the lens in the $x-z$ plane, and $\alpha = \sin\theta$ ($\cos\phi = 1$). N is the number of elements in the $x-z$ plane, and FOD_x and FOD_y are the focal length-to-length ratios for the two principal planes. Factors of $1/(4\pi f^2)$ are introduced as a result of the isotropic spreading of the wave in the region of space between the feed and pickup aperture. Note that this expression is only approximate because it assumes (for amplitude purposes) that the distance from the feed to each pickup element is f. The reflection coefficient of both aperture elements is assumed to be r_r and, therefore, the factor of t_r^8 occurs because of two-way transmission through two apertures.

The feed should be well matched and, therefore, the reflection coefficient r_r will be small. Furthermore, the RCS falls off very rapidly because of the A_N^4. (Note that, when there are no Bragg lobes, A_N is essentially a sinc function.) When the array is scanned, $\sin\theta$ in the argument of the pattern function must be replaced with ($\sin\theta - \sin\theta_s$), where θ_s is the scan angle.

Example 6.11: Radar Cross Section of a Large Scanning Array with a Space Feed

A 50-element array has a space-fed network, and the elements are spaced 0.4λ at the operating frequency, which is also the threat frequency. Figure 6.35 shows the array RCS for $f/D = 1$ and the beam scanned to 45 deg. All reflection coefficients are 0.2, and the feed is assumed to be perfectly matched ($r_f = 0$).

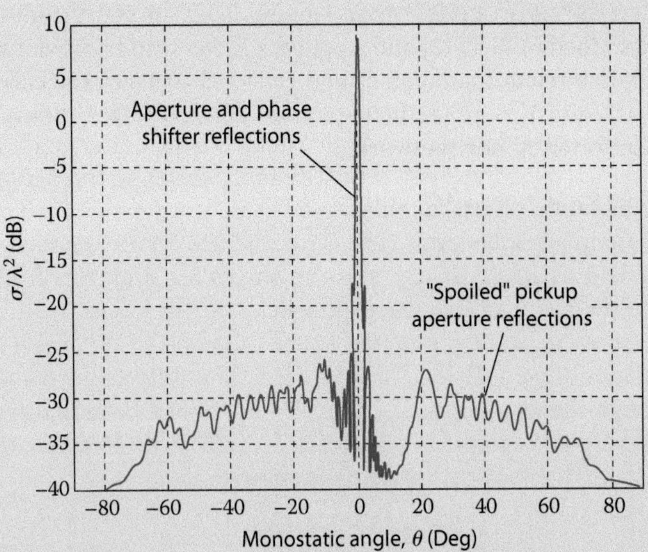

Fig. 6.35 Radar cross section of a space-fed array for a threat frequency equal to the operating frequency ($\theta_s = 45$ deg and a matched feed is assumed).

In Fig. 6.35, the lobe associated with the pickup aperture reflection has disappeared because of the quadratic phase errors introduced by the phase-shifter spherical wave corrections. The aperture and phase-shifter input reflection yield a sharp spike at 0 deg that falls off rapidly. The pickup aperture scattering is smeared out or "spoiled" over a wide angular region and begins to dominate where $|\theta| > \pm 20$ deg. If the pickup elements and feed aperture were perfectly matched, the level in these regions would be well below -40 dB. In practice, this may not be a problem because random errors due to manufacturing and assembly imperfections will result in comparable average levels.

The disadvantage of the space feed with regard to RCS performance is the presence of the pickup aperture reflections, which are significant over a wide angular region. This problem is avoided mainly by matching the pickup aperture so that the reflection coefficient is as small is possible.

6.9 Rigorous Calculation of Feed Radar Cross Section

Our previous analysis of antenna and feed scattering has neglected multiple reflections within the feed and interactions between the feed and the aperture. In most cases, these approximations are sufficient but, in some instances, these second-order effects are significant. A rigorous solution based on integral or differential equations must be used. This consists of a finite difference or MM technique used on the antenna surfaces and a network matrix method applied to the feed. The two sets of equations are related by *continuity* or *joining* relations based on the conservation of energy.

As described in Ref. [10] and Appendix E, the scattering parameters of an N-port device relate the incident and reflected voltages (or currents) at all ports. If V_n^+ and V_n^- are the incident and reflected waves at port n, respectively, then for an N-port network,

$$V_n^- = S_{n1}V_1^+ + S_{n2}V_2^+ + \cdots + S_{NN}V_N^+, \qquad n = 1, \ldots, N, \tag{6.66}$$

or in matrix form: $\mathbf{V}^- = \mathbf{S}\mathbf{V}^+$.

The S_{nm} are the scattering parameters that give the signal at port n due to an input at port m with all other ports terminated in matched loads. Thus, S_{nm} is given by

$$S_{nm} = \left.\frac{V_n^-}{V_m^+}\right|_{V_k^+ = 0 \text{ for } k \neq m} \tag{6.67}$$

Hence, S_{nn} is the reflection coefficient seen looking into port n when all other ports are terminated in matched loads, and S_{nm} is the transmission coefficient from port m to port n when all other ports are terminated in matched loads.

For a rigorous calculation of antenna RCS, the scattering parameters can be combined with the MM impedance matrix through the joining equations.

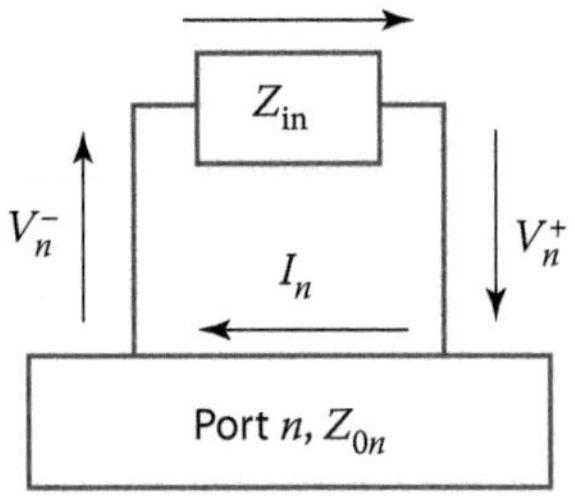

Fig. 6.36 Incoming and outgoing voltage waves at a loaded port.

The magnitude of the total voltage V_n and current I_n at port n must be equal to the magnitude of the voltage V_a and the MM expansion current coefficient I_a at the dipole terminal, respectively. Currents I_n and I_a must be in antiphase based on the reference system defined in Fig. 6.36. The two joining equations for voltage and current are

$$V_n = V_n^+ + V_n^- = V_a \tag{6.68}$$

and

$$I_n = \frac{V_n^+ - V_n^-}{Z_{on}} = -I_a \tag{6.69}$$

where Z_{on} is the characteristic impedance of port n.

A center-fed dipole with M method of moments subdomains connected to a transmission line is shown in Fig. 6.37. The scattering matrix of a two-port device is

$$S = \begin{bmatrix} S_{11} & S_{12} \\ S_{21} & S_{22} \end{bmatrix} \tag{6.70}$$

Hence, there are two scattering equations:

$$\begin{bmatrix} c_1 \\ c_3 \end{bmatrix} = \begin{bmatrix} S_{11} & S_{12} \\ S_{21} & S_{22} \end{bmatrix} \begin{bmatrix} c_2 \\ V_{in} \end{bmatrix} \tag{6.71}$$

or, rearranging

$$\begin{aligned} c_1 - S_{11}c_2 &= S_{12}V_{\text{in}} \\ c_3 - S_{21}c_2 &= S_{22}V_{\text{in}} \end{aligned} \tag{6.72}$$

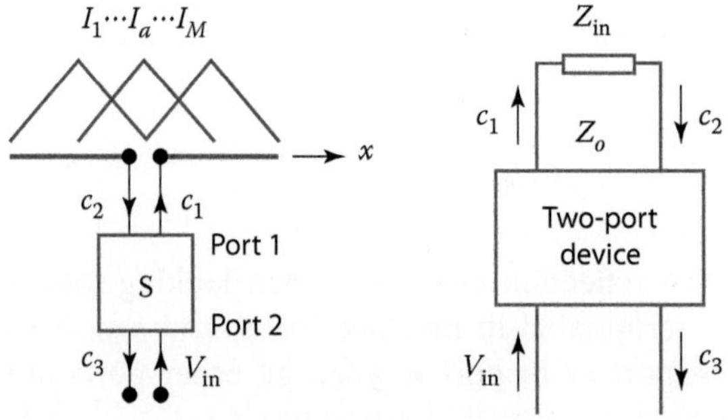

Fig. 6.37 Dipole connected to a transmission line and its equivalent circuit.

The unknown signals c_1, c_2, and c_3, in the feed network need to be determined along with the MM expansion coefficients. Using (6.68) and (6.69), the two joining equations are

$$Z_{a1}I_1 + Z_{a2}I_2 + \cdots + Z_{aM}I_M - c_1 - c_2 = 0 \tag{6.73}$$

and

$$I_a - \frac{c_1}{Z_o} + \frac{c_2}{Z_o} = 0 \tag{6.74}$$

where a is the feeding subdomain number and Z_o is the characteristic impedance of the transmission line.

The scattering parameters are then combined with the MM impedance matrix through the two joining equations to obtain one matrix that completely describes the entire antenna and feed network

$$\begin{bmatrix} Z_{11} & \cdots & \cdots & Z_{1a} & \cdots & \cdots & Z_{1M} & 0 & 0 & 0 \\ \vdots & & & \vdots & & & \vdots & \vdots & \vdots & \vdots \\ \vdots & & & \vdots & & & \vdots & 0 & 0 & 0 \\ Z_{a1} & \cdots & \cdots & Z_{aa} & \cdots & \cdots & Z_{aM} & -1 & -1 & 0 \\ \vdots & & & \vdots & & & \vdots & 0 & 0 & 0 \\ \vdots & & & \vdots & & & \vdots & \vdots & \vdots & \vdots \\ Z_{M1} & \cdots & \cdots & Z_{Ma} & \cdots & \cdots & Z_{MM} & 0 & 0 & 0 \\ 0 & \cdots & 0 & 1 & 0 & \cdots & 0 & -1/Z_0 & +1/Z_0 & 0 \\ 0 & \cdots & 0 & 0 & 0 & \cdots & 0 & 1 & -S_{11} & 0 \\ 0 & \cdots & 0 & 0 & 0 & \cdots & 0 & 0 & -S_{21} & 1 \end{bmatrix} \begin{bmatrix} I_1 \\ \vdots \\ \vdots \\ I_a \\ \vdots \\ \vdots \\ I_M \\ c_1 \\ c_2 \\ c_3 \end{bmatrix} = \begin{bmatrix} V_1 \\ \vdots \\ \vdots \\ V_a \\ \vdots \\ \vdots \\ V_M \\ 0 \\ S_{12}V_{in} \\ S_{22}V_{in} \end{bmatrix}$$

which, in matrix form is

$$\mathbf{ZI} = \mathbf{V} \tag{6.76}$$

The complete system of equations is solved using $\mathbf{I} = \mathbf{Z}^{-1}\mathbf{V}$.

In the transmit mode, V_{in} is the applied voltage. In the receive mode, the excitation is a plane wave of amplitude 1 V/m, and thus the applied feed voltage V_{in} is set to zero. The feed network interaction with the array elements is included; there is no approximation other than the numerical evaluation of the matrix elements. Once the vector $\mathbf{I}$ is determined, the current on the antenna surfaces and the voltage and current at all points in the feed are known. Thus, the current density, power density, reflection coefficients, voltage standing wave ratio (VSWR), signal amplitude and phase, or any other related quantity can be computed.

Example 6.12: Radar Cross Section of a Linear Array of Dipoles with a Corporate Feed

Figure 6.38 shows a linear array of eight dipoles fed by a corporate feed [15]. The matrix solution was used to compute the RCS patterns of this array as a function of various feed device parameters. The dipoles are 0.5λ long and have a radius of 0.05λ. Normally, a ground plane would be located under the dipoles but, to reduce the number of unknowns, the ground plane is approximated by a row of shorted dipoles. These are crude approximations of the images that an infinite ground plane would provide.

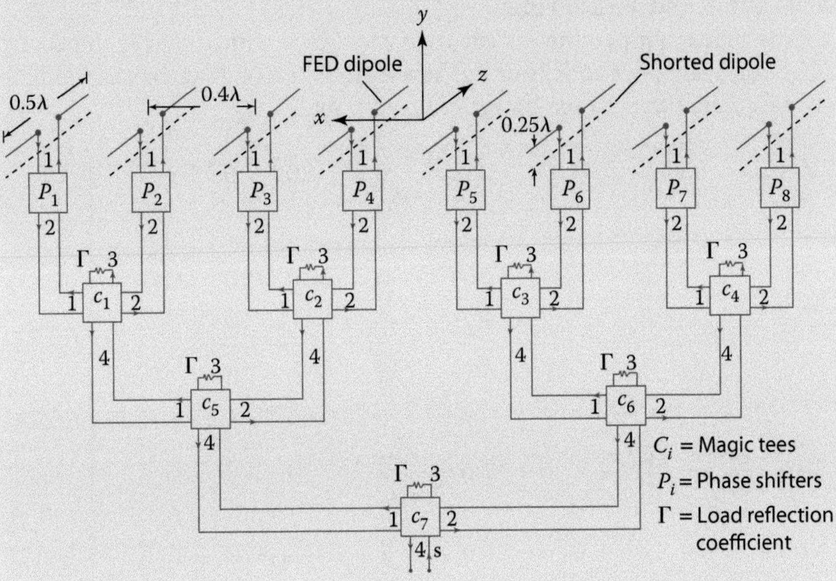

Fig. 6.38 Scattering matrix model for a linear array of dipoles with a parallel feed.

The feed consists of seven couplers denoted by c_i ($i = 1, \ldots, 7$) and interconnected by transmission lines. The coupler scattering matrix for equal power dividers can be written

$$\mathbf{C} = \begin{bmatrix} r_c & 0 & t_c & t_c \\ 0 & r_c & -t_c & t_c \\ t_c & -t_c & -r_c & 0 \\ t_c & t_c & 0 & -r_c \end{bmatrix}$$

where $2|t_c|^2 + |r_c|^2 = 1$. The input reflection coefficient r_c can be different for each coupler in the feed. Similarly, the coupler loads and the insertion loss and phase of the interconnecting lines can be individually specified.

(Continued)

Example 6.12: Radar Cross Section of a Linear Array of Dipoles with a Corporate Feed *(Continued)*

Behind each element is a phase shifter denoted by P_i ($i = 1, \ldots, 8$), with scattering matrix

$$\boldsymbol{P} = \begin{bmatrix} r_p e^{j\xi_p} & t_p e^{j\chi_p} \\ t_p e^{j\chi_p} & r_p e^{j\xi_p} \end{bmatrix}$$

where $|t_p|^2 + |r_p|^2 = 1$. The quantity r_p is the input reflection coefficient of the phase shifter, and χ_p is the interelement phase shift required to provide a linear phase shift for beam scanning. There is a restriction on the transmission and reflection phases if the device is reciprocal as discussed in Appendix E:

$$\xi_p = \chi_p - \ell\pi/2, \qquad \ell = \pm 1, \ \pm 2, \ \ldots$$

For this example, there are 44 scattering equations, 8 continuity equations, and 7 triangles per dipole, for a total number of 164 unknowns. Figure 6.39 shows the RCS as a function of several feed-line impedances.

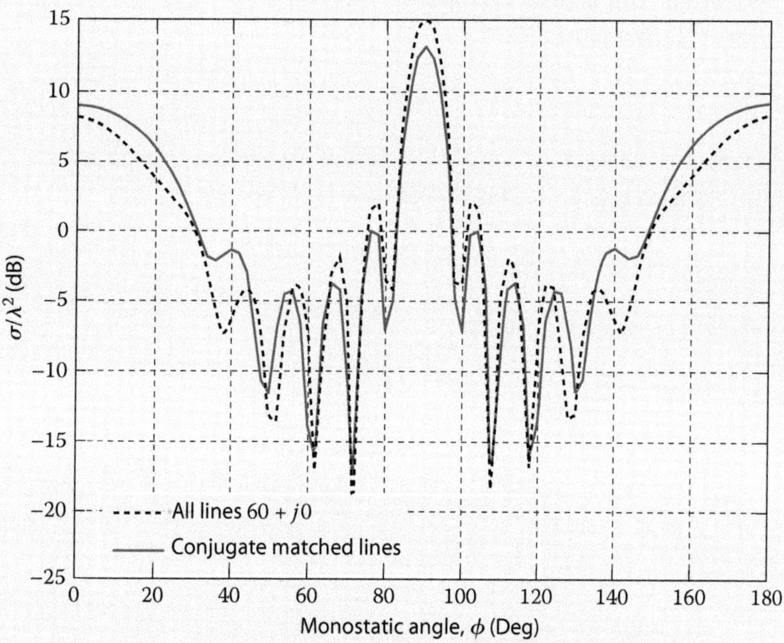

Fig. 6.39 Radar cross section of a linear array of dipoles with a parallel feed computed using MM and scattering parameters. Two line impedance conditions are shown.

6.10 Frequency Selective Surfaces

The RCS of a complicated object like an antenna is difficult or impossible to control over a wide frequency range. The most efficient and cost-effective approach in these situations is to shield the scattering object from the threat radar. This can be accomplished by retracting the object and covering the cavity in which it is housed with a *shutter.* If the object is an antenna, then, obviously, the system served by this antenna cannot operate when it is stowed. An alternate approach is to cover the antenna with an FSS, that is, one that is transparent at the antenna operating frequency yet opaque at the threat radar frequency [16–18].

6.10.1 Frequency Characteristics

A simple example of an FSS is shown in Fig. 6.40. The antenna is covered by a conducting plate with rectangular holes of dimension a by b. Consider a normally incident wave polarized along the y axis. If the holes behave like open-ended waveguides, their reflection coefficient will be small if the incident wave is above the cutoff frequency. Conversely, the reflection coefficient will be large if the threat frequency is below cutoff. The lowest frequency that can penetrate the screen is approximately given by the dominant mode (TE_{10}) cutoff frequency [10],

$$f_c = \frac{c}{2a} \tag{6.77}$$

The frequency characteristics of this surface are similar to a high-pass filter, as illustrated in Fig. 6.41. Frequencies above cutoff are passed, whereas frequencies below cutoff are reflected. The region below f_c is the *stopband* and the region above f_c the *passband.*

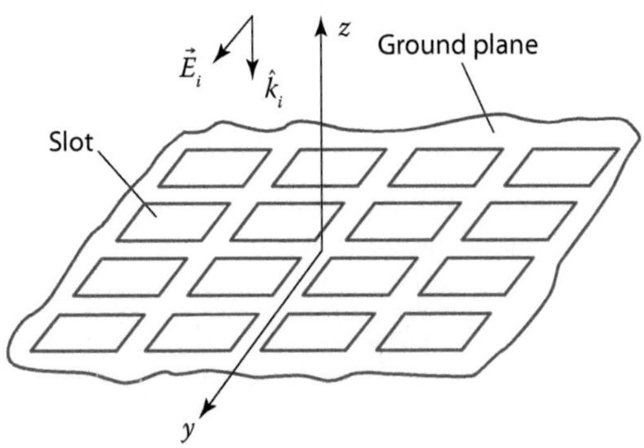

Fig. 6.40 Array of slots in a ground plane.

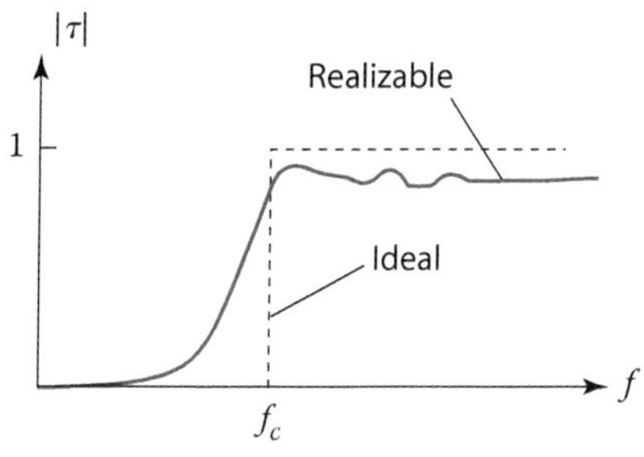

Fig. 6.41 Frequency characteristic of a high-pass FSS.

The cutoff frequency of an actual surface is not as abrupt as indicated in Fig. 6.41. The rolloff rate is more gradual and, at very high frequencies (near harmonics of the cutoff frequency), the surface is partially transparent again. Surfaces can be layered to increase the falloff rate, as shown in Fig. 6.42. This is equivalent to cascading filter sections, and the same methods used in filter design can be applied [10]. Generally, increasing the number of layers makes the frequency dependence a stronger function of angle.

As depicted in Fig. 6.43, complementary structures have "complementary" frequency responses. Slots in a ground plane provide a high-pass response; patches suspended in free space provide a low-pass response. The latter case has already been encountered in the study of dipole antenna RCS. At frequencies below the dipole's resonance (i.e., threat wavelengths $\gg 0.5\lambda_0$, where λ_0 is the operating wavelength), its effective height is so small that essentially no current is induced, and the scattered field is negligible. The incident wave passes by the dipoles and is reflected by the ground plane.

A *bandpass* characteristic can be synthesized using combinations of these complementary structures, although it is not obvious what they would look like. Most of the successful elements were developed empirically and tuned in the laboratory, more or less by trial and error. Some commonly encountered FSS structures are shown in Fig. 6.44.

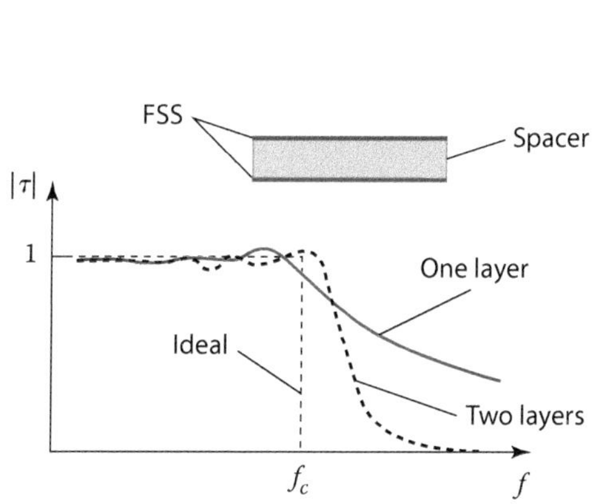

Fig. 6.42 Multilayer FSS and its frequency characteristic.

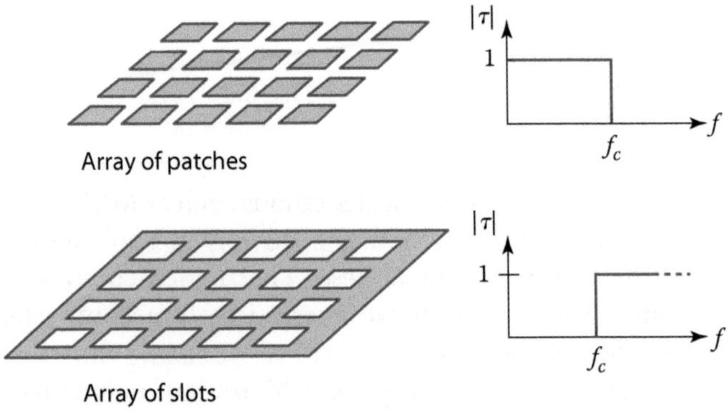

Fig. 6.43 Complementary structures and their ideal frequency responses.

6.10.2 Frequency Selective Surface Analysis Methods

All of the previously discussed numerical methods can be used to analyze frequency selective surfaces. Because MM and the FD methods are applied to finite computational domains, the FSS grids must be terminated. On the other hand, they must be large enough and contain a sufficient number of elements so that edge effects are not important.

The analysis of an infinite planar FSS can be handled by applying Floquet's theorem [17,19]. For an infinite periodic structure, the fields and currents are periodic as well as the boundary conditions. It is necessary to solve

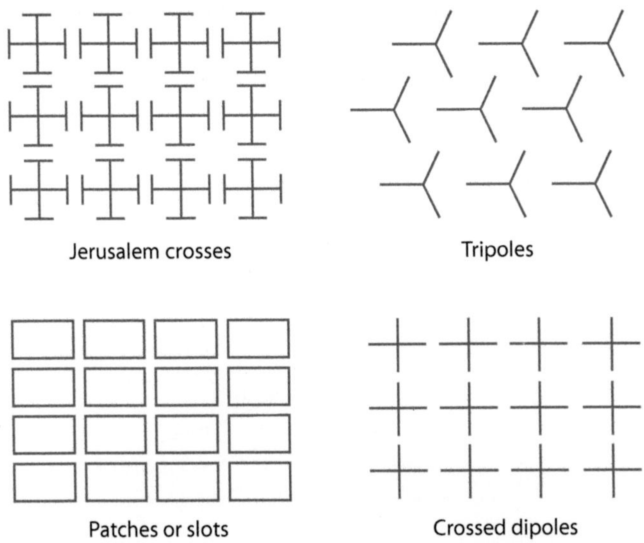

Fig. 6.44 Some common FSS configurations.

Maxwell's equations for only one typical *unit cell.* The fields and currents at any point on the structure outside of the unit cell will be simply a complex constant (i.e., phase shift) times the values at the corresponding point in the unit cell. The quantities in the reference cell can be computed using either numerical or analytical techniques. The main advantage is computational speed. When using numerical methods such as MM and FEM in conjunction with the Floquet method, they need only be applied to the unit cell, which is typically a wavelength or less. The implementation of Floquet's theorem in numerical simulation packages is usually accomplished by introducing *periodic boundary conditions.*

Finite FSSs can be modeled using the MM. Integral equations for the electric and magnetic currents are formulated as described in Sec. 3.7. It is necessary to include interactions between the currents as well. Thus, basis functions must be defined on all surfaces and apertures. At the junctions of multilegged elements, Kirchhoff's current law must be satisfied.

6.10.3 Waveguide Simulators

Experimental verification of FSS performance can be obtained by building a sufficiently large surface and measuring the transmission through it. The dimensions must be large enough so that leakage around the edges and edge diffraction do not corrupt the measurement. The surface must be in the far field of the transmit and receive antennas to provide a plane wave. These requirements can result in a significantly large surface that can be difficult to measure and expensive to fabricate. An alternative is to use a small sample of the surface and measure its transmission when it is inserted in a waveguide. A fixture designed to make this measurement is called a *waveguide simulator.*

A basic simulator using a rectangular waveguide is shown in Fig. 6.45. It is based on the fact that a TE_{10} mode in a rectangular waveguide can be

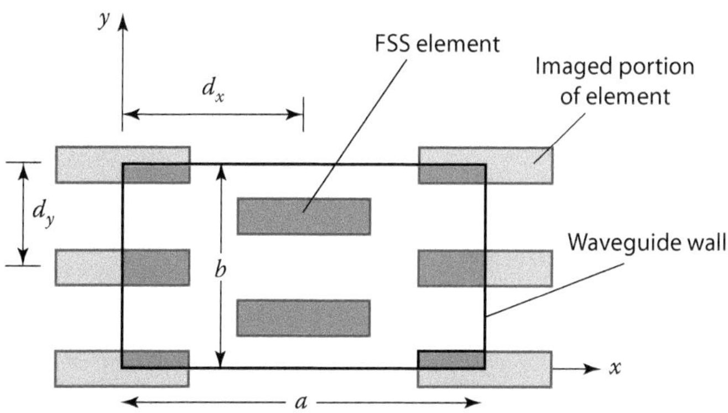

Fig. 6.45 Waveguide simulator for measuring FSS or radome transmission and reflection.

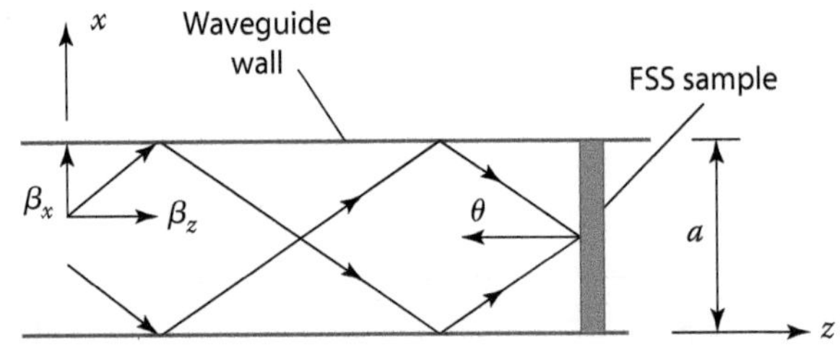

Fig. 6.46 Plane wave incident on a FSS sample in a waveguide simulator.

decomposed into a sum of plane waves traveling down the guide. The electric field for the TE_{10} mode is

$$E_z = E_0 \cos\left(\frac{\pi x}{a}\right) e^{-j\beta z}$$
$$= 2E_0[\exp(j\pi x/a - j\beta z) + \exp(-j\pi x/a - j\beta z) \qquad (6.78)$$

which corresponds to two waves, as shown in Fig. 6.46. The angle of incidence is given by

$$\theta_i = \tan^{-1}[\pi/(\beta a)]$$

This particular mode simulates a wave approaching in the $x-z$ plane and is called an *H-plane simulator.* The angle of incidence for an *E-plane simulator* is determined by the waveguide height b [20]. Multimode waveguides can be used to simulate several incidence angles simultaneously.

Not only do the dimensions of the waveguide determine the incidence angle, but they must also be chosen to provide proper imaging of the FSS elements. Consequently, the angles that can be simulated are limited. However, two or three angles are usually sufficient to verify the basic performance of the FSS design.

Figure 6.47 shows the transmission properties of an FSS incorporated into an aerodynamic missile radome [21]. Note that the transmission is significantly different for the two polarizations, especially the dependence on angle of incidence.

6.11 Cavities and Ducts

Complex targets frequently contain partially closed structures that act like microwave cavities. An example is the intake duct on a jet aircraft. If the cavity opening is large, signficant RCS can occur, even at wide angles off the cavity axis. This is analogous to the glowing of a cat's eye when it is illuminated. The RCS characteristics depend on the Q of the cavity which, in turn, depends on materials and objects inside.

6.11.1 Waveguide Modes

If the cavity is electrically small to moderate in size at the threat frequency, the MM can be used to compute the RCS. Rotationally symmetric

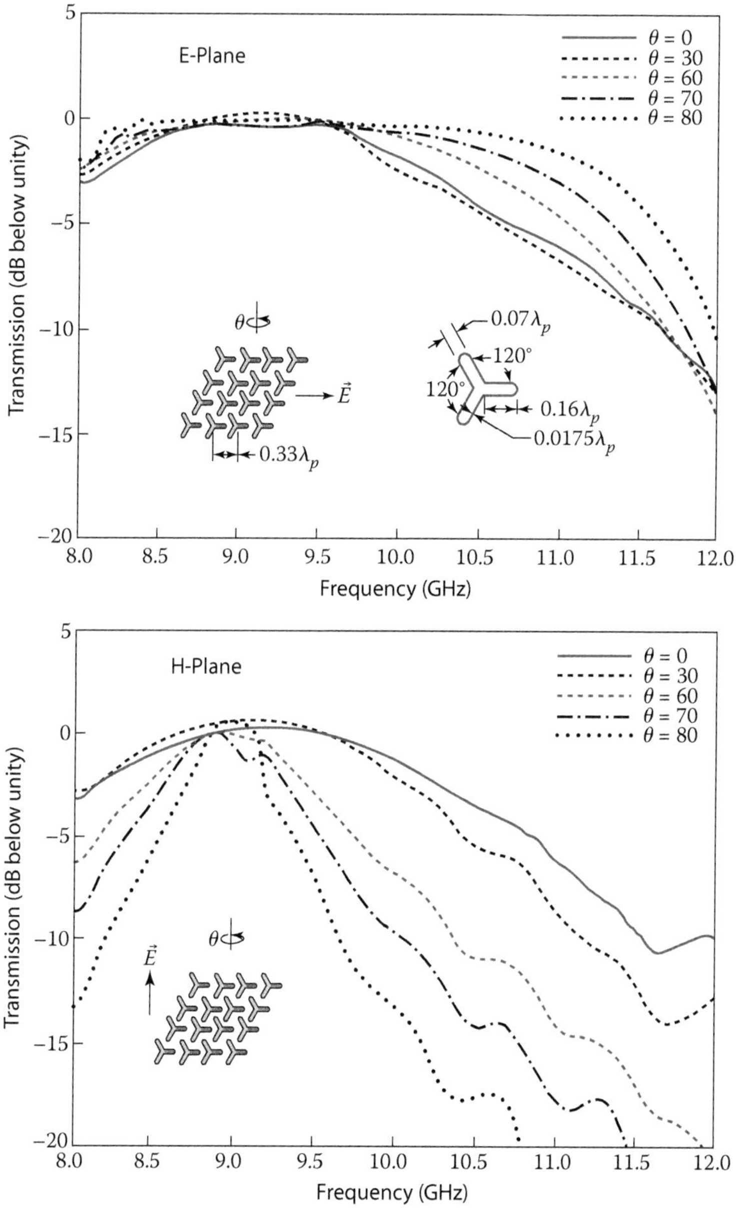

Fig. 6.47 Transmission through an aerodynamic radome incorporating FSS elements (from Ref. [20], courtesy IEEE).

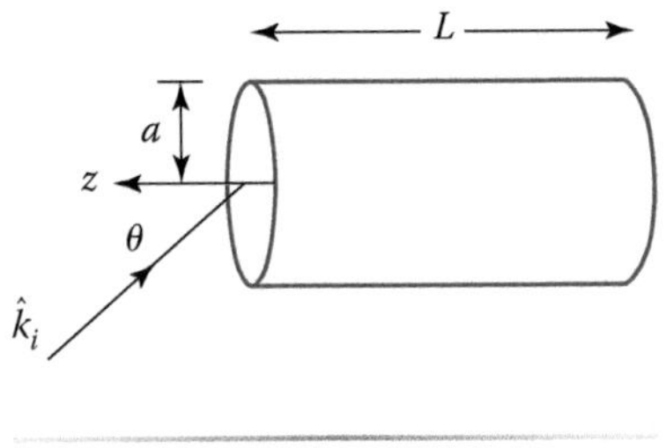

Fig. 6.48 Cylindrical cavity.

cavities can take advantage of the BOR formulation described in Sec. 3.6.2. For large cavities whose surfaces conform to coordinate system axes, modal expansions of the internal fields based on the method of separation of variables can be applied. This is the same procedure used to compute the fields inside a waveguide. In practice, this implies circular or rectangular cavity cross sections. The fields are expanded into a sum of discrete modes that satisfy the boundary conditions. The coefficients of the modes are determined using the orthogonality properties of the mode functions.

Several authors have applied this method to circular cavities [22–25], such as the one shown in Fig. 6.48. Based on the results derived in Ref. [22], the total field scattered from a cavity can be expressed as

$$\begin{bmatrix} E_{s\theta} \\ E_{s\theta} \end{bmatrix} = \begin{bmatrix} S^I_{\theta\theta} + S^R_{\theta\theta} & S^I_{\theta\phi} + S^R_{\theta\phi} \\ S^I_{\phi\theta} + S^R_{\phi\theta} & S^I_{\phi\phi} + S^R_{\phi\phi} \end{bmatrix} \begin{bmatrix} E_{i\theta} \\ E_{i\phi} \end{bmatrix} \tag{6.79}$$

The superscripts I and R refer to the words *internal* and *rim*, respectively. The internal scattering coefficients are given by

$$S^I_{\phi\phi} = \sum_m \sum_n \frac{j(-1)^m J_m^2(\alpha)}{\epsilon_{0m} \sin^2\theta} \left\{ \frac{[(k_{zmn}/k) + \cos\theta]^2 e^{-j2k_{zmn}L}}{k_{zmn}[1 - (\xi_{mn}/\alpha)^2]^2} + \frac{m^2[1 + (k'_{zmn}\cos\theta)/k]^2 e^{-j2k'_{zmn}L}}{k'_{zmn}[(\xi'_{mn})^2 - m^2]} \right\} \tag{6.80}$$

$$S^I_{\phi\phi} = \sum_m \sum_n \frac{j(-1)^m [kaJ'_m(\alpha)]^2 [(k'_{zmn}/k) + \cos\theta]^2 e^{-j2k'_{zmn}L}}{\epsilon_{0m} k'_{zmn}[(\xi'_{mn})^2 - m^2][1 - (\alpha/\xi'_{mn})]^2} \tag{6.81}$$

where $\alpha = ka\sin\theta$. In Eqs. (6.80) and (6.81), the following quantities are required:

1. Neuman's number

$$\epsilon_{0m} = \begin{cases} 2 & \text{if } m = 0 \\ 1 & \text{if } m > 0, \end{cases}$$

2. ξ_{mn} = the nth root of the Bessel function $J_m(\alpha)$
3. ξ'_{mn} = the nth root of the derivative of the Bessel function $J'_m(\alpha)$
4. k_{zmn} = the longitudinal component of the propagation constant

The sums in the preceding equations are taken over only the propagating modes. The cutoff frequencies and wave numbers for a circular waveguide are given by

$$(f_c)_{mn} = \begin{cases} \dfrac{\xi'_{mn}}{2\pi a\sqrt{\mu\epsilon}} & \text{TE modes} \\ \dfrac{\xi_{mn}}{2\pi a\sqrt{\mu\epsilon}} & \text{TM modes} \end{cases} \tag{6.82}$$

and

$$k_{zmn} = k\sqrt{1 - \left[\frac{(f_c)_{mn}}{f}\right]^2} \tag{6.83}$$

The expressions for the off-diagonal terms are based on the GTD diffraction coefficients and being rather involved, these expressions will not be presented here. (They can be found in Ref. [21].) The off-diagonal terms in Eq. (6.78) are zero if rim diffraction is neglected.

Example 6.13: Radar Cross Section of a Cylindrical Cavity Using Modal Analysis

The RCS of the cavity shown in Fig. 6.48 is computed using the modal technique (with rim diffraction neglected) and compared to the MM solution [26]. The radius is $a = 0.037$ m, the length $L = 0.22$ m, and the frequency is 9.13 GHz. The results are shown in Fig. 6.49.

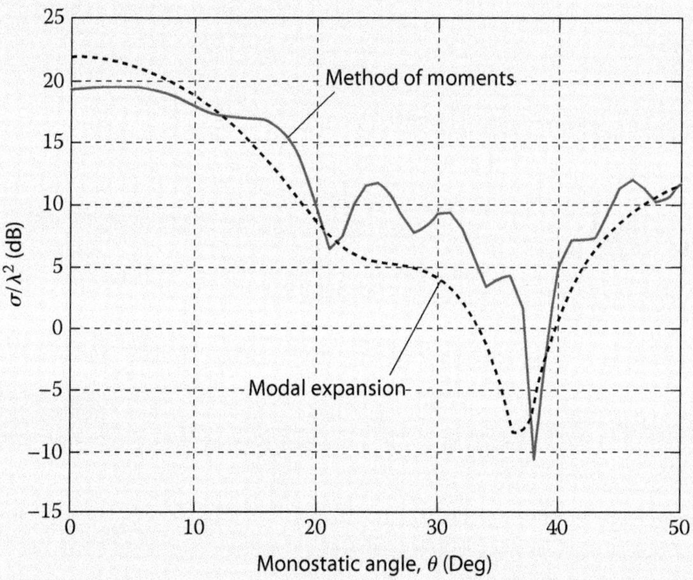

Fig. 6.49 Radar cross section of a cylindrical cavity using modal analysis (no rim diffraction) compared to the MM.

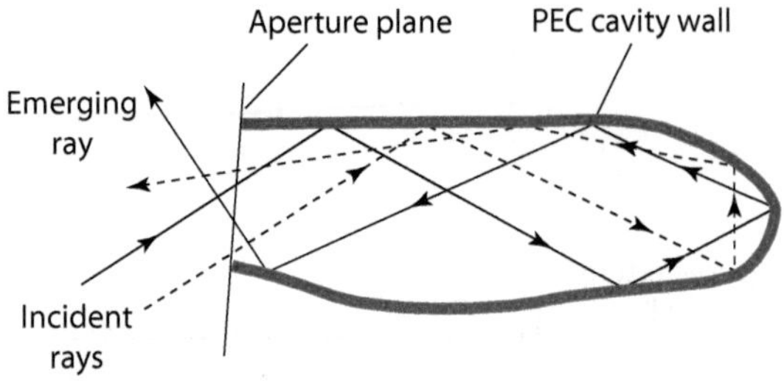

Fig. 6.50 Shooting and bouncing ray concept.

6.11.2 Shooting and Bouncing Rays

Another analysis method applicable to cavities of arbitrary shape is called *shooting and bouncing rays* (SBR) [27]. As illustrated in Fig. 6.50, a bundle of rays is propagated into the cavity and the path of each ray traced until it emerges from the cavity. This geometrical optics field in the aperture, E_i, is used to find an equivalent magnetic current that can be substituted into the radiation integrals:

$$\vec{J}_{ms} = \begin{cases} 2\vec{E}_i(x', y', z') \times \hat{n}, & \text{on } S_A \\ 0 & \text{else} \end{cases} \tag{6.84}$$

If the aperture lies in the $x - y$ plane, the monostatic scattered field is (Appendix C)

$$\vec{E}(P) = \frac{jke^{-jkR}}{2\pi R}\{\hat{\theta}[E_{xi}\cos\phi + E_{yi}\sin\phi] + \hat{\phi}[(-E_{xi}\sin\phi + E_{yi}\cos\phi)\cos\theta]\} \times \iint_{S_A} \exp[j2k(x'u + y'v)]\,dx'\,dy' \tag{6.85}$$

A simple example illustrates the details.

Example 6.14: Radar Cross Section of a Rectangular Cavity Using the SBR Technique

The side view ($x - z$ plane cut) of a rectangular cavity of dimension L_x by L_y is shown in Fig. 6.51. The length-to-height ratio in this plane is 7:4, and the angle of incidence is conveniently chosen to be -26.56 deg (a 2:1 ratio for z to x). The negative sign denotes that the wave is incident from the -x direction. The wave is TM_z-polarized and, for simplicity, only four rays are considered and the problem is treated as two-dimensional.

(Continued)

Example 6.14: Radar Cross Section of a Rectangular Cavity Using the SBR Technique *(Continued)*

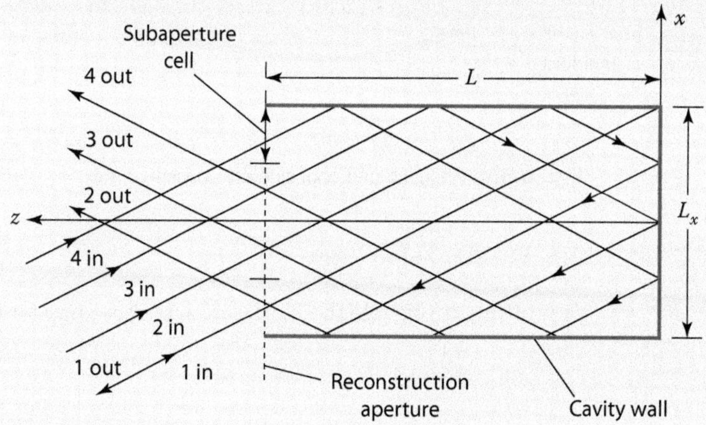

Fig. 6.51 Cross section of a rectangular cavity with four incident rays.

First, the aperture is divided equally into N subapertures of size $L_y \times d_s$ $(=L_x/N)$, with each ray entering the cavity in the center of a subaperture. The path of each ray is traced using Snell's law at reflection points. It is found that each ray travels exactly the same distance inside the cavity and, for this particular angle, has an odd number of reflections. This is not true for arbitrary geometries and angles of incidence. The number of reflections is important because a 180-deg phase reversal occurs at each bounce, and the relative phase of each emerging ray must be known precisely to construct an aperture distribution.

Ray 1 is a unique case because it hits a corner. Somewhat arbitrarily, we model the direction of the reflected ray as directly back toward the source (similar to a corner reflector). Table 6.2 summarizes the ray behavior based on a unit plane wave incident. Ray 1's entry point is used as the phase reference. The phase $\alpha = kd_s \sin\theta$ accounts for the path delay on incidence between adjacent subapertures. Note that in Fig. 6.51 no ray emerges from subaperture 4.

Table 6.2 Shooting and Bouncing Rays for Example 6.14

Ray Number	Entry Number	Exit Number	Amp∠ Phase a_n	θ_n, Exit Direction Deg
1	1	1	$1\angle 0$	−26.56
2	2	1	$1\angle\alpha$	+26.56
3	3	2	$1\angle 2\alpha$	+26.56
4	4	3	$1\angle 2\alpha$	+26.56

(Continued)

Example 6.14: Radar Cross Section of a Rectangular Cavity Using the SBR Technique *(Continued)*

The amplitude and phase of each subaperture is constant and, therefore, the integral in Eq. (6.85) reduces to a sum of sinc functions weighted by the ray amplitude and phase. Because $E_{yi} = 0$ and $\phi = 0$ in this case, Eq. 6.85 becomes

$$E_{\theta}(P) = \frac{j\,k\,L_y\,d_s\,e^{-jkR}}{2\pi R} E_{\theta i} \cos\theta \sum_{n=1}^{4} a_n \mathrm{sinc}[kd_s(u - u_n)/2] \tag{6.86}$$

where $u_n = \sin\theta_n$ is obtained from column 5 in Table 6.2. This term essentially scans the sinc function for each ray so that its maximum corresponds to the exit direction of its ray. This is not essential and is less signifcant as the subaperture becomes smaller. Note that the phase delay for travel through the cavity and reversals due to reflections have been dropped because they are common to all rays and only introduce a phase constant. Finally, the total RCS is

$$\sigma = \frac{4\pi A^2}{\lambda^2} \frac{|1/4E_{\theta}(P)|^2}{|E_{\theta i}|^2} \cos^2\theta \tag{6.87}$$

The factor of $1/4$ provides the proper normalization, so that the RCS can be expressed in terms of the physical area. For $L_x = L_y = 1\lambda$ $(L = 2\lambda)$, the RCS is $\sigma/\lambda^2 \approx 6$ dB at 30 deg. This is about 10 dB higher than a conducting plate of the same area at the same angle.

As more rays are used, d_s decreases and the sinc function approaches an isotropic scattering pattern. This is expected based on Huygen's principle. Obviously, the SBR method will be much more complicated for cavities with bends and circular cross sections. The cross section of a ray tube "shot" into the cavity will not be the same when it exits. Thus, the density of rays in the aperture plane will not be uniform, and the path lengths in the cavity will not be the same. When the cavity surface is curved, the appropriate divergence factor must also be included. All these details present a significant bookkeeping task for large cavities that require many rays to provide adequate resolution for the aperture field construction.

To improve the SBR calculation, diffraction from the rim of the cavity opening can be added. In most cases single diffraction from the two edges in the plane of incidence (defined by $\hat{k}_i$ and $\hat{n}$) is sufficient. Second order diffraction includes multiply diffracted rays between edges on opposite sides of the rim. Higher order terms are due to diffracted rays that enter the cavity and then return to the aperture after several bounces. These should be included in the reconstructed aperture distribution, but they are generally much smaller than the primary SBR rays discussed previously.

6.12 Errors and Imperfections

6.12.1 General Comments

Previous formulations for RCS have dealt with ideal target shapes and materials. Surfaces were considered perfectly smooth and their material composition (μ, ε, and σ) known precisely. In a real world, perfection is impossible, and attempts to approach it are costly. The physical dimensions of target components will be held to some appropriate *tolerance*. For instance, a plate surface roughness might be specified as 0.05 inch. At first glance, this appears to be such a small number that the plate could be approximated accurately by a perfectly smooth surface. However, this is not the case if the wavelength of the incident wave is small or if the plate area is very large. An example of the first case is a millimeter wave radar illuminating the plate at 35 GHz ($\lambda = 0.34$ in.). The surface roughness is 0.15λ, which corresponds to path differences of about 53 deg.

Imperfections in the manufacturing and assembly processes of a target tend to behave in a random fashion. Thus, the quantities in the expressions for RCS that represent these parameters can be modeled as random variables. The scattered field, and thus RCS, are functions of these random variables. The effect of random errors on the RCS pattern is to extract energy from the specular reflection and convert it to random scattering that is distributed nearly equally in all directions. This is called *diffuse* scattering. As the size of the random error is increased, the specular scattering level decreases and the diffuse scattering increases.

6.12.2 Sources of Imperfections

Errors are an expected part of manufacturing and assembly. Physical dimensions of machined and fabricated parts will not be precise, and the electrical characteristics of materials will not be uniform from piece to piece. The errors can be controlled to some extent but at the expense of increased cost. Frequently, clever engineering can be used to minimize the impact of imperfections, even in the presence of rather “sloppy” tolerances.

The imperfections can manifest themselves in the RCS pattern in several ways. In many cases, pattern features can be used to track down the source of an error. For instance, periodic errors result in coherent lobes, and the height, spacing, and lobe width are known functions of the intensity and spacing of the source scatterers. On the other hand, random errors yield a “noise floor” similar to noise added to a voltage signal. The noise level can be predicted from the statistics of the error sources.

Figure 6.52 illustrates three approaches to assembling a large plate from smaller tiles. This might be desirable if the plate is very large, and extremely close tolerances must be maintained on surface roughness. Machining large surfaces is very expensive and time consuming. In the first method shown, each tile has its own support mount. It could be argued that each tile location

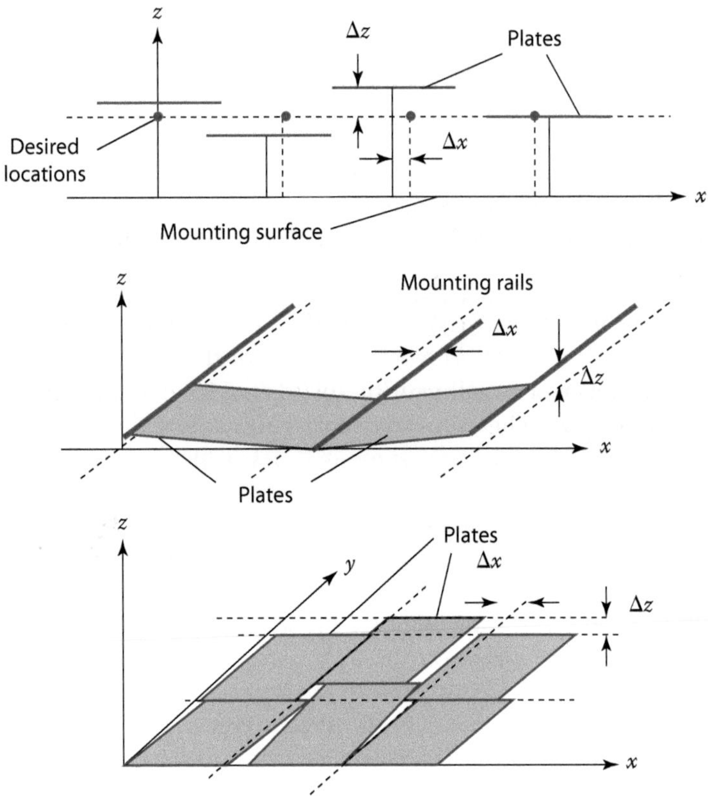

Fig. 6.52 Illustration of three methods for assembling a plate that result in different error behavior.

is random and uncorrelated with the errors of other tiles. In the second case, tiles in the same row lie on common rails. Thus, errors along the rows are perfectly correlated from tile to tile, whereas the location of the individual rails is random and uncorrelated. Finally, if the tiles are simply butted against each other, errors will accumulate. In each method, assumptions must be made about the statistics of the error: mean, variance, and probability density function (PDF). In almost all cases, it is adequate to model errors as normally distributed, with the error-free quantity as the mean and a variance determined by the tolerance on the quantity.

6.13 Random Errors

6.13.1 Array of Discrete Scatterers

An estimate of the average RCS level due to random errors can be obtained from the *mean power pattern* $\overline{P}$ if the statistics of the errors are known. As an example, consider the plate of tiles discussed in the preceding section. Assume that each tile has dimensions $\ell_x \times \ell_y$ and lies in the $z = 0$ plane. The location of tile (m, n) should be

$$x_{mn} = (m - 1/2)\ell_x, \qquad y_{mn} = (n - 1/2)\ell_y, \qquad z_{mn} = 0$$

Assume that the tiles will be accurately aligned in the x and y coordinates but possibly misaligned in the z dimension. The errors are zero mean and normally distributed with variance $\overline{\delta^2}$.

The monostatic scattered field (omitting polarization factors) is given by

$$E_{s\theta} = F(u, v) \cos\theta \sum_{m=1}^{M} \sum_{n=1}^{N} \exp[j2k(ux_{mn} + vy_{mn} + w\delta_{mn})] \tag{6.88}$$

where δ_{mn} is the error at element (m, n); it is modeled as a random variable with a Gaussian PDF. The element factor for this array, $F(u, v)$, is the scattering pattern for a single plate of size $\ell_x \times \ell_y$ (a product of sinc functions), and u, v, and w are the x, y, and z direction cosines, respectively.

To find the average RCS we follow the technique used by Ruze [28] for analyzing the surface errors of reflector antennas. The mean power pattern is used in the definition of RCS, where

$$\overline{\mathcal{P}} = \langle \vec{E}_s \cdot \vec{E}_s^* \rangle \tag{6.89}$$

The asterisk (*) denotes complex conjugation and $\langle\cdot\rangle$ denotes expectation. The factor of $1/\eta$ has been neglected. Substituting in the summations [with the factor $F(u, v)$ omitted] gives

$$\overline{\mathcal{P}} \sim \cos^2\theta \Bigg\langle \sum_{m=1}^{M} \sum_{n=1}^{N} \exp[j2k(ux_{mn} + vy_{mn} + w\delta_{mn})]$$
$$\times \sum_{p=1}^{M} \sum_{q=1}^{N} \exp[-j2k(ux_{pq} + vy_{pq} + w\delta_{pq})] \Bigg\rangle$$

or

$$\overline{\mathcal{P}} \sim \cos^2 \Bigg\langle \sum_{m=1}^{M} \sum_{n=1}^{N} \sum_{p=1}^{M} \sum_{q=1}^{N} \exp\{j2k[u(x_{mn} - x_{pq})$$
$$+ v(y_{mn} - y_{pq}) + w(\delta_{mn} - \delta_{pq})]\} \Bigg\rangle \tag{6.90}$$

Only the last term in the exponent contains a random variable. Thus

$$\overline{\mathcal{P}} \sim \cos^2\theta \sum_{m=1}^{M} \sum_{n=1}^{N} \sum_{p=1}^{M} \sum_{q=1}^{N} \exp\{j2k[u(x_{mn} - x_{pq})$$
$$+ v(y_{mn} - y_{pq})]\overline{\exp[j2kw(\delta_{mn} - \delta_{pq})]}\} \tag{6.91}$$

where the overbar denotes expected value.

Define a new random variable $\Delta = 2kw(\delta_{mn} - \delta_{pq})$, which is generated as the difference of two samples from a normal distribution. From statistics,

$$\overline{e^{j\Delta}} = \overline{\cos\Delta} + j\overline{\sin\Delta} = \begin{cases} e^{-\overline{\Delta^2}}, & m \neq p \text{ and } n \neq q \\ 1, & m = p \text{ and } n = q \end{cases} \tag{6.92}$$

where $\overline{\Delta^2} = 4(kw)^2\overline{\delta^2}$. Thus,

$$\overline{\mathcal{P}} \sim \cos^2\theta \left(e^{-\overline{\Delta^2}} \sum_{\substack{m=1 \\ m\neq p}}^{M} \sum_{\substack{n=1 \\ \text{and}}}^{N} \sum_{\substack{p=1 \\ n\neq q}}^{M} \sum_{q=1}^{N} \exp\left\{ j2k[u(x_{mn} - x_{pq}) + v(y_{mn} - y_{pq})] \right\} + \sum_{\substack{m=1 \\ m=p}}^{M} \sum_{\substack{n=1 \\ \text{and}}}^{N} \sum_{\substack{p=1 \\ n=q}}^{M} \sum_{q=1}^{N} (1) \right) \tag{6.93}$$

Now add and subtract the quantity

$$e^{-\overline{\Delta^2}} \sum_{\substack{m=1 \\ m=p \text{ and } n=q}}^{M} \sum_{n=1}^{N} \sum_{p=1}^{M} \sum_{q=1}^{N} (1) = (MN) e^{-\overline{\Delta^2}}$$

from the quantity inside the boldface parentheses in Eq. (6.93). After a bit of manipulation, the mean power pattern becomes

$$\overline{\mathcal{P}} \sim \cos^2\theta \, MN\left(1 - e^{-\overline{\Delta^2}}\right) + e^{-\overline{\Delta^2}}\mathcal{P}_0 \tag{6.94}$$

where $\mathcal{P}_0$ represents the error-free power pattern ($\delta_{mn} = 0$).

The maximum value of $\mathcal{P}_0$ is $(MN)^2$. The normalized mean power pattern can be written by dividing through by $(MN)^2 e^{-\Delta^2}$

$$\overline{\mathcal{P}}_{\text{norm}} \approx \mathcal{P}_{0_{\text{norm}}} + \frac{\cos^2\theta}{MN}\left(1 - e^{-\overline{\Delta^2}}\right)e^{\overline{\Delta^2}} \tag{6.95}$$

and, if the errors are small,

$$(1 - e^{-\overline{\Delta^2}})e^{\overline{\Delta^2}} \approx \overline{\Delta^2} \tag{6.96}$$

yielding a final result

$$\overline{\mathcal{P}}_{\text{norm}} \approx \mathcal{P}_{0_{\text{norm}}} + \frac{\overline{\Delta^2}\cos^2\theta}{MN} \tag{6.97}$$

The final quantity of interest is the RCS

$$\sigma = \frac{4\pi A^2 \cos^2\theta}{\lambda^2} |F_n(u, v)|^2 \overline{\mathcal{P}}_{\text{norm}} \tag{6.98}$$

where F_n is the normalized element factor.

6.13.2 Characteristics of Random Scattering

Several important properties of the RCS are noted:

1. The $\cos^2\theta$ factor is the usual projected aperture factor. There is an additional $\cos^2\theta$ factor embedded in Δ^2. As illustrated in Fig. 6.53, this second factor accounts for the projection of the error distance δ_{mn} in the direction of the observer. When viewed along the z axis, the phase change introduced by the position error is $k\delta_{mn}$; when viewed from $\theta = 90$ deg, there is no error introduced because the z direction is transverse to the propagation vector.
2. If errors are present in the x and y directions in addition to the z direction, Eq. (6.97) holds with

$$\overline{\Delta^2} = 4k^2\left[u^2\overline{\delta_x^2} + v^2\overline{\delta_y^2} + w^2\overline{\delta_z^2}\right] \tag{6.99}$$

3. If the errors are angle independent (that is, $\overline{\Delta^2}$ is independent of θ), the second term would represent an isotropic RCS "noise floor."
4. Inclusion of the element factor causes the RCS to decay even faster at wide angles. The element factor may also include a polarization dependency.
5. In its nonnormalized form, $\mathcal{P}_0$ is proportional to $(M, N)^2 \sim \text{AREA}^2$, whereas the noise term varies as $(M, N) \sim \text{AREA}$.

The last point is particularly important. A common method for suppressing wide-angle RCS lobes is edge tapering, as described in Problem 2.5. If more area is added to a surface to increase the taper region, $\mathcal{P}_0$ will indeed be reduced at wide angles but, simultaneously, the noise term increases because the area is increased. In practice, there is a limit to wide-angle RCS reduction via tapering.

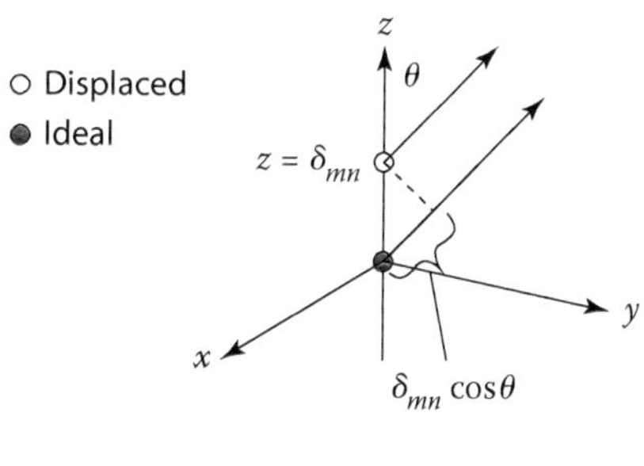

Fig. 6.53 Angle dependence of displacement errors.

6.13.3 Monte Carlo Method

Another method of analyzing the effects of tolerances on RCS is the *Monte Carlo technique*, also known as the *method of repeated trials*. Equation (6.88) is evaluated explicitly many times using a random number generator to provide values for δ_{mn}. Then, $E_{s\theta}$ is used to compute the RCS, and the results of a large number of trials provide an average. The advantage of this technique is that a pattern is obtained based on the chosen set of δ_{mn} that allows the effects of the tolerances to be observed directly. A large number of trials must be performed, however, to determine exactly how much degradation is "average." If only one calculation is performed, the sidelobe increase could be much smaller than the average; on the other hand, it might be much larger than the average if a particularly large set of δ_{mn} was chosen.

Example 6.15: Monte Carlo Simulation of a Plate of Tiles

The Monte Carlo method is to be used to simulate the effect of random errors in the tile example. Equation (6.88) is computer programmed and a set of δ_{mn} is chosen from a random number generator. The errors are scaled to represent samples from a normal distribution with zero mean and specified standard deviation. A typical result is shown in Fig. 6.54 for a 50 × 50 array of tiles, each tile being 0.25λ square. The pattern in shown in the diagonal plane of the plate ($\phi = 45$ deg), and the errors are assumed to be uncorrelated with a rms error of 0.11λ.

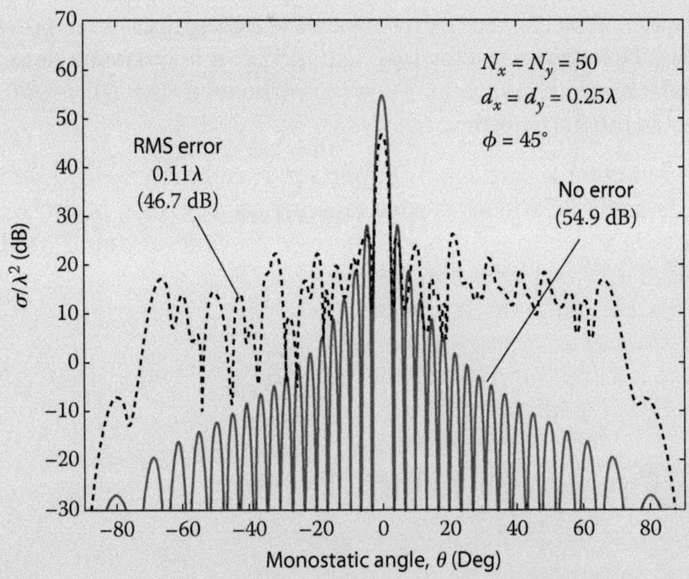

Fig. 6.54 Effect of random vertical displacement errors on RCS computed using the Monte Carlo method.

6.13.4 Solid Scattering Surfaces

The statistical approach used to compute the mean power pattern in the tile example can be applied to solid surfaces as well. Ruze first investigated the effect of random surface errors on the gain of reflector antennas using this method [29]. For rough solid surfaces, the deviation of points in the same neighborhood is clearly correlated. For instance, a dent in a plate will probably have a hemispherical shape, or something close to it. If the dent diameter is 1 in., the deviation of points less than 0.5 in. from the center will be correlated to the deviation at the center.

The statistical method can lead to closed-form expressions for the mean power pattern just as it did in the discrete case if several assumptions and approximations are made. First, assume that the surface deviations resemble Gaussian dents, as shown in Fig. 6.55. The extent of the Gaussian shape is characterized by a *correlation interval c*, which is the average distance at which the deviations become uncorrelated. As illustrated in Fig. 6.56, a large correlation interval implies a slowly varying surface error, whereas a small correlation interval is associated with a rapidly varying error. The variance of the deviations is denoted as $\overline{\delta^2}$.

To compute the mean power pattern, physical optics is first used to find the scattered field. For the moment, assume a flat surface in the $x-y$ plane. Because of the dents, the surface will not lie completely in the $x-y$ plane but will have a tangential surface vector component parallel to z in some spots. Thus, the tangential surface current will also have a z component in these spots but, if the dents are kept small, this component can be ignored. Based on this argument, there will be only x and y components of the surface current.

For a TM_z incident wave

$$
\begin{aligned}
J_x &= -\frac{2e^{jkg}}{\eta_0} E_{0\theta} \cos\phi \\
J_y &= -\frac{2e^{jkg}}{\eta_0} E_{0\theta} \sin\phi
\end{aligned}
\tag{6.100}
$$

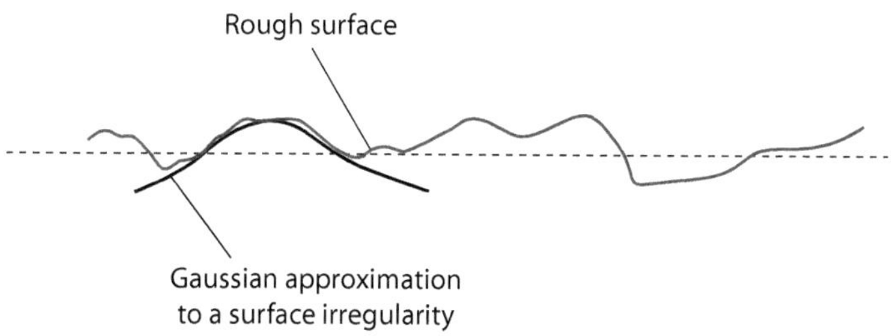

Fig. 6.55 Dent in a surface approximated by a Gaussian shape.

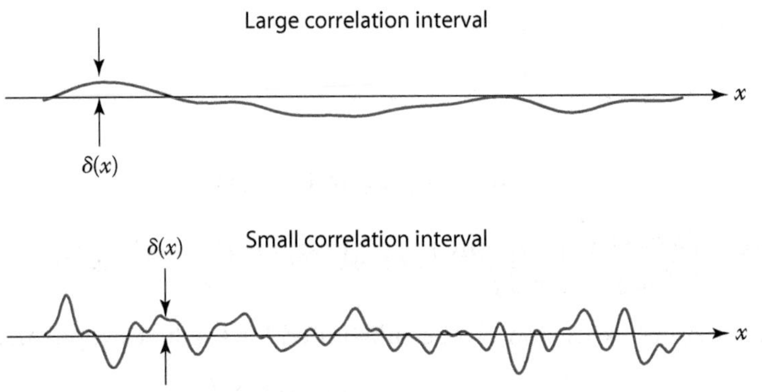

Fig. 6.56 Illustration of large and small correlation intervals.

where $g = -\hat{k} \cdot \vec{r}'$ and $\vec{r}' = \hat{x}x' + \hat{y}y'$. For a unit incident field ($E_{0\theta} = 1$), the scattered field is

$$E_{s\theta} = \frac{-jk \cos \theta e^{-jkr_0}}{2\pi r_0} \iint_S e^{j2\vec{k}\cdot\vec{r}'} e^{j2k\delta(\vec{r}')} ds' \tag{6.101}$$

where r_0 represents the distance from the surface origin to the observation point. Using Eq. (6.101) yields a mean power pattern of

$$\overline{\mathcal{P}} = \left\langle \vec{E}_s \cdot \vec{E}_s^* \right\rangle = \left(\frac{k \cos \theta}{2\pi r_0} \right)^2$$
$$\times \iint_S \iint_S \exp \left[j2\vec{k} \cdot (\vec{r} - \vec{r}')\right] \overline{\exp \{j2k[\delta(\vec{r}) - \delta(\vec{r}')]\}} ds' ds \tag{6.102}$$

Now, perform a substitution of variables with $\vec{\tau} = \vec{r} - \vec{r}'$:

$$\overline{\mathcal{P}} = \left(\frac{k \cos \theta}{2\pi r_0} \right)^2 \iint_S \iint_S e^{\{j2\vec{k}\cdot\vec{\tau}\}} \overline{\exp \{j2k[\delta(\vec{r}' + \vec{\tau}) - \delta(\vec{r}')]\}} ds' ds \tag{6.103}$$

Further evaluation requires knowledge of the variance of the random variable in the exponent, $\Delta(\vec{\tau}) \equiv \delta(\vec{r}' + \vec{\tau}) - \delta(\vec{r}')$. It is formed by the difference of two independent normally distributed random variables with variance $\overline{\delta^2}$. If the separation of two points on the surface $\vec{\tau}$ is much greater than the correlation interval, then,

$$\overline{\Delta(\vec{\tau})^2} = 2\overline{\delta^2}$$

On the other hand, when two surface points coincide, $\vec{\tau} = 0$, and $\overline{\Delta(\vec{\tau})^2} = 0$. Thus, a suitable form for the variance is

$$\overline{\Delta^2} = 2\overline{\delta^2} \left[1 - e^{-\tau^2/c^2}\right] \tag{6.104}$$

The expected value of the exponential becomes

$$\overline{\cos\Delta} = \exp\left[-\overline{\delta^2}(1 - e^{-\tau^2/c^2})\right]$$
$$\overline{\sin\Delta} = 0$$

Returning to Eq. (6.103) and using Eq. (6.104) give

$$\overline{\mathcal{P}} = \left(\frac{k\cos\theta}{2\pi r_0}\right)^2 \iint_S \iint_S e^{j2\vec{k}\cdot\vec{\tau}} \exp\left(4k^2\overline{\delta^2}e^{-\tau^2/c^2}\right) e^{-4k^2\overline{\delta^2}}\, ds'\, ds \tag{6.105}$$

A Taylor series expansion can be substituted for the first exponential:

$$\exp\left(4k^2\overline{\delta^2}e^{-\tau^2/c^2}\right) = \sum_{n=0}^{\infty} \frac{\left(4k^2\overline{\delta^2}\right)^n}{n!} e^{-n\tau^2/c^2} \tag{6.106}$$

The first term in the sum can be broken out and, given that the error-free power term is

$$\mathcal{P}_0 = \left(\frac{k\cos\theta}{2\pi r_0}\right)^2 \underbrace{\iint_S \iint_S e^{j\vec{k}\cdot\vec{\tau}}\, ds'\, ds}_{\equiv \Omega_0} \tag{6.107}$$

the mean power pattern can be expressed as Ω_0

$$\overline{\mathcal{P}} = \left(\frac{k\cos\theta}{2\pi r_0}\right)^2 e^{-4k^2\overline{\delta^2}} \left\{ \Omega_0 + \iint_S \sum_{n=1}^{\infty} \iint_S \frac{\left(4k^2\overline{\delta^2}\right)^n}{n!} e^{-(n\tau^2/c^2)} e^{j2\vec{k}\cdot\vec{\tau}}\, ds'\, ds \right\} \tag{6.108}$$

For convenience, assume that the surface is a circular disk with radius a, and polar coordinates can be used. For each point on the plate (ρ, ϕ), an integration is performed over all possible distances τ. Note that the exponential in the integrand decays rapidly when the distance is greater than the correlation length. The limits on the integral can be extended to infinity with little error:

$$\iint_S ds' \to \int_0^a \int_0^{2\pi} \rho'\, d\rho'\, d\phi' \to \int_0^{\infty} \int_0^{2\pi} \rho'\, d\rho'\, d\phi' \tag{6.109}$$

When polar coordinates are used and the integration limits extended, Eq. (6.108) becomes

$$\overline{\mathcal{P}} = \left(\frac{k\cos\theta}{2\pi r_0}\right)^2 e^{-4k^2\overline{\delta^2}} \left\{ \Omega_0 + \int_0^a \int_0^{2\pi} \rho\, d\rho'\, d\phi' \right. \\ \left. \times \sum_{n=1}^{\infty} \int_0^{\infty} \int_0^{2\pi} e^{j2\vec{k}\cdot\vec{\rho}} \frac{\left(4k^2\overline{\delta^2}\right)^n}{n!} e^{-n(\rho')^2/c^2)} \rho'\, d\rho'\, d\phi' \right\} \tag{6.110}$$

In polar coordinates, the dot product in the exponent can be written

$$\vec{k} \cdot \vec{\rho}' = k\rho' \sin\theta \cos(\phi - \phi')$$

and the ϕ' integral takes the form of a Bessel function (see Problem 2.4):

$$\overline{\mathcal{P}} = \left(\frac{k\cos\theta}{2\pi r_0}\right)^2 e^{-4k^2\overline{\delta^2}} \left\{ \Omega_0 + \int_0^a \int_0^{2\pi} \rho d\rho' d\phi' \sum_{n=1}^{\infty} \frac{\left(4k^2\overline{\delta^2}\right)^n}{n!} \right.$$
$$\left. \times\, 2\pi \left[\int_0^{\infty} J_0(k\rho' \sin\theta) e^{-[n(\rho')^2/c^2]} \rho' d\rho' \right] \right\} \quad (6.111)$$

The integral in square brackets can be written in Schlaefli's contour integral form [30] and evaluated using the method of residues

$$\int_0^{\infty} J_0(k\rho' \sin\theta) e^{-[n(\rho')^2/c^2]} \rho' d\rho' = \frac{c^2}{2n} \exp\left(-\frac{c^2\pi^2 \sin^2\theta}{\lambda^2 n}\right) \quad (6.112)$$

Combining all the preceding relationships gives

$$\overline{\mathcal{P}} = \left(\frac{k\cos\theta}{2\pi r_0}\right)^2 e^{-4k^2\overline{\delta^2}} \left\{ \Omega_0 + \left[\int_0^a \int_0^{2\pi} \rho d\rho' d\phi' \right] \cdot \right.$$
$$\left. \times\, 4k^2\overline{\delta^2} c^2 \pi \sum_{n=0}^{\infty} \left[\frac{\left(4k^2\overline{\delta^2}\right)^{n-1}}{n!n} \exp\left(-\frac{c^2\pi^2 \sin^2\theta}{\lambda^2 n}\right) \right] \right\} \quad (6.113)$$

The remaining integral is recognized as the plate area A. Finally, keeping only the first term in the sum yields

$$\overline{\mathcal{P}} = \left(\frac{k\cos\theta}{2\pi r_0}\right)^2 e^{-4k^2\overline{\delta^2}} \left\{ \Omega_0 + 4k^2\overline{\delta^2} c^2 \pi A \exp\left(-\frac{c^2\pi^2 \sin^2\theta}{\lambda^2}\right) \right\} \quad (6.114)$$

Because the incident plane wave has unit amplitude, the RCS is

$$\sigma = 4\pi r_0^2 \overline{\mathcal{P}}$$
$$= \frac{4\pi A^2 \cos^2\theta}{\lambda^2} e^{-4k^2\overline{\delta^2}} \left\{ \mathcal{P}_{0_{\mathrm{norm}}} + \frac{4k^2\overline{\delta^2} c^2 \pi}{A} \exp\left(-\frac{c^2\pi^2 \sin^2\theta}{\lambda^2}\right) \right\} \quad (6.115)$$

The error-free power term Ω_0 has been normalized by the area squared:

$$\mathcal{P}_{0_{\mathrm{norm}}} = \frac{\Omega_0}{A^2} \to 1 \quad (6.116)$$

which has a maximum value for normal incidence.

The second term in Eq. (6.115), the average RCS sidelobe level, emphasizes the importance of maintaining tight surface tolerances. Furthermore, any unavoidable errors should have a small correlation interval. This will result in lower, more uniformly dispersed scattering than that from errors with a large correlation interval.

The distribution of sidelobes follows a modified Rayleigh distribution [31] and, hence, it is expected that 98% of the sidelobes will be less than the average plus 6 dB.

Example 6.16: Scattering from a Rough Plate [32]

Figure 6.57 shows a 3λ by 3λ plate that has been made rough by displacing the nodes of a triangular mesh randomly in the z direction. The location errors are uniformly distributed over the interval $[0, 0.1\lambda]$ (the variance of the errors is $0.82 \times 10^{-3}\lambda$). The RCS in direction cosine space is shown in Fig. 6.58 for a smooth plate and in Fig. 6.59 for the rough plate. The effects of the errors are clearly evident, especially off of the principal planes.

The average edge length of the triangular facets of the plate is about 0.15λ. If we estimate the correlation interval to be approximately half of the edge length, 0.075λ, then the average RCS can be computed from the second term of Eq. (6.115)

$$\sigma_{\text{ave}} \approx \frac{4\pi A}{\lambda^2}\cos^2\theta e^{-4k^2\overline{\delta^2}}(4\pi c^2 k^2 \overline{\delta^2})\exp\left(-\frac{c^2\pi^2\sin^2\theta}{\lambda^2}\right)$$

where $k^2\overline{\delta^2} = (2\pi/\lambda)^2(0.1\lambda)^2 \approx 0.395$. At $\theta = 45°$ this gives $\sigma_{\text{ave}} \approx -4.9$ dBsm.

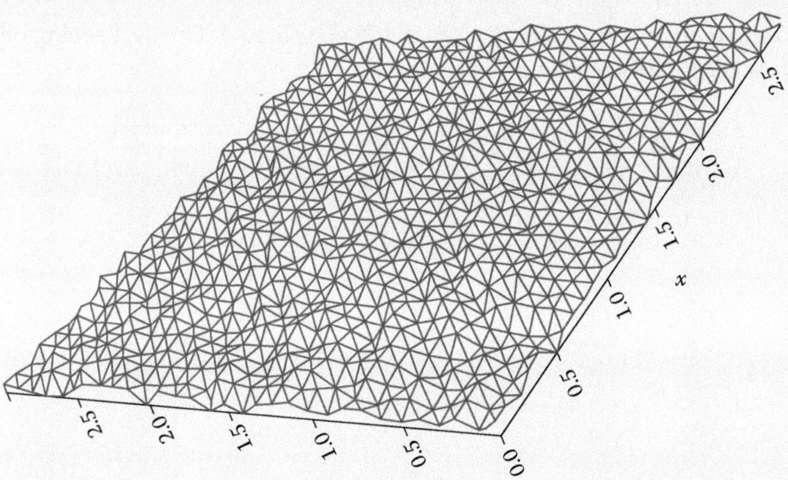

Fig. 6.57 Rough PEC plate generated by randomly displacing the triangle nodes of the MM mesh.

(Continued)

Example 6.16: Scattering from a Rough Plate [32] *(Continued)*

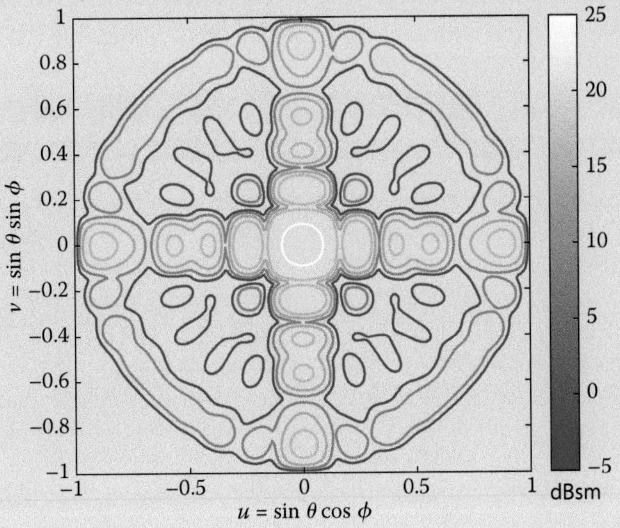

Fig. 6.58 Radar cross section of a 3λ by 3λ rough PEC plate using the method of moments and plotted in direction cosine space (from Ref. [31]).

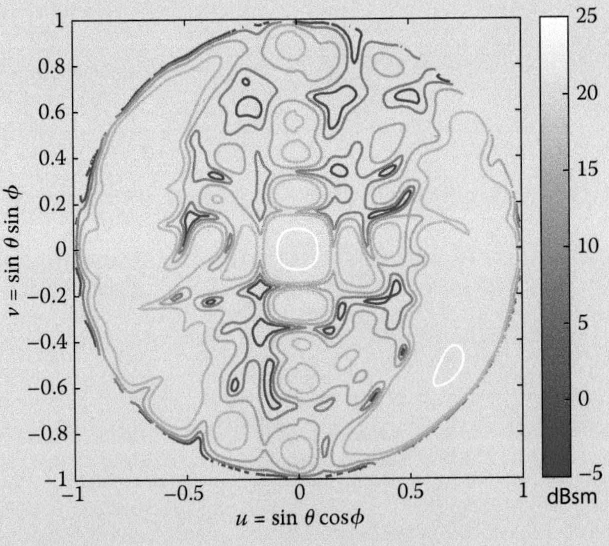

Fig. 6.59 Radar cross section of a 3λ by 3λ flat PEC plate using the method of moments and plotted in direction cosine space. The mesh nodes are randomly distributed in the *z* with a uniform distribution with limits [0, 0.1λ] (from Ref. [31]).

6.14 Periodic Errors

Periodic errors frequently occur in the manufacture of grids and lattices for frequency-selective surfaces and antennas. For example, in the approximate analysis of parallel feed scattering, it was assumed that the errors at every level of the feed were identical. However, if an error was introduced at only one coupler in the second level shown in Fig. 6.22, it would affect only four elements in the array (a *block error*). As another example, photolithography is used to fabricate aperture elements. If several steps or layers are involved, a photoresist misalignment can yield a periodic error. These types of errors can also arise in the assembly of surfaces, as the tile example illustrates.

Example 6.17: Grating Lobes Due to Periodic Errors

Consider an $N_x \times N_y$ array of isotropic scatterers in the $x-y$ plane. Let every other row be displaced from the $z = 0$ plane by a distance δ. Thus, the period of the error is twice the element spacing in x. Figure 6.60 shows the RCS of an $(N_x = 25) \times (N_y = 1)$ array of elements spaced $d_x = 0.3\lambda$. The observation

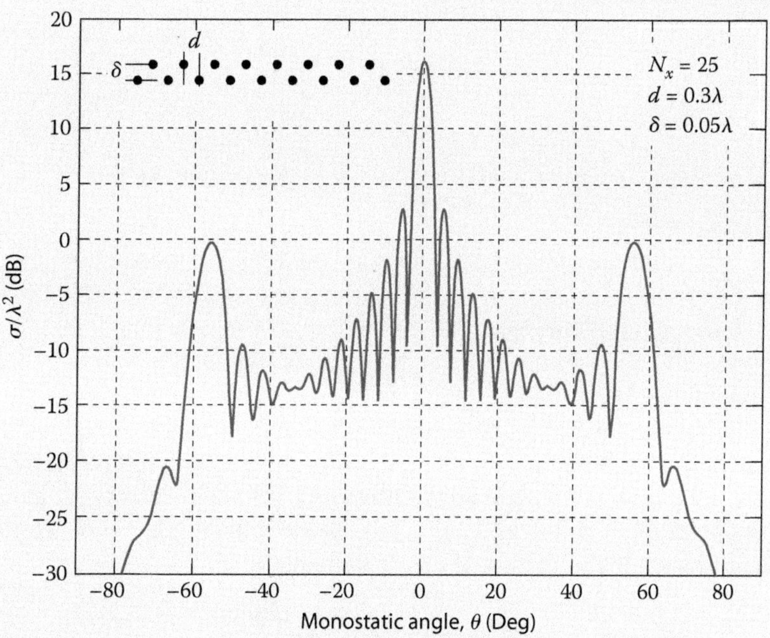

Fig. 6.60 Radar cross section of an array with a periodic displacement error (odd number of elements).

(Continued)

Example 6.17: Grating Lobes Due to Periodic Errors (Continued)

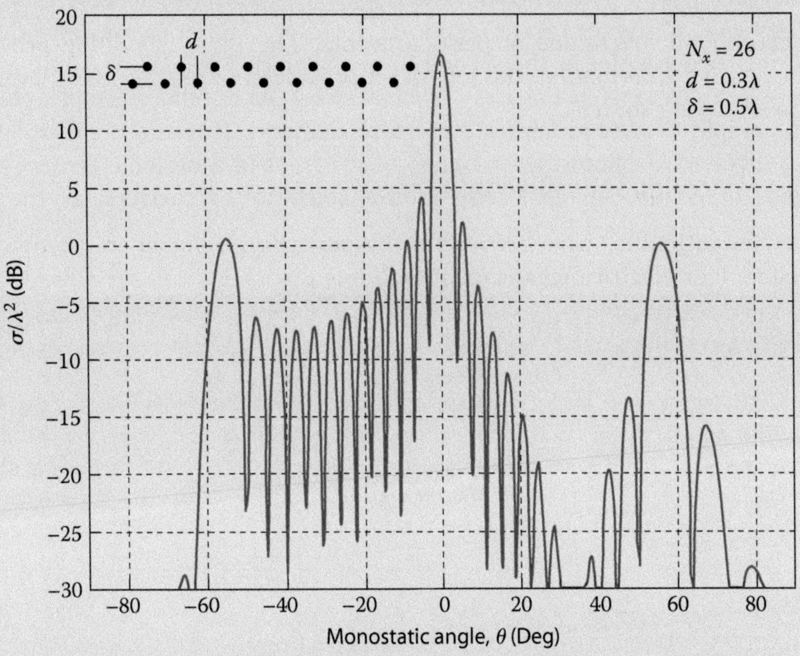

Fig. 6.61 Radar cross section of an array with a periodic displacement error (even number of elements).

point is in the $\phi = 0$ deg plane. Notice that the pattern has symmetry about $\theta = 0$ deg because, given that N_x is odd, the array itself is symmetric about the z axis. On the other hand, if N_x is even, the element locations are asymmetrical with respect to the z axis, and the RCS of Fig. 6.61 results.

The pattern characteristics observed in Figs. 6.60 and 6.61 are easily predicted based on the principle of pattern multiplication (see Problem 2.8). If N_y is even, the array can be decomposed into two elements: an $N_x/2$ element array at $z = 0$ and an $N_x/2$ element array at $z = \delta$. Thus, the element factor is the pattern of $N_x/2$ elements spaced $2d_x$:

$$\mathrm{EF} = \frac{\sin(N_x k d_x \sin\theta \cos\phi)}{\sin(k d_x \sin\theta \cos\phi)} \tag{6.117}$$

The array factor is determined by the displacements of the two elements in x and z:

$$\mathrm{AF} = 1 + \exp[jk(d_x u + \delta w)] \Rightarrow \cos[k(d_x u + \delta w)] \tag{6.118}$$

The final RCS pattern is given by the product (AF)(EF). When $\delta = 0$, the array factor nulls are coincident with the Bragg lobe locations for the element factor in Eq. (6.118). As δ increases, the nulls in the array factor are scanned away from the Bragg lobe peaks, allowing them to increase in magnitude.

This simple concept can be used as a diagnostic tool to determine the location and magnitude of periodic errors. The Bragg condition provides the spacing information; the lobe level provides the relative error levels of the scattering sources.

6.15 Miscellaneous Discontinuities

Discontinuities have been defined as any abrupt change in geometrical, material, or electrical characteristics. Some discontinuities are a product of the design process (windows in aircraft), whereas others are unavoidable consequences of the assembly process (noncontacting joints). If the individual discontinuities are large in number and small in extent, their behavior can be considered random and the methods of the last section can be applied. This section takes up the effect on RCS of three important types of discontinuities: 1) gaps and cracks, 2) holes, and 3) thick or rounded edges.

6.15.1 Gaps and Cracks

Gaps and cracks refer to *small electrical discontinuities* that occur when two discrete objects are joined. *Small* implies that the width (but not necessarily the depth) is much less than a wavelength. Note that the appearance of physical contact does not ensure that the two pieces are making good electrical contact. A gap between two parts may be filled with nonconducting epoxy and, thus, the parts may look as though they are in good contact. Similarly, two pieces can be touching yet not make good contact because of rust, paint, or corrosion.

The primary effect of a crack in a flat surface is the disruption of surface waves flowing along the interface. From the view of microwave optics, edge diffraction occurs at the breaks. Figure 6.62 illustrates the effect of a break in a wire. This is easily simulated with the MM code for a solid wire when overlapping basis functions are used. First, the MM solution is applied to a continuous wire, with a segment present where the gap should be. Next, the rows and columns of the resulting MM impedance matrix that correspond to the segment in the gap are nulled. This essentially forces the current to zero at the gap and also negates continuity of the current.

Referring to Fig. 6.62, we see that the maximum traveling wave lobe is essentially unchanged but that the RCS at intermediate angles has been increased by as much as 10 dB. This is bad news if the wire has been rotated so that it is viewed at these aspects, and this is the logical design choice because the threat is away from the large specular and traveling wave lobes. Note that the break has essentially no effect on the specular

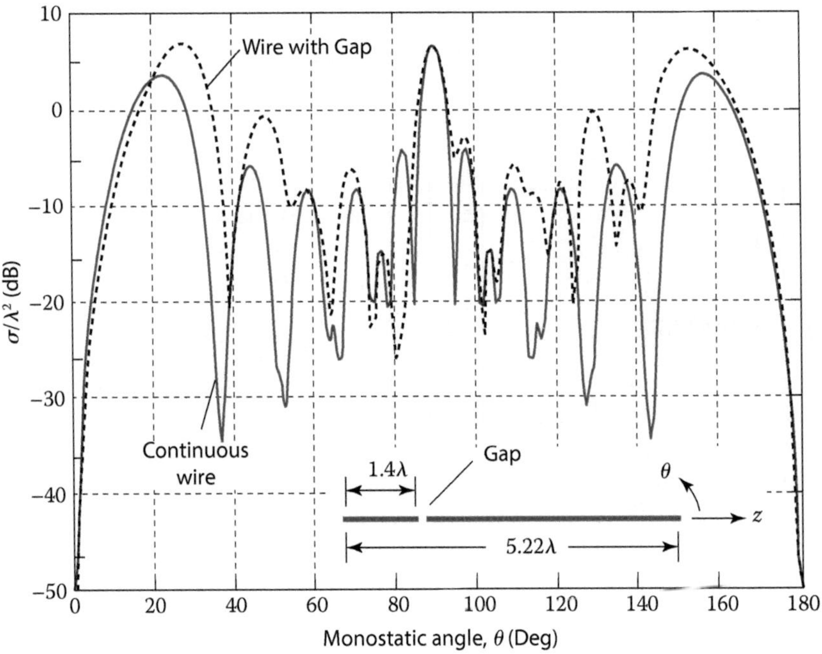

Fig. 6.62 The effect of disrupting the current along a wire on its RCS.

component ($\theta = 90$ deg). Some other aspects of gaps and cracks are discussed in Ref. [33].

6.15.2 Holes

Holes are another potential RCS problem. A simple example along the line of Problem 2.3 illustrates the effect. Figure 6.64 presents the RCS of a square plate with a square hole cut in it. The hole edges are rotated 45 deg with respect to the plate edges. In the diagonal plane of the plate, the hole has increased the RCS by almost 10 dB. This calculation is based on the physical optics approximation and, therefore, does not include the traveling wave component. (The traveling wave contribution will be small in this plane.)

6.15.3 Blunt Edges

Another approximation that has been used throughout the development of the PO, MM, and GTD formulas is that of "infinitely thin" surfaces. Again, the classification of a slab of material of thickness t as thick or thin will depend on the threat frequency. At millimeter wavelengths, most structures will be sufficiently thick so that the scattering from the edge will differ appreciably from that of a knife edge. For thick edges, the question arises as to the optimum shape: flat, rounded, or wedged.

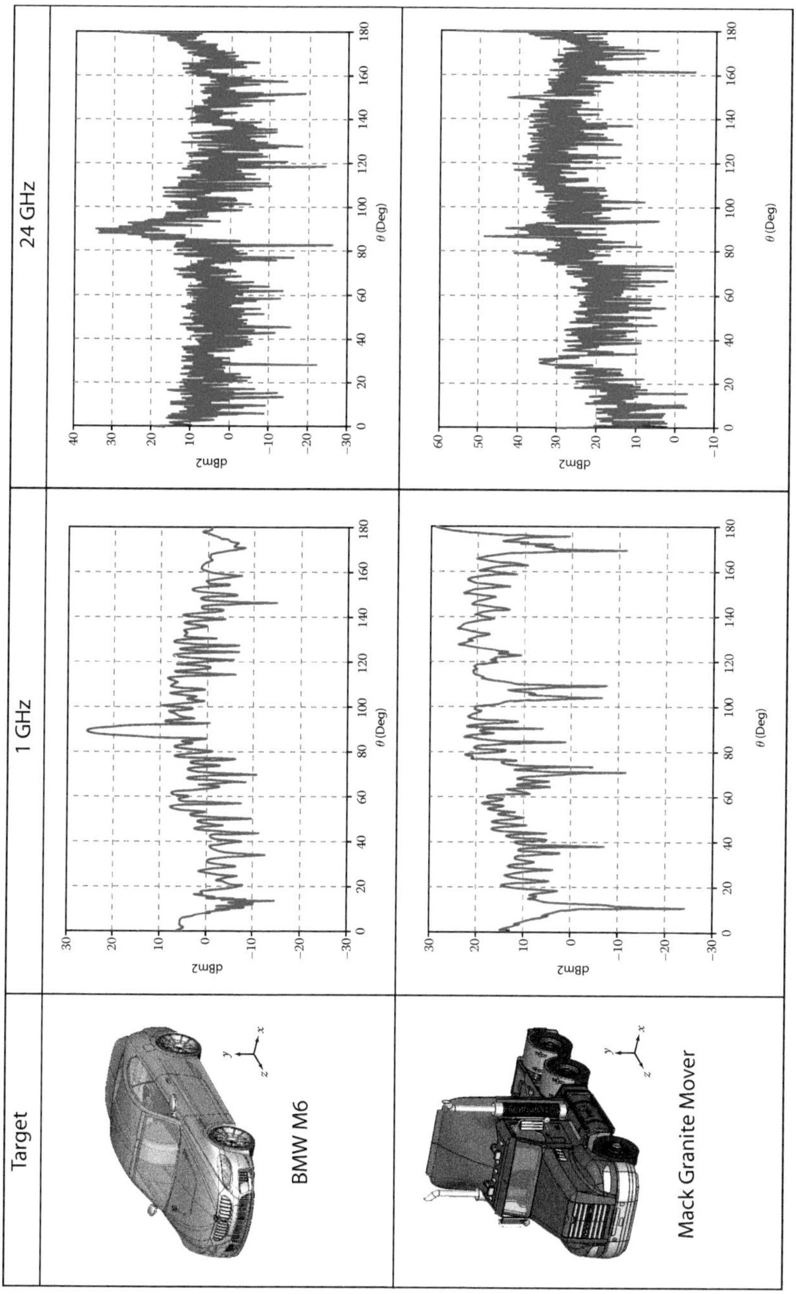

Fig. 6.63 RCS of two motor vehicles at two frequencies.

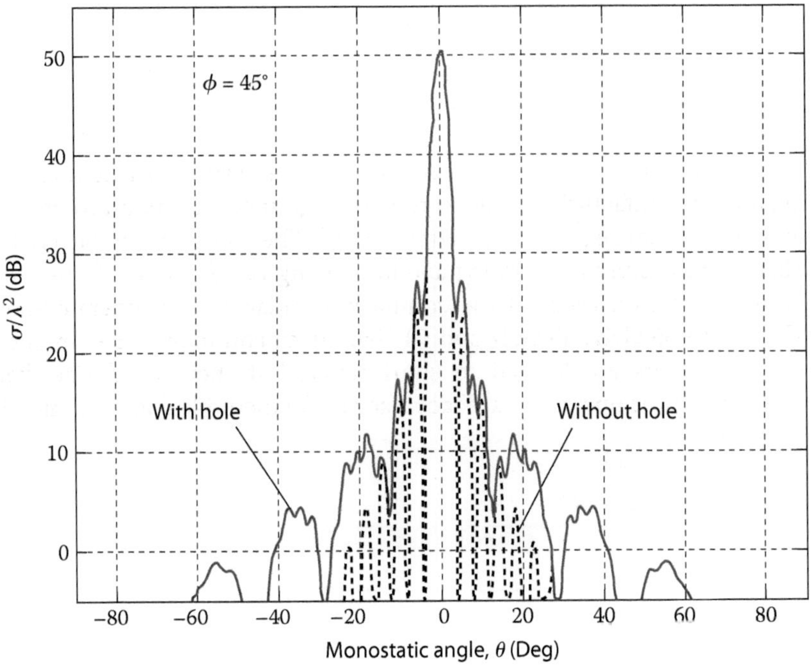

Fig. 6.64 Radar cross section of a plate with a hole computed using physical optics.

The scattering mechanism for a blunt edge (radius, $a > \lambda$) is specular whereas, in the sharp-edge case ($a < \lambda$), diffraction is the primary scattering mechanism. In the transition region, the field is given approximately by the empirical formula [34]

$$|E| \approx \sqrt{E_{\text{high}}^2 + E_{\text{low}}^2} \tag{6.119}$$

where E_{high} is the high-frequency physical optics solution and E_{low} the GTD result for a wedge with its edge at the specular point on the rounded edge. Thus, the RCS for the blunt edge of length L is approximately

$$\sigma_{\text{high}} \approx kaL^2 \tag{6.120}$$

and it depends on both the radius and length. On the other hand, the RCS of the sharp edge depends only on the length:

$$\sigma_{\text{low}} \approx L^2/\pi \tag{6.121}$$

and does not scatter the normal component of the incident field. As to airfoils, subsonic leading edges are more rounded than those for supersonic or hypersonic aircraft.

6.16 Non-Traditional Radar Targets

Until recently, military applications drove research interests in the area of RCS. As radar sensors become more common, knowledge of the RCS of a variety of targets and objects in the environment has become important in predicting sensor performance. Radar sensors are used in self-driving cars, automated manufacturing, safety monitoring, medical monitoring, and unmanned air vehicle (UAV) operation [35,36]. The scattering characteristics of objects are useful in separating and identifying valued targets from background and clutter. Most of the commercial radar sensors operate above 20 GHz, with 35 GHz, 60 GHz and 94 GHz being commonly used frequencies. The sensors are limited to short range, but their small size, light weight, low power operation, and accuracy make them ideal for commercial applications

The methods of RCS prediction and reduction are the same as those discussed for military targets. Figure 6.63 shows several vehicles and their RCS at two frequencies (1 GHz and 24 GHz)[§]. Note that at the high frequency there are few well defined lobes even though the surfaces are much larger electrically. The curved surfaces introduce greater phase error at the higher frequency, and therefore currents induced on the entire surface are never in phase.

Rarely is there a need to reduce the RCS, in fact, the opposite may be true. For example, it might be advantageous for safety reasons to increase the RCS of a pedestrian or cyclist.

References

[1] Skolnik, M. I., *Introduction to Radar Systems*, McGraw-Hill, New York, 1980.
[2] Tittensor, P. J., and Newton, M. L., "Prediction of the Radar Cross Section of an Array Antenna," *Proceedings of the 6th International Conference on Antennas and Propagation*, London, April 1989, pp. 258–262.
[3] Williams, N., "The RCS of Antennas—An Appraisal," *Military Microwaves Conference Proceedings*, Brighton, England, June 1986, pp. 505–508.
[4] Hansen, R. C., "Relationships Between Antennas as Scatterers and as Radiators," *Proceedings of the IEEE*, Vol. 77, No. 5, May 1989, pp. 659–667.
[5] King, R. W. P., and Harrison, C. W., "The Receiving Antenna," *Proceedings of the IRE*, No. 32, Jan. 1944, pp. 18–49.
[6] Stevenson, A. E, "Relations Between Transmitting and Receiving Properties of Antennas," *Quarterly of Applied Mathematics*, Vol. 5, Jan. 1948, pp. 369–384.
[7] Hu, Y. Y., "Back-Scattering Cross Section of a Center-Loaded Antenna," *IRE Transactions on Antennas and Propagation*, Vol. AP-6, No. 1, Jan. 1958, pp. 140–148.
[8] Balanis, C. A., *Antenna Theory Analysis and Design*, Harper and Row, New York, 1982.
[9] Green, R. B., "Scattering from Conjugate-Matched Antennas," *IEEE Transactions on Antennas and Propagation*, Vol. AP-14, No. 1, Jan. 1966, pp. 17–21.
[10] Pozar, D. M., *Microwave Engineering*, Addison-Wesley, Reading, MA, 1990.

[§]The dimensions of these vehicles are readily available on the internet. The models were purchased at www.3dcadbrowser.com. The detailed material composition of the targets is not accurately modelled.)

[11] Chu, R.-S., Jenn, D. C., and Wong, N. S., "Scattering From a Finite Phased Array of Dipoles Over a Finite Ground Plane," *1987 IEEE AP-S International Symposium Digest*, Vol. II, June 1987, pp. 722–726.
[12] Jenn, D. C., Fletcher, J. E., and Prata, A., "Radar Cross Section of Symmetric Parabolic Reflectors with Cavity-Backed Dipole Feeds," *IEEE Transactions on Antennas and Propagation*, Vol. 41, No. 7, July 1993, pp. 992–998.
[13] Jenn, D. C., and Flokas, V., "In-band Scattering from Arrays with Parallel Feed Networks," *IEEE Transactions on Antennas and Propagation*, Vol. 44, No. 2, 1996, pp. 172–178.
[14] Jenn, D. C., and Lee, S., "Inband Scattering from Arrays with Series Feed Networks," *IEEE Transactions on Antennas and Propagation*, Vol. 43, No. 8, 1995, pp. 867–873.
[15] Jenn, D. C., "A Complete Matrix Method for Antenna Analysis," *1989 IEEE AP-S International Symposium Digest*, June 1989, pp. 126–130.
[16] Cwik, T., Mittra, R., Lang, K. C., and Wu, T. K., "Frequency Selective Screens," *IEEE Antennas and Propagation Society Newsletter*, April 1987, pp. 6–10.
[17] Munk, B. A., *Frequency Selective Surfaces: Theory and Design*, Wiley, New York, 2000.
[18] Vardaxoglou, J. C., *Frequency Selective Surfaces: Analysis and Design*, Wiley, New York, 1997.
[19] Bhattacharyya, A. K., *Phased Array Antennas*, Wiley, Hoboken, NJ, 2006.
[20] Hannon, P. W., and Balfour, M. A., "Simulation of a Phased-Array Antenna in Waveguide," *IEEE Transactions on Antennas and Propagation*, Vol. 13, No. 3, May 1965, pp. 342–353.
[21] Pelton, E., and Munk, B., "A Streamlined Metallic Radome," *IEEE Transactions on Antennas and Propagation*, Vol. 22, No. 6, Nov. 1974, pp. 799–803.
[22] Lee, C.-S., and Lee, S.-W., "RCS of a Coated Circular Waveguide Terminated by a Perfect Conductor," *IEEE Transactions on Antennas and Propagation*, Vol. 35, No. 4, April 1987, pp. 391–398.
[23] Huang, C.-C., "Simple Formula for the RCS of a Finite Hollow Circular Cylinder," *Electronics Letters*, Vol. 19, No. 20, Sept. 1983, pp. 854–856.
[24] Crispin, J. W., and Maffett, A. L., "Estimating the Radar Cross Section of a Cavity," *IEEE Transactions on Aerospace and Electronic Systems*, Vol. AES-6, No. 5, Sept. 1970, pp. 672–674.
[25] Moll, J. W., and Seecamp, R. G., "Calculation of Radar Reflecting Properties of Jet Engine Intakes Using a Waveguide Model," *IEEE Transactions on Aerospace and Electronic Systems*, Vol. AES-6, No. 5, Sept. 1970, pp. 675–683.
[26] Mautz, J. R., and Harrington, R. F., "An Improved E-Field Solution for a Conduction Body of Revolution," Syracuse Univ., Syracuse, NY, TR-80-1, 1980.
[27] Ling, H., Chou, R.-C., and Lee, S.-W., "Shooting and Bouncing Rays: Calculating the RCS of an Arbitrarily Shaped Cavity," *IEEE Transactions on Antennas and Propagation*, Vol. AP-37, No. 2, Feb. 1989, pp. 194–205.
[28] Ruze, J., "Physical Limitations on Antennas," MIT Research Laboratory of Electronics, Cambridge, MA, Tech. Rept. 248, Oct. 1952.
[29] Ruze, J., "Antenna Tolerance Theory—A Review," *Proceedings of the IEEE*, Vol. 54, No. 4, April 1966, pp. 633–640.
[30] Copson, E. T., *Theory of Functions of a Complex Variable*, Oxford Univ. Press, New York, 1944.
[31] Steinberg, B. D., *Principles of Aperture and Array System Design: Including Random and Adaptive Arrays*, Wiley-Interscience, New York, 1976.
[32] Waddell, J. M., *Scattering from Rough Plates*, M.S. Thesis, Naval Postgraduate School, Monterey, CA, Sept. 1995.
[33] Senior, T. B. A., and Volakis, J. L., "Scattering by Gaps and Cracks," *IEEE Transactions on Antennas and Propagation*, Vol. AP-37, No. 6, June 1989, pp. 744–750.
[34] Mitzner, K. M., and Kaplin, K. J., "Frequency Dependence of Scattering from a Wedge with a Rounded Edge," *1986 National Radio Science Symposium Digest*, June 1986, p. 71.

[35] Schipper, T., Fortuny-Guasch, J., Tarchi, D., Reichardt, L., and Zwick, T., "RCS Measurements Results for Automotive Related Objects at 23-27 GHz," *Proc. of the 5th European Conf. on Antennas and Prop. (EUCAP)*, 11–15 April 2011, pp. 683–686.

[36] Belgiovane, D., Chen, C-C., Chien, S. Y.-P., and Sherony, R., "Surrogate Bicycle Design for Millimeter-Wave Automotive Radar Pre-Collision Testing," *IEEE Trans. on Intelligent Transportation Systems*, Vol. 18, No. 9, Sept. 2017, 2413–2422.

Problems

6.1 Derive the expression for the variance of the RCS given in Eq. (6.9).

6.2 Two isotropic scatterers are spaced a distance d along the z axis. Derive an expression for the RCS of this two-element array and find the beamwidth between first nulls as a function of d. Repeat the calculation for the case in which one of the two elements only scatters one-half of the incident signal.

6.3 A linearly polarized antenna has an effective height $\vec{h}$. Show that the monostatic antenna RCS is given by

$$\sigma = \lim_{r\to\infty} 4\pi r^2 \frac{|\vec{h}\cdot\vec{E}_s|^2}{|\vec{h}|^2|\vec{E}_i|^2}$$

6.4 The load impedance of a receiving antenna is switched from Z_{L1} to Z_{L2}. Show that the incremental change in monostatic RCS can be expressed as

$$\Delta\sigma = \frac{\lambda^2}{4\pi} G^2(\mathrm{PLF})^2|\Gamma_1 - \Gamma_2|^2$$

where G is the power gain of the scattering antenna in the direction of the transmitter and receiver, and PLF is the polarization loss factor (refer to Appendix C).

6.5 A three-element array with a series feed is shown in Fig P6.5. Assume that the elements are lossless isotropic sources with a reflection coefficient of r_r. The coupler scattering matrices all have the form

$$\mathbf{C}_n = \begin{bmatrix} 0 & t_n & jc_n & 0 \\ t_n & 0 & 0 & -jc_n \\ jc_n & 0 & 0 & t_n \\ 0 & -jc_n & t_n & 0 \end{bmatrix}$$

where t is the transmission coefficient.

a) Find the coupling coefficients c_n, $n = 1, 2, 3$ to provide equal power to each element when transmitting.

b) For a unit plane wave incident from an angle θ, find the total RCS of the aperture and feed by tracing signals. Neglect multiple reflections, and use r_L as the reflection coefficient of the loads. Also, neglect any signal that reflects more than twice from any combination of loads; that is, apply Eqs. (6.54–6.56) to a three-element array.

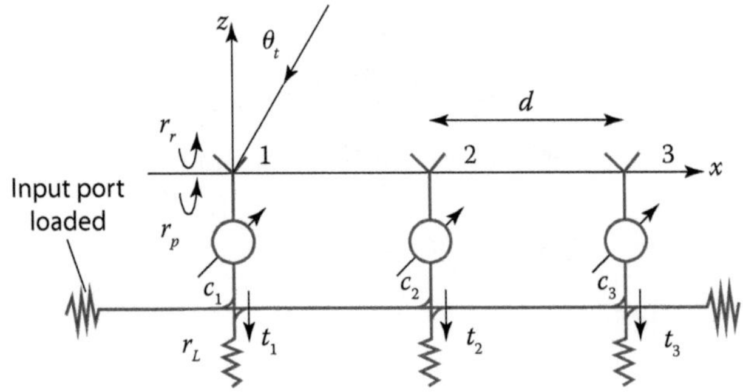

Fig. P6.5

6.6 A four-element array has a parallel feed, as shown in Fig P6.6. Each coupling device is a magic tee with the fourth port loaded. The reflection coefficients are r_r, r_p, r_Σ, and r_Δ for the radiating elements, phase shifters, and coupler sum and difference ports, respectively. Derive expressions for the in-band RCS of the antenna. Neglect multiple reflections; that is, apply Eq. (6.50).

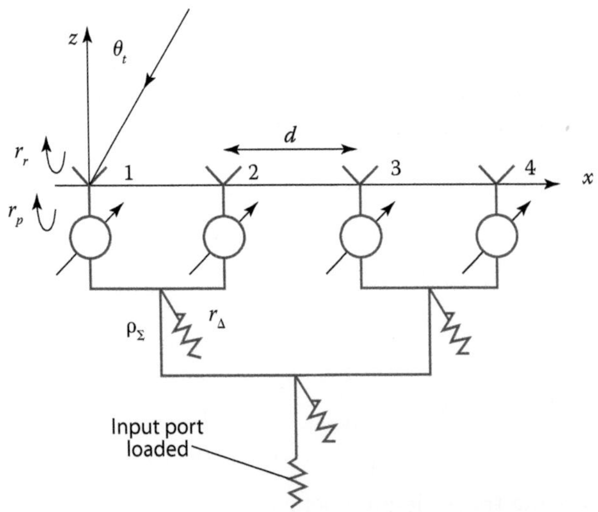

Fig. P6.6

6.7 A 2 × 2 array of isotropic elements is fed by a corporate feed, as shown in Fig P6.7. The reflection coefficients are r_r, r_p, r_Σ, and $r\Delta$ for the radiating elements, phase shifters, and coupler sum and difference ports, respectively. Derive expressions for the in-band RCS of the antenna. Neglect multiple reflections.

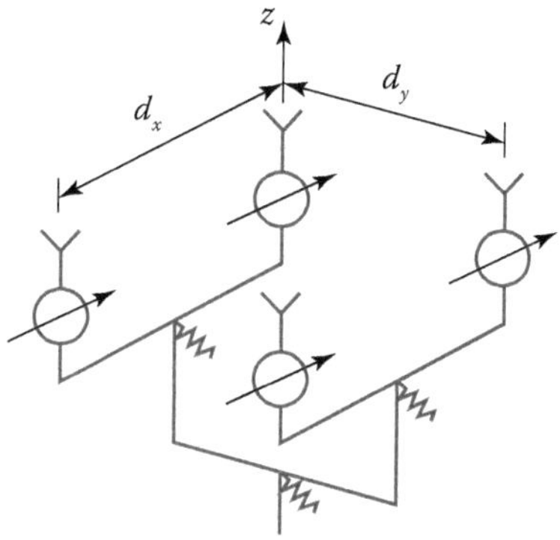

Fig. P6.7

6.8 An array of N elements is fed by the series feed network. Using the model described in Fig 6.31, derive expressions for the forward, backward, and self-beams given in Eqs. (6.58–6.60). Neglect reflections at transmission line and device junctions (i.e., consider only reflections from loads), and also neglect multiple reflections. Assume an arbitrary aperture distribution with coupling coefficients given by c_n, $n = 1, 2, \ldots, N$. All line lengths from the main line to the elements are equal, and the interelement line length is L ($\psi_0 = kL$).

6.9 A *Luneburg lens* is a solid inhomogeneous dielectric sphere with a radially symmetric dielectric constant:

$$\epsilon(r) = 2 - (a/r)^2$$

where a is the radius and r the distance from the center of the lens, as shown in Fig. P6.9. If a reflecting plate is placed behind the focal point,

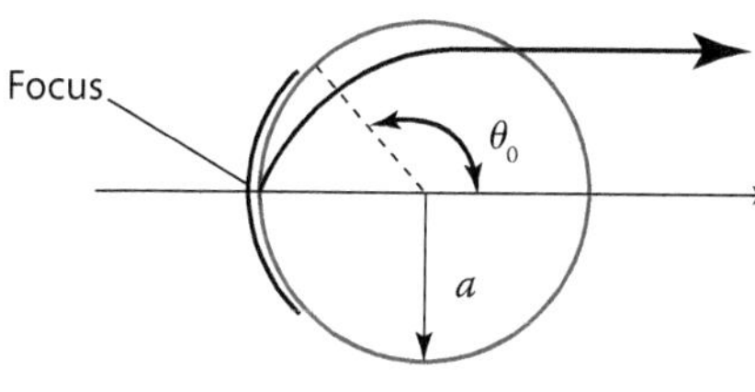

Fig. P6.9

all the power collected by the lens is reradiated. Using the same approach as in Example 1.1, find an expression for the maximum RCS. If the reflecting plate subtends a half angle $\pi - \theta_0$, sketch the RCS as a function of θ.

6.10 A 10-ft × 10-ft plate lies in the $x-y$ plane and is composed of 1-ft × 10-ft panels. The short sides of the panels are aligned with the x axis. If every other panel is displaced by a constant amount from the $z = 0$ plane, at what angles will lobes occur in the $\phi = 0$ deg plane? Verify your result using the data in Fig. 6.60.

6.11 An array lies in the $x-y$ plane and occupies a 10-ft × 10-ft area. The array elements are laid out in a square grid of 0.5λ on a side, and the array operates at 4 GHz.

a) At a threat frequency equal to the operating frequency, what tolerances must be maintained to keep the average RCS less than −40 dBsm?

b) If the values obtained in part a are used, what will thc average level be at a threat frequency of 10 GHz?

6.12 Derive the feed contribution to the space-fed array in Fig. 6.34, Eq. (6.65). Assume that the feed horn has a reflection coefficient r_f and gain G_f. The feed horn radiates in three-dimensional space, and all the lens elements are isotropic scatterers with an effective area $A_e \approx d\ell$ (ℓ is the element length in the y direction, and d the element spacing). Show that the maximum value of σ_f is

$$\sigma_f \approx t_r^8 t_p^4 r_f^2 \frac{4\pi A^2}{\lambda^2} \left[\frac{A_f^2}{\mathrm{FOD}_x \mathrm{FOD}_y} \right]$$

6.13 Compute the RCS of the rectangular cavity of Example 6.14 for a monostatic angle of −45deg ($\phi = 0$). Use the SBR method with four rays.

6.14 Compute the approximate RCS of the cavity in Example 6.14 and Problem 6.13 by considering the cavity opening to be a conducting ground plane of the same area.

6.15 The isotropic scattering elements of a 500 × 500 array are spaced 0.25λ apart on a square grid in the $x-y$ plane. The RCS of the array must satisfy $\sigma \leq -40$ dBsm for $\theta > 40$ deg. The threat frequency is 1 GHz.

a) Determine the envelope of the RCS vs the angle for the array. Use the fact that, because the spacing is small and there are no Bragg lobes, the array factor can be approximated by a sinc function.

b) Use a linear taper to reduce the wide-angle RCS lobes. The taper region consists of adding an additional 50-element strip around the perimeter of the array (i.e., the array dimensions are now 550 × 550, with a linear taper extending over the 50 elements closest to the edges). Find the rate of falloff of the sidelobe envelope for this tapered array.

c) What is the maximum allowable variance in the surface error $\overline{\Delta^2}$ to achieve the desired RCS levels?

6.16 Compute the monostatic RCS of a cylindrical paraboloid (i.e., a surface generated by translating a parabola along the y axis). Let $f/D = 0.4$ and $D = 10\lambda$. The length of the antenna in the y direction is 20λ and the frequency is 3 GHz. Approximate the parabolic shape by eight plates of equal size and use PO on each plate. (In other words, the paraboloid is represented by an array of identical plates tangent to the parabola at their centers.)

6.17 Starting with the antenna mode term of the basic equation of antenna scattering [square bracket term of Eq. (6.17)], derive Eq. (6.24).

6.18 For a series feed, the backward waves from all elements sum at the input port, yielding a total field $E_{b_{\text{in}}}$. If a reflection occurs at the receiver with coefficient r_{in}, it appears as if the signal $r_{\text{in}}E_{b_{\text{in}}}$ is injected into the antenna in the transmit mode. Show that the scattered field due to the reflection at the input port is given by

$$E_{\text{in}} \sim r_{\text{in}} t_r^2 t_p^2 \left\{ \sum_{n=1}^{N} c_n e^{j(n-1)(\zeta+\psi)} \left(\prod_{m=1}^{n-1} t_m \right) \right\}^2$$

With this term included, Eq. (6.61) becomes

$$|E_s^s|^2 \approx |E_f|^2 + |E_b|^2 + |E_{\text{in}}|^2 + |E_{\text{self}}|^2$$

6.19 A unit amplitude plane wave is normally incident from the -x direction on the thin half-wave dipole antenna shown in Fig. P6.19. Assume that the total wire length is $\ell = 0.5$ m and the frequency is 300 MHz.

a) What is the radiation resistance of the dipole (assume resonance and no ohmic loss)?

b) What is the effective area of the dipole?

c) What is the effective height of the dipole?

d) Calculate the dipole's radiation mode RCS if a conjugate matched load is connected across the terminals. (Assume that the dipole is at resonance and its impedance is purely real.)

e) If the terminals are shorted, calculate the dipole's radiation mode RCS using Equation (6.17). (Observe that the RCS is the directivity times the effective area.)

f) Estimate the structural mode RCS from Fig. 6.12. Note that for the data in the figure ℓ is the half length.

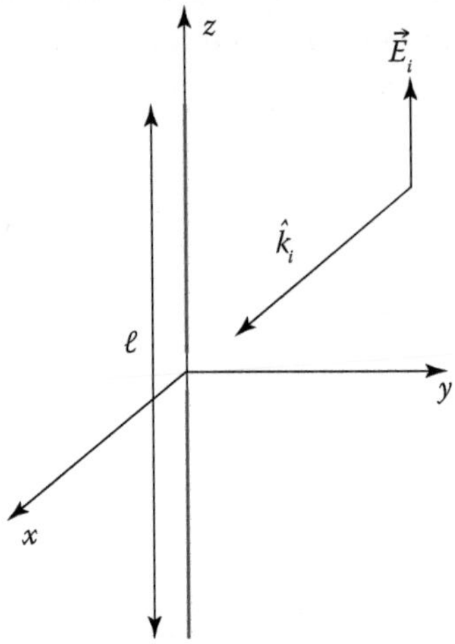

Fig. P6.19

6.20 What is the maximum rms surface error allowable if the average RCS of a plate with area A and correlation interval $c = 1\lambda$ cannot exceed -60 dB relative to the peak RCS? Calculate the allowable values for areas of 1 m^2 and 10 m^2.

6.21 Using commercial software such as CST Microwave Studio® or Altair FEKO® compute the RCS of a tuned dipole with several load conditions. Typical values for a dipole tuned at 300 MHz are length = 0.44 m, radius = 0.005 m, and feed gap = 0.01 m. This should give an input impedance of approximately $Z_{in} = 75 + j0\ \Omega$. Run RCS simulations with terminal loads of $Z_L = 0$, 200, and $10^4\ \Omega$.

Chapter 7 Radar Cross Section Reduction

7.1 Introduction

This chapter examines methods of controlling RCS and the tradeoffs involved in implementing these methods. Radar cross section reduction techniques generally fall into one of four categories: 1) target shaping, 2) materials selection and coatings, 3) passive cancellation, and 4) active cancellation.

Application of each of these methods involves a compromise in performance in other areas. For instance, there are limitations to modification of an aircraft's shape from the aerodynamic optimum. Sharply angled facets may be desirable from an RCS perspective, but they degrade the aircraft's maneuverability and handling characteristics. Until recently, reduction methods also tended to be narrowband and effective only over limited spatial regions. They must be chosen based on the platform's missions and expected threats. Reduction methods are applied to maintain the RCS below a specified threshold level over a range of frequencies and angles.

The next several sections examine the advantages and disadvantages of each of the four methods. The final section discusses RCS synthesis and inverse scattering; that is, given a target's scattered field, what must its composition be to achieve the specified RCS? The prescription for this problem is the RCS designer's dream. In principle, a solution to this problem can be formulated in terms of integral or differential equations. Unfortunately, solving the integral equations is practical in only the simplest cases.

7.2 Target Shaping

7.2.1 Shaping Philosophy

Shaping is considered by many to be the first line of RCS control. It refers to the tilting and contouring of surfaces to direct scattered energy away from high-priority "quiet zones." This leads to the unconventional platform shapes that have become associated with the term stealth. Shaping is a wideband approach to RCS reduction, in that an angle of reflection is independent of frequency (i.e., Snell's law is satisfied at all frequencies). However, in practice it is usually applied to electrically large, flat surfaces for the primary purpose of preventing the specular "flash" from hitting the radar. In this case, it is best to keep large surfaces as flat and smooth as possible so that the specular flash is confined to a very narrow angular region.

For angles out of the RCS mainlobe, the effectiveness of shaping diminishes. This point is illustrated in Fig. 7.1, which plots the principal plane RCS of plates of various lengths, all with a constant width of 10 m. Equation (2.49) is used, and the frequency is 300 MHz. Note that at wide angles well into the "skirt" of the sinc function, the RCS lobes are essentially the same level for all plate lengths. Physical optics was used to generate these patterns and, therefore, no surface waves are included.

Obviously, to present the smallest possible RCS, a plate should be oriented so that it is viewed off of the principal planes to avoid the locations of high RCS. The principal planes are determined by the orientation of the plate edges; the principal planes are transverse to the edges of a flat plate. This is illustrated in Fig. 7.2 for square and diamond-shaped plates. The plate shape controls not only the principal planes but also the characteristics of its traveling wave and edge-diffracted radiation. Therefore, it is necessary to choose edge alignments that simultaneously direct traveling waves and diffracted waves as well as specular RCS.

Multiple reflections are also controlled or eliminated by proper shaping. A 90-deg corner between two flat surfaces (a *dihedral*) results in a strong multiple reflection, and therefore 90-deg angles are to be avoided. Figure 7.3 shows that by simply changing the corner angle, the RCS in the angular region between the two plate surface normals can be reduced dramatically.

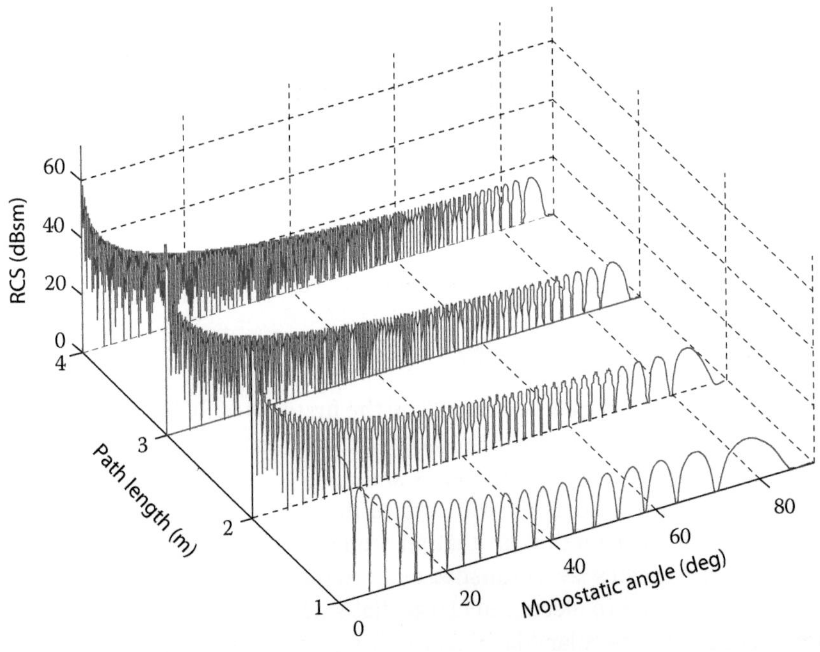

Fig. 7.1 RCS of plates of constant width ($b = 10$ m) vs length a ($\phi = 0$ deg cut).

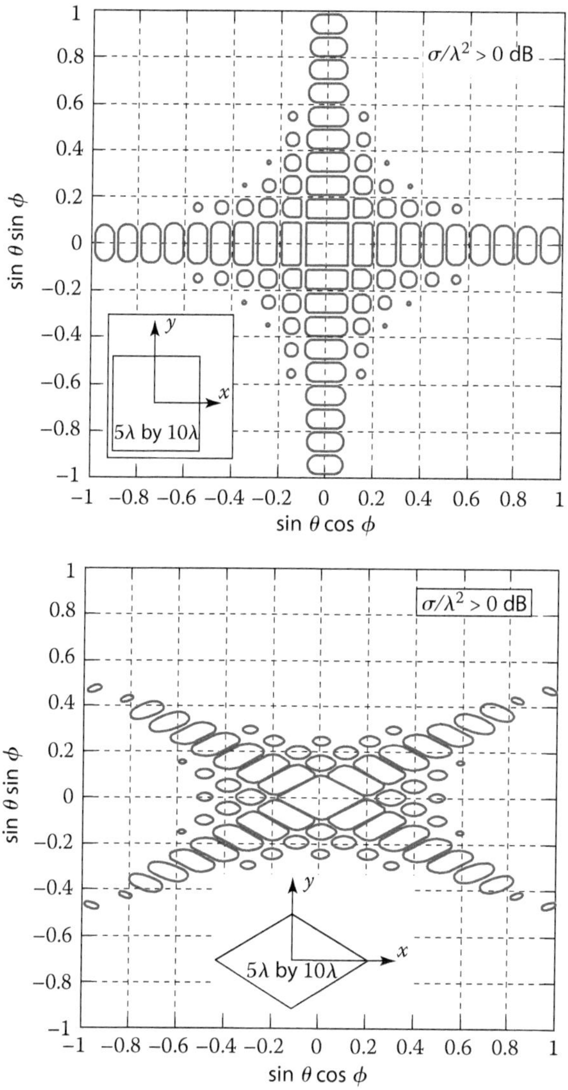

Fig. 7.2 Radar cross section contours for a square plate (top) and diamond-shaped plate (bottom) having the same area. The PO approximation has been used.

The two plates are 0.5 m by 1 m and connected at the short edge. The frequency is 5 GHz.

7.2.2 Shaping Examples

All modern military air and surface platforms incorporate some aspect of shaping. Obvious examples are the now retired F-117A fighter [1] and the

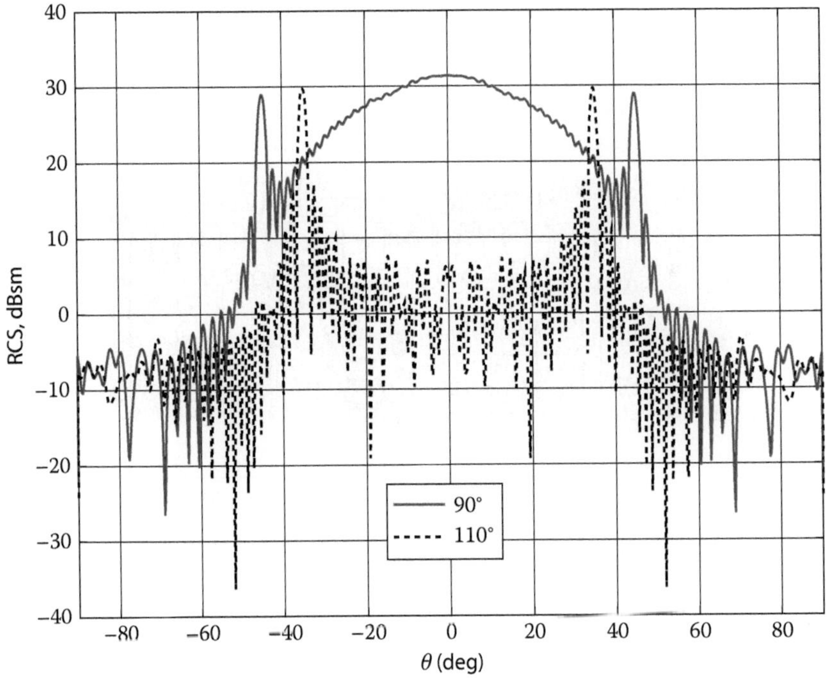

Fig. 7.3 Corner reflector RCS for corner angles of 90 deg and 110 deg (0.5 m by 1.0 m plates, $f = 5$ GHz).

B-2 bomber. Neither aircraft has the standard vertical and horizontal tail surfaces. In plan view, it is clear that edges lie at a few common angles. Basic shaping concepts have continued to be applied to recent generations of fighter aircraft such as the F-22 (Fig. 7.4) and Joint Strike Fighter (JSF). Unmanned combat air vehicles (UCAVs), like the X-47B shown in Fig. 7.5, are also taking on stealthy forms.

Shaping principles have been applied to ships as well. Two examples are shown in Fig. 7.6. The United States Navy's *Sea Shadow* demonstrated the feasibility of shaping for ocean going vessels. Stealth and signature control are sure to be high priorities in the design of the next generation of warships as evidenced by the U.S. Navy's *Zumwalt* class destroyer. Examples of currently deployed stealth-inspired ships include the Swedish *Visby* and the French *LaFayette*.

7.3 Materials Selection

7.3.1 Introduction

The second method of RCS reduction is the use of radar-absorbing materials (RAM). Although the term RAM suggests the selection of materials

Fig. 7.4 Lockheed Martin F-22 (U.S. Air Force photo).

that absorb and attenuate waves, it actually has come to encompass a wide range of techniques and materials, including coatings, *composites*, and most recently, *artificial materials*. These will be described in later sections.

At radar frequencies, there are two primary approaches to reducing reflections from a structure. The first is *absorption*, which refers to the transfer of energy from the wave to the material as it passes through. The physical mechanisms of absorption include *ohmic loss*, *dielectric loss*, and *magnetic loss*. Ohmic loss is due to the motion of free charges (usually electrons) in an imperfect conductor with conductivity σ_c. Dielectric and magnetic losses are due to the imaginary parts of the complex relative permittivity

Fig. 7.5 X-47B (U.S. Navy photo).

Fig. 7.6 Examples of shaping applied to ships: (a) Sea Shadow and (b) the DDG-1000 *Zumwalt* destroyer (U.S. Navy photos).

$\epsilon_r' - j\epsilon_r''$ and permeability $\mu_r' - j\mu_r''$, respectively. In place of ϵ_r'' and μ_r'', the loss can be specified by giving the material's electric and magnetic *loss tangents*

$$\begin{aligned} \tan \delta_\epsilon &= \epsilon_r''/\epsilon_r' \\ \tan \delta_\mu &= \mu_r''/\mu_r' \end{aligned} \tag{7.1}$$

Second, coatings may also be used to reduce radar cross section by cancellation of multiple reflections, or *destructive interference.* The coating properties are selected so that reflections from the front and back faces of a layer cancel. This type of coating is also referred to as a *resonant absorber* because cancellation is a narrowband technique that relies on the layer being the proper thickness in wavelengths, so that the necessary round-trip phase is introduced into the back face reflection. Materials that attenuate waves via the dielectric and magnetic loss mechanisms can generally be made to operate over a wider frequency range than resonant absorbers and thus are categorized as *broadband absorbers.*

7.3.2 Interaction of Materials and Waves

The dielectric and magnetic losses can be explained using the simple *Debye model* [2–5]. Atoms and molecules are represented by nuclei (with positive charge centers) and electron clouds, which have circular orbits about negative charge centers. In the absence of an external field, the positive and negative charge centers are coincident. When an external field is applied, the positive and negative charge centers separate. The separation of charge induces a polarization electric field. The response of the molecule is expressed in terms of a polarization vector, $\vec{P}(t)$. If the external field is time harmonic, then each molecule of material is essentially an oscillating dipole, as depicted in Fig. 7.7.

The separation is referred to as *electronic* polarization and χ_e is the *electric susceptibility*. The permittivity is defined in terms of this susceptibility by

$$\vec{D} = \epsilon\vec{E} = \epsilon_o\epsilon_r\vec{E} = \epsilon_0(1 + \chi_e)\vec{E}$$

It takes time for the molecules to respond to the impressed field. The time-dependent form of the polarization is

$$P(t) = \underbrace{\epsilon_o\chi_e(0)E_{\text{ext}}}_{P_o}\, e^{-t/\tau}$$

where τ is the relaxation constant (having a value of about 10^{-15} s).

The Debye model is never seen in real materials, but it can be approached for single particle noninteracting systems like gases. The assumptions are that all of the dipoles are identical and independent, and relaxation times are the same. In fact, dipoles are spatially and temporally coupled, relaxation times vary, and other types of polarization exist. They include *ionic polarization,* due to the displacement of positive and negative ions of a molecule (with a relaxation constant of about 10^{-13} s), and *orientational polarization,* which arises from the rotation of the molecule with a permanent dipole moment (with a relaxation constant of about 10^{-11} s).

Typical behavior of the dielectric constant with frequency is shown in Fig. 7.8. The relationship between the real and imaginary parts is not independent, but is governed by the *Kramers–Kronig* formula [3].

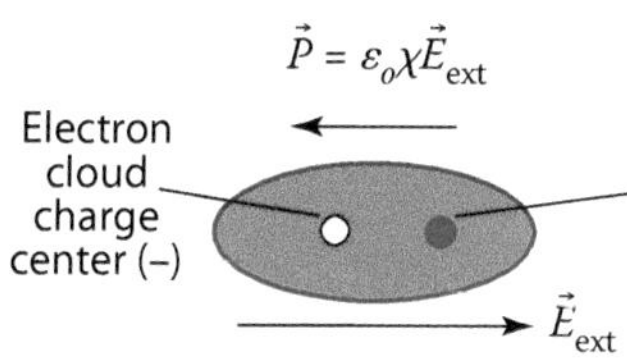

Fig. 7.7 Electronic polarization.

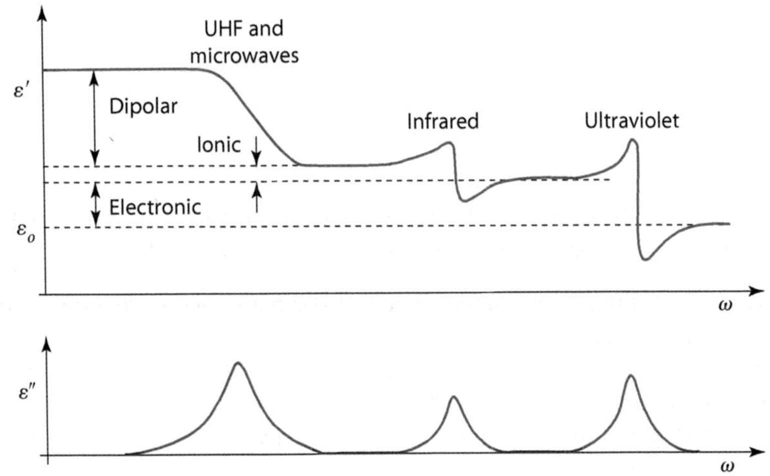

Fig. 7.8 Behavior of the real and imaginary parts of the permittivity of a typical material.

At characteristic (resonant) frequencies there is a rapid change in ϵ_r' and a sharp peak in ϵ_r''. High peaks in ϵ_r' correspond to frequencies where the material is a strong absorber.

To explain magnetic losses, the orbiting electrons can be modeled as small current loops, each with a *magnetic moment,* $\vec{m}$ [4]. The magnetic moment is caused by 1) electron orbit (electron orbiting the nucleus), 2) electron spin (electron spinning about its axis), and 3) nuclear spin (nucleus spinning about its axis). As illustrated in Fig. 7.9, an external magnetic field, $\vec{B}_{\text{ext}}$, puts a torque on the atomic loops causing the dipoles to align with or against the external field. The magnetization $\vec{M}$ is a measure of how the external magnetic field aligns the internal dipole moments. In a linear medium $\vec{M} = \chi_m \vec{H}$ where χ_m is the *magnetic susceptibility.* The permeability is defined in terms of the susceptibility

$$\vec{B} = \mu\vec{H} = \mu_o\mu_r\vec{H} = \mu_o(1 + \chi_m)\vec{H} \quad (7.2)$$

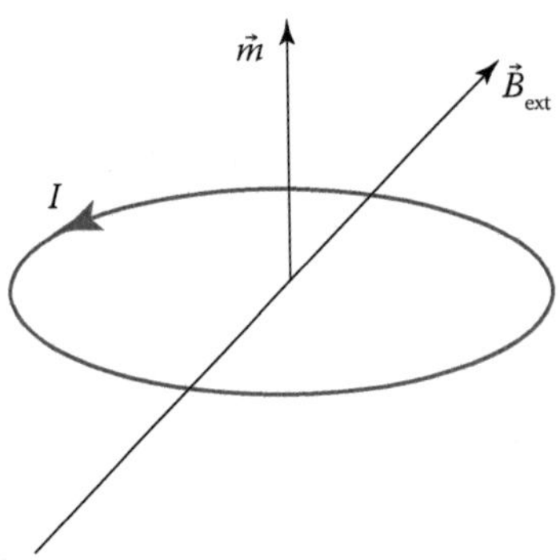

Fig. 7.9 An orbiting electron in an external magnetic field can be modeled as an equivalent current loop.

A broad classification of the magnetic properties of materials is possible based on the range of χ_m: 1) *diamagnetic,* small negative χ_m; 2) *paramagnetic,* small positive χ_m; and 3) *ferromagnetic,* large positive χ_m.

Most natural materials have a very weak magnetization and can be considered nonmagnetic (thus $\mu_r = 1$). Exceptions are materials such as iron, which have a very strong magnetization and exhibit hysteresis [4]. Exposing iron to a very high external magnetic field under the right conditions can cause a permanent magnetization that remains even after the external field is removed.

7.3.3 General Constitutive Parameters

Materials can be categorized with regard to several aspects of their behavior through their constitutive parameters, σ, μ, and ϵ.

1. Linear or nonlinear: for a linear medium the parameters are independent of the field strengths.
2. Conducting or nonconducting: conducting materials generally have a high conductivity, σ_C.
3. Dispersive or nondispersive: for a nondispersive material the parameters are independent of frequency.
4. Homogeneous or inhomogeneous: the parameters of a homogeneous medium are independent of position in the medium.
5. Isotropic, anisotropic, or *bianisotropic**: an isotropic medium has no preferred directionality; the field vectors are parallel to the flux density vectors.

A generalized representation of matrix constitutive relations that covers all of these cases is the *Tellegen representation* [6,7]. The fields are related to the flux densities by

$$\begin{bmatrix} \mathbf{D} \\ \mathbf{B} \end{bmatrix} = \begin{bmatrix} \epsilon & \xi \\ \zeta & \mu \end{bmatrix} \begin{bmatrix} \mathbf{E} \\ \mathbf{H} \end{bmatrix} \tag{7.3}$$

For example, in the Cartesian system the vectors and matrices, respectively, have the forms

$$\mathbf{D} = \begin{bmatrix} D_x \\ D_y \\ D_z \end{bmatrix} \quad \text{and} \quad \epsilon = \begin{bmatrix} \epsilon_{xx} & \epsilon_{xy} & \epsilon_{xz} \\ \epsilon_{yx} & \epsilon_{yy} & \epsilon_{yz} \\ \epsilon_{zx} & \epsilon_{zy} & \epsilon_{zz} \end{bmatrix} \tag{7.4}$$

In the time-harmonic case (i.e., phasor representation) these quantities are complex and may be frequency dependent.

The most naturally occurring materials are isotropic or perhaps have limited anisotropy (for example, some crystals in one or two dimensions). Almost any medium that is in motion is bianisotropic, and therefore much of the past research has dealt with wave propagation in uniformly moving media [7]. Examples are plasmas created by hot jet exhaust or weapon explosions. However, in recent years artificial materials have been constructed with complex behaviors, some even with negative permittivity and permeability.

---

*The prefix *bi* means D depends on the two fields E and H (and similar for B). *Anisotropic* signifies that D is not parallel to E, and B is not parallel to H.

Because E and B are the fundamental quantities (as illustrated by duality), often the *Boys–Post representation* is more useful

$$\begin{bmatrix} D \\ H \end{bmatrix} = \begin{bmatrix} \epsilon_p & \alpha_p \\ \beta_p & \mu_p^{-1} \end{bmatrix} \begin{bmatrix} E \\ B \end{bmatrix} \tag{7.5}$$

These new constitutive parameters (with subscript p) are related to the original ones as follows:

$$\boldsymbol{\epsilon} = \boldsymbol{\epsilon}_p - \boldsymbol{\alpha}_p \boldsymbol{\mu}_p \boldsymbol{\beta}_p, \quad \boldsymbol{\mu} = \boldsymbol{\mu}_p, \quad \boldsymbol{\xi} = \boldsymbol{\alpha}_p \boldsymbol{\mu}_p, \quad \text{and} \quad \boldsymbol{\zeta} = -\boldsymbol{\mu}_p \boldsymbol{\beta}_p \tag{7.6}$$

Maxwell's equations require that the constitutive parameters satisfy the following relationship [6]:

$$\text{Trace}(\boldsymbol{\xi}\boldsymbol{\mu}^{-1} + \boldsymbol{\mu}^{-1}\boldsymbol{\zeta}) = 0 \tag{7.7}$$

With regard to the matrix notation for the constitutive parameters, an *isotropic* medium has scalar permittivity and permeability. In a *homogeneous* medium, the parameters are independent of position, and in an *anisotropic* medium, either or both the permittivity and permeability can be a 3-by-3 matrix. If all four submatrices are diagonal (i.e., reduce to scalars), the medium is *bi-isotropic.*

Three further material classifications are now defined.

1. A *simple* medium is linear, isotropic, and homogeneous. Therefore

$$\begin{bmatrix} D \\ B \end{bmatrix} = \begin{bmatrix} \boldsymbol{\epsilon} & \mathbf{0} \\ \mathbf{0} & \boldsymbol{\mu} \end{bmatrix} \begin{bmatrix} E \\ H \end{bmatrix} \tag{7.8}$$

where $\mathbf{0}$ is a 3-by-3 matrix of zeros and

$$\boldsymbol{\epsilon} = \begin{bmatrix} \epsilon_r & 0 & 0 \\ 0 & \epsilon_r & 0 \\ 0 & 0 & \epsilon_r \end{bmatrix} \epsilon_o \quad \text{and} \quad \boldsymbol{\mu} = \begin{bmatrix} \mu_r & 0 & 0 \\ 0 & \mu_r & 0 \\ 0 & 0 & \mu_r \end{bmatrix} \mu_o \tag{7.9}$$

2. A *uniaxial* medium has two of the three diagonal elements the same (e.g., $\epsilon_{xx} = \epsilon_{yy}$); for example:

$$\boldsymbol{\epsilon} = \begin{bmatrix} \epsilon_{xx} & 0 & 0 \\ 0 & \epsilon_{xx} & 0 \\ 0 & 0 & \epsilon_{xx} \end{bmatrix} \epsilon_o \tag{7.10}$$

Media with three different diagonal elements are called *biaxial.*

3. For a *bi-isotropic* or *chiral* medium [8,9], the 3-by-3 matrices are diagonal. Hence the constitutive relations can be written with scalars

$$\begin{bmatrix} D \\ B \end{bmatrix} = \begin{bmatrix} \epsilon & \xi \\ \zeta & \mu \end{bmatrix} \begin{bmatrix} E \\ H \end{bmatrix} \rightarrow \begin{array}{l} \vec{D} = \epsilon\vec{E} + \xi\vec{H} \\ \vec{B} = \zeta\vec{E} + \mu\vec{H} \end{array} \tag{7.11}$$

Furthermore, the elements of the off-diagonal blocks can be expressed as $\xi = \chi - j\kappa$ and $\zeta = \chi + j\kappa$. The quantity κ is the *chirality parameter*, and it measures the degree of "handedness" of the medium (κ is real for a lossless medium); χ is the *magneto-dielectric parameter*, and if $\chi \neq 0$ the medium is *nonreciprocal*. In a nonreciprocal medium, a permanent electric dipole is tied to a permanent magnetic dipole by a nonelectromagnetic force. In the Russiacn literature the term *gyrotropy*, referring to the *gyromagnetic* characteristics of the medium, is often used instead of chirality.

There are several important issues in dealing with EM waves interacting with any material. The fundamental ones are the symmetry conditions that must be satisfied by the constitutive relations; the behavior of propagating waves in the medium; time reversal and spatial inversion; and the applicability of reciprocity, image theory, and duality.

The wave equation determines the characteristics of propagating waves in a medium. To take a basic case, assume that the medium is nonmagnetic ($\mu = \mu_o$) and that both E and H vary as exp $(-j\vec{k} \cdot \vec{r})$. Therefore, the curl operator $\nabla\times$ reduces to $-j\vec{k}\times$, and Maxwell's first and second equations become

$$\vec{k} \times \vec{E} = \omega\mu_o\vec{H} \tag{7.12}$$

$$\vec{k} \times \vec{H} = -\omega\vec{D} \tag{7.13}$$

Now the wave equation can be written as

$$\vec{k} \times \vec{k} \times \vec{E} + \omega^2\mu_o\vec{D} = 0 \tag{7.14}$$

Equation (7.14) is equivalent to three equations in Cartesian coordinates that can be cast into matrix form:

$$\begin{bmatrix} k_x^2 - k^2 + \omega^2\mu_o\epsilon_{xx} & k_xk_y + \omega\mu_o\epsilon_{xy} & k_xk_z + \omega\mu_o\epsilon_{xz} \\ k_xk_y + \omega\mu_o\epsilon_{yx} & k_y^2 - k^2 + \omega^2\mu_o\epsilon_{yy} & k_zk_y + \omega\mu_o\epsilon_{yz} \\ k_xk_z + \omega\mu_o\epsilon_{zx} & k_zk_y + \omega\mu_o\epsilon_{zy} & k_z^2 - k^2 + \omega^2\mu_o\epsilon_{zz} \end{bmatrix} \begin{bmatrix} E_x \\ E_y \\ E_z \end{bmatrix} = 0 \tag{7.15}$$

This system must be solved for k_x, k_y, and k_z in terms of ω. The resulting relationship is called the *dispersion equation*.

If a medium is lossless, not all of the elements of ϵ are independent. To investigate the properties of a lossless anisotropic dielectric, the Poynting theorem must be examined. It is derived in the same manner as for the isotropic case (see Problem 7.10) and can be written

$$\nabla \cdot (\vec{E} \times \vec{H}^*) = j\omega\mu_o|\vec{H}|^2 - j\omega\vec{D}^* \cdot \vec{E} \tag{7.16}$$

where

$$\vec{D}^* \cdot \vec{E} = E_x(\epsilon_{xx}E_x + \epsilon_{xy}E_y + \epsilon_{xz}E_z)^* + E_y(\epsilon_{yx}E_x + \epsilon_{yy}E_y + \epsilon_{yz}E_z)^* + E_z(\epsilon_{zx}E_x + \epsilon_{zy}E_y + \epsilon_{zz}E_z)^* \tag{7.17}$$

For a lossless medium and real ω

$$\mathrm{Re}\{j\omega\vec{D}^* \cdot \vec{E}\} = \mathrm{Im}\{\vec{D}^* \cdot \vec{E}\} \tag{7.18}$$

From this condition, it is found that ϵ_{xx}, ϵ_{yy}, and ϵ_{zz} are real and that $\epsilon_{xy} = \epsilon_{yx}^*$, $\epsilon_{xz} = \epsilon_{zx}^*$, and $\epsilon_{yz} = \epsilon_{zy}^*$.

Example 7.1: Uniaxial Medium

A uniaxial medium has the permittivity matrix

$$\boldsymbol{\epsilon} = \begin{bmatrix} \epsilon_{xx} & 0 & 0 \\ 0 & \epsilon_{xx} & 0 \\ 0 & 0 & \epsilon_{zz} \end{bmatrix}$$

Assume that a wave is propagating in the x direction, $\vec{k} = k_x\hat{x}$. The wave equation simplifies to

$$\begin{bmatrix} \omega^2\mu_0\epsilon_{xx} & 0 & 0 \\ 0 & \omega^2\mu_0\epsilon_{yy} - k_x^2 & 0 \\ 0 & 0 & \omega^2\mu_0\epsilon_{zz} - k_x^2 \end{bmatrix} \begin{bmatrix} E_x \\ E_y \\ E_z \end{bmatrix} = 0$$

which can be separated into three equations. The first gives

$$\omega^2\mu_0\epsilon_{xx}E_x = 0 \Rightarrow E_x = 0$$

which is the requirement that the field components be transverse to the direction of propagation. The remaining two equations give

$$(\omega^2\mu_0\epsilon_{xx} - k_x^2)E_y = 0 \Rightarrow k_{x_1} = \omega\sqrt{\mu_0\epsilon_{xx}}$$

and

$$(\omega^2\mu_0\epsilon_{zz} - k_x^2)E_z = 0 \Rightarrow k_{x_2} = \omega\sqrt{\mu_0\epsilon_{zz}}$$

Thus, the allowable values of k_x have been determined.

Figure 7.8 indicates that at frequencies just past a resonance, the real part of the dielectric constant decreases rapidly. Is it possible to force ϵ_r' negative, and if so, what does it mean to have negative permeability or permittivity? Mathematically the requirement is that the induced polarization and magnetization vectors must be anti-parallel (opposite) to the original definitions and the susceptibilities must be sufficiently large to drive the permittivity and permeability negative.

Let us examine a plane wave propagating in an isotropic medium (ϵ_r, μ_r are scalars):

$$\vec{E}(z) = \hat{x}E_o e^{-\gamma z} \tag{7.19}$$

where

$$\begin{aligned}\gamma &= j\omega\sqrt{\mu\epsilon} = j\omega\sqrt{\mu_o\mu_r\epsilon_o\epsilon_r} \\ &= jk_o\sqrt{\mu_r\epsilon_r} = jk_o\sqrt{(\mu_c' - j\mu_r'')(\epsilon_c' - j\epsilon_r'')} \equiv \alpha + j\beta\end{aligned} \tag{7.20}$$

Materials with $\epsilon_r > 0$ and $\mu_r > 0$ are referred to as *right-handed (RH) materials* because the direction of power flow is according to the right-hand rule: $\vec{W} = \vec{E} \times \vec{H}$ (W/m^2). The propagation vector is also in the direction of $\vec{W}$:

$$\vec{k} = \hat{k}k_o\sqrt{\mu_r\epsilon_r} \tag{7.21}$$

If it were possible to have both negative μ_r and ϵ_r [*double negative,* (DNG)], then the direction of propagation would be given by

$$\vec{k} = -\hat{k}k_o\sqrt{|\mu_r||\epsilon_r|} \tag{7.22}$$

which is a left-hand rule. Therefore, this is called a *left-handed (LH) material.*† The "handedness" parameter p of a medium is given by the determinant

$$p = \begin{vmatrix} \hat{x}\cdot\hat{e} & \hat{y}\cdot\hat{e} & \hat{z}\cdot\hat{e} \\ \hat{x}\cdot\hat{h} & \hat{y}\cdot\hat{h} & \hat{z}\cdot\hat{h} \\ \hat{x}\cdot\hat{k} & \hat{y}\cdot\hat{k} & \hat{z}\cdot\hat{k} \end{vmatrix} = \begin{cases} +1 & \text{for RH materials} \\ -1 & \text{for LH materials} \end{cases} \tag{7.23}$$

where $\hat{e} = \vec{E}/|\vec{E}|$ and $\hat{h} = \vec{H}/|\vec{H}|$ are unit vectors in the directions of the fields.

The implications of a LH material are as follows [10–12]:

1. $\vec{W}$ and $\vec{k}$ are in opposite directions.
2. The group velocity is opposite the phase velocity.
3. The index of refraction $n = \sqrt{\mu_r\epsilon_r}$ is negative; the impedance $\eta = \eta_o\sqrt{\mu_r/\epsilon_r}$ is positive.
4. The Doppler shift is reversed (an approaching source has a negative Doppler shift).
5. Snell's law must be amended:

$$\frac{\sin\theta_i}{\sin\theta_t} = \frac{p_2}{p_1}\left|\sqrt{\frac{\mu_{r_2}\epsilon_{r_2}}{\mu_{r_1}\epsilon_{r_1}}}\right| \tag{7.24}$$

6. Convex and concave lenses change roles when rays impinge from infinity.

†They have also been referred to as *metamaterials* and *backward wave materials.*

Example 7.2: Plane Wave Incident on a DNG Medium

Consider a plane wave incident on a plane interface between two media where $\mu_{r1}, \epsilon_{r1} > 0$, $\epsilon_{r2} = -\epsilon_{r1}$, and $\mu_{r2} = -\mu_{r1}$ (i.e., medium 2 is DNG). The handedness parameters are

$$p_1 = -p_2 = \begin{vmatrix} \cos\theta_i & 0 & -\sin\theta_i \\ 0 & 1 & 0 \\ -\sin\theta_i & 0 & \cos\theta_i \end{vmatrix} = 1$$

Snell's law gives $\sin\theta_t = -\sin\theta_i$, with the negative sign signifying that the transmitted ray is on the opposite side of the normal. The reflection coefficients at the boundary are zero for any angle because $\cos\theta_t = \cos\theta_i$. As depicted in Fig. 7.10, the boundary has no reflection and consequently it has no RCS.

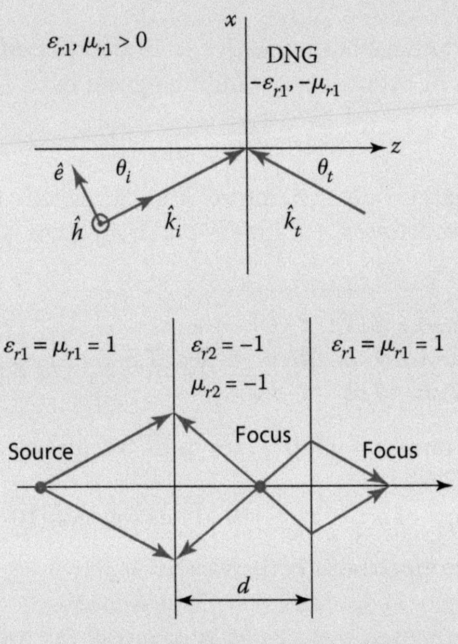

Fig. 7.10 Plane wave incident on a boundary between RH and LH media (top), and a point source illuminating a DNG panel (bottom).

The opposite direction of the arrow on the phase vector in the DNG material implies that propagation through the LH can restore the phase change of a RH medium. In addition, evanescent waves increase in amplitude away from the interface in the LH medium. This does not violate conservation of energy because evanescent waves do not contribute to energy transport. The fact that the decay in evanescent wave fields can be cancelled by a

DNG has led to the concept of *Pendry's perfect lens*, which is illustrated in Fig. 7.10. Further investigation has shown that the DNG must be nondispersive ($\epsilon_{r2}(\omega) = \mu_{r2}(\omega) = -1$), which is not physically realizable. Simulations of realistic DNG materials using the FDTD have verified that the foci do not develop exactly as indicated by the ray diagram in Fig. 7.10, but it was found that the DNG panel channelizes the energy as it passes through the material, behaving somewhat as a *beam waveguide* [11].

The perfect lens images a point source as a point of the same size, not a larger "blur circle" as for a conventional lens. Conventional optical devices focus only the propagating modes. The loss of information contained in the evanescent modes results in a degraded image. The conventional diffraction limit is approximately a wavelength. That is, if multiple image features are separated by less than a wavelength, then they appear as a blur that is about one wavelength in diameter in the image (this is called the *diffraction limit*). Therefore, theoretically, LH materials allow the possibility of subdiffraction limit imaging.

The interface between RH and LH media are capable of supporting a wide range of surface waves, both fast waves (radiating) and slow waves (nonradiating). The reversal in phase delay has led to the term *backward wave*. Traveling wave antennas built with LH materials have demonstrated the capability to radiate a beam over 180 deg, not just 90 deg as the case for a RH material. Undoubtedly this capability will have implications on how traveling waves are excited on targets with DNG materials and on how one might treat or exploit them from a RCS perspective.

The general problem of reflection from multiple layers of LH materials has been presented in Ref. [12]. In Sec. 7.5 the design and fabrication of LH materials is discussed.

7.3.4 Plasmas

A *plasma* [2,3] can be generated from neutral molecules that are separated into negative electrons and positive ions by an ionization process (e.g., laser heating or spark discharge). A *Lorentz plasma* is a simple model in which the electrons interact with each other only through collective space-charge forces. The positive ions and neutral particles are much heavier than the electrons, and therefore the electrons can be considered as moving through a continuous stationary fluid of ions and neutrals with some viscous friction.

The propagation characteristics of electromagnetic waves in a uniform ionized medium can be inferred from the equation of motion of a single "typical" electron. This model would be rigorous if the ionized medium was composed entirely of electrons that do not interact with the background particles (neutrals and ions) and possess thermal speeds that are negligible with respect to the phase velocity of the EM wave. Such a medium is called a *cold plasma*.

In the absence of a magnetic field, the important parameters for a cold plasma are the *electron density* N_e (electrons/m^3) and the *collision frequency* ν (/m^3). The complex relative dielectric constant of the plasma is given by [2]

$$\epsilon_r = \epsilon_r' - j\epsilon_r'' = 1 - \frac{X}{(1 - jZ)} = 1 - \frac{\omega_p^2}{\omega(\omega - j\nu)} \tag{7.25}$$

where

$$\omega_p = \sqrt{\frac{N_e e^2}{m\epsilon_o}}$$

is the *plasma frequency*, and

$$X = \left(\frac{\omega_p}{\omega}\right)^2, \quad Z = \frac{\nu}{\omega}, \quad m = 9.0 \times 10^{-31}\,\text{kg (electron mass)}$$

and $e = 1.59 \times 10^{-19}$ C (electron charge).

Referring to Eq. (7.20), the real and imaginary parts of the propagation constant are the attenuation and phase constants, respectively, $\gamma \equiv \alpha + j\beta = jk_o\sqrt{\mu_r\epsilon_r}$. For a plasma $\mu_r = 1$. Separating Eq. (7.25) into real and imaginary parts gives

$$\epsilon_r' = 1 - \frac{N_e e^2}{\epsilon_o m(\nu^2 + \omega^2)} \quad \text{and} \quad \epsilon_r'' = \frac{N_e e^2 \nu}{\omega\epsilon_o m(\nu^2 + \omega^2)} \tag{7.26}$$

For the special case of negligible collisions, $\nu \approx 0$, the corresponding propagation constant is

$$\gamma = jk_o\sqrt{1 - X} = jk_o\sqrt{1 - \frac{\omega_p^2}{\omega^2}} \tag{7.27}$$

There are three special cases of interest.

1. $\omega > \omega_p$: γ is imaginary and $e^{-j\beta z}$ is a propagating wave.
2. $\omega < \omega_p$: γ is real and $e^{-\alpha z}$ is an evanescent wave.
3. $\omega = \omega_p$: $\gamma = 0$ and this value of ω is called the *critical frequency* because it defines the boundary between propagation and attenuation of the EM wave.

The intrinsic impedance of the plasma medium is

$$\eta = \sqrt{\frac{\mu_o}{\epsilon_o(\epsilon_r' - j\epsilon_r'')}} \tag{7.28}$$

The magnitude of the reflection coefficient at an infinite plane boundary between plasma and free space is given by the standard formula $|\Gamma| = |(\eta - \eta_0)/(\eta + \eta_0)|$.

Example 7.3: Reflection from a Plasma Boundary

The reflection coefficient as a function of frequency for a plane wave normally incident on a sharp plasma/air boundary with $N_e = 1 \times 10^{12}/\text{m}^3$ and $\nu = 0$ is

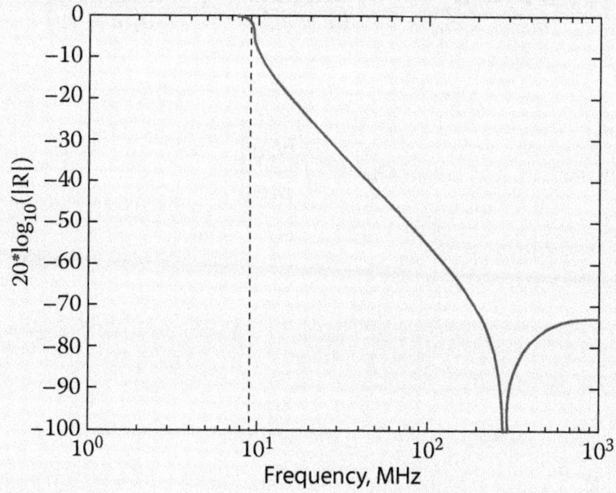

Fig. 7.11 Reflection from a plasma boundary (Example 7.3).

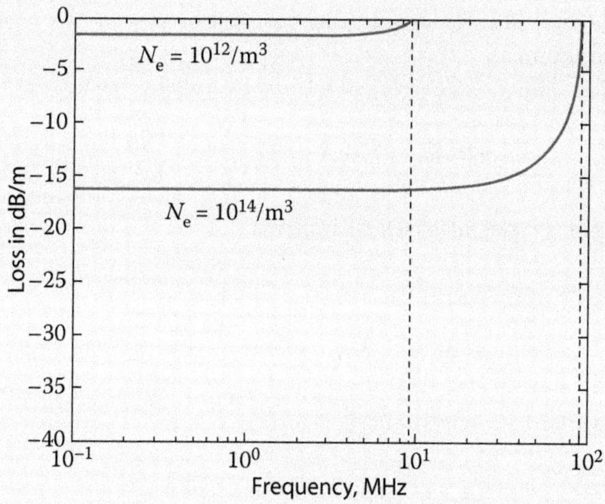

Fig. 7.12 Attenuation per meter of waves propagating in a plasma having the electron density indicated.

(Continued)

Example 7.3: Reflection from a Plasma Boundary *(Continued)*

shown in Fig. 7.11. The dashed line denotes the plasma frequency, $f_p = 8.9$ MHz. It is evident that at frequencies below the plasma frequency, the plasma is a good reflector. A plasma has the characteristics of an FSS, as described in Chapter 6. EM waves below the plasma frequency ($\omega < \omega_p$) are attenuated by the plasma at a rate determined by the attenuation constant:- $|E(z)| \sim e^{-\alpha z} = \exp\left(-zk_o\sqrt{X-1}\right)$. The loss in decibels per meter (dB/m) is $20\log_{10}\{\exp(-k_o\sqrt{X-1})\}$. In Fig. 7.12 the loss per meter is plotted, showing that the plasma can be a good absorber once the EM wave enters the plasma medium. The electron density controls the absorption and cutoff frequency.

Example 7.4: Plasma Loss as a Function of Collision Frequency

For an EM wave frequency of $f = 1$ GHz, the dielectric constant and attenuation in dB/m of a plasma with densities $N_e = 10^{15}/\text{m}^3$ and $10^{16}/\text{m}^3$ is computed for collision frequencies of $\nu = 0/\text{s}$, $10^7/\text{s}$, and $10^9/\text{s}$. The results are given in Table 7.1. For reference, a standard fluorescent bulb has an electron density of about $N_e \approx 10^{11}/\text{cm}^3$. It is clear that a plasma sheet can operate as an effective radar absorbing material.

Table. 7.1 Dielectric Constant and Loss Per Meter for a Range of Plasma Densities and Collision Frequencies

N_e	$\nu = 0/s$	$\nu = 10^7/s$	$\nu = 10^9/s$
$10^{15}/\text{m}^3$	$0.9196 - j0$	$0.9196 - j0.00010.01$	$0.9196 - j0.01251.18$
($f_p = 284$ MHz)	0 dB/m	2 dB/m	3 dB/m
$10^{16}/\text{m}^3$	$0.196 - j0$	$0.196 - j0.0013$	$0.2159 - j0.124823.5$
($f_p = 897$ MHz)	0 dB/m	0.263 dB/m	23.5 dB/m

If there is a static magnetic field present, the plasma medium becomes anisotropic. The nonzero elements of the permittivity matrix of a plasma in the presence of a magnetic field $\vec{B} = B_o\hat{z}$ are [2]

$$\begin{aligned}
\epsilon_{xx} &= \epsilon_{yy} = \epsilon_o\left[1 + \frac{\omega_p}{\omega_c^2 - \omega^2}\right] \\
\epsilon_{xy} &= \epsilon_{yx}^* = \frac{j\omega_p^2(\omega_c/\omega)\epsilon_o}{\omega_c^2 - \omega^2} \\
\epsilon_{zz} &= \epsilon_o\left[1 - \frac{\omega_p^2}{\omega^2}\right]
\end{aligned} \tag{7.29}$$

where $\omega_c = -eB_0/m$ is called the *cyclotron frequency*. A moving electron in a static magnetic field rotates with an angular velocity ω_c, even in the absence of an EM wave. If a wave at frequency ω_c enters the medium, it is synchronized with the electron motion and will continue to push the electrons to higher velocities. All energy is extracted from the wave and no propagation occurs. The magnetic field provides another means of controlling the plasma absorption, cutoff, and reflection characteristics.

It has been reported that plasma techniques have been used in the design of stealthy Russian cruise missiles [13].

7.3.5 Theorems Regarding Absorbers

There are two interesting theorems regarding absorbers due to Weston [14]. Theorem 1 states: *If a plane electromagnetic wave is incident on a body composed of material such that $\mu/\mu_0 = \epsilon/\epsilon_0$ at each point, then the backscattered field is zero, provided that the incidence direction is parallel to an axis of the body about which a rotation of 90 deg leaves the shape of the body, together with its material medium invariant.*

A wave incident along the axis of symmetry of a body of revolution satisfies the geometric requirements of the theorem. If the body has constant permeability and permittivity, they must be equal: $\mu_r = \epsilon_r$.

Theorem 2, which deals with surface impedance, states: *If a plane wave is incident on a body composed of material such that the total field components satisfy the impedance boundary condition [Eq. (2.98)] and if the surface is invariant under a 90-deg rotation, the backscattered field is zero if the direction of incidence is along the axis of symmetry and $Z_s = 1$.*

These theorems are derived directly from Maxwell's equations and the boundary conditions. They apply to any frequency and include all possible scattering and loss mechanisms. Unfortunately, practical radar targets do not satisfy the geometry limitations imposed by the theorems. However, the condition that $\epsilon_r = \mu_r$ does shed some insight into the reduction problem. Because the impedance of a medium relative to free space is essentially determined by $\sqrt{\mu_r/\epsilon_r}$, equal permeability and permittivity yield an impedance in the medium of Z_0. Consequently, the reflection coefficient between free space and the medium is zero. Note that this applies only to the specular field and, thus, diffractions from edges and surface waves could still yield a nonzero scattered field for a finite interface. The nonspecular contributions to scattering will be zero only if the above-mentioned geometrical conditions are satisfied as well.

7.4 RCS Reduction Techniques Employing Materials

Over the years many RCS reduction techniques have been investigated. In the early days of radar, the aircraft were made mostly of metal and glass. Topical applications of materials and paints were used in an attempt to reduce RCS without adding substantial weight or disturbing the aerodynamic performance of the aircraft. Two approaches to reducing the RCS of a metal structure that met with some success were the Salisbury screen and Dallenbach layer, discussed in the following sections.

Modern aircraft and ships contain a substantial amount of composite material, which may not have a high conductivity. In fact, it may be possible to engineer a material having a very high attenuation constant by introducing another constituent to the mix, as described in Sec. 7.5.1. However, this solves only part of the problem. In addition to high loss, the material must be matched to free space. That is, the reflection coefficient must be negligible so that most of the incident radar wave enters the material.

7.4.1 Salisbury Screen

For the Salisbury screen, a resistive film is introduced to cancel reflections from the target surface (Fig. 7.13). If the resistive film is $\lambda/4$ from the conductor, the short is transformed to an open at the screen location. Thus,

$$\frac{1}{Z_{\text{in}}} = \frac{1}{R_S} + \frac{1}{\infty} = \frac{1}{R_S} \tag{7.30}$$

To force the input reflection coefficient to zero requires $R_S = 377\ \Omega$. For other values of resistivity or spacing, the cancellation is not complete. In fact, for a finite screen size, the cancellation is not complete, even for ideal materials, because edge diffractions from the screen and ground plane exist and do not cancel.

7.4.2 Dallenbach Layer

As in the case of the Salisbury screen, the thickness and material properties of a Dallenbach layer (Fig. 7.14) are chosen so that the reflection from the outer

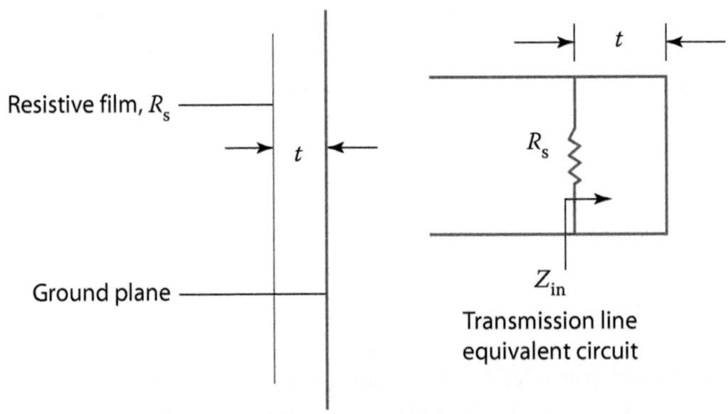

Fig. 7.13 Salisbury screen and its equivalent circuit.

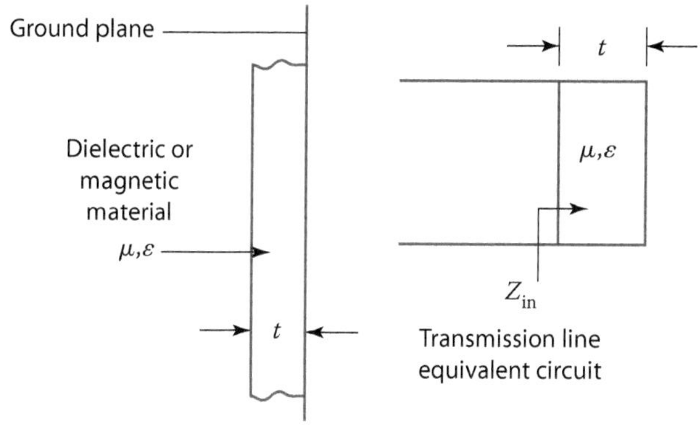

Fig. 7.14 Dallenbach layer and its equivalent circuit.

surface cancels the reflection from the back surface. For normal incidence, the reflection coefficient at the outer surface is (see Problem 5.8)

$$\Gamma = \frac{Z_{\text{in}} - Z_0}{Z_{\text{in}} + Z_0} \tag{7.31}$$

where

$$Z_{\text{in}} = Z_d \frac{Z_L + Z_d \tanh(\gamma t)}{Z_d + Z_L \tanh(\gamma t)} \tag{7.32}$$

with γ determined by Eq. (7.20), $Z_d = \sqrt{\mu/\epsilon}$, and $Z_L = 0$ for a perfectly conducting back plane.

Figure 7.15 shows the reflection coefficient of Dallenbach layers as a function of thickness for several values of permittivity and permeability [15]. The first notch in the reflection coefficient occurs at $0.25\lambda_0$ for electric materials ($\mu_r = 1$, $\epsilon_r > 1$) and at $0.5\lambda_0$ for magnetic materials ($\epsilon_r = 1$, $\mu_r > 1$). The thickness should be as small as possible to minimize the additional size and weight of the coating. At low radar frequencies (e.g., UHF), a quarter of a wavelength is unacceptable, and much thinner layers must be used. The effectiveness of thin layers is determined primarily by the electric and magnetic loss tangents.

7.4.3 Multiple Layers and Wave Matrices

Both the Salisbury and Dallenbach concepts have been extended to multiple layers [15] in an attempt to increase the bandwidth of the structure. When dealing with several layers of dielectrics, the multiple reflections and transmissions that occur within each layer must be considered. In principle, if a layer is lossless, there are an infinite number of reflections as discussed in Problem 2.15. The multiple reflections can be summed to yield a closed form

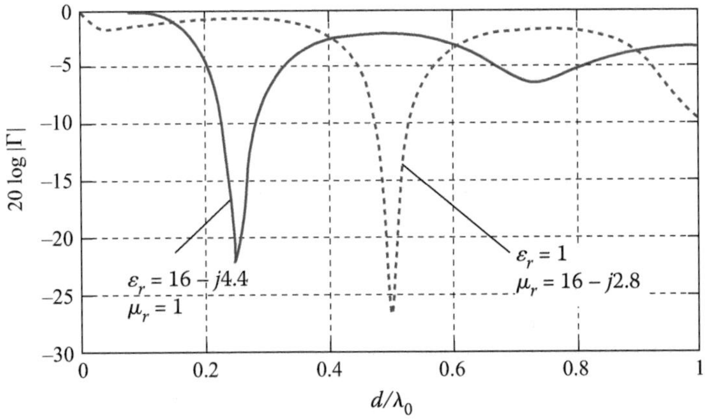

Fig. 7.15 Performance of Dallenbach layers (after Ref. [15]).

result in some cases, but the ray tracing process becomes overwhelming as the number of layers is increased.

For multilayered (stratified) media, a matrix formulation can be used to determine the net transmitted and reflected fields [16]. Figure 7.16 shows incident and reflected waves at the boundary between two media. The positive z traveling waves are denoted c and the negative z traveling waves b. Incident waves from both sides are allowed simultaneously. Therefore,

$$\begin{aligned} b_1 &= \Gamma_1 c_1 + \tau_{21} b_2 \\ c_2 &= \Gamma_2 b_2 + \tau_{12} c_1 \end{aligned} \tag{7.33}$$

where Γ and τ are the appropriate Fresnel reflection and transmission coefficients at the boundary.

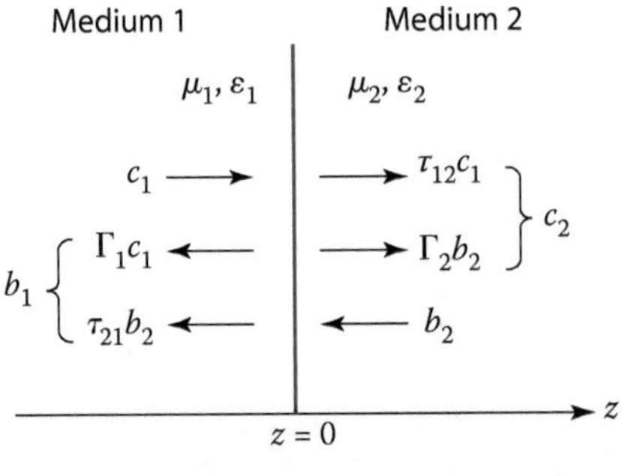

Fig. 7.16 Incident and reflected waves at an interface.

Rearranging the two equations gives

$$\begin{aligned} b_1 &= \left(\tau_{21} - \frac{\Gamma_1\Gamma_2}{\tau_{12}}\right) b_2 + \frac{\Gamma_1}{\tau_{12}} c_2 \\ c_1 &= \frac{c_2}{\tau_{12}} - \frac{\Gamma_2 b_2}{\tau_{12}} \end{aligned} \tag{7.34}$$

which can be written in matrix form as

$$\begin{bmatrix} c_1 \\ b_1 \end{bmatrix} = \frac{1}{\tau_{12}} \begin{bmatrix} 1 & -\Gamma_2 \\ \Gamma_1 & \tau_{12}\tau_{21} - \Gamma_1\Gamma_2 \end{bmatrix} \begin{bmatrix} c_2 \\ b_2 \end{bmatrix} \tag{7.35}$$

The 2-by-2 matrix is called the *wave transmission matrix.* It relates the forward and backward propagating waves on the two sides of the boundary.

As defined, c and b are the waves incident on the boundary $z = 0$. At some other location $z = z_1$ the forward traveling wave becomes $c_1 e^{-j\beta z_1}$ and the backward wave becomes $b_1 e^{j\beta z_1}$. Referring to Fig. 7.17, a plane wave incident from free space onto N layers of different materials (μ_n, ϵ_n) and thickness (t_n), each boundary is represented by a 2-by-2 matrix. The wave matrices can be cascaded to give the total reflection and transmission through the layers:

$$\begin{bmatrix} c_1 \\ b_1 \end{bmatrix} = \prod_{n=1}^{N} \left\{ \frac{1}{\tau_n} \begin{bmatrix} e^{j\Phi_n} & \Gamma_n e^{-j\Phi_n} \\ \Gamma_n e^{j\Phi_n} & e^{-j\Phi_n} \end{bmatrix} \right\} \begin{bmatrix} c_{N+1} \\ b_{N+1} \end{bmatrix} \equiv \begin{bmatrix} A_{11} & A_{12} \\ A_{21} & A_{22} \end{bmatrix} \begin{bmatrix} c_{N+1} \\ b_{N+1} \end{bmatrix} \tag{7.36}$$

For normal incidence, $\Phi_n = \beta_n t_n$ is the electrical length of layer n. If the last layer extends to $z \to \infty$ then $b_{N+1} = 0$, and we can use $\Phi_N = 0$. The overall transmission coefficient of the layers (i.e., the transmission into layer N when $b_{N+1} = 0$) is $c_{N+1}/c_1 = 1/A_{11}$. The overall reflection coefficient is $b_1/c_1 = A_{21}/A_{11}$.

If the incidence angle in region 1 is not normal, then Φ_n must be determined by taking into account the refraction in all of the previous $n - 1$ layers. As shown in Fig. 7.18, the transmission angle for layer n becomes the incidence angle for layer $n + 1$, and they must be related by Snell's law:

$$\beta_o \sin\theta_i = \beta_1 \sin\theta_{i_1} = \beta_2 \sin\theta_{i_2} = \cdots = \beta_{N-1} \sin\theta_{i_{N-1}} \tag{7.37}$$

Thus oblique incidence can be handled by modifying the transmission and reflection formulas to include the dependence on θ_i, the incidence angle at

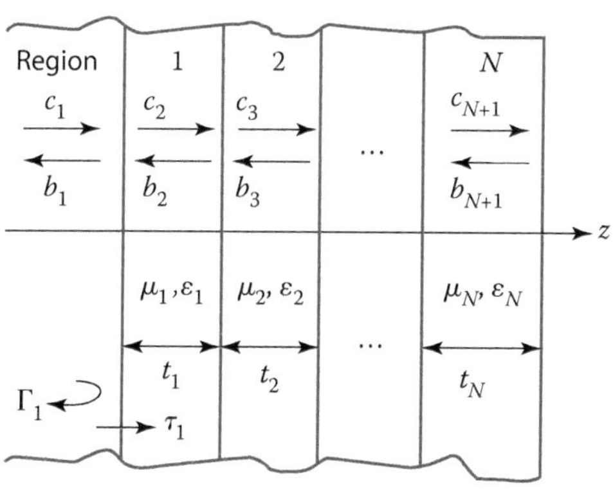

Fig. 7.17 Application of the wave matrix technique to multiple layers.

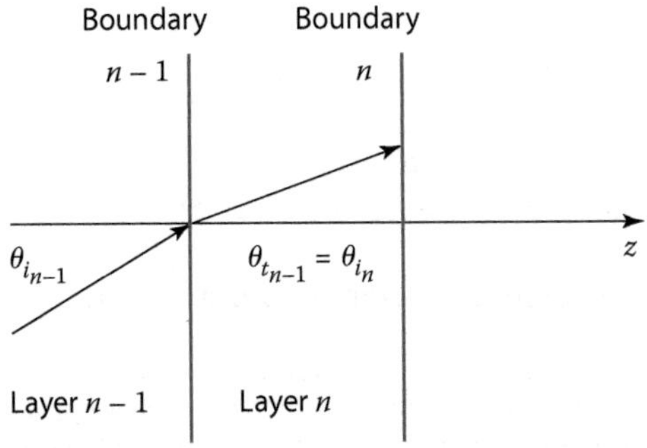

Fig. 7.18 Snell's law applied to a boundary within multiple layers.

the first boundary:

$$
\begin{aligned}
\Phi_n &= \beta_o t_n \left(\epsilon_{r_n} - \sin^2\theta_i\right)^{1/2} \\
\Gamma_n &= \frac{Z_n - Z_{n-1}}{Z_n + Z_{n-1}} \\
\tau_n &= 1 + \Gamma_n \\
\frac{Z_n}{Z_o} &= \frac{\sqrt{\epsilon_{r_n} - \sin^2\theta_i}}{\epsilon_{r_n} - \cos\theta_i} \quad \text{for parallel polarization} \\
\frac{Z_n}{Z_o} &= \frac{\cos\theta_i}{\sqrt{\epsilon_{r_n} - \sin^2\theta_i}} \quad \text{for perpendicular polarization}
\end{aligned}
\tag{7.38}
$$

For lossy materials the real dielectric constant is replaced by a complex value that is determined using the conductivity of the layer:

$$\epsilon_{r_n} \rightarrow \epsilon_{r_n} - j\frac{\sigma_{c_n}}{\omega\epsilon_o}$$

The wave matrix model holds for LH materials if the proper sign conventions are observed [12].

7.5 Composite and Artificial Materials

7.5.1 Composite Materials

Composites refer to a wide range of materials that can include everything from textiles to steel, depending on the breadth of the definition one accepts. The definition of a composite can be made at three levels.

1. *Elemental*: In this sense, a composite is any material composed of two or more different atoms. By this definition, most materials are composites; only pure elements are excluded.

2. *Microstructural*: At this level, a composite is defined as a material that consists of two or more different crystals, molecular structures, or phases. According to this definition, steel is considered a composite.
3. *Macrostructural*: A composite is a material composed of different gross constituents (matrices, particles, and fibers).

A commonly accepted definition is found in Ref. [17]:

> *A composite is a material brought about by combining materials differing in composition or form on a macroscale for the purpose of obtaining specific characteristics and properties. The constituents retain their identity such that they can be physically identified and exhibit an interface between one another.*

Composite materials have replaced metals in most applications, especially in the aerospace and military industries. They include plastics, polymers, resins, and epoxies. Their advantages are primarily mechanical: increased strength, reduced weight, resistance to the environment, increased fatigue life, thermal stability, and ease of manufacture.

A composite material is composed primarily of a *body* constituent, or *matrix*, which gives it its bulk form. Additional *structural* constituents determine the internal structure. These include flakes, fibers, particles, laminates, and fillers, as depicted in Fig. 7.19. The resultant properties of a combination of materials are determined in one of three ways. First, the net characteristics can be a simple *summation* of the individual properties. An obvious example is a laminar composite, in which case electrical and thermal conductivity are given by a summation rule. A second way in which the properties of a composite can differ from those of the original materials is *complementation*. Each component contributes separate and distinct properties. Laminates and clad materials are examples. Finally, the third way is *interaction*. Each constituent supplements the others, and the final result is a material that has properties intermediate between those of the constituents or higher than those of both.

The four most common composites used in the aircraft industry are graphite/epoxy, boron/epoxy, aramid/epoxy, and glass/epoxy [17]. Extensive use of graphite/epoxy is incorporated into the F/A-18 wing skins, horizontal and vertical tail fins, fuselage dorsal cover and avionics bay door, and speed brake. Graphite/epoxy is in widespread use in commercial aircraft as well. The trend toward composites is not new; even older designs, such as

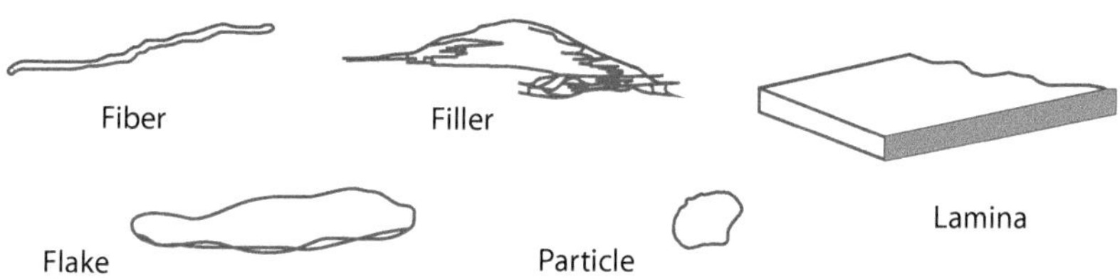

Fig. 7.19 Structural constituents of composites.

the Lockheed L-1011, Boeing 727, and McDonnell-Douglas DC-10, made use of graphite/epoxy. Essentially, the entire structure of the Lear Fan 2100 is graphite/epoxy.

Scattering from a composite is determined as it is for any "pure" material; it is simply a matter of specifying μ_r, ϵ_r, tan $\delta\varepsilon$, and tan δ_μ. The complexity of the composite structure and constituents usually results in constitutive parameters that are at least mildly anisotropic and inhomogeneous. For example, fiber-reinforced materials have a dielectric constant that is slightly different for the electric field vector parallel to the fiber compared to the electric field vector perpendicular to the fiber. Likewise, the dielectric constant of each layer in a laminar composite can vary. Frequently, both these cases are treated by using an *equivalent* or *effective* permittivity and permeability. The electrical characteristics of several composite materials are summarized in Appendix F.

The use of composites has both advantages and disadvantages insofar as the RCS is concerned. The (specular) reflection coefficients of these materials are usually less than those of metals. Traveling waves are also less of a problem because the surface impedance has a larger real part than a good conductor and the reactive part is not inductive. A disadvantage is that composites are *penetrable*; significant field strengths can exist in the interior of composite bodies. Interior metal structures that would normally be shielded by a conducting enclosure will scatter.

The penetration of waves through composites not only affects the RCS but also presents the potential for electromagnetic interference (EMI) between electronic systems. Care must be taken to shield and ground components properly, not only to reduce EMI but also for lightning protection. The latter usually requires embedding a mesh or wire grid inside the composite, which will also affect the RCS.

7.5.2 Magnetic Materials

Magnetic materials refer to those whose electrical characteristics are determined primarily by the permeability and magnetic loss tangent. Iron, the most common magnetic material, is unsuitable for RCS applications because it is a fairly good conductor. Thus, the loss is limited to the depth of penetration of the wave into the material.

The first microwave applications of magnetic materials were primarily devices and components and, therefore, low losses were desirable. There is a fundamental limit to the initial susceptibility at a given frequency for which small magnetic losses occur. *Snoek's law* states that the product of the initial susceptibility and the resonant frequency (defined by the peak of the μ_r curve) is a constant. This has limited the use of low-loss ferrites to relatively low frequencies.

For RCS applications, large losses are desired. The first magnetic radar absorbers were composed of ferrite or iron particles suspended in a dielectric.

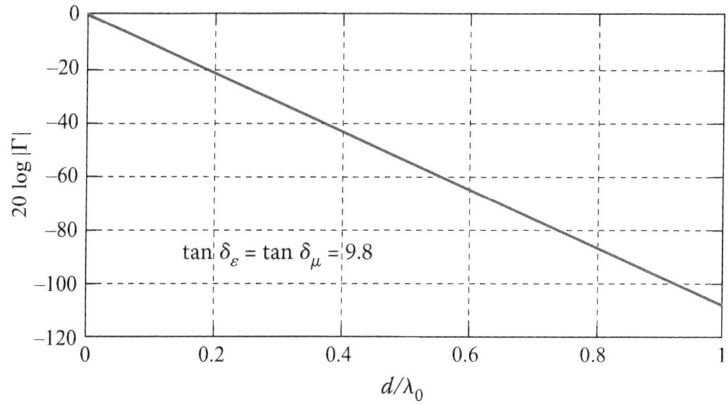

Fig. 7.20 Dallenbach layer with equal electric and magnetic loss tangents.

Figure 7.15 shows that there is no fundamental advantage in electrical performance between electric and magnetic materials. The primary difference is the thickness required for the first resonance or notch. A survey of composites (Appendix F) shows that most materials have relative dielectric constants in the 2 – 10 range and small electric loss tangents.

Figure 7.20 shows the reflection coefficient from a material having $\mu_r = \epsilon_r$ and $\tan\delta_\epsilon = \tan\delta_\mu$ as a function of thickness. When $\mu_r = \varepsilon_r$, the front face reflection coefficient is very small (as expected from the first theorem of Weston), and the attenuation is determined by the loss tangents of the material. The advantage of magnetic material is that large μ_r is accompanied by relatively large ϵ_r, so that $\sqrt{\mu_r/\epsilon_r}$ is on the order of 1. Because the permeability of most materials is frequency-dependent, behavior of the type shown in Fig. 7.20 occurs only over a relatively small frequency band. Broadbanding can be achieved by using multiple layers such that the permeability of a single layer dominates over a portion of the design bandwidth.

7.5.3 Artificial Materials

In general, *artificial materials* refer to materials that do not occur naturally. Depending on the breadth of the definition, composites may or may not be considered as artificial materials. Artificial materials are built around inclusions (added structures or elements) that are small in scale compared to the wavelengths at which the material is designed to operate. Inclusions are generally man-made structures like rings, helices, wires, spheres, discs, etc. They may be distributed periodically or randomly, depending upon the desired electromagnetic properties. On a macroscopic scale, the material can be characterized by an effective permittivity ϵ_{eff} and effective permeability μ_{eff}. One of the first types of artificial materials was the thin-film magnetodielectrics (TFM) [18].

Collective oscillations of electrons (*plasmons*) occur in conductors as well as plasmas. A plasma can be simulated by a three-dimensional array of wires. Confining electrons to thin wires effectively enhances their mass by a factor of 10^4. A three-dimensional grid of thin wires approximates a plasma. The effective dielectric constant is [19]

$$\epsilon_{r_{\text{eff}}} = 1 - \frac{\omega_p^2}{\omega\left(\omega + \dfrac{j\epsilon_o a^2 \omega_p^2}{\pi \sigma_c r^2}\right)} \tag{7.39}$$

where

$$\omega_p = \frac{2\pi c^2}{a^2 ln(a/r)}$$

and r is wire radius, a is the grid spacing, and σ_c is the wire conductivity.

Example 7.5: Effective Dielectric Constant of a Three-Dimensional Grid of Wires

Consider a wire grid with the following specifications: wire radius, $r = 10^{-6}$ m; grid wire spacing, $a = 5$ mm; and conductivity, $\sigma_c = 3.65 \times 10^7$ S/m. Figure 7.21 shows that a plot of Eq. (7.39) for the above quantities yields negative values of ϵ_r' in the gigaHertz frequency region.

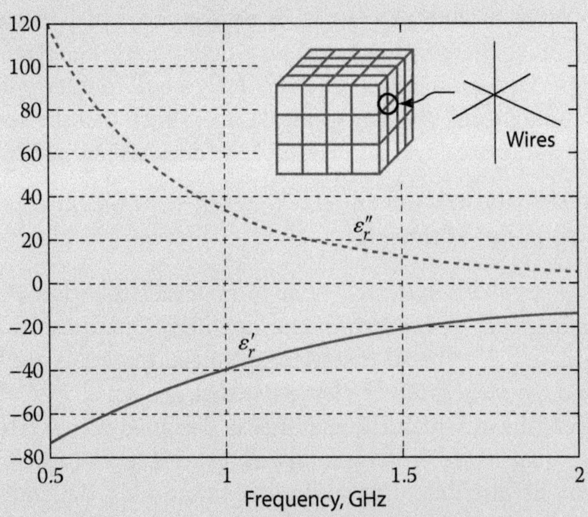

Fig. 7.21 Effective dielectric constant of a 3-D grid of conducting wires.

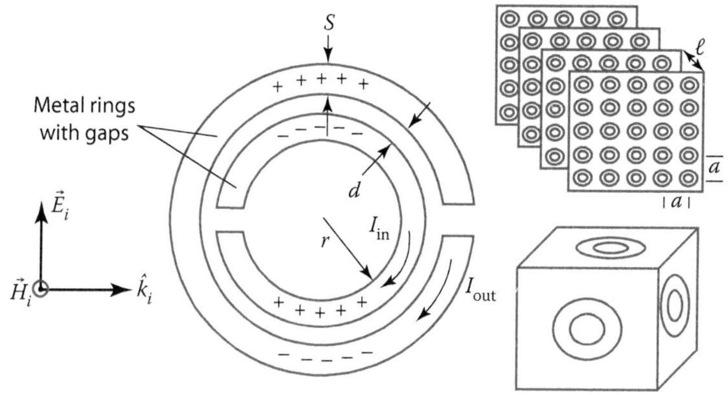

Fig. 7.22 Coplanar ring geometry (left), assembled in sheets (top right), and a unit cell (bottom right).

A coplanar ring (CPR) such as the one shown in Fig. 7.22 is an example of an inclusion that influences both the effective permittivity and permeability of a material. The magnetic field induces currents on the rings. According to Lentz's law, the induced currents oppose the external field, which is diamagnetic behavior. The electric field causes charge separation, which results in a polarization vector, thus changing the permittivity. If the rings are arrayed in one dimension (laid out on a plane), the material will have an anisotropic behavior. Isotropic properties are achieved by having three-dimensional inclusions, as shown in Fig. 7.22.

The effective permeability of the CPR is [20]

$$\mu_{\text{eff}} = 1 - \frac{\pi r^2/a^2}{1 + j\dfrac{2\ell\rho_c}{\omega r \mu_o} - \dfrac{3\ell}{\pi^2\omega^2\mu_o C r^3}} \tag{7.40}$$

where $\rho_c =$ resistivity of the metal (ohms/m), $C = (\epsilon_0/\pi) ln(2s/d) =$ capacitance/m of two parallel strips, $a =$ lattice spacing in plane of rings, $d =$ ring separation, $s =$ ring width, and $\ell =$ spacing between sheets of rings.

Example 7.6: Effective Permeability of a Three-Dimensional Array of Coplanar Rings

Consider the following values for Eq. (7.40): $a = 10^{-2}$ m, $\ell = 2 \times 10^{-3}$ m, $d = 10^{-4}$ m, $s = 10^{-3}$ m, and $r = 2 \times 10^{-3}$ m. The components of the relative permeability are shown in Fig. 7.23, with negative values of μ_r' in the vicinity of 13.5 GHz.

(Continued)

Example 7.6: Effective Permeability of a Three-Dimensional Array of Coplanar Rings *(Continued)*

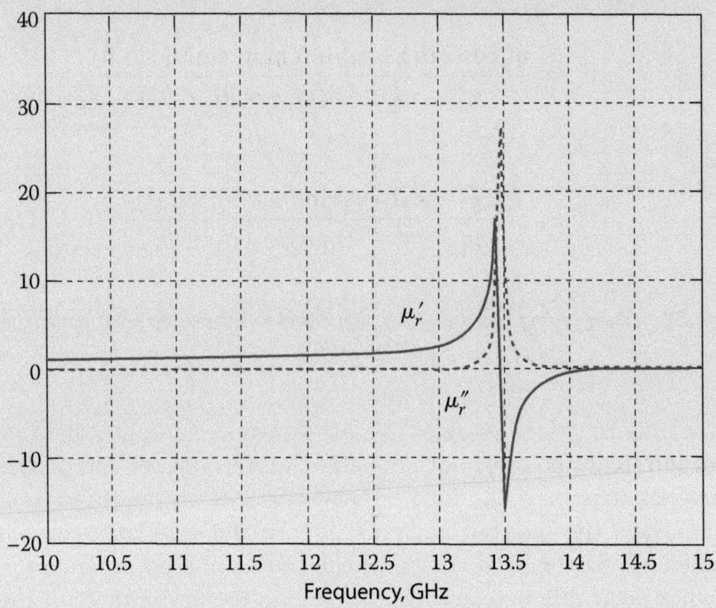

Fig. 7.23 Relative permeability of a grid of coplanar rings.

The helix is another element used in artificial materials. The magnetic response is maximum when the magnetic field is parallel to the helix axis. If large numbers of randomly oriented small helices are added to a material, its macroscopic properties will be isotropic (i.e., no preferred direction). For the sample shown in Fig. 7.24, $\chi = 0$ (reciprocal) and $\kappa = 0.44$. Often the

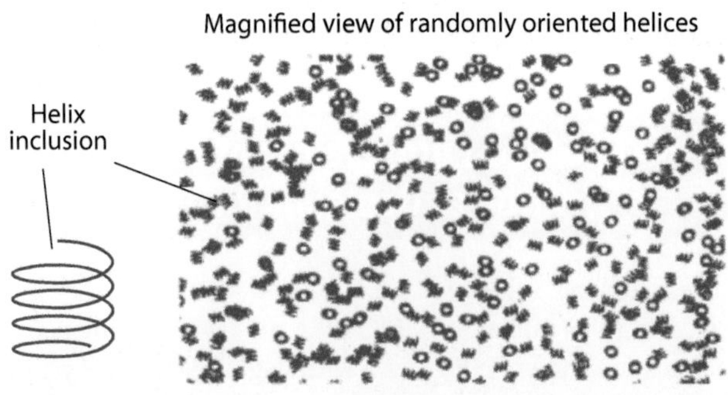

Fig. 7.24 Magnified view of an isotropic chiral material (from Ref. [8]).

magneto-dielectric parameter χ in Eq. (7.11) is expressed in terms of a new parameter ϑ such that $\chi = \sin\vartheta$. In terms of this new parameter, the reflection coefficients at normal incidence for the co- and cross-polarized waves reflected from a plane boundary between two bi-isotropic materials with complex intrinsic impedances η_1 and η_2, are [8]

$$\Gamma_c = \frac{\eta_2^2\cos(2\vartheta_1) - \eta_1^2 + 2\eta_1\eta_2\sin\vartheta_1\sin\vartheta_2}{\eta_1^2 + \eta_2^2 + 2\eta_1\eta_2\cos(\vartheta_1 + \vartheta_2)} \tag{7.41}$$

and

$$\Gamma_x = \frac{2\eta_2\cos\vartheta_1(\eta_2\sin\vartheta_1 - \eta_1\sin\vartheta_2)}{\eta_1^2 + \eta_2^2 + 2\eta_1\eta_2\cos(\vartheta_1 + \vartheta_2)} \tag{7.42}$$

Example 7.7: Reflection from a Chiral Boundary

Medium 1 is free space ($\eta_1 = \eta_0, \chi_1 = \sin\vartheta_1 = 0$) and medium 2 is a reciprocal chiral medium ($k_2 \neq 0$ and $\chi_2 = \sin\vartheta_2 = 0$); the equations reduce to the Fresnel formulas for isotropic media

$$\Gamma_c = \frac{\eta_2^2 - \eta_o^2}{\eta_o^2 + \eta_2^2 + 2\eta_o\eta_2} = \frac{(\eta_2 - \eta_o)^2}{(\eta_2 - \eta_o)(\eta_2 + \eta_o)} = \frac{\eta_2 - \eta_o}{\eta_2 + \eta_o} \tag{7.43}$$

$$\Gamma_x = 0$$

For a Dallenbach layer that employs a bi-isotropic material of thickness t, the reflection coefficients at normal incidence for the co- and cross-polarized cases are [8]

$$\Gamma_c = \frac{\left(\eta_o^2 - \eta_2^2\right)\sin^2(k_2 t\cos\vartheta_2) - \eta_o^2\cos^2\vartheta_2}{\eta_o^2\cos^2\vartheta_2 - \left(\eta_o^2 + \eta_2^2\right)\sin^2(k_2 t\cos\vartheta_2) + j\eta_o\eta_2\cos\vartheta_2\sin(2k_2 t\cos\vartheta_2)} \tag{7.44}$$

$$\Gamma_x = \frac{2\eta_o\eta_2\sin\vartheta_2\sin^2(k_2 t\cos\vartheta_2)}{\eta_o^2\cos^2\vartheta_2 - \left(\eta_o^2 + \eta_2^2\right)\sin^2(k_2 t\cos\vartheta_2) + j\eta_o\eta_2\cos\vartheta_2\sin(2k_2 t\cos\vartheta_2)} \tag{7.45}$$

where the intrinsic impedance of the bi-isotropic medium is η_2, and k_2 is the propagation constant. It can be shown that if the thickness satisfies

$$\sin(k_2 t\cos\vartheta_2) = \cos\vartheta_2/\sqrt{1 - (\eta_2/\eta_1)^2} \tag{7.46}$$

then $\Gamma_c = 0$ and $\Gamma_x = 1$. The layer acts as a "twist polarizer." The boundary completely reflects the incident wave in the cross-polarized component. This effectively reduces the RCS of the target when the threat radar is linearly polarized.

7.5.4 Universal Charts for Materials Selection

Radar absorbing material (RAM) is an effective method of reducing the RCS of targets. A thin layer of RAM is applied to the target skin, which is usually a metal or composite. The circuit model of RAM is a lossy transmission line; it attenuates the plane wave as it propagates through the material. To achieve a large attenuation (20 to 50 dB), the material must have a large attenuation constant, or the layer must be very thick. Clearly a thick layer is not desirable or acceptable on a ship or aircraft, and so it is necessary to select materials that provide a large attenuation constant.

There are many possible RAM configurations. Consider the standard Dallenbach layer of Fig. 7.14. Assume that the plane wave is normally incident. The design parameters are the thickness, t, and constitutive parameters of the medium μ_r and ϵ_r. Note that we are assuming an isotropic material. More specialized and advanced designs may be composed of anisotropic materials.

There are two approaches to selecting the layer parameters:

1. *Matched characteristic impedance method*: Make $\mu_r = \epsilon_r$ everywhere in the material and μ_r'' and ϵ_r'' large enough so that the attenuation constant α provides sufficient round-trip attenuation.
2. *Matched wave impedance method*: Make the wave impedance at the input of the equivalent transmission line circuit equal to that of free space. That is, make the net reflection coefficient at the front face of the layer zero

$$\Gamma = \frac{Z_{\text{in}} - Z_o}{Z_{\text{in}} + Z_o} = 0 \tag{7.47}$$

where

$$Z_{\text{in}} = Z_d \frac{Z_L + Z_d \tanh(\gamma t)}{Z_d + Z_L \tanh(\gamma t)} \tag{7.48}$$

and $Z_d = \sqrt{\mu/\epsilon}$ is the impedance of the coating layer. If the backing material is a PEC then $Z_L = 0$ and $Z_{\text{in}} = Z_d \tanh(\gamma t)$ so that

$$|\Gamma| = 0 \rightarrow Z_d \tanh(\gamma t) - Z_o = 0 \rightarrow \frac{Z_d}{Z_o} \tanh(\gamma t) = 1 \tag{7.49}$$

The matched wave impedance method can serve as the basis for a set of *universal curves* for RAM design [21]. There are six degrees of freedom in designing the Dallenbach layer: t, λ, μ_r', μ_r'', ϵ_r', and ϵ_r''. The six parameters can be reduced to four by normalizing the permittivity and permeability by t/λ

$$a = \frac{t}{\lambda}\epsilon_r' \quad b = \frac{t}{\lambda}\epsilon_r'' \quad x = \frac{t}{\lambda}\mu_r' \quad y = \frac{t}{\lambda}\mu_r'' \tag{7.50}$$

In terms of the new variables, the transcendental equation (7.49) becomes

$$
\begin{aligned}
&\sqrt{\frac{\mu_r' - j\mu_r''}{\epsilon_r' - j\epsilon_r''}}\tanh\left(j\frac{2\pi}{\lambda}t\sqrt{(\epsilon_r' - j\epsilon_r'')(\mu_r' - j\mu'')}\right) = 1\\
&\sqrt{\frac{x - jy}{a - jb}}\tanh\left(j2\pi\sqrt{(a - jb)(x - jy)}\right) = 1\\
&j\sqrt{\frac{x - jy}{a - jb}}\tan\left(2\pi\sqrt{(a - jb)(x - jy)}\right) = 1
\end{aligned}
\tag{7.51}
$$

Now a set of curves can be drawn up where the abscissa is x and the ordinate is y, for constant values of a and b. Figure 7.25 shows such a curve taken from Ref. [21].

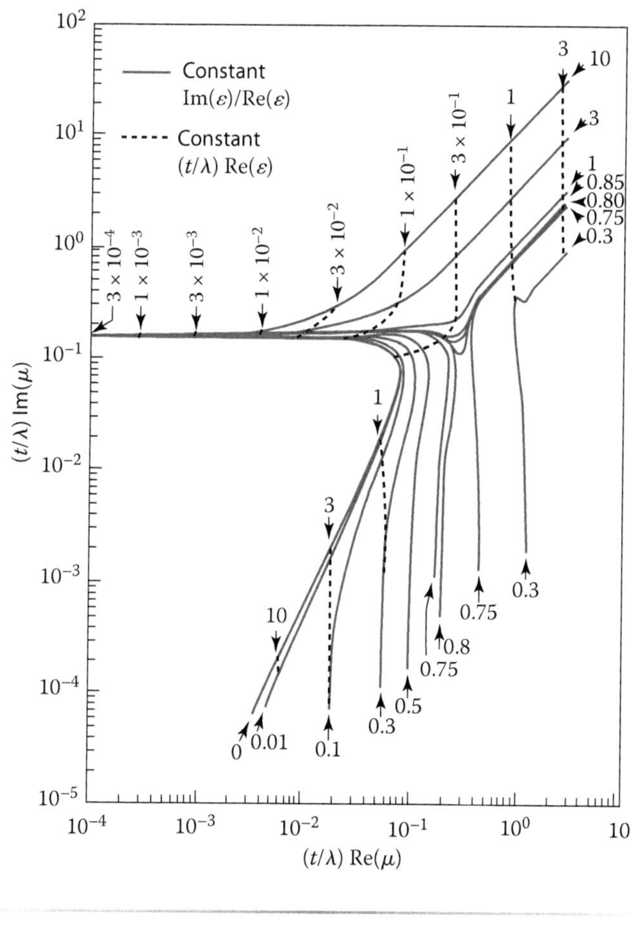

Fig. 7.25 Universal curve of points with zero reflection based on the matched wave impedance transcendental equation (from Ref. [20]).

Example 7.8: Selection of RAM Material Properties

As an example, a RAM layer must be designed at 3 GHz and with a maximum thickness of 1 cm ($t/\lambda = 0.1$). If the layer is composed of a dielectric with $\epsilon_r' = 10$ and $\epsilon_r'' = 100$, we will use the universal chart to determine μ_r' and μ_r'' for zero specular reflection. To enter the chart first find $a = (t/\lambda)\epsilon_r' = (0.1)(10) = 1$ and the electric loss tangent, $\tan\delta_e = \epsilon_r''/\epsilon_r' = 100/10 = 10$. Therefore $b = a \tan\delta_e = 10$.

Next, go to the intersection of the curves for these values of a (dashed lines) and $\tan\delta_e$ (solid lines). Finally, drop down to read x and over to the left to read y:

$$x = \frac{t}{\lambda}\mu_r' \approx 1 \rightarrow \mu_r' \approx 10$$

$$y = \frac{t}{\lambda}\mu_r'' \approx 10 \rightarrow \mu_r'' \approx 100$$

It is also possible to solve the transcendental equation numerically rather than using the chart. The MATLAB® "solve" routine can be used in some cases:

```
Syms y
u=solve(sqrt (y/(1-j*10))*tanh(6.28*sqrt((1-j*10)*y))-1)
```

It returns the result `u=1.000005197-10.000140985i`.

Note that in this case $\epsilon_r = \mu_r$, which is the definition of a matched characteristic impedance absorber. Curves in the upper part of the chart correspond to matched characteristic impedance absorbers.

Example 7.9: Simulation of a RAM Coated Plate

In this example [22], a PEC plate with dimensions 3-by-2 m in the y and z directions, respectively, is coated with a 0.1 m layer of absorbing material. The frequency is 300 MHz ($t/\lambda = 0.1$) and the material parameters are (arbitrarily) chosen for a case of zero specular reflection at normal incidence:

$$\tan\delta_\epsilon = 0.3$$

$$t/\lambda \,\mathrm{Re}[\epsilon_r] = 0.3$$

$$t/\lambda \,\mathrm{Re}[\mu_r] = 0.1103$$

$$t/\lambda \,\mathrm{Im}[\mu_r] = 0.1266$$

The magnetic loss tangent is calculated as

$$\tan\delta_\mu = \frac{t/\lambda \,\mathrm{Im}[\mu_r]}{t/\lambda \,\mathrm{Re}[\mu_r]} = \frac{0.1266}{0.1103} = 1.147$$

(Continued)

Example 7.9: Simulation of a RAM Coated Plate *(Continued)*

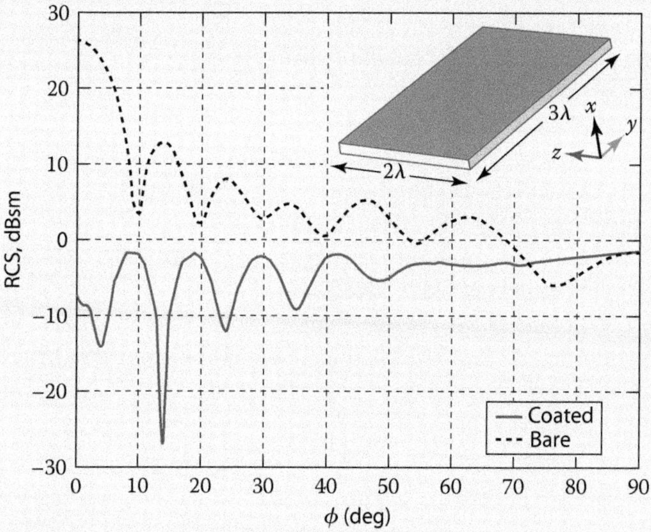

Fig. 7.26 Simulated monostatic RCS of the coated plate in Example 7.9.

The real parts of the permeability and permittivity of the coating material are calculated as

$$\mathrm{Re}[\mu_r] = 1.103$$
$$\mathrm{Re}[\epsilon_r] = 3$$

Simulations of the coated plate were run using *CST Microwave Studio.* The monostatic RCS patterns for the bare and coated plates are compared in Fig. 7.26. Notice that at normal incidence we do not achieve "zero specular reflection" even though the parameters have been chosen from the chart. The reduction from the PEC RCS is approximately 33 dB. The residual scattered field is mainly due to edge diffraction from the finite plate but may also include numerical errors in the computation (as discussed in Chapter 4). For larger angles the reduction decreases because the path length through the coating increases, and the thickness is no longer 0.1λ.

It should be pointed out that even though Eq. (7.51) gives the required constitutive parameters for no specular reflection it does not ensure that such a material exists. Mathematically it is possible to calculate the required permittivity and permeability for a perfectly matched layer (PML) that gives zero reflection at all frequencies and incidence angles. Problems 7.20 and 7.21 examine the conditions for a PML.

7.5.5 Bandwidth of RAM Treatments

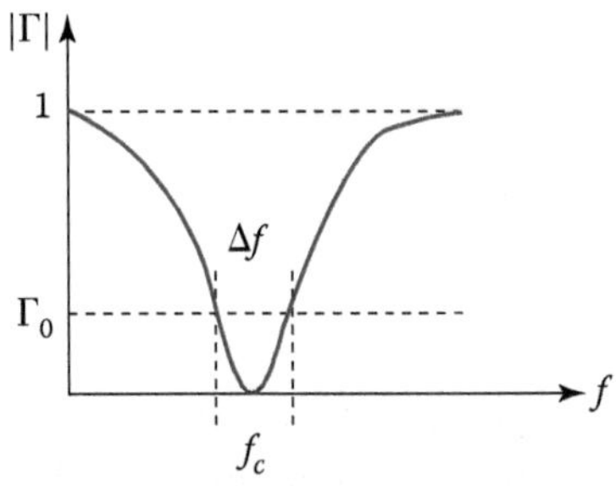

Fig. 7.27 Bandwidth of a RAM treatment.

The reflection coefficient for a Dallenbach layer over a PEC is given by Eq. (7.48) with $Z_L = 0$. For no reflection the numerator of Eq. (7.47) must be zero. The question arises: What are the conditions on the dielectric constant and permeability to maximize the bandwidth of the structure, Δf? [23–25]. The bandwidth is defined as the maximum contiguous range of frequencies over which the reflection coefficient is no greater than a specified value, Γ_o, as depicted in Fig. 7.27. If the layer medium is lossless then $\alpha = 0$ and $\gamma = j\beta$, so that

$$jZ_d \tan(\beta t) - Z_o = 0 \tag{7.52}$$

where $\beta = 2\pi/\lambda_c$. Therefore, at the center frequency f_c we can write

$$Z_o = jZ_d \tan(\beta t) \tag{7.53}$$

Now consider a second frequency $f' > f_c$. The reflection coefficient of the Dallenbach layer will be smallest when $Z_d \tan(\beta' t) - Z_o \approx 0$. Substituting in Eq. (7.53) in the reflection coefficient formula at f' gives

$$|\Gamma| = \left|\frac{\tan(\beta' t) - \tan(\beta t)}{\tan(\beta' t) + \tan(\beta t)}\right| = \left|\frac{\tan(B) - \tan(A)}{\tan(B) + \tan(A)}\right| \tag{7.54}$$

For a small change in frequency $B \approx A$, which can be used in the denominator. Next, make use of the identities

$$\begin{aligned} \tan B - \tan A &= \tan(B - A)(1 + \tan B \tan A) \\ \tan B + \tan A &= \tan(B + A)(1 - \tan B \tan A) \\ \sin B &= \tan B/(1 + \tan^2 B). \end{aligned} \tag{7.55}$$

Equation (7.54) becomes

$$\begin{aligned} |\Gamma_o| &= \left|\frac{B - A}{2\tan(B)}\sqrt{1 + \tan^2(B)}\right| = \left|\frac{j(B - A)\sqrt{1 - (\epsilon_r/\mu_r)}}{2\sqrt{\epsilon_r/\mu_r}}\right| \\ &= \frac{2\pi(f' - f_c)t|\epsilon_r - \mu_r|}{2\lambda'} \end{aligned} \tag{7.56}$$

Solving this for the bandwidth and normalizing to the center frequency,

$$\mathrm{BW} = \frac{2(f' - f_c)}{f_c} = \frac{\Delta f}{f_c} = \frac{\lambda_c}{\Delta\lambda} = \frac{2|\Gamma_o|}{\pi|\epsilon_r - \mu_r|t/\lambda'} \tag{7.57}$$

Now assume that 1) the layer is thin $t \ll \lambda_c$; 2) the bandwidth is small compared to the center frequency $\Delta\lambda \ll \lambda_c$; 3) $|\epsilon_r| \gg |\mu_r|$, which is true for a wide range of natural materials; and 4) $2\pi t\sqrt{\epsilon_r \mu_r}/\lambda_c \approx \pi/2$. These assumptions give the approximation

$$\Delta\lambda \approx \frac{32\ \mathrm{Re}(\mu_r)t}{\pi}|\Gamma_o| \tag{7.58}$$

Equation (7.57) illustrates that the bandwidth of a single-layer Dallenbach treatment can be increased by increasing the permeability of the layer. The conclusion is that, generally, wideband absorbers require a high permeability. This has led to the development of artificial materials that meet the permeability requirements of RAM.

7.5.6 Geometric Absorbers and Edge Treatments

Control of a target's constitutive parameters is the primary means of reducing RCS. Additional reduction in the low RCS levels due to multiple reflections and diffraction may be possible by the geometric control of discontinuities. For example, scattering from a knife edge can increase the RCS in the plane transverse to the edge. A serrated edge causes diffractions to be dispersed in all directions, thereby reducing the RCS in the transverse plane, as in Fig. 7.28.

The first use of geometric absorbers was to reduce reflections in anechoic chambers, as discussed in Chapter 8. The pyramidal shape breaks up the incident wave front, partially absorbing the transmitted energy and partially scattering it. The transmitted signal is attenuated by the carbon particles embedded in the foam. The reflected signal is directed toward adjacent pyramids, where further reflection and attenuation occur. Reflection from conductor-backed absorbers is typically in the range of 20–40 dB down from the incident intensity.

Serration is a geometric treatment (not an absorber per se) that can be applied to the leading and trailing edges of wings. Obvious examples are the B-2 and F-117A, where the trailing-edge zig zags are clearly visible.

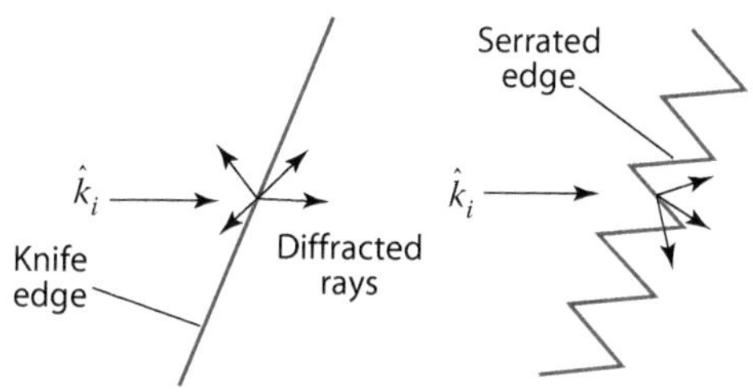

Fig. 7.28 Diffraction from a straight edge (left) versus serrated (jagged) edge.

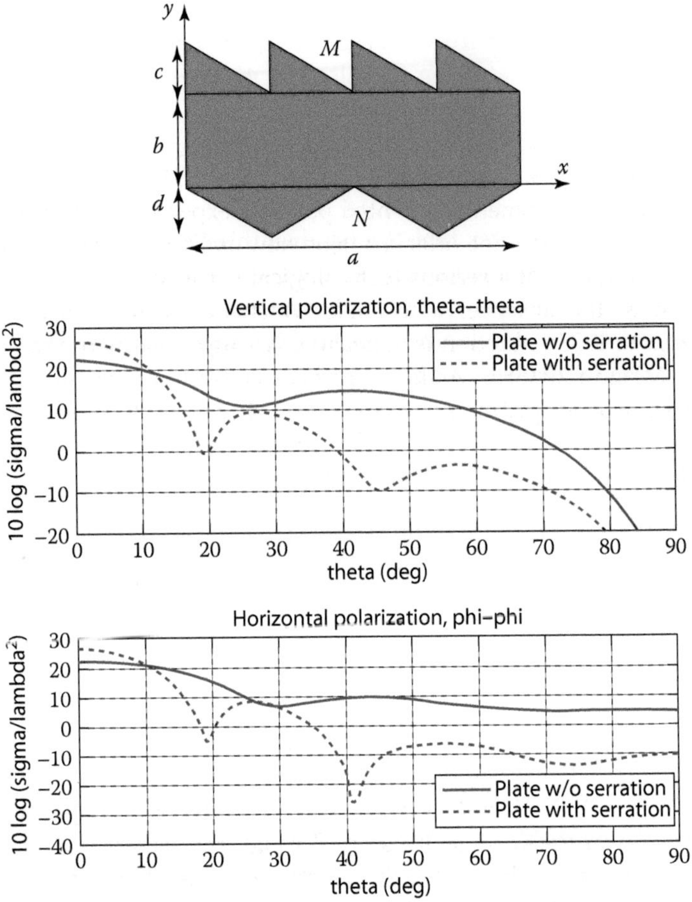

Fig. 7.29 Improvement in wide angle RCS provided by edge serrations ($a = 4\lambda$, $b = \lambda$, $c = d = 0.5774\lambda$, $M = 4$, and $N = 0.5\lambda$, from Ref. [27]).

In other applications, such as the SR-71 Blackbird [26], conducting foils cover composite structures, and the type of surface treatment employed is not obvious from the wing's physical shape.

Figure 7.29 shows how serrations can reduce the wide angle RCS of a simple plate [27]. The RCS of the plain rectangular plate is shown along with the RCS with the triangles added. Note that the wide angle RCS has been reduced even though peak RCS has increased because the total area is larger with the triangles.

7.5.7 Ultrawideband Effects

The trend in radar has been toward increased bandwidth. This in turn has prompted research into wideband RCS reduction methods. *Ultrawideband*

(UWB) radar (also known as impulse radar) and communication systems have now been deployed, as a quick search of the Internet shows. These systems use extremely short pulses (1 ns or less) that contain significant energy over a wide range of frequencies.‡

In the past, many extraordinary capabilities of UWB radar have been alleged, most notably, the ability to defeat stealth [28]. After extensive investigation by a government-appointed panel of experts, it was concluded that "impulse radar does not offer a major new military capability nor does it present the threat of a serious technological surprise [29]."

Even in the absence of a single wideband radar threat, a wideband approach to RCS reduction is appealing. A single treatment that is effective from 1 MHz to 20 GHz would be more desirable than having to apply multiple narrowband treatments to cover the many individual radar threats over this extremely wide frequency range. Most likely such a treatment would rely on special material properties. As alluded to in the previous discussion of materials and waves, media have a far more complex EM relaxation behavior than predicted by the simple Debye model. Awareness of the model shortcomings has arisen from research involved with ultrashort pulse lasers interacting with materials. New theories have been devised such as the *Dissado–Hill* model [30, 31] that takes all of the spatial and temporal factors into account. The atomic dipoles have both a homogeneous and an inhomogeneous lifetime. The inhomogeneous lifetime depends on the number of neighboring dipoles and their distances as well as their relaxation times.

A predicted behavior of some materials at high frequencies is *self-induced transparency* [32]. If the signal frequencies are high enough, then energy extracted from the wave as it passes through the material can be returned back to the wave. The wave can penetrate the medium without loss and therefore any RAM would be ineffective.

New insight into the behavior of materials has given rise to the concept of designing waveforms for specific materials, and vice versa. For example, a radar might transmit a waveform based on the type of target that it is trying to detect. The waveform would be designed to efficiently scatter from the specific material of which that particular target is made. This one of the considerations in *cognitive radar* design [33].

Another possible effect of UWB waveforms is the phenomenon of *precursors*. When a conventional waveform passes through a nondispersive material, the waveform out of the material is a time-delayed replica of the waveform at the input. (We assume that the waveform has a long pulse width compared to the relaxation time of the material.) The group velocity u_g is usually taken as the velocity of energy propagation in the material (neglecting any distortions due to dispersion.) The group velocity is less than the phase velocity, which in

‡At this time there are no formal definitions of narrow band (NB), wide band (WB), and UWB. Generally systems with absolute bandwidths of at least 1 GHz and relative bandwidths of 25% or more can be considered UWB. Narrow band systems are those with less than about 5% relative bandwidth, and wide band spans the region in between the two.

turn is less than the velocity of light in a vacuum (except for anomalous cases). From Fig. 7.8, it can be seen that high frequencies travel faster than low frequencies because $u_p = 1/\sqrt{\mu\epsilon'}$. The high frequency components of an ultra-wideband waveform propagate faster than the low frequency components, and thus the high frequency components arrive at an observation point first. These precursors are features in waves transmitted through media due the ultrafast rise and fall times of the pulse envelope. They occur because the transfer of energy is not instantaneous.

7.6 Passive Cancellation

Passive cancellation refers to RCS reduction by introducing a secondary scatterer to cancel with the reflection of the primary target. The target with the scattering element is called the *loaded body,* as opposed to the bare target, which is the *unloaded body.* Consequently, this method is also known as *impedance loading,* and it is essentially the same approach as that used in the design of the Salisbury screen and Dallenbach layer. Several approaches to passive cancellation are illustrated in Fig. 7.30.

As with any cancellation or tuning method, this technique is effective over only a narrow frequency band and is usually limited to a small spatial

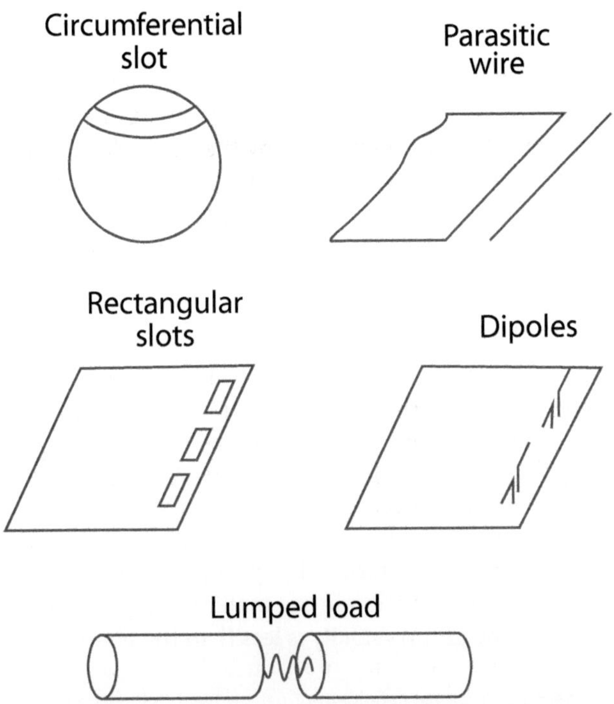

Fig. 7.30 Parasitic elements used for passive RCS cancellation.

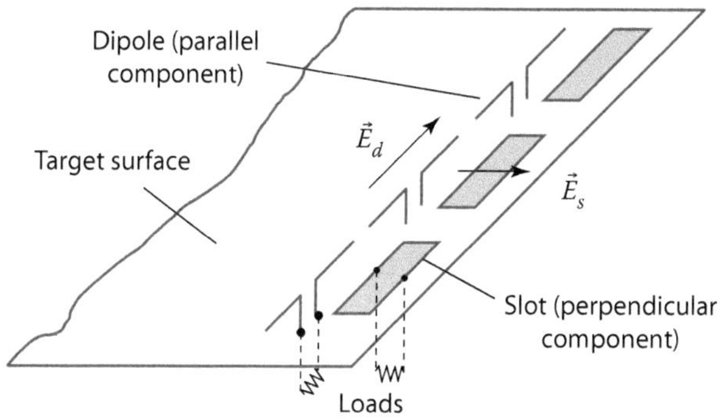

Fig. 7.31 Radar cross section cancellation using a combination of slots and dipoles.

sector. If large parasitic elements are to be avoided, the magnitude of RCS that can be canceled is relatively small. Thus, passive cancellation is used to supplement shaping and absorbers. The exception to this is the treatment of traveling waves, in which case passive cancellation by parasitic structures is often the primary means of RCS reduction (Sec. 7.8).

A simple application for a solid surface is shown in Fig. 7.31. Two orthogonal elements are required to cancel an incident wave of arbitrary polarization. These elements could be slots, dipoles, or a combination of both. The terminals are shorted, and the transmission lines are adjusted to provide the proper phase for the field reflected at the short. The number of elements is chosen to provide sufficient gain to cancel the RCS in the direction of interest. Thus, the field reradiated by the network cancels the RCS of the bare target. It is evident that a fair amount of information about the threat and the target itself is required: the threat radar frequency, direction, and polarization and the target RCS in the direction of the threat.

7.7 Active Cancellation

Active cancellation is the extension of passive cancellation to handle dynamic threat scenarios. In the context used here, it does not include *deception,* that is, techniques for modifying and retransmitting a signal. For the present discussion, the source of rf energy is the threat radar's signal. Two levels of sophistication are considered.

1. *Fully active*: The cancellation network receives, amplifies, and retransmits the threat signal such that it is out of phase with the static RCS of the target. The transmitted signal amplitude, phase, frequency, and polarization can be adjusted to compensate for changing threat parameters.
2. *Semiactive*: No boost in threat signal energy is provided by the cancellation network, but adjustable devices in the network allow the reradiated signal to

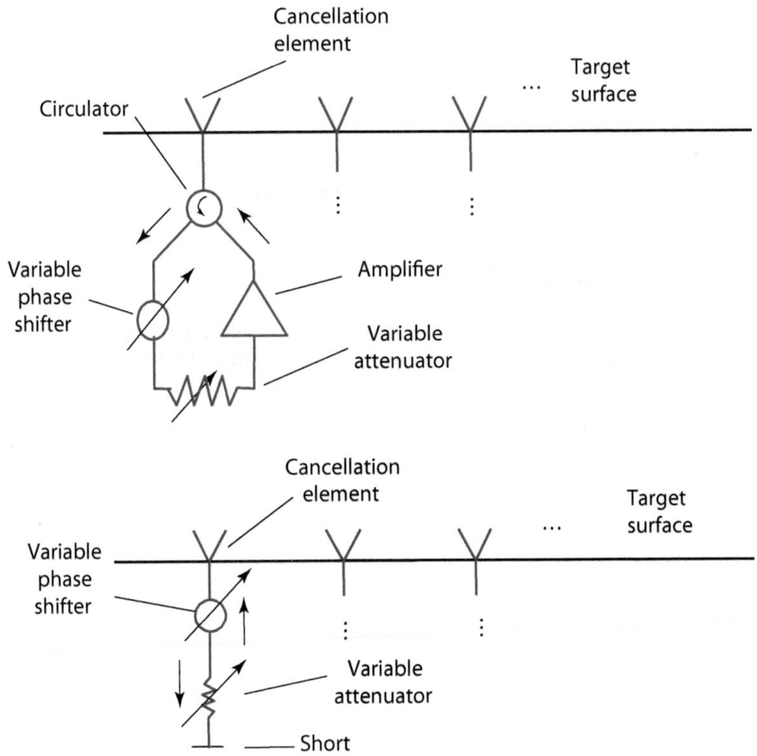

Fig. 7.32 Networks for fully active and semi-active RCS cancellation.

compensate for limited changes in the threat signal parameters. Networks for implementing active cancellation are illustrated in Fig. 7.32.

The demands for a fully active system are almost always so severe as to make it impractical. It requires a transmitter and antennas that cover the anticipated threat angles, frequencies, incident power densities, and polarizations. A knowledge of the threat direction is required as well as the target's own RCS. A semiactive system is not as complicated in terms of hardware, but the use of adjustable devices still requires bias lines, controller units, and a computer with the appropriate databases.

7.8 Treatments for Traveling Waves

Traveling waves present a special challenge as far as RCS reduction is concerned. The target acts as a combination of antenna and transmission line, collecting incident energy and guiding it along the surface until a discontinuity or load is encountered. As illustrated in Fig. 2.41, a dielectric-coated ground plane can support a surface wave for TM_z incidence. Thus,

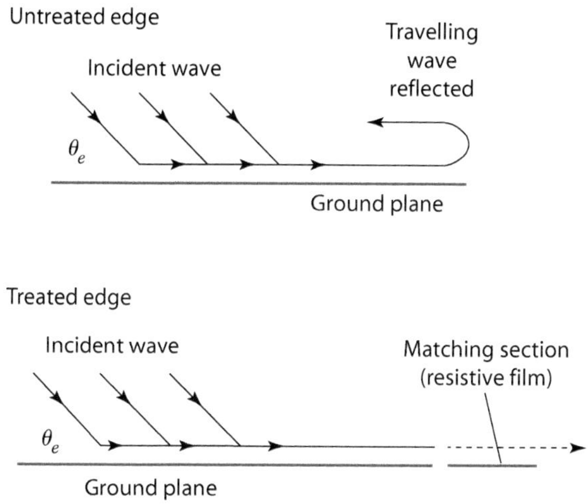

Fig. 7.33 Matching an edge with a resistive strip.

a Dallenbach layer, which is used effectively to reduce specular RCS, has the potential for supporting a surface wave.

From Eq. (2.171), the propagation constant for the wave traveling along the z axis is

$$\gamma_z = k_0\sqrt{(R_s')^2 - \left[1 + (X_s')^2\right] + 2jX'R_s'} \tag{7.59}$$

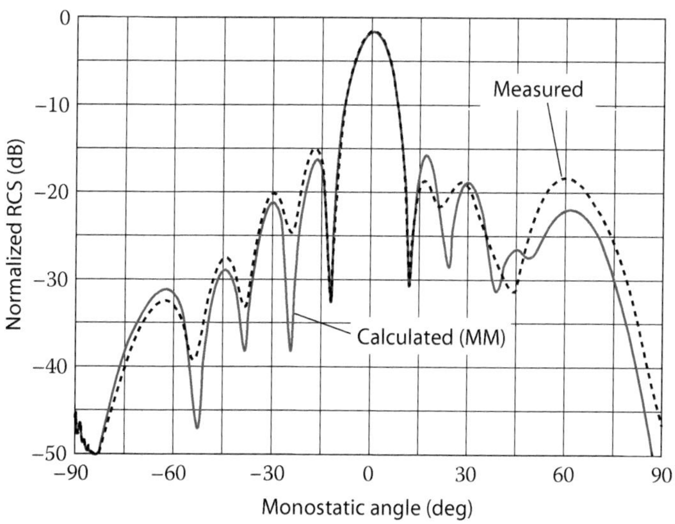

Fig. 7.34 Comparison of calculated (MM) and measured RCS data for a plate with one edge treated.

For strong attenuation, the product $X_s'R_s'$ must be large and, to confine most of the field near the interface, X_s' must be large. The surface resistance R_s' depends on the conductivity and, hence, loss tangent. The implication of Eq. (7.59) is that sufficient attenuation of surface waves cannot be achieved with short lengths of a thin layer of material having a moderate loss tangent.

In light of these constraints, RCS reduction of traveling wave components often focuses on redirecting the wave rather than attenuating it. Options include:

1. Adjusting the length of the structure L to obtain θ_e in Fig. 2.43 so that the lobe is out of quiet zones or the level is acceptably low.
2. Reorienting or curving the edge to direct the lobe out of quiet zones.
3. Matching the terminating discontinuity so that reflection and re-radiation in the direction of the threat does not occur.

Option 3 is a narrowband approach based on the impedance transformer concept used in transmission-line matching. (Actually, it has been used in the design of Salisbury screens and Dallenbach layers as well.) A second discontinuity is introduced and its characteristics are adjusted so that its scattered field cancels with that of the original discontinuity. The application to a strip is shown in Fig. 7.33. From the transmission-line analogy, the length of the matching section should be about 0.25λ at the threat frequency. Its resistivity value is determined by measurement or is calculated using MM.

Figure 7.34 compares measured and calculated data for a 2.25λ square plate with one side loaded with $R_s = 22\ \Omega/\text{square}$. The RCS of the travelling wave lobe has been reduced about 15 dB. If a two-section transformer is used a reduction of more than 25 dB can be achieved. The technique just described can be extended to more complex doubly curved surfaces.

7.9 Antenna Radar Cross Section Reduction

Antenna RCS was shown to be distinctly different for frequencies in the operating band as compared to those out of the operating band. Thus, effective control of antenna RCS must address the in-band and out-of-band frequencies separately. Unfortunately, methods that are effective out of the operating band impact the antenna performance (both gain and RCS) in its operating band.

A high-performance antenna that maintains low RCS over a wide frequency band is most likely to be a phased array. Out of the operating band, an FSS is employed to shield the array. The spacing of the FSS elements must be very small in terms of the threat wavelength to avoid Bragg diffraction. The in-band RCS of the array is controlled by feed network design and by minimizing feed component scattering. The brute-force approach is to keep the VSWR of every device in the feed as close to 1 as possible; but,

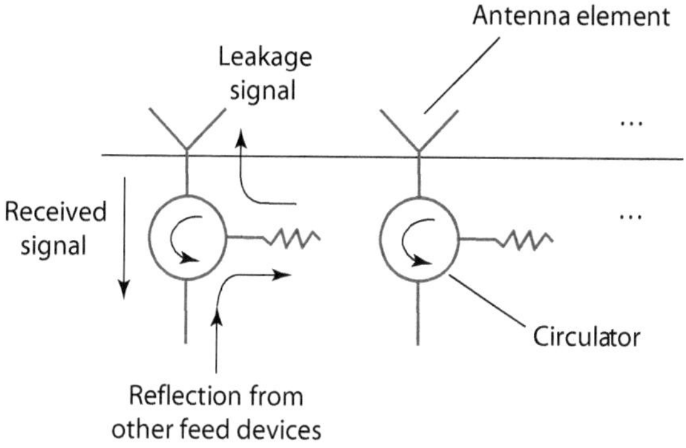

Fig. 7.35 Array antenna with a circulator behind each element.

given the sensitivity of VSWR to manufacturing and assembly tolerances, this approach is expensive and impractical.

Another reduction method that can be applied to an array that only transmits or receives is the use of nonreciprocal devices (Fig. 7.35). Circulators, nonreciprocal phase shifters, and amplifiers attenuate reflected signals traveling in the reverse direction. These types of devices have become more common with the increase in the so-called *solid-state arrays*, which have the entire feed and radiating elements printed on a common substrate.

Example 7.10: In-Band Radar Cross Section Reduction Using Circulators

The RCS of the array in Example 6.9 is recomputed for the case of a circulator located behind each element [34]. Any reflection occurring from devices behind the circulator is reduced by its isolation, which can be anywhere from 10 to 40 dB. Figure 7.36 shows the array RCS for $N = 50$, $d = 0.5\lambda_0$, all device reflection coefficients 0.2, and the beam scanned to 45 deg as in Fig. 6.25. The circulator isolation is 15 dB. Note that the reflection at the circulator input contributes a sinc pattern with maximum at 0 deg. Thus, control of the circulator input VSWR is important in controlling the level of this lobe. However, scattering from the radiating element also contributes to this maximum, and its level is usually much greater than that of the circulator.

Circulators are effective in suppressing reflections from points inside the feed network, but they add complexity and therefore increase cost. Circulators usually contain ferrites, which are heavy, and the size of the device increases with the amount of isolation provided.

(Continued)

Example 7.10: In-Band Radar Cross Section Reduction Using Circulators *(Continued)*

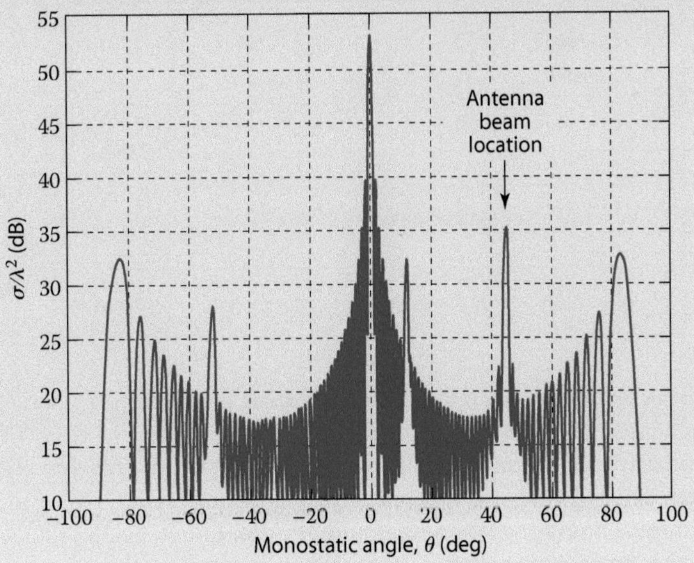

Fig. 7.36 Radar cross section of an array with circulators behind each element.

Nonreciprocal phase shifters can be used for limited control of the RCS of a receive antenna. Signals reflected from points behind the phase shifter pass through unattenuated, but the phase shift in the reverse direction is not the same as the phase shift in the forward direction. Recall, from the formulas of Sec. 6.7, that the peak RCS lobe for the scanning mismatches occurs when

$$2\alpha + 2\chi_0 = 0 \tag{7.60}$$

where $\chi_0 = kd \sin\theta_s$ is the beam-scanning phase [Eq. (6.40)]. Thus, the phase introduced by the shifter compensates for the path difference between elements, yielding a maximum RCS in the direction θ_s. For a nonreciprocal phase shifter, the shift in the reverse direction is independent of χ_0. Consequently, the phase shifter adds nothing to the phase slope of the reflected wave, and Eq. (7.60) becomes

$$2\alpha = 0 \tag{7.61}$$

Thus, the RCS has a maximum in the direction of θ_s only when $\theta_s = 0$. Retrodirective scanning lobes have been eliminated.

Reflectionless filters [35] are another potential antenna RCS reduction tool. Conventional filters reflect threat signals outside of the passband, but reflectionless filters absorb the signals. The out-of-band RCS contributions from the feed network (or any device behind the filter) are reduced by an amount determined by the filter's round-trip insertion loss at the threat frequency.

7.10 Cavity Cross Section Reduction

Cavities formed by aircraft cockpits and engine intakes can enhance RCS because of the complex wave scattering that occurs internally, as described in Sec. 6.11. A cavity inlet should be shielded from the threat as much as possible. When the inlet is visible, RCS can be reduced somewhat by bending the cavity or lining it with an absorbing material.

Application of an absorbing coating approximately reduces the RCS in direct proportion to the reduction in reflection coefficient of the material. Referring to Example 6.14, the reflected ray intensity for a lossy wall is reduced relative to that for a perfect conductor (which is -1). For example, if a ray traced through the cavity reflects three times, as shown in Fig. 6.51, the RCS is reduced by a factor of $(|\Gamma|^2)^3$. The 2 in the exponent occurs because Γ is a voltage coefficient and the 3 because there are three reflections. For example, if the reflection coefficient of the lining is 0.5, the RCS is reduced by approximately 18 dB.

Some metal matrix composites like boron/aluminum are used in engine components, but they are relatively good conductors. Any candidate material

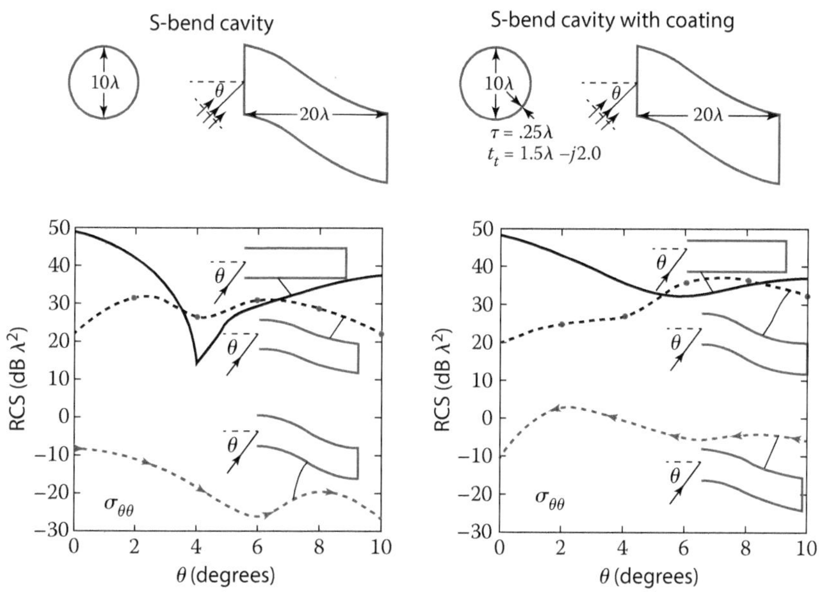

Fig. 7.37 Radar cross section of jet engine cavities: straight and S-bend, with and without coating (from Ref. [36]).

for the wall coating must be capable of maintaining structural integrity at the high temperatures of a jet engine.

For a cavity with straight walls, the characteristics of the back wall primarily determine the RCS (except at wide angles). Significant RCS reductions can be achieved by simply coating the back wall, reshaping it, or removing it altogether. Jet cavities are more complicated in that they contain compressors, turbine blades, diffusers, and so forth, which cannot be reshaped or removed. S-bend cavities have been used to hide these obstacles, as shown in Fig. 7.37 (see Ref. 36). The greatest RCS reduction occurs when both shaping and coating are applied. This approach has been used in the design of the B-2 engine cavity [37].

The cockpit of an aircraft also resembles a cavity. Waves that penetrate the canopy can be scattered by internal objects. Direct illumination of the cockpit is not a problem in most situations; the radar must be above the aircraft looking down. If it is necessary to prevent a wave from entering the cavity, however, a thin metallic film can be printed on the window material. A simple example of this is window tinting of the type used in automobiles. When the canopy becomes more reflective, the use of a low-profile shape is effective.

7.11 Cloaking

A *cloak* is a cover or shield that isolates a region of space from an illuminating field. The concept has been a favorite of science fiction writers for years, and has been demonstrated recently using metamaterials [38, 39].

There are current-based and ray-based approaches to cloaking. Using Love's equivalence principle, it may be possible to find electric and magnetic currents that can be impressed on the cloak surface to give zero fields inside and some specified fields outside. The outside fields are those in the absence of the object, at least for some limited region of space.

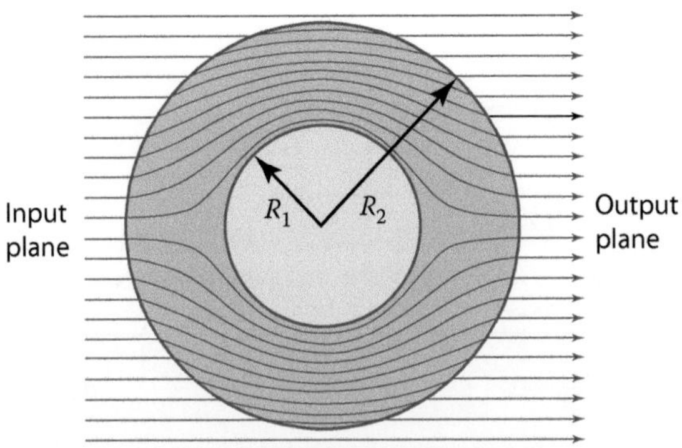

Fig. 7.38 Ray-based approach to cloak design (from Ref. [38]).

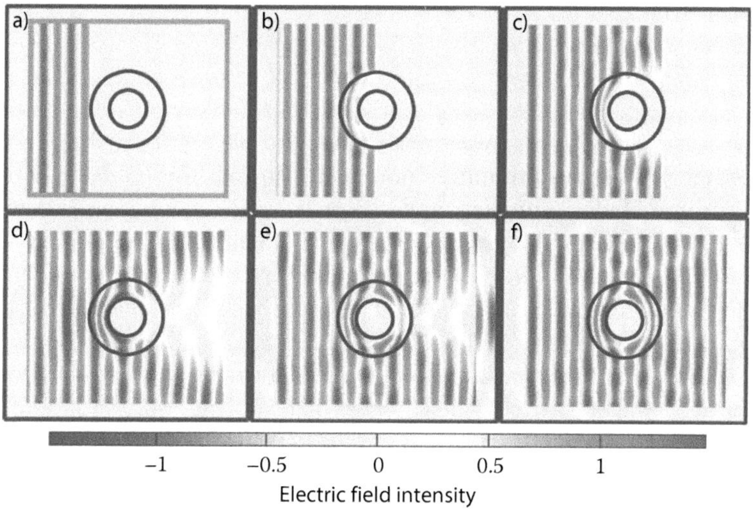

Fig. 7.39 Field around an electromagnetic cloak at various time steps (a) through (f) (from Ref. [39]).

For a ray-based approach, a wavefront at the input plane in the direction of incidence must be transferred to the output plane in the direction of the observer. A solution for all angles is the spherical structure shown in Fig. 7.38. Equal optical path lengths are needed in spite of the differing physical lengths. This requires a medium where phase advance is possible. DNG metamaterials and plasmas have this property for the phase velocity. Figure 7.39 shows a simulated time sequence of wave propagation around a metamaterial cloak [39].

Carpet cloaks [40] have been proposed to hide the presence of ground based objects. In Fig. 7.40 the carpet cloak produces the specular reflection

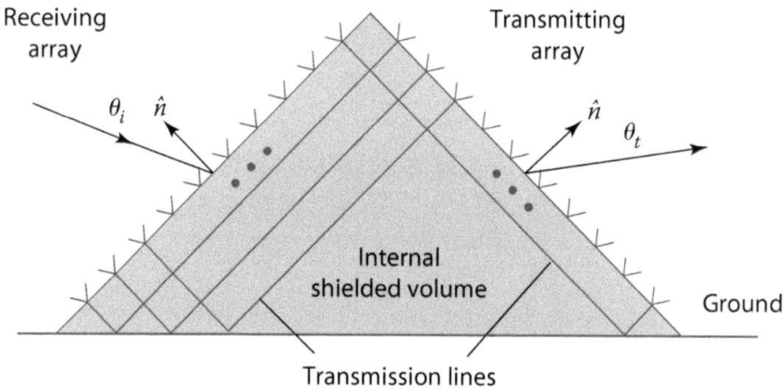

Fig. 7.40 Carpet cloak using transmission lines.

characteristics of a planar surface, thereby hiding an internal object from a threat sensor. The proper phases are obtained using transmission lines, and so these structures are also referred to as *transmission-line cloaks.*

A major obstacle in realizing a practical cloak is the differing phase and group velocities in a metamaterial. The phase velocity can be increased, but the group velocity cannot be increased using passive media and devices [41]. In a practical application other interesting and fundamental problems arise, such as how an EM sensor enclosed in a cloak would perform.

7.12 Summary: Radar Cross Section Design Guidelines

For any system or platform, RCS reduction is achieved at the expense of almost every other performance measure. In the case of an aircraft, stealth has resulted in decreased aerodynamic performance and increased complexity. Complexity, of course, translates into increased cost. The trend in the design of new platforms has been to integrate all the engineering disciplines into a common process. Thermal, mechanical, aerodynamic, and RCS simulations are carried out in parallel, using a common database. Changes made to the structure or materials for the purpose of decreasing RCS are automatically recorded in the databases for all other design simulations. This approach is called *concurrent engineering.*

The previous sections of this chapter illustrate the complexity of the RCS reduction problem. Treating the most intense sources of scattering is the easiest, and the payoff in decibels is greatest. For example, just tilting a surface a few degrees can reduce the RCS presented to a monostatic radar by 20 or 30 dB. Further reduction is much more difficult because second-order scattering mechanisms (multiple reflections, diffractions, surface waves, etc.) become important. Therefore, it can be more costly to drop the RCS the next 5 dB than it was for the first 30 dB.

With that point in mind, it is evident that the guidelines for designing a low-RCS vehicle will not be as extensive as those for an ultralow-RCS vehicle. The guidelines for both vehicles have several basic points in common, however, and these are summarized as follows:

1. Design for specific threats when possible to minimize cost. Keep in mind the threat radar frequency, whether it is monostatic or bistatic, and the target aspect angles that will be presented to the radar.
2. Orient large, flat surfaces away from high-priority quiet zones.
3. Use lossy materials or coatings to reduce specular/traveling wave reflections.
4. Maintain tolerances on large surfaces and materials.
5. Treat trailing edges to avoid traveling wave lobes.
6. Avoid corner reflectors (dihedrals or trihedrals).
7. Do not expose cavity inlets; use a mesh cover, or locate the inlets out of view of the radar.

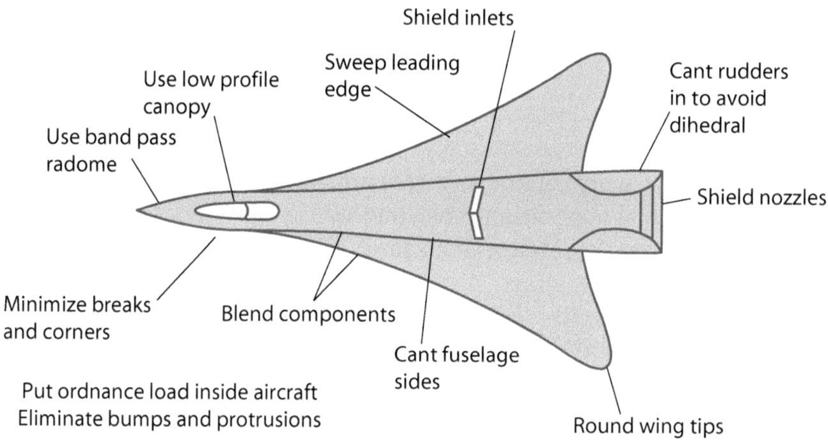

Fig. 7.41 Conceptual stealthy jet fighter with RCS reduction guidelines incorporated (from Ref. [1], courtesy AIAA).

8. Shield high-gain antennas from out-of-band threats.
9. Avoid discontinuities in geometry and materials to minimize diffraction and traveling wave radiation.

As an example of the application of these points, consider Fig. 7.41, which shows a generic stealthy design that has incorporated many of the points just listed [1].

7.13 Inverse Scattering and Radar Cross Section Synthesis

7.13.1 General Comments

The methods of RCS reduction just described are based primarily on a combination of experience, intuition, and trial and error. According to the uniqueness and equivalence theorems, there is a one-to-one relationship between the shape of a target, the boundary conditions on its surface, and its RCS. Thus, in principle, it is possible to determine the target's material composition and shape to achieve a specified RCS. This is RCS *synthesis* and is part of the more general problem of *inverse scattering* [42]. Inverse scattering attempts to extract information about an unknown target from its scattered field. The name is derived from the fact that this is the inverse of the usual *direct problem*, which involves finding the scattered field of a known target.

Inverse-scattering problems are categorized as *general* or *restricted* [42]. In the first case, nothing is known about the target; only the most general conditions on continuity and regularity hold. In the restricted case, some characteristic of the target is known, perhaps shape or material composition, but not both. Radar cross section synthesis generally falls into the second category. The target shape is known (e.g., a missile), but the surface materials (boundary conditions) are to be determined.

7.13.2 Time Domain

Inverse problems can be solved in either the time or frequency domain. In the time domain, the target is characterized by its *complex resonances.* Figure 7.42 shows a target illuminated by a plane wave. The received signal $r(t)$ is a delayed version of the transmitted signal $x(t)$ plus the target scattered signal $y(t)$. Therefore, the scattered signal is

$$y(t) = r(t) - x(t - \tau) \tag{7.62}$$

When a linear systems approach is used, the system transfer function is

$$H(s) = \frac{Y(s)}{X(s)} \tag{7.63}$$

and the system impulse response can be cast into the form

$$h(t) = \sum_{j=1}^{\infty} R_j e^{s_j t} + \Phi(t) \tag{7.64}$$

where R_j are the residues, s_j are the poles, and $\Phi(t)$ is a time-limited response.

Because the time response is observed over a finite time interval, the sum must be truncated at some finite value. Usually, the time-limited response is

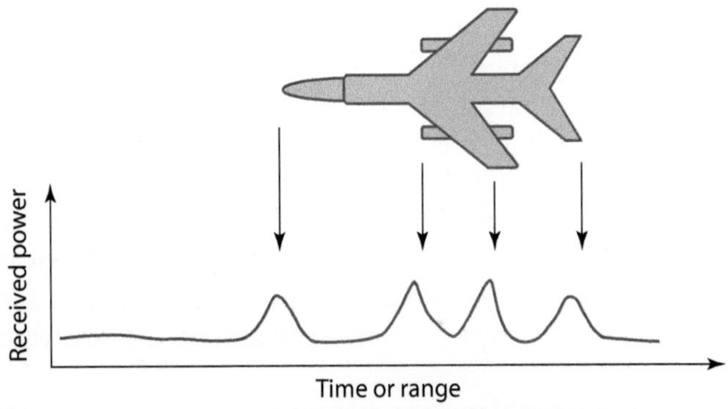

Fig. 7.42 Time domain scattering from a target.

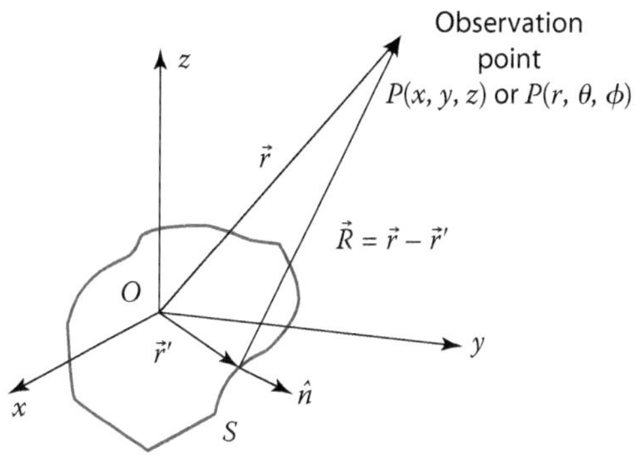

Fig. 7.43 Geometry for the generalized synthesis problem.

ignored, even though it may not be justified. Equation (7.64) becomes

$$h_N(t_k) = \sum_{j=1}^{N} R_j e^{s_j t_k} \tag{7.65}$$

Thus, the problem is reduced to identifying the poles and residues for a given input and output. This is a common system identification problem, and numerous techniques exist to solve it [43].

To date, this approach in extracting complex resonances from data has not been very successful for several reasons. The resonances in the frequency domain must be identifiable in the presence of noise and separable from each other. For targets whose response decays rapidly, early time data are crucial and $\Phi(t)$ is important.

7.13.3 Frequency Domain

In the frequency domain, the integral equations derived in Chapter 3 provide a basis for obtaining the surface impedance information from a target's RCS. Assume that the shape of the body is known, and let its surface be denoted S as shown in Fig. 7.43. The electrical properties of the body will be approximated by the normalized surface impedance Z_s, defined in Chapter 2. In general, both electric and magnetic currents can exist on the surface. The two are related by the surface impedance, as given in Eq. (2.101):

$$\vec{J}_{ms} = -Z_s \eta_0 \hat{n} \times \vec{J}_s \tag{7.66}$$

At an observation point P, the scattered electric field that arises from the currents $\vec{J}_s$ and $\vec{J}_{ms}$ radiating in free space is given by the radiation integral

Eq. (2.14):

$$\vec{E}_s(P) = -jk\eta_o \iint_S \left[\vec{J}_s G + \frac{\vec{J}_{ms} \times \nabla' G}{j\omega\mu} - \frac{\nabla' \cdot \vec{J}_s}{k^2} \nabla' G \right] ds' \tag{7.67}$$

where

$$G = \frac{e^{-jkR}}{4\pi R}$$

$$R = \sqrt{(x-x')^2 + (y-y')^2 + (z-z')^2}$$

As usual, the primed quantities refer to a current source point and the unprimed quantities to the observation point.

Using Eq. (7.66) in Eq. (7.67) eliminates the magnetic current from the integrand:

$$\vec{E}_s(P) = -jk\eta_0 \iint_S \left[\vec{J}_s G + \frac{jZ_s}{k} (\hat{n}' \times \vec{J}_s) \times \nabla' G - \frac{\nabla' \cdot \vec{J}_s}{k^2} \nabla' G \right] ds' \tag{7.68}$$

For the synthesis problem, $\vec{E}_s(P)$ is known and Z_s is unknown. The current $\vec{J}_s$ is also unknown. There are two approaches to solving for Z_s: 1) Provide an estimate of $\vec{J}_s$ at every point on the surface and solve the integral equation for Z_s, or 2) provide additional boundary conditions and simultaneously solve two integral equations for both unknowns, $\vec{J}_s$ and Z_s. In the following paragraphs, these are referred to as the *approximate method* and the *rigorous method*, respectively.

The additional information needed to solve the problem rigorously is available from Eq. (3.94) and is repeated here for convenience:

$$\begin{aligned} \frac{\vec{E}_i(\vec{r})|_{\tan}}{\eta_0} - Z_s(\vec{r})\hat{n}(\vec{r}) \times \vec{H}_i(\vec{r}) &= Z_s(\vec{r})\hat{n}(\vec{r}) \times \left[\nabla \times \iint_S \vec{J}_s G ds' \right. \\ &\left. + jk \iint_S Z_s (\hat{n}' \times \vec{J}_s) G ds' + \frac{j}{k} \iint_S \nabla' \cdot [Z_s \hat{n}' \times \vec{J}_s] \nabla G ds' \right] \\ &+ \left[jk \iint_S \vec{J}_s G ds' + \frac{j}{k} \iint_S (\nabla' \cdot \vec{J}_s) \nabla G ds' - \nabla \times \iint_S Z_s (\hat{n}' \times \vec{J}_s) G ds' \right]_{\tan} \end{aligned} \tag{7.69}$$

Recall that the quantities $\hat{n}'$, Z_s, G, and $\vec{J}_s$ in the integrands are functions of the source point coordinates even though the $\vec{r}'$ dependence has not been explicitly written. Thus, a rigorous solution requires that Eqs. (7.68) and (7.69) be solved simultaneously. In general, solving the two integral equations is a formidable task and, in many cases, it is impractical, if not impossible.

In the monostatic case, the current induced on the target depends on the angle of incidence and, therefore, the set of coupled integral equations must be solved at each aspect angle. Using this approach, the bistatic RCS needs to

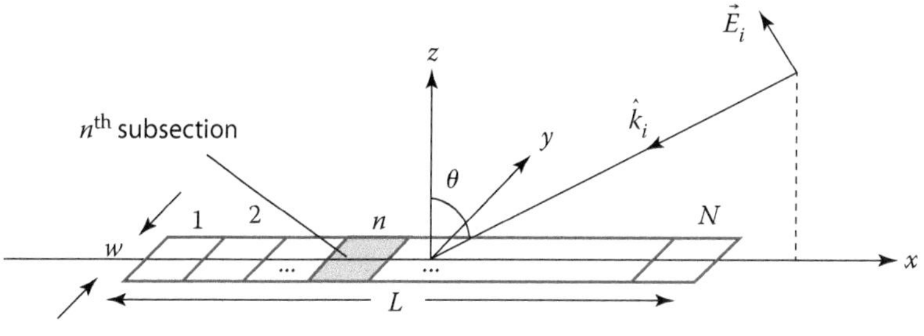

Fig. 7.44 Geometry for a thin strip synthesis example.

be specified at each incidence angle. For monostatic synthesis, however, only one scattering angle is of interest at each angle of incidence, that is, the one back in the direction of the radar. For each incidence angle, only one synthesis point is provided and all others are arbitrary. Any arbitrary value of RCS could be used for the remaining angles,[§] but this would result in a solution for Z_s that depends on incidence angle. In other words, the target's material composition must change with the angle of incidence, which is not possible for a passive target. This problem can be avoided by using an approximate solution with a current function that is independent of incidence angle.

A thin strip, as shown in Fig. 7.44, is one special case where the equations simplify considerably. Let us apply an approximate method to find the surface resistivity distribution along the strip that is required to achieve a given RCS pattern. Let the wave be incident in the $\phi = 0$ plane and TM_z polarized:

$$\vec{H}_i = \hat{\phi}\frac{E_o}{\eta_o}e^{-j\vec{k}\cdot\vec{r}} \tag{7.70}$$

For the approximate solution, an estimate of the resistivity is obtained by specifying $\vec{J}_s$ and solving Eq. (7.68) for Z_s. It is possible, by judicious choice of a current function, to make the solution angle independent. Let the current be given by

$$\vec{J}_s = 2\cos\theta_i(1 - R_s')\,\hat{n} \times \vec{H}_i \tag{7.71}$$

This form was arrived at empirically by comparing the monostatic RCS computed using this current to the actual RCS computed using the method of moments. When $R_s' \ll 1$ and $\cos\theta_i \approx 1$, Eq. (7.71) approaches the conventional PO current approximation. However, for the strip, the approximation gives excellent results over a much larger range of values.

[§]In practice the specified RCS values cannot be completely arbitrary in that they are constrained by the laws of physics.

For monostatic scattering, the far-field form of Eq. (7.67) becomes

$$E_\theta(\theta) = \frac{-jke^{-jkr}}{2\pi r} E_o w \cos\theta \int_{-L/2}^{L/2} (1 - R_s')e^{j2kx'\sin\theta} dx' \tag{7.72}$$

Now we use an MM approach and expand the resistivity into a series with pulse basis functions:

$$R_s'(x') = \sum_{n=1}^{N} a_n p_n(x') \tag{7.73}$$

Substituting Eq. (7.73) into Eq. (7.72) and using the properties of the pulse basis functions to evaluate the integral gives

$$E_\theta(\theta) = F(\theta) \sum_{n=1}^{N} b_n e^{j2kx_n \sin\theta} \tag{7.74}$$

where $b_n = 1 - a_n$ and

$$F(\theta) = \cos\theta \frac{w\Delta \sin(k\Delta \sin\theta)}{k\Delta \sin\theta} \tag{7.75}$$

Now Eq. (7.74) can be enforced in M directions for which the RCS is specified $(\theta_1, \theta_2, \ldots \theta_M)$, yielding M equations that can be written in matrix form as

$$\mathbf{E} = \mathbf{Ab} \tag{7.76}$$

where $E_m = E_\theta(\theta_m)$ With $F(\theta_m)$ denoted by F_m, the elements of $\mathbf{A}$ are

$$A_{mn} = F_m e^{j2kx_n \sin\theta_m}. \tag{7.77}$$

The vector of coefficients, **b**, can be solved for using standard matrix methods if $M = N$ or using a generalized inverse if $M < N$ [44].

Example 7.11: Approximate Monostatic RCS Synthesis for a Resistive Strip

An application of the approximate method for monostatic RCS synthesis is illustrated beginning with Fig. 7.45, which shows the monostatic RCS of a perfectly conducting strip (dashed) and the desired low-RCS pattern (solid). The synthesis procedure is initiated by sampling the desired pattern and using the samples to form the vector **E** in Eq. (7.76). The matrix equation can be solved for **a**. The normalized synthesized resistivity distribution using 17 far-field points is shown in Fig. 7.46. (A linearly tapered normalized resistivity from

(Continued)

Example 7.11: Approximate Monostatic RCS Synthesis for a Resistive Strip *(Continued)*

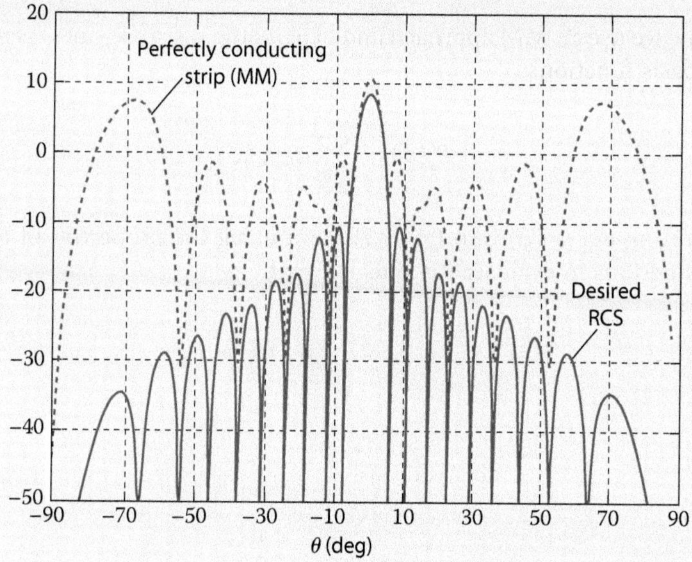

Fig. 7.45 Monostatic RCS of a PEC strip and the desired RCS to be synthesized in Example 7.11.

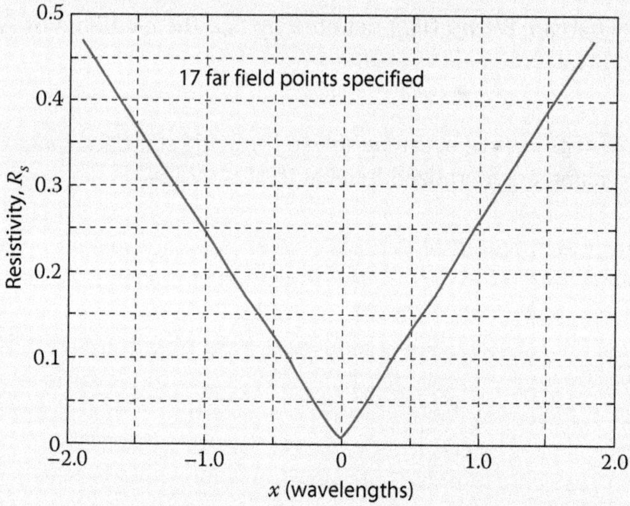

Fig. 7.46 Synthesized resistivity for the strip (Example 7.11).

(Continued)

Example 7.11: Approximate Monostatic RCS Synthesis for a Resistive Strip *(Continued)*

0.5 at the edge to 0 at the center was used to generate the desired pattern.) When the synthesized distribution was reinserted into the MM code, the dashed pattern of Fig. 7.47 was obtained. The original RCS is also plotted for reference.

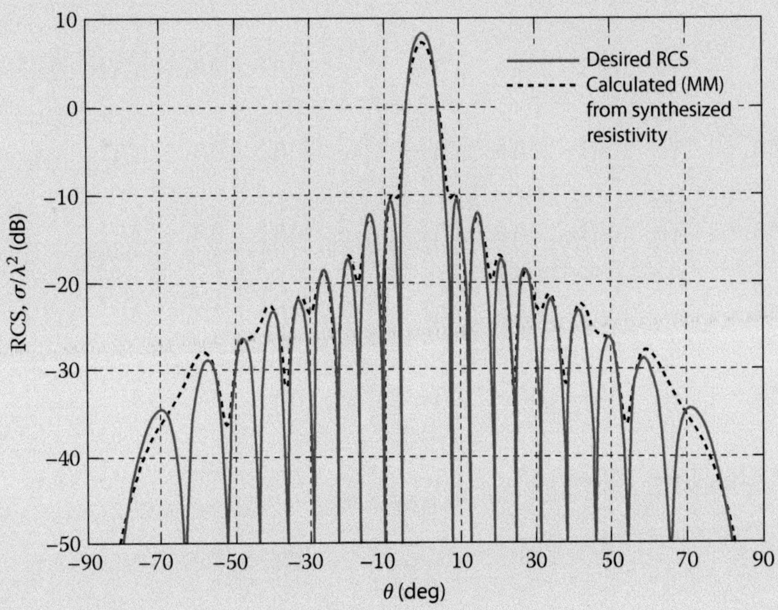

Fig. 7.47 Comparison of the desired RCS and synthesized RCS for Example 7.11.

The synthesis procedure generally gives accurate results for the simple strip geometry but, as expected, the synthesized distribution is sensitive to several parameters. The Nyquist sampling rate for the particular RCS pattern must be satisfied. For a sinc(kua) type of pattern, this is generally in the range of one sample per half-wavelength for bistatic scattering or per quarter-wavelength for monostatic scattering. (This is approximately the lobe spacing of the scattering pattern.) Sampling periodically in direction cosine u is more effective than sampling in θ. For other pattern shapes, the sampling interval must be increased or decreased, depending on the rate of variation of the pattern with angle.

Numerical problems can occur for several reasons. One is the possibility of a large spread in the elements of **E**. For patterns with regions of low and high RCS, the spread can lead to inaccurate results due to numerical roundoff. In the approximate solution, the assumed current distribution affects the accuracy. In cases in which the synthesized impedance distribution does not support surface waves, the PO approximation for the current gives excellent results.

The usual MM issues apply to this synthesis procedure, in particular, the choice of basis functions and the impact on convergence. It is well known that, in the conventional MM solution for the current, overlapping subsectional basis functions converge faster than pulses. For the rigorous solution, however, it is more convenient to avoid overlap between subsections so that the equations simplify. The final step in implementing a synthesized impedance distribution is selecting the corresponding material or coating that can be applied to the body.

References

[1] Fuhs, A. E., *RCS Lectures*, AIAA, New York, 1984.
[2] Inan, U. S., and Inan, A. S., *Electromagnetic Waves*, Prentice-Hall, Upper Saddle River, NJ, 2000.
[3] Jackson, J. D., *Classical Electrodynamics*, 2nd ed., Wiley, New York, 1975.
[4] Cheng, D. C., *Field and Wave Electromagnetics*, 2nd ed., Addison Wesley, New York, 1992.
[5] Taylor, J. D., (ed.), *Ultra – Wideband Radar*, CRC Press, Boca Raton, FL, 1991.
[6] Ishimaru, A., Lee, S. W., Kuga, Y., and Jandhyala, V., "Generalized Constitutive Relations for Metamaterials Based on the Quasi – Static Lorentz Theory," *IEEE Transactions on Antennas and Propagation*, Vol. 51, No. 10, 2003, pp. 2250 – 2557.
[7] Kong, J. A., "Theorems of Bianisotopic Media," *Proceedings of the IEEE*, Vol. 60, No. 9, 1972, pp. 1036 – 1046.
[8] Lindell, I. V., Sihvola, A. H., Tretyakov, S. A., and Viitanen, A. J., *Electromagnetic Waves in Chiral and Bi – isotropic Media*, Artech House, Norwood, MA, 1994.
[9] Engheta, N., and Jaggard, D. L., "Electromagnetic Chirality and Its Applications," *IEEE Antennas and Propagation Society Newsletter*, Oct. 1968, pp. 6 – 12.
[10] Smith, D. R., Padilla, W. J., Vier, D. C., Nemat – Nasser, S. C., and Schultz, S., "Composite Medium with Simultaneously Negative Permeability and Permittivity," *Physical Review Letters*, Vol. 84, No. 18, 2000, pp. 4184 – 4187.
[11] Ziolkowski, R. W., and Heyman, E., "Wave Propagation in Media Having Negative Permittivity and Permeability," *Physical Review E*, Vol. 64, pp.056625-1 – 056625-15.
[12] Cory, H., and Zach, C., "Wave Propagation in Metamaterial Multi – layered Structures," *Microwave and Optical Technology Letters*, Vol. 40, No. 6, 2004, pp. 460 – 465.
[13] Barrie, D., "LO and Behold," *Aviation Week and Space Technology*, 11 Aug. 2003, pp. 50 – 53.
[14] Weston, V. H., "Theory of Absorbers in Scattering," *IEEE Transactions on Antennas and Propagation*, Vol. AP-11, No. 5, Sept. 1963, pp. 578 – 584.
[15] Knott, E., Shaeffer, J., and Tuley, T., *Radar Cross Section*, 2nd ed., Artech House, Norwood, MA, 1993, Chap. 8.
[16] Collin, R. E., *Field Theory of Guided Waves*, McGraw-Hill, New York, 1960.
[17] Schwartz, M., *Composite Materials Handbook*, 2nd ed., McGraw-Hill, New York, 1992.
[18] Walser, R. M., *A Study of Thin Film Magnetodielectrics*, Ph.D. Dissertation, Dept. of Electrical Engineering, Univ. of Michigan, Univ. Microfilms, Inc., Ann Arbor, MI, 1967.
[19] Pendry, J. B., Holden, A. J., Stewart, W. J., and Youngs, I., "Extremely Low Frequency Plasmons in Metallic Mesostructures," *Physical Review Letters*, Vol. 76, No. 25, 1996, p. 4773.

[20] Pendry, J. B., Holden, Robbins, D. J., and A. J., Stewart, W. J., "Magnetism from Conductors and Enhanced Nonlinear Phenomena," *IEEE Trans. on Microwave Theory and Techniques,* Vol. 47, No. 11, Nov. 1999, pp. 2075–2084.

[21] Musal, S., and Smith, S., "Universal Design Chart for Specular Absorbers," *IEEE Transactions on Magnetics,* Vol. 26, No. 5, 1990, p. 1462.

[22] Yucelik, K., *Radar Absorbing Material Design*, M.S. Thesis, Naval Postgraduate School, Monterey, CA, Sept. 2003.

[23] Rozanov, K. N., "Ultimate Thickness to Bandwidth Ratio of Radar Absorbers," *IEEE Transactions on Antennas and Propagation,* Vol. 48, No. 8, 2000, p. 1230.

[24] Ruck, G., Barrick, D., Stuart, W., and Kirchbaum, C., *Radar Cross Section Handbook,* Plenum Press, New York, 1970.

[25] Wallace, J., "Broadband Magnetic Microwave Absorbers: Fundamental Limitations," *IEEE Transactions on Magnetics,* Vol. 29, No. 6, 1993, p. 4209.

[26] Stealth, *Great Fighting Jets*, series, Time–Life Video (V648-01).

[27] Yong, K. M., *Radar Cross Section Reduction: Geometric Control of Discontinuities Using Serrated Edges,* M.S. Thesis, Monterey, CA, Naval Postgraduate School, March 1998.

[28] Davis, C. W., Tomljanovich, N. M., Kramer, D. R., and Poirier, J. L., "Characteristics of Ultra Wide Band Radars Pertinent to Air Defense," *Proceedings of the 1st Los Alamos Symposium on UWB Radar,* CRC Press, Boston, 1991, pp. 501–518.

[29] Fowler, C. A., "The UWB (Impulse) Radar Caper," *IEEE AES Systems Magazine,* Dec. 1992, pp. 3–5.

[30] Dissado, L. A., and Hill, R. M., "A Cluster Approach to the Structure of Imperfect Materials and Their Relaxation Spectroscopy," *Proceedings of the Royal Society of London, A390,* 1983, pp. 131–180.

[31] Barrett, T. W., "Energy Transfer and Propagation and the Dielectrics of Materials: Transient Versus Steady State Effects," *Proceedings of the 1st Los Alamos Symposium on UWB Radar,* edited by B. Noel , CRC Press, Boston, 1991, pp. 1–19.

[32] Barrett, T. W., "Energy Transfer Through Media and Sensing of the Media," *Introduction to Ultra–Wideband Radar Systems,* edited by J. D. Taylor , CRC Press, Boca Raton, FL, 1995, pp. 365–434.

[33] Haykin, S., "Cognitive Radar: A Way of the Future," *IEEE Signal Processing Magazine,* Vol. 23, No. 1, June 2006, pp. 30–40.

[34] Knop, C. M., "Radar Echo Reduction Using Circulators," *IEEE Transactions on Antennas and Propagation,* Vol. AP–14, No. 6, Nov. 1966, pp. 789–791.

[35] Morgan, M. A., *Reflectionless Filters,* Artech House, Norwood, MA, 2017.

[36] Ling, H., Chou, R.-C., and Lee, S.-W., "Shooting and Bouncing Rays: Calculating the RCS of an Arbitrarily Shaped Cavity," *IEEE Transactions on Antennas and Propagation,* Vol. AP–37, No. 2, Feb. 1989, pp. 194–205.

[37] Ashley, S., and Gilmore, C., "Finally: Stealth," *Popular Science,* July 1988, p. 460.

[38] Pendry, J. B., Schurig, J. B., and Smith, D. R., "Controlling Electromagnetic Fields," *Science,* Vol. 312, 23 June 2006, pp. 1780–1782.

[39] Liang, Z., Yao, P., Sun, X., and Jiang, X., "The Physical Picture and the Essential Elements of the Dynamical Process for Dispersive Cloaking Structures, *Applied Physics Letters,* Vol. 92, 2008, 131118.

[40] Ramaccia, D., Tobia, A., Allessandro, T., and Bilotti, F. "Antenna Arrays Emulate Metamaterial-Based Carpet Cloak Over a Wide Angular and Frequency Bandwidth," *IEEE Transactions on Antennas and Propagation,* Vol. 66, No. 5, May 2018, pp. 2346–2352.

[41] Alitalo, P., and Tretyakov, S., "On Electromagnetic Cloaking: General Principles, Problems and Recent Advances Using the Transmission-line Approach," XXIX URSI General Assembly, 7–16 Aug. 2008, Chicago.

[42] Dudley, D. G., "Progress in Identification of Electromagnetic Systems," *IEEE Antennas and Propagation Society Newsletter,* Aug. 1988, pp. 5–11.

[43] Ljung, L., *Systems Identification, Theory for the User,* Prentice-Hall, Englewood Cliffs, NJ, 1987.

[44] Rao, C., and Mittra, S. K., *Generalized Inverse of Matrices and Its Applications,* Wiley, New York, 1971.

[45] Gedney, S. D., "An Anisotropic Perfectly Matched Layer-Absorbing Medium for the Truncation of FDTD Lattices," *IEEE Transactions on Antennas and Propagation,* Vol. 44, No. 12, 1996, pp. 1630–1639.

[46] Sacks, Z., Kingsland, D., Lee, R., and Lee, J., "A Perfectly Matched Anisotropic Absorber for Use as an Absorbing Boundary Condition," *IEEE Transactions on Antennas and Propagation,* Vol. 43, No. 12, Dec. 1995, pp. 1460–1463.

[47] Caloz, C., Lee, C. J., Smith, D., Pendry, J., and Itoh, T., "Existence and Properties of Microwave Surface Plasmons at the Interface Between a Right–Handed and Left–Handed Media," *2004 IEEE International Symposium on Antennas and Propagation Digest,* Vol. 3, pp. 3151–3154.

[48] Chambers, B., and Tennant, A., "A Smart Radar Absorber Based on the Phased–Switched Screen," *IEEE Transactions on Antennas and Propagation,* Vol. 53, No. 1, Jan. 2005, pp. 394–403.

[49] Galarregui, J. C. I., Pereda, A., de Falcon, J. L. M., Ederra, I., Gonzalo, R., and de Maagt, P. B., "Broadband Radar Cross-Section Reduction Using AMC Technology," *IEEE Transactions on Antennas and Propagation,* Vol. 61, No. 12, Dec. 2013, pp. 6136–6143.

Problems

7.1 Prove Weston's first theorem.

7.2 Prove Weston's second theorem.

7.3 Plot the reflection coefficient of a Salisbury screen as a function of normalized frequency, f/f_0, where f_0 is the design frequency at which the film/ground plane spacing is $\lambda_0/4$. Use values of $R_s = 377$, 320, and 250 Ω/square.

7.4 A radome panel consists of honeycomb sandwiched between two dielectric layers as shown in the Fig. P7.4. Using a transmission-line

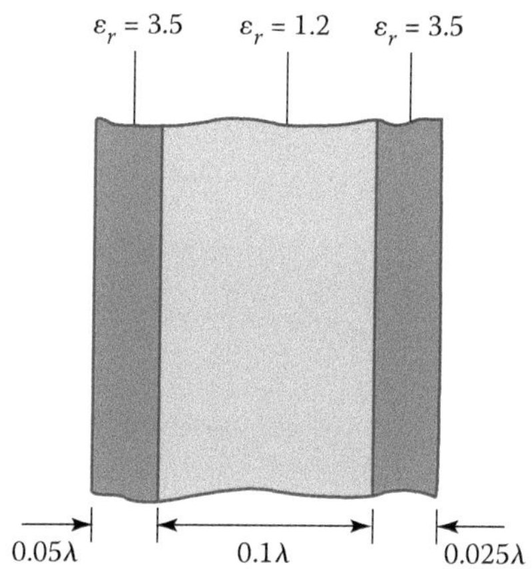

Fig. P7.4

equivalent, find the plane wave reflection coefficient and surface impedance for a wave normally incident from the left. Repeat the calculation for a wave incident from 45 deg off the interface normal. The thicknesses are in free-space wavelengths.

7.5 Given that a Dallenbach layer uses a material with $\epsilon_r = 5$ and $\mu_r = 1$, find the electric loss tangent $\tan\delta_\varepsilon$ that provides a 10-dB reduction in RCS. Assume that the magnetic loss tangent is zero.

7.6 The monostatic RCS (σ/λ^2) of the square plate shown in Fig. P7.6 (with sides of length $a = 5\lambda$) is to be canceled at an angle $\theta = 48.5$ deg, $\phi = 0$ deg, using an array of two shorted slots. The slots have a length $l = 0.5\lambda$ and are spaced a distance $d = 0.6\lambda$. They are oriented as shown in the figure so that a TM-polarized wave can be canceled.

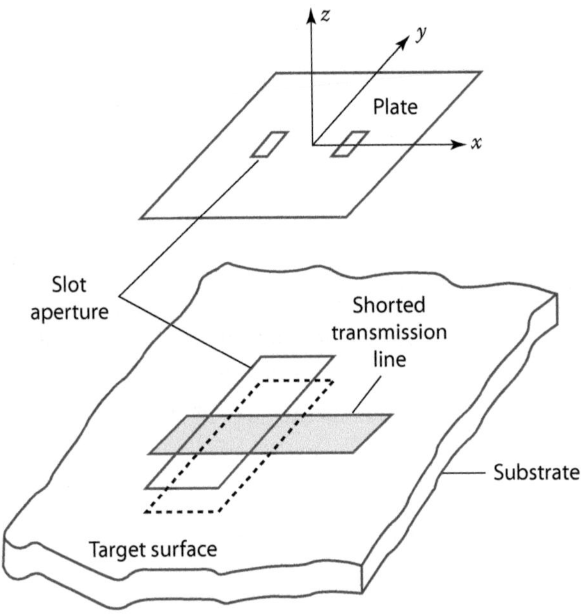

Fig. P7.6

a) If physical optics gives an accurate estimate of the plate RCS, find σ/λ^2 at the angle of interest. Include the phase of the RCS power pattern referenced to the plate center in your result.

b) Each slot is terminated with a shorted transmission line. Assuming that any reflection at the slot opening is negligible, the incident wave will travel down the transmission line, be reflected, and then be reradiated. The characteristics of the transmission line can be chosen to give a specific reflected wave amplitude and phase back at the slots. The phase ψ can be changed by adjusting the location and length of the transmission line at the back of each slot. The magnitude of the reflection Γ_0 can be changed by increasing or decreasing the loss in the material that fills the slot. Because this is a passive device (no amplifiers), $\Gamma_0 \leq 1$.

When viewed in the $x-z$ plane, the effective area of each slot is $A_e = ld\Gamma$, where $\Gamma = \Gamma_0 e^{j\psi}$. What is the monostatic RCS of the two-slot array? (Assume that the radiation mode dominates, and neglect mutual coupling.)

c) Specify the value of ψ for each slot to scan the RCS from the plate normal at $\theta = 0$ deg to the cancellation angle $\theta = 48.5$ deg. (Note that there are several possible answers because only the relative phase between the two slots is important.)

d) What value of Γ_0 is required to cancel the plate RCS?

e) Will there be any Bragg lobes in the visible region? If so, at what angles?

f) How would it be possible to cancel larger plate RCS values that require $\Gamma_0 > 1$?

7.7 A linear edge taper is to be used to reduce the wide angle RCS of a plate. Using the result of Problem 2.5, find the rate of decrease in RCS sidelobe height in the principal plane for angular regions away from the maximum. Express the answer in decibels per octave (*octave* refers to a doubling of direction cosine u). Discuss the requirements of the material that might be used to achieve the taper.

7.8 The array in Example 6.9 is to be used on receive only, and a low-noise amplifier (LNA) will be located behind each element. What reverse isolation is needed to drop the retrodirective lobe at 45 deg to a level below the specular skirt of the aperture reflections (i.e., those with a RCS maximum at $\theta = 0$ deg)?

7.9 Derive the wave equation for an anisotropic dielectric Eq. (7.15).

7.10 Derive the Poynting theorem for an anisotropic dielectric [Eqs. (7.16) and (7.17)].

7.11 An optical wave retarder is made of mica ($\sqrt{\epsilon_{xx}} = 1.599$ and $\sqrt{\epsilon_{zz}} = 1.594$) and operates at a wavelength of 63.3 μm. What thickness is required to give a phase difference of 180 deg between the two normal modes traveling along the x direction?

7.12 A square resistive sheet lies in the $x-y$ plane.

a) Use PO to calculate the bistatic RCS of the sheet for a wave incident from ($\theta = 30$ deg, $\phi = 45$ deg). Assume that the normalized resistivity of the sheet is zero at the center and tapers linearly to 0.5 at the edges along the x and y axes:

$$R_s'(x, y) = 0.5\frac{2x}{L}\frac{2y}{L}$$

where $L = 2\lambda$ is the length of the plate.

b) Use the result in part (a) and the approximate synthesis method with pulse basis functions to recover the resistivity distribution.

7.13 Consider a rectangular plate (Fig. 2.12) with reflection characteristics that are represented by a normalized anisotropic surface impedance matrix:

$$\mathbf{Z}_s = \begin{bmatrix} Z_{xx} & Z_{xy} \\ Z_{yx} & Z_{yy} \end{bmatrix}$$

The incident wave can have both TE and TM components as described by Eq. (2.38) and, therefore, the PO surface current is given by Eq. (2.41). For convenience, the rectangular components of $\vec{J}_s$ are defined as

$$J_{sx} = E_{0\theta} \cos\phi - E_{0\phi} \cos\theta \sin\phi$$

$$J_{sy} = E_{0\theta} \sin\phi + E_{0\phi} \cos\theta \cos\phi$$

a) Apply the definition of surface impedance [Eq. (2.111)], and show that the magnetic current, in matrix form, is given by

$$\begin{bmatrix} J_{msx} \\ J_{msy} \end{bmatrix} = 2\begin{bmatrix} Z_{xx} & Z_{xy} \\ Z_{yx} & Z_{yy} \end{bmatrix}\begin{bmatrix} J_{sx} \\ J_{sy} \end{bmatrix} e^{-j\vec{k_i}\cdot\vec{r}}$$

b) If the elements of $\mathbf{Z}_s$ are constant over the entire plate, show that the monostatic scattered field components at a far observation point P are

$$\begin{aligned} E_\theta(P) = &-E_M\{J_{sx}(Z_{xx} \sin\phi - Z_{yx} \cos\phi + \cos\theta \cos\phi) \\ &+ J_{sy}(Z_{xy} \sin\phi - Z_{yy} \cos\phi + \cos\theta \sin\phi)\} \\ E_\phi(P) = &E_M\{J_{sx}(-Z_{xx} \cos\theta \cos\phi - Z_{yx} \cos\theta \sin\phi - \sin\phi) \\ &+ J_{sy}(-Z_{xy} \cos\theta \cos\phi - Z_{yy} \cos\theta \sin\phi + \cos\phi)\} \end{aligned}$$

where

$$E_M = \frac{jk}{2\pi r} e^{-jkr} ab \,\text{sinc}(kua)\text{sinc}(kvb)$$

7.14 a) Using the universal chart, construct a RAM layer over PEC that uses a material with $\epsilon_r = 20 - j10$ and is effective against a threat radar operating at $f = 10$ GHz. Find μ_r assuming a maximum layer thickness of 15 mm. Verify the result from the chart using the Mathcad "find" command.

b) Find the propagation constant γ and intrinsic impedance η in the layer material of Part a. What is the loss of the material in dB/m? That is, calculate 20log $(\exp(-\alpha \cdot 1))$.

c) If the backing material is a composite with intrinsic impedance of 60 + j0 ohms, write the transcendental equation that must be solved for zero specular reflection.

7.15 Repeat Problem 7.4 using wave matrices.

7.16 Given a collision frequency of $10^3/s$, an electron density $N_e = 0.55 \times 10^{11}/\text{m}^3$, and wave frequency of 10 MHz, calculate the following for the plasma:

a) relative dielectric constant

b) phase velocity, u_p

c) attenuation in dB/km

d) intrinsic impedance

e) reflection coefficient at a planar boundary between the plasma and air.

7.17 (After Ref. [2]) During a spacecraft's reentry into the atmosphere there is a 1 m plasma sheath surrounding the vehicle with $N_e = 10^{13}/\text{cm}^3$ and $\nu = 10^{11}/s$. What communications frequency should be used for a minimum power loss of 10 dB through the plasma? You can neglect the presence of the earth's magnetic field. (The answer is 343 GHz, a very high frequency compared to the plasma frequency. Use this fact to simplify the formulas that need to be solved.)

7.18 A plasma has $N_e = 10^{15}/\text{m}^3$ and $\nu = 0/s$. For a plane wave of frequency 500 MHz:

a) What is the phase velocity in the plasma?

b) What is the phase shift of the plane wave after it has traveled 1 m in free space?

c) What is the phase shift of the plane wave after it has traveled 1 m in the plasma?

7.19 The B-52 *Stratofortress*, shown in Fig. P7.19, has been described as "how not to make a stealthy aircraft." Give a brief explanation with specific reasons why the B-52's RCS is high.

Fig. P7.19 (USAF Photo)

7.20 The objective of this problem is to design a *perfectly matched layer* (PML) [45]. It is a material that has no reflection for any polarization, frequency, and angle of incidence. Consider a plane wave that is incident at an angle θ_i onto a boundary between free space and a material with ε and μ equal everywhere. Let the permittivity and permeability matrices be

$$\boldsymbol{\epsilon} = \begin{bmatrix} a & 0 & 0 \\ 0 & a & 0 \\ 0 & 0 & b \end{bmatrix}, \quad \boldsymbol{\mu} = \begin{bmatrix} c & 0 & 0 \\ 0 & c & 0 \\ 0 & 0 & d \end{bmatrix}$$

Find the conditions on the complex quantities a, b, c and d so that there is no reflection for all frequencies, angles, and polarizations. (You should find that $c = a$, $b = 1/a$ and $d = 1/c$.)

7.21 For the PML of Problem 7.20 [46] show that Snell's law at the boundary can be written as $\sqrt{bc}\sin\theta_t = \sin\theta_i$ and the Fresnel reflection coefficients for parallel and perpendicular polarizations can be expressed as

$$\Gamma_\perp = \Gamma_{\mathrm{TE}} = \frac{\cos\theta_i - \sqrt{b/a}\cos\theta_t}{\cos\theta_i + \sqrt{b/a}\cos\theta_t}$$

$$\Gamma_\parallel = \Gamma_{\mathrm{TM}} = \frac{\sqrt{b/a}\cos\theta_t - \cos\theta_i}{\cos\theta_i + \sqrt{b/a}\cos\theta_t}$$

Let $a = b = A - jB$ and $c = 1/(A - jB)$ where $A = B = 1$. (Note that B controls the attenuation of the layer.)

a) Verify that the reflection coefficients are zero.

b) How would you classify this material?

c) Is this material physically realizable?

7.22 Repeat Example 7.9 using POFACETS as the simulation software.

7.23 A plane wave is incident onto a DNG half space ($\epsilon_{r1} = \mu_{r1} = 1$, $\epsilon_{r2} = -4$, $\mu_{r2} = -1$) at an angle of 29 deg from the surface normal, as depicted in Fig. P7.23. Find the impedance of the DNG material, the transmission angle, and the reflection coefficients for both parallel and perpendicular polarizations.

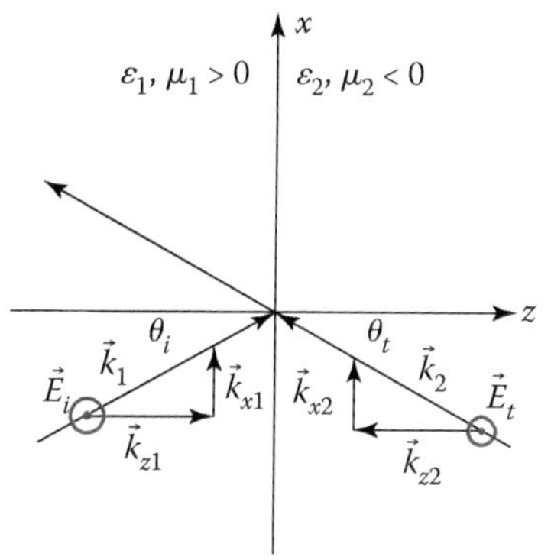

Fig. P7.23

7.24 Referring to Fig. P7.23, where region 1 is not necessarily free space, consider the following:

a) Use the separation equations (i.e., $k_n^2 = k_{xn}^2 + k_{zn}^2$, for regions $n = 1, 2$) to show that the reflection coefficient can be expressed as [47]

$$\Gamma_\perp = \frac{\mu_{r2}k_{z1} - \mu_{r1}k_{z2}}{\mu_{r2}k_{z1} + \mu_{r1}k_{z2}}$$

b) When the reflection coefficient is zero (i.e., at Brewster's angle), surface plasmons are excited. Show that for this condition

$$k_{x_\perp} = \frac{\omega}{c}\sqrt{\frac{\mu_{r1}\mu_{r2}(\epsilon_{r1}\mu_{r2} - \mu_{r1}\epsilon_{r2})}{\mu_{r2}^2 - \mu_{r1}^2}}$$

Note that if one were to repeat the exercise for parallel polarization, the result would be

$$k_{x_\parallel} = \frac{\omega}{c}\sqrt{\frac{\epsilon_{r1}\epsilon_{r2}(\mu_{r1}\epsilon_{r2} - \epsilon_{r1}\mu_{r2})}{\epsilon_{r2}^2 - \epsilon_{r1}^2}}$$

c) Let $\epsilon_{r1} = \mu_{r1} = 1$, $\epsilon_{r2} = -1.3783$, $\mu_{r2} = -2.428$, and $f = 2.5$ GHz. Find $k_{x_\perp}$ and determine whether this a radiating (fast wave) or nonradiating (slow wave) condition.

d) Repeat Part c for $\epsilon_{r1} = \mu_{r1} = 1$, $\epsilon_{r2} = -0.5907$, $\mu_{r2} = -1.0545$, and $f = 3.3$ GHz.

7.25 A general form for the permittivity and permeability of a DNG material is

$$\epsilon_r(f) = 1 - \frac{A_e}{B_e[B_e - jC_e]}$$

$$\mu_r(f) = 1 - \frac{A_m}{B_m[B_m - jC_m]}$$

The B and C parameters (omitting the subscripts e and m) can be expressed as $\mathrm{B} = \sqrt{c^2 - b^2 - 1}$ and $C = -2bc/B$. Consider a material that has $A_e = 1$, $A_m = 0.8$, $b_e = b_m = 0.01$, $c_e = f/f_e$, $c_m = f/f_m$, $f_e = 1000$ GHz, and $f_m = 937.5$ GHz.

a) Plot the real and imaginary parts of the permittivity and permeability from 750 GHz to 3000 GHz.

b) Plot the reflection coefficient for a plane wave normally incident on the boundary between free space and the DNG for the same frequency range.

7.26 *A phase switched screen* (PSS) [48] can be thought of as an active version of a combination of a Salisbury screen and Dallenbach layer. It consists of a shunt conductance that can take on two values (say G_1 and G_2) located a distance d from a PEC. The region between the conductance and PEC has a characteristic admittance $Y_c = 1/Z_c$. By switching between G_1 and G_2 the input admittance changes from Y_{in_1} to Y_{in_2}. The input reflection coefficient changes from

$$\Gamma_1 = \frac{Y - Y_{\text{in}_1}}{Y + Y_{\text{in}_1}}$$

to

$$\Gamma_2 = \frac{Y - Y_{\text{in}_2}}{Y + Y_{\text{in}_2}}$$

where $Y = Y_0/\cos\,\theta$ for parallel polarization and $Y = Y_o\,\cos\,\theta$ for perpendicular polarization. Let the switching between the conductance values be a square wave in time:

$$G(t) = \begin{cases} G_1, & 0 \le t < \tau \\ G_2, & \tau \le t < T \end{cases}$$

Assume that the screen is operated at a quarter wave condition ($d\cos\theta = \lambda/4$) and that $\tau = T/2$. Find the relationship between G_1 and G_2 so that the time-averaged reflection coefficient is zero; that is,

$$\Gamma_{\mathrm{av}} = \frac{1}{T}\int_0^T \Gamma(t)\mathrm{d}t = 0$$

7.27 Derive the wave matrix equations for the case where $\mu_r \neq 1$.

7.28 A checkerboard arrangement of alternating PMC ($\Gamma = +1$) and PEC ($\Gamma = -1$) tiles can be used to reduce the RCS of a surface for near normal incidence [49]. Consider a square of edge length $L = 5\lambda$ consisting of N by N tiles (N even) in the $x-y$ plane. Calculate the monostatic RCS for $\phi = 20$ deg and 45 deg with tile sizes of 0.1λ, 0.5λ, and λ. Explain the behavior from the perspective of array theory.

7.29 For the carpet cloak shown in Fig. 7.40 find the requirement on the transmission line lengths to give an apparent specular reflection from a flat ground ($\theta_t = \theta_i$) when the vertex angle is 90 deg.

Chapter 8 Measurement of Radar Cross Section

8.1 Introduction

To establish the stealthiness of a platform, it could simply be built and then tested against the threat radar. The obvious danger in this approach is that deficiencies occurring in the early stages of the design would not be uncovered until completion and, thus, could not be corrected. Another complication is that the effective measured RCS will depend on the radar parameters and test environment. For these reasons, the precise quantity of RCS as defined in Chapter 1 (i.e., σ) is measured in a controlled environment. In general, measured RCS data can serve three purposes:

1. To establish a target's RCS for design verification.
2. For comparison with calculated RCS, to evaluate numerical methods, approximations, and so forth.
3. As a diagnostic or troubleshooting tool.

Measurements play an important part in the design of a stealthy platform. During the conceptual stage, scale models of candidate designs might be built and tested to verify that the RCS levels are in the ballpark and that no unexpected scattering arises. At various stages in the development of the full-scale components (such as wings and antennas), measurements can be used to verify assembly tolerances, RCS treatments, and other manufacturing processes. Problem areas can be reworked and then remeasured. Finally, if the target is not too large, a measurement can be made on the completed product. This step may serve as the official fulfillment of a contract.

Measurement facilities can be categorized by a wide variety of attributes, which include physical characteristics, instrumentation capabilities, and data analysis and presentation. Table 8.1 summarizes some commonly used descriptors.

Most systems measure monostatic RCS by using a fixed transmit/receive antenna and rotating the target [1]. First, a known calibration target like a sphere is measured, and its received power is recorded as a reference. Next, the sphere is replaced by the object with unknown RCS, and the process is repeated as the target is rotated. By comparing the received power to the reference power, the RCS of the unknown target is obtained. In older systems the scaled power difference is sent to a plotter that is slewed to the target mount. The final result is a plot of RCS vs aspect angle.

The classical far-field continuous wave (CW) RCS measurement is illustrated in Fig. 8.1. For a true far-field measurement, the target must be

Table 8.1 Descriptors Used in the Classification of RCS Measurement Facilities and Data Processing Capabilities

System Element	Descriptor
Physical configuration	Indoor/outdoor
	Near field
	Far field
	Compact
	Tapered
Instrumentation	Time domain
	Frequency domain
	Continuous wave (CW)
	Pulsed CW
Data analysis and presentation	Fixed frequency/variable aspect
	Fixed aspect/frequency sweep
	Two-dimensional frequency and aspect
	Time-domain trace
	Target imaging
	Polar or rectangular

illuminated by a plane wave. This condition is never rigorously satisfied in practice because the need for a directive transmit antenna usually results in a nonuniform amplitude illumination of the target. Furthermore, the wave incident on the target is always spherical, not planar. Finally, because the system components are near the surface, reflections from the ground and other obstacles will affect the measurement.

Several measures of chamber performance are 1) pattern symmetry, 2) noise floor (noise level), 3) quiet zone, 4) repeatability, 5) allowable

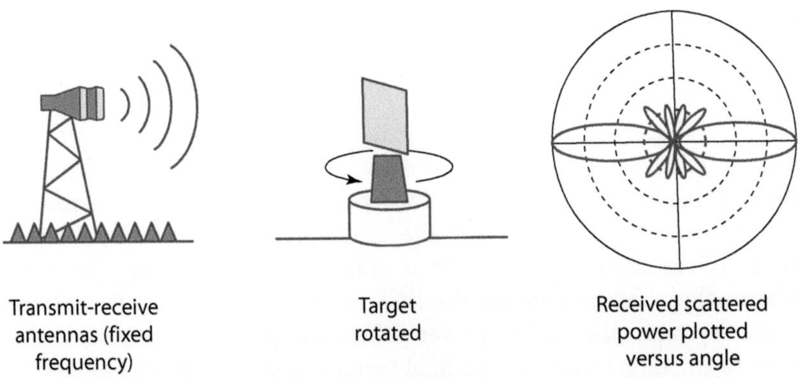

Fig. 8.1 Monostatic continuous wave (CW) RCS measurement.

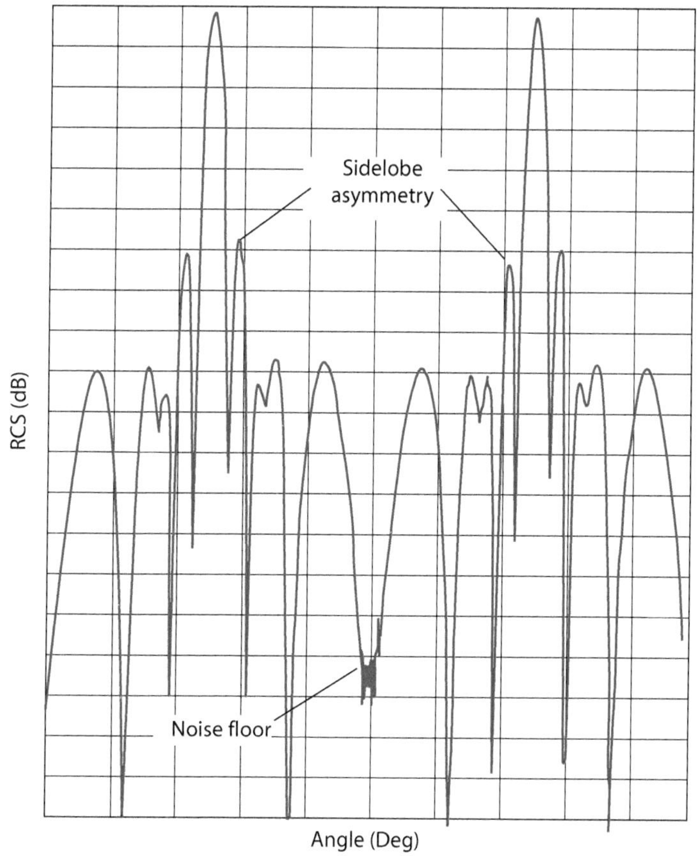

Fig. 8.2 Typical CW measurement of RCS vs aspect angle for a plate (principal plane).

target size and weight, 6) allowable range of target motion, 7) instrumentation capabilities, 8) data processing capabilities, and 9) data presentation capabilities. A sample plot from a CW measurement is shown in Fig. 8.2. It illustrates the types of inaccuracies that occur, such as sidelobe asymmetry and the existence of a *noise floor.* The noise floor sets a lower limit on the RCS than can be reliably measured. These will be discussed in some detail in the following sections.

8.2 Chamber Configurations

The majority of all RCS measurement facilities are completely enclosed (*indoor ranges*), providing isolation from the external environment, which is desirable for both security and technical reasons. An indoor chamber has reduced or eliminated interference from other radio frequency (rf) sources and a controlled temperature for increased instrumentation stability. The end result is more accurate and repeatable measurements than are possible in an outdoor facility. The obvious limitation to an indoor chamber is the size of the building required for large targets.

8.2.1 Conditions for Plane Wave Incidence

Figure 8.3 shows three types of indoor chambers. The first is the standard far-field configuration. The separation of the antenna and target must be large enough so that the difference between an ideal plane wave and the actual spherical wave is negligible. The required distance is a function of target size and the frequency. If the distance is R and the target length L, as indicated in Fig. 8.4, the phase error at the edge of the target relative to

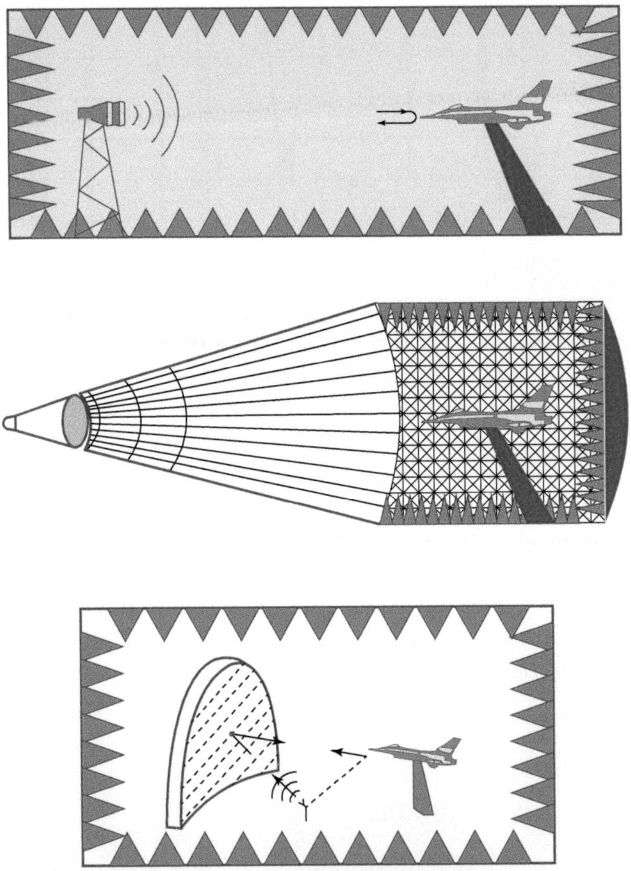

Fig. 8.3 Common indoor chamber configurations: farfield (top), tapered (center), and compact (bottom).

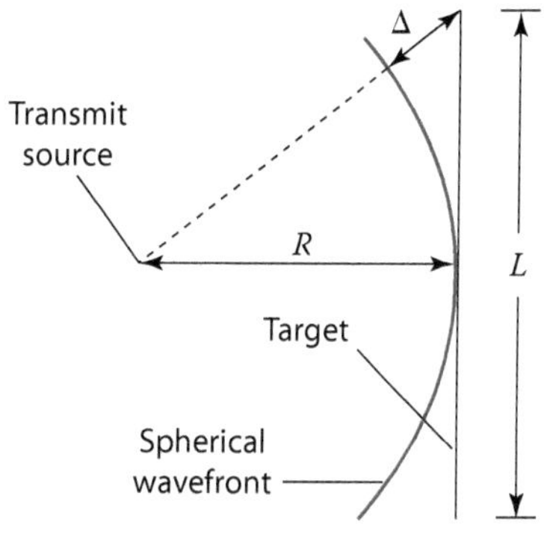

Fig. 8.4 Phase error limitation for a far-field measurement.

the center is

$$k\Delta_{\max} = k[\sqrt{R^2 + (L/2)^2} - R] \approx \frac{\pi L^2}{4R\lambda} = \frac{kL^2}{8R} \tag{8.1}$$

The right-hand side of Eq. (8.1) holds for $R \gg L$. Based on the convention used for the definition of the far field of a radiating antenna, we might limit the round-trip phase error to $\pi/8$, so that Eq. (8.1) becomes

$$2k\Delta_{\max} \approx \frac{\pi L^2}{4R\lambda} \equiv \frac{\pi}{8} \tag{8.2}$$

Solving for R, the minimum distance is

$$R_{\min} = \frac{(4L)^2}{\lambda} \tag{8.3}$$

For example, at 6 GHz, a target with length $L = 1$ m should be at least 320 m from the source. This distance turns out to be overly restrictive and is usually not adhered to in practice.

Equation (8.1) assumes that all of the target surface contributes coherently to the total RCS, for example, a smooth plate. On the other hand, if the target is composed of many independent scatterers, the far-field distance is determined primarily by the dominant scatterers and their distribution in space. Thus, when the dominant scatterers are "clustered," a much smaller range will suffice [2].

In addition to a plane phase front at the target, the amplitude of the incident wave across the target must be constant. The amplitude distribution is controlled by the antenna beamwidth and the ratio L/R. An isotropic antenna provides the most uniform wave front but also results in stronger illumination of the chamber walls and other obstacles. The transmitted power level must also be higher if the antenna gain is low. Thus, there is a tradeoff necessary to determine the optimum antenna beamwidth; this is investigated in more detail in Problem 8.2.

8.2.2 Compact Ranges

The demands on chamber size required for accurate low-RCS measurements have resulted in the development of *compact ranges.* The layout of a typical compact range is shown in Fig. 8.5. The parent reflector configuration is a Cassegrain, and the reflector surfaces have been arranged so that no blockage occurs. The transmit/receive antenna is located at the focus, and

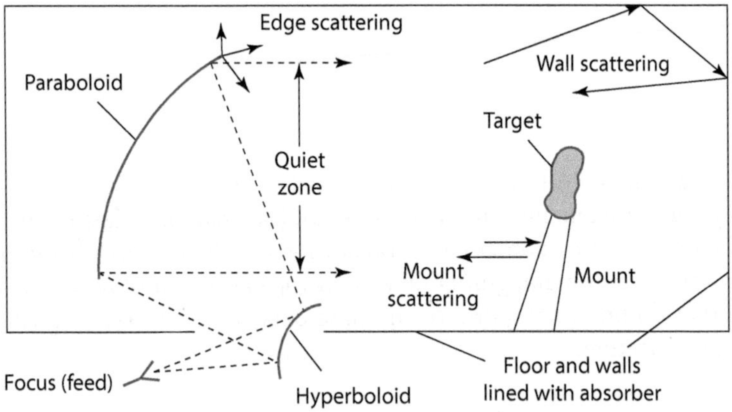

Fig. 8.5 Illustration of some sources of RCS measurement error for a compact chamber.

a plane wave field is formed a short distance in front of the paraboloid. The total volume of a compact range is much less than that of an equivalent far-field configuration. Figure 8.6 shows a compact range at the U.S. Navy's facility in Point Mugu, California.

The compact range is not without its own problems. To obtain uniform amplitude over the target requires a shaped feed pattern or some subreflector

Fig. 8.6 Photo of the compact range at the Pacific Missile Test Range (Point Mugu, California, U.S. Navy photo).

shaping. Cross polarization and edge diffraction can be significant. Because the target is close to the reflector edges, the effects of edge scattering are especially important at low RCS levels. Much effort has gone into the design of rolled or serrated edges that control the level and direction of edge diffraction.

The state-of-the-art instrumentation used in RCS measurement is essentially a programmable multimode radar [3,4]. Data can be collected in either the time or frequency domain and transformed back and forth between the two. In the time domain, *gating* or *windowing* can be employed to discard signals that do not correspond to the time delay of a scattered signal from the target's location.

8.3 Sources of Measurement Error

The received signal in a measurement is not due entirely to scattering from the target. It also includes noise introduced by the receiving system and extraneous scattered signals originating from the chamber walls and other obstacles, as illustrated in Fig. 8.5. Some of the transmit signal couples directly to the receive antenna. This *leakage* is strongest when a common antenna is used for both transmit and receive.

Chambers that are used to measure low or very low RCS must be carefully designed, maintained, and calibrated. Great care is taken to reduce the error signals in the vicinity of the target. The volume of the chamber that is essentially free of extraneous signals and meets the plane wave illumination conditions is referred to as the *quiet zone.* It is usually specified in feet or meters and is perhaps the most important figure of merit of an RCS facility. It represents the largest diameter of a sphere in which a centered target can be rotated while maintaining the phase and amplitude error specifications of the chamber. The quiet zone and the operating frequency range effectively define the chamber performance.

If the test range is indoors, then ideally the walls should be perfect absorbers. From the discussion in Chapter 7 it is clear that the walls cannot be made perfect absorbers and, therefore, error signals corrupt the RCS pattern measurements. A technique for reducing the reflection from the front face of a flat absorber is to produce a material whose intrinsic impedance is very close to unity. Achieving the impedance gradient by geometrical shaping of a medium of constant impedance provides a more complete transition from free space to the loading medium. Geometrical shaping means replacing the front face of the surface with shaped pointed elements such as cones or pyramids, where the axis of individual elements is oriented transverse to the plane of the absorber. A wave entering such a medium encounters a smoothly changing ratio of medium to the adjacent free space. This is similar to the case where the actual properties are changing gradually. Two examples are the "horse-hair" type and the carbon-loaded low-density foam absorbers. The most common one is the pyramidal

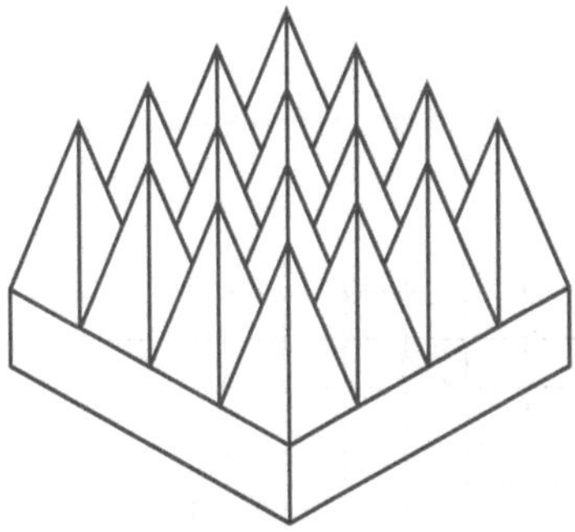

Fig. 8.7 A pyramidal absorber (after Ref. [5]).

absorber, illustrated in Fig. 8.7. The faceted faces of the pyramidal absorber cause dispersion of the incident plane wave by causing reflections and diffractions in all directions, thereby reducing the RCS in the transverse plane. This type of absorber can provide reflectivity reductions in excess of 50 dB but may require a thickness in excess of 10λ to achieve such a level [5].

Measurement errors are mitigated using *background subtraction.* First, a measurement is taken with the target absent. Ideally, the received signal should be zero if the chamber walls are perfect absorbers and there is no leakage. In practice, a nonzero signal is present, stemming from the sources indicated in Fig. 8.5, and this value is stored as a reference. Next, the measurement is repeated with the target present. The measured value will include the target's RCS as well as the error signals. If the interaction between the target and chamber is negligible, vector subtraction of the background signal should yield a close approximation to the isolated target's RCS.

The subtraction method requires that the characteristics of the measurement equipment remain stable during the calibration and measurement runs. If the power level or frequency drifts, the background reference is no longer valid, and the difference is erroneously attributed to the target. Even small changes in phase caused by cable movement can affect low-RCS measurements.

A potential source of large errors is the target mount. If at all possible, the pedestal should be made of low-density material that does not scatter. Frequently, the target is suspended and held in tension using thin polyester filaments similar to fishing line. For large, heavy targets that must be tilted and rotated, there is no alternative but to use metal mounts. In this case, specially shaped fixtures are used to shield the pedestal. Absorbing material can also be used in conjunction with shaping to reduce mount scattering.

8.4 Resolved Radar Cross Section and Target Imaging

Returns from individual scattering elements of a complex target can be isolated or *resolved.* The strength of the return is plotted as a function of *downrange (x)* and *cross range (y)* in Fig. 8.8. If the cell size becomes small enough, a high-resolution two-dimensional image of the target can be constructed.

Resolving the individual scattering sources on a target requires that accurate angle and range information be available as the returns are received. Range resolution is achieved by using a short pulse of duration τ that

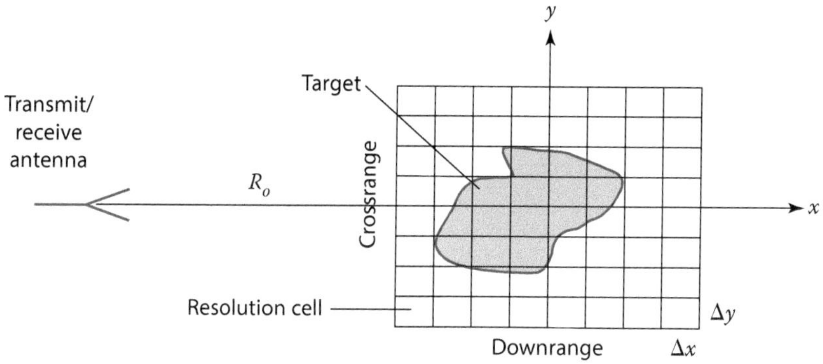

Fig. 8.8 Simplified resolution grid for target imaging. The actual grid is polar but approaches rectangular for large R_O.

illuminates only a small slice of the target, as shown in Fig. 8.9. Knowledge of the time elapsed after the leading edge of the pulse hits the target is equivalent to knowledge of the downrange coordinate x.

One method of obtaining the cross-range value y is by accurately pointing a narrow antenna beam as shown in Fig. 8.9. The beam is scanned over the entire solid angle of the target, thereby providing angle data that can be processed along with the range data to construct an image. This approach is not efficient at microwave frequencies, however, because of the large antenna size needed for fine angle resolution. It is more practical at millimeter wavelengths and smaller and is highly successful at laser wavelengths.

A more practical method for obtaining cross-range takes advantage of the *Doppler shift* [5]. The target is rotated at an angular frequency ω_r, as shown in Fig. 8.10. Target surface points rotate with a linear velocity that depends on their distance from the center of rotation. The component of velocity parallel to R gives rise to a Doppler shift.

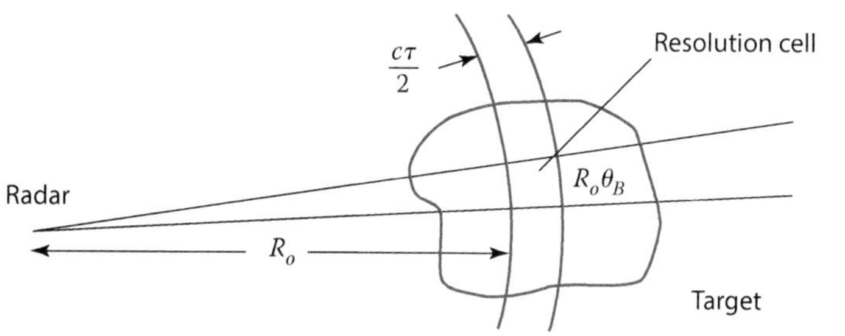

Fig. 8.9 Resolution cell of a conventional radar (based on antenna beamwidth and pulse width).

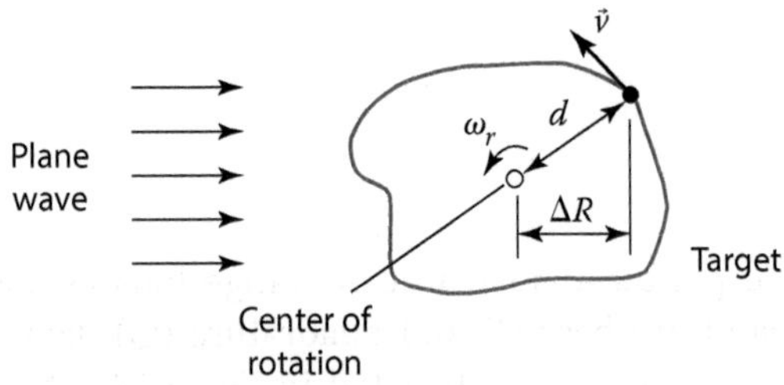

Fig. 8.10 Doppler shift from target rotation.

Referring to the coordinate system shown in Fig. 8.11, let R_0 be the range to the target center of rotation along the y coordinate and r the distance to a scattering point at a distance d from the center. When the law of cosines is used,

$$r^2 = R_0^2 + d^2 - 2R_0 d\cos(\pi - \phi) \tag{8.4}$$

which simplifies to

$$r = R_0\sqrt{1 + \frac{2d\cos\phi}{R_0}} \approx R_0 + d\cos\phi \tag{8.5}$$

when $R_0 \gg d$. The instantaneous value of the monostatic scattered electric field varies as

$$E_s(t) \sim \cos(\omega t - 2kR_0 - 2kd\cos\phi) \tag{8.6}$$

The term ωt in the argument is associated with the carrier and $-2kR_0$ with the round-trip time delay to and from the target center; $2kd\sin\phi$ is the Doppler shift due to rotation. Since $\phi = \omega_r t$, the Doppler frequency in hertz is

$$f_d = \frac{1}{2\pi}\frac{\mathrm{d}}{\mathrm{d}t}(-2kR_0 - 2kd\cos(\omega_r t)) = \frac{2\omega_r}{\lambda}y \tag{8.7}$$

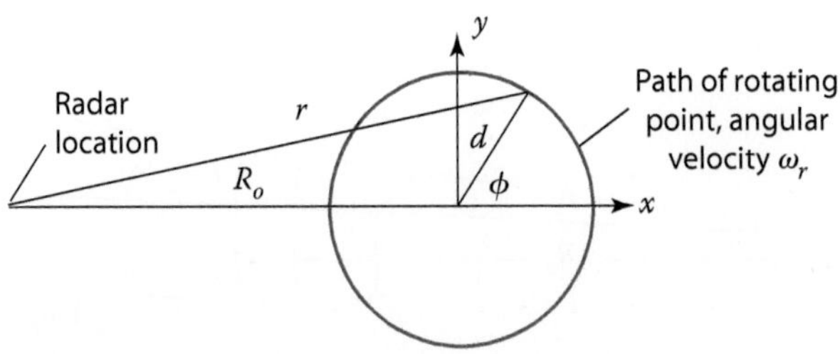

Fig. 8.11 Obtaining cross range information from the Doppler shift.

This method of rotating the target to obtain cross-range information is one form of *inverse synthetic aperture radar* (ISAR).

There are a variety of waveforms and processing approaches that can be applied to target imaging [9]. One of the most fundamental is Fourier-based image processing. A train of N pulses (a dwell) of width τ is transmitted and the interpulse period divided into M range bins as shown previously in Fig. 1.7. Baseband in-phase (I) and quadrature (Q) data are collected in each bin forming the data block depicted in Fig. 8.12. The knowledge of I and Q is equivalent to knowing the complex target scattered field. If the target stays in a single range bin for the entire dwell (for example, the shaded cells in the figure), a Fourier transform of the time samples gives the frequency content, and thus target Doppler and crossrange.

In practice the Doppler shift is not constant over the observation time and the target can move out of its range cell. The consequence is a blurred image. In order to sharpen the image motion compensation must be applied. Motion compensation is very effective in a measurement scenario where the distances are known precisely and the target motion can be controlled.

The image data can be analyzed and the scattering centers extracted [10]. Generally they correspond to the high intensity pixels and indicate strong scattering sources on the target. They can be used to focus RCS reduction efforts such as RAM application or shaping modifications. Scattering centers are also used to build computationally efficient target models for rapid near real time radar system simulations.

Figure 8.13 shows an image based on a 11-point scattering model of an F-35 aircraft at 10 GHz using a 500 MHz bandwidth. The image processing

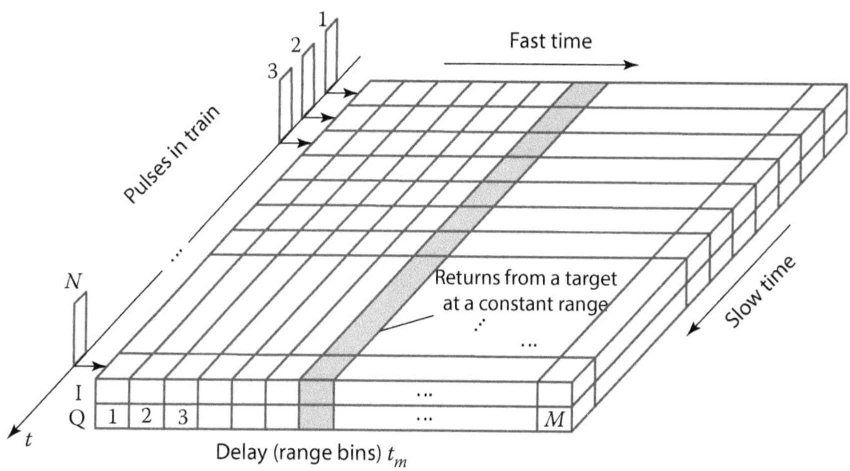

Fig. 8.12 Baseband I and Q data collected for image processing of target returns.

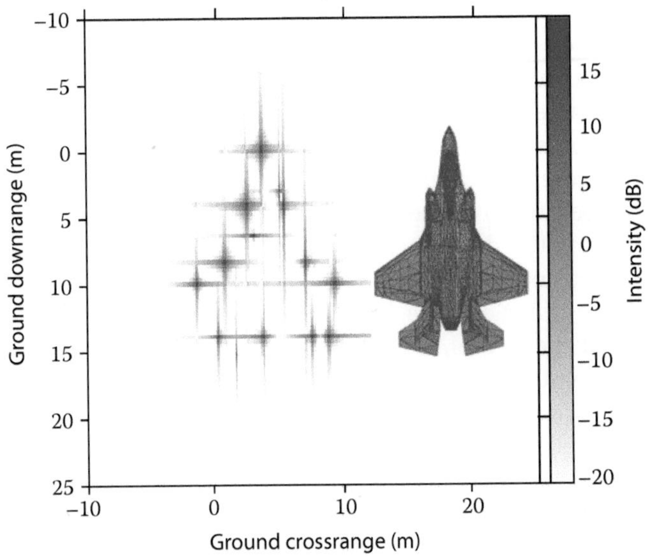

Fig. 8.13 Radar image of a 11 point scattering model of a F-35 aircraft.

algorithm described in [11] has been used to successfully recover the scattering points, although some are slightly displaced or blurred.

8.5 Diagnostic Techniques

Diagnostic techniques are used to isolate and identify scattering sources and their mechanisms. The techniques can be used with scale models in the design phase to refine RCS performance, or in later phases for verification purposes. The presentation mode of the data can aid in diagnostics. An example is shown in Fig. 8.14, which plots RCS of an aircraft over azimuth angle for a band of frequencies. The frequency varies with radial distance. A plot such as this can be used to identify specular reflections which give high RCS that is relatively independent of frequency (the long radial lines). The plot also can be used to identify the constructive and destructive patterns (high and low RCS patches) caused by the interaction of multiple scatterers. The frequency spacing of the high and low spots can be linked to the separation of the scatterers.

In Sec. 6.12, the importance of limiting the phase and amplitude errors due to imperfections was discussed. Additional errors are introduced in the operational environment. Aircraft examples include dust and dirt deposited over the surface, loosening and displacement of parts due to turbulence and hard landings, and wearing away of surface material from rain and ice impact. A diagnostic system is required to determine when the platform's RCS is sufficiently degraded to the point where maintenance is

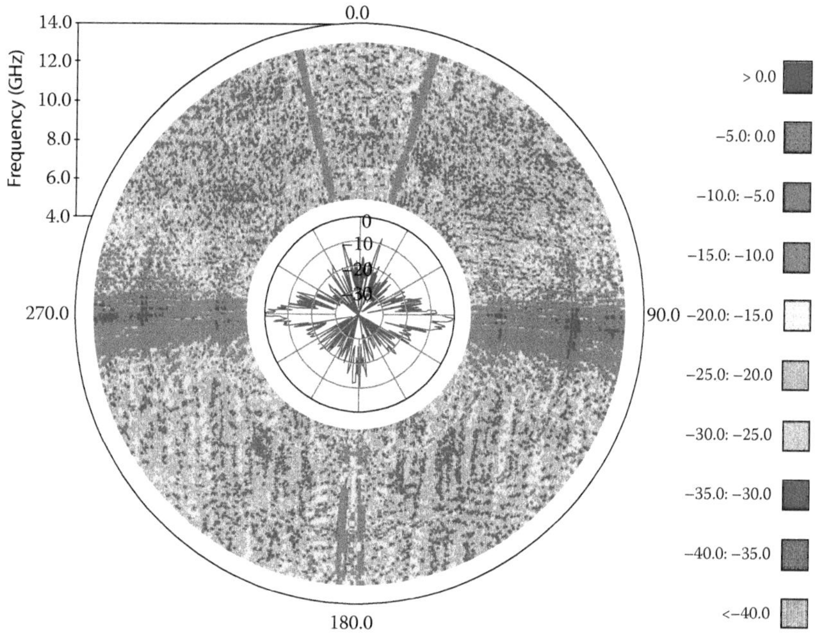

Fig. 8.14 Example of measured data presentation for multiple frequencies and angles [From Ref. 7].

needed. Some of these systems operate "in the field" to provide a quick look at the LO platform and give an indication whether it meets the RCS specifications before a mission. Still other diagnostic systems are needed to verify the integrity of repairs that are done in the field or at the maintenance depot.

For localized repairs (i.e., patches or fixes that affect a small area of the platform), handheld devices are used. They are essentially duplications of the so-called *NRL* (Naval Research Laboratory) *arch*. Physically, miniature versions of the NRL arch resemble a radar gun. The instrument is held close to the surface and reflections are measured.

For more extensive repairs a *diagnostic imaging radar* (DIR) can be used. It is a specialized, scaled-down portable radar system. It may view the platform over only a limited range of azimuth and elevation angles. The collected data are processed and may be compared to a set of reference data that serves as threshold to determine whether the platform is within specification.

References

[1] Blacksmith, P., Hiatt, R. E., and Mack, R. B., "Introduction to Radar Cross-Section Measurements," *Proceedings of the IEEE*, Vol. 53, No. 8, Aug. 1965, pp. 901–920.

[2] Welsh, B. M., "A Minimum Range Criterion for RCS Measurements of a Target Dominated by Point Scatterers," *1984 AP-S/URSI Symposium*, Vol. APS-17-3, pp. 666–669.

[3] *Short Pulse Radar Cross Section Measurement Systems*, Hughes Aircraft Co., Microwave Products Div., Torrance, CA.

[4] Tavormina, J., "Instrumentation Radars Fulfill Role in RCS Measurement," *Microwave Systems News (MSN & CT)*, Feb. 1985, pp. 75–96.

[5] Knott, E. F., Shaeffer, J. F., and Tuley, M. T., *Radar Cross Section*, 2nd ed. Artech House, Norwood, MA, 1993.

[6] Mensa, D., *High Resolution Radar Cross-Section Imaging*, Artech House, Norwood, MA, 1981.

[7] *RCS Measurement Capabilities at the Pacific Missile Test Center*, Radar Signature Branch, Pt. Mugu, CA.

[8] Johansson, M., Holloway, C., and Kuester, E., "Effective Electromagnetic Properties of Honeycomb, Composites, and Hollow-Pyramidal and Alternating-Wedge Absorbers," *IEEE Transactions on Antennas and Propagation*, Vol. 53, No. 2, Feb. 2005, pp. 728–736.

[9] Chen, V. C., and Ling, H. *Time Frequency Transforms for Radar Imaging and Signal Analysis*, Artech House, 2002.

[10] Rajan Bhalla, John Moore, and Hao Ling, "A Global Scattering Center Representation of Complex Targets Using the Shooting and Bouncing Ray Technique," *IEEE Trans. on Ant. and Prop.*, Vol. 45, No. 12, Dec. 1997, pp. 1850–1856.

[11] Garren, D. A., "Theory of Two-Dimensional Signature Morphology for Arbitrarily Moving Surface Targets in Squinted Spotlight Synthetic Aperture Radar," *IEEE Transactions on Geoscience and Remote Sensing*; Vol. 53, No. 9, Sept. 2015, pp. 4997–5008; DOI: 10.1109/TGRS.2015.2416066; Date of Publication to IEEE Xplore as an Early Access Article: 17 April 2015

Problems

8.1 A CW measurement is made at 4 GHz on a target of unknown RCS, and the received power due to target scattering is −20 dBm. Calibration was performed using a 6-in. sphere, and the measured power in this case was −15 dBm. Compute the RCS of the target.

8.2 An RCS chamber will be used in the frequency range 500 MHz–1 GHz. Targets will be as long as 6 ft.

a) What is the transmitter/target far-field separation R required to maintain a round-trip phase error of $\pi/4$?

b) What is the largest antenna diameter that can be used if the amplitude of the wave over the target cannot vary by more than 1 dB? (Assume a uniformly excited circular antenna aperture.)

8.3 To examine the effects of incident wave phase deviation from that of a plane wave, consider a circular plate of radius a $(=L/2)$ and distance to the transmitter R, as shown in Fig. 8.4.

a) Show that the path difference Δ is a quadratic function of distance from the center ρ that it can be expressed as

$$\Delta(\rho) = \Delta_m(\rho/\alpha)^2$$

Find Δ_m in terms of R.

b) Using the physical optics approximation, find a closed-form expression for the backscattered RCS of the plate as a function of Δ_m. Plot the peak RCS error as a function of Δ_m.

8.4 Time gating is to be used to eliminate chamber reflections. The range to the target is 60 ft, and typical target lengths are 6 ft.

a) What is the required gate time for a typical target?

b) What pulse width is needed to resolve a 2-in. length of the target?

8.5 Discuss the relative advantages and disadvantages of using a sphere vs a square plate as calibration targets.

8.6 Plot the relative amplitude error (i.e., the amplitude at the target edge divided by the amplitude at the target center) as a function of target extent L if the transmitter is located 19 ft from the target. Assume a Gaussian beam shape, which can be expressed as

$$G(\theta) = Ge^{-2.776(\theta/\theta_B)^2}$$

where G is the main beam antenna gain (pointed at the center of the target), θ is the half-angle subtended by the test object, and θ_B is the half-power beamwidth (HPBW). Plot curves for half-power beamwidths of 5, 10, and 20 deg.

8.7 Formulas are available for absorbers of various shapes that are composed of multiple dielectric materials [8]. One model for a wedge absorber made with a foam having a dielectric constant ϵ_a that fills a volume fraction of space g gives an effective dielectric constant that is bounded by

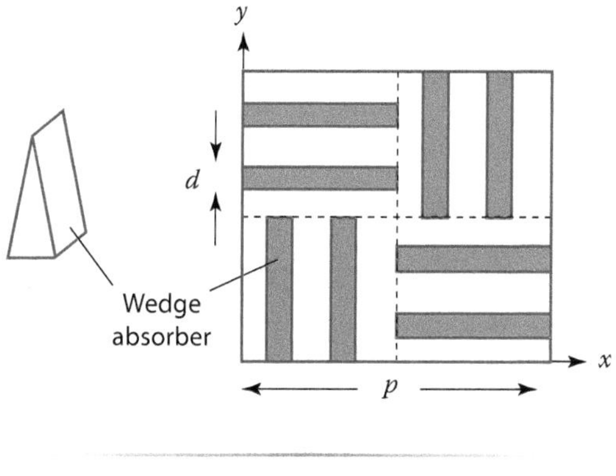

Fig. P8.7 Ref. [8].

$\epsilon_1 \leq \epsilon_{eff} \leq \epsilon_2$ where

$$\epsilon_1 = \epsilon_o \frac{(1+g)\epsilon_a + (1-g)\epsilon_o}{(1-g)\epsilon_a + (1+g)\epsilon_o}$$

$$\epsilon_2 = \epsilon_a \frac{(2-g)\epsilon_o + g\epsilon_a}{g\epsilon_o + (2-g)\epsilon_a}$$

a) What is the fill factor g for the alternating wedge arrangement shown in Fig. P8.7?

b) Plot the upper and lower limits ϵ_1 and ϵ_2 of the arrangement shown for fill factors in the range $0.1 \leq g \leq 1.0$ when $\epsilon_a = 2\epsilon_0$ and $\epsilon_a = 50\epsilon_0$.

Chapter 9 Laser Cross Section

9.1 Introduction

Laser radars are commonly referred to as *ladar* for *laser radar* or as *lidar* for *light detection and ranging.* They operate on the same principles as microwave radar [1,2]. Ranging is accomplished by measuring the time delay to and from the target. Angular information is obtained from the antenna beam-pointing direction or from the target's Doppler shift. Laser radars are capable of extremely accurate angular measurement because of the small beam diameters of lasers (on transmit) and narrow fields of view (on receive). On the negative side, the detection and tracking ranges are much shorter than microwave radar because of lower transmitter power and higher atmospheric attenuation.

Atmospheric transmittance as a function of wavelength is shown in Fig. 9.1. Ladars usually operate in the 10.6-μm* wavelength region in the far infrared and the 1.06-μm wavelength region in the visible [3]. The former use CO_2 lasers and the latter neodymium YAG (yttrium aluminum garnet) crystal lasers with typical efficiencies of 10% and 3%, respectively [4–6].

9.2 Scattering and Propagation of Light

The scattering and propagation of light obey the same set of laws as radar waves, that is, those set forth by Maxwell's equations and the boundary conditions. However, the wavelength of laser light is so small that minute particles and even molecules are significant scatterers. Target surfaces are very rough and, consequently, the random or *diffuse* scattering component dominates. In fact, there may not be any significant specular component to the laser cross section unless the surface is highly polished.

The methods of guiding radar waves are no longer practical at laser wavelengths. For example, a hollow metal waveguide of the type used for microwave transmission typically has a radius of 0.5λ. Cutting a hole with a radius of one-millionth of a meter is an impossible task. Optical fibers are used at these frequencies. They take advantage of the existence of a critical angle and total internal reflection to confine the light (see Appendix A). Over short distances, *beam waveguides* can be used. As shown in Fig. 9.2, a beam waveguide uses a system of lenses and mirrors to keep the beam focused and directed.

*The standard length unit is *micrometers* or *microns*, which is 10^{-6} m. It is sometimes denoted by μ, but μm will be used here to avoid confusion with permeability.

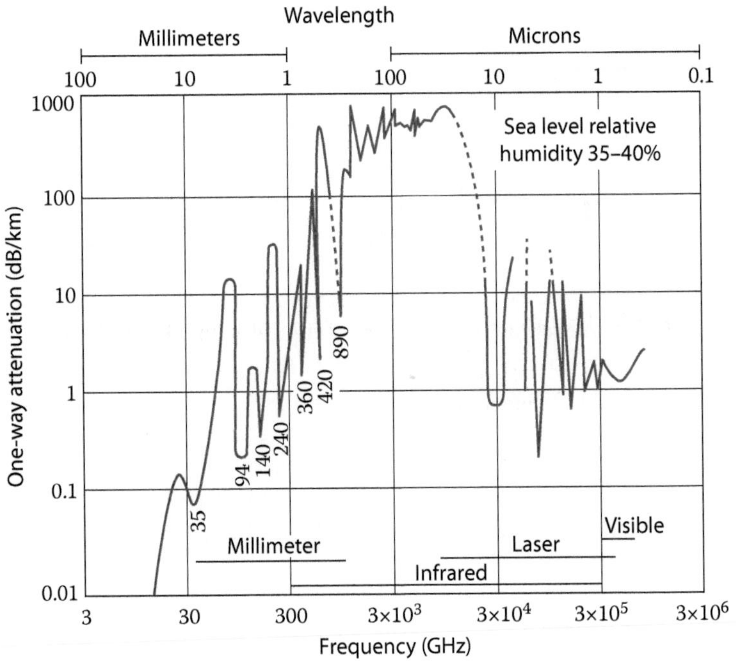

Fig. 9.1 Atmospheric transmittance (from Ref. [4], courtesy Artech House, www.artechhouse.com).

Geometrical optics (GO) is applicable to most problems in this frequency range. Optical systems are not perfect, however, and caustics and foci are not points, as GO assumes. Figure 9.3 compares the GO rays in a beam waveguide to the actual beam diameter. The intensity of the beam is essentially constant inside the two rays shown and decays rapidly outside the edge rays. Frequently a Gaussian shape is used to represent the intensity distribution across the beam. The narrowest beam diameter w_o is called the *beam waist.* Figure 9.4 shows the Gaussian beam intensity as a function of distance from the beam waist. The spreading of the beam is referred to as *divergence* as depicted in Fig. 9.5.

A *beam expander* is used to increase the diameter of a laser beam. Large beam diameters are not as hazardous to personnel as small beam diameters,

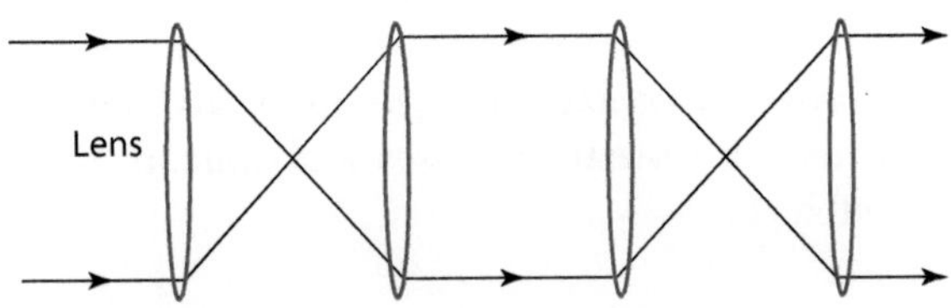

Fig. 9.2 Section of a beam waveguide.

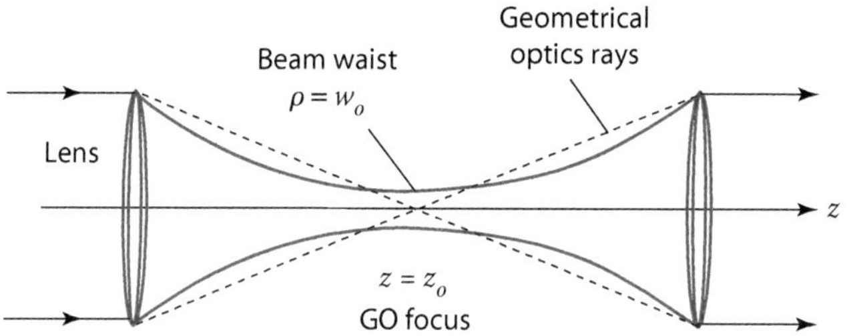

Fig. 9.3 Focused beam and its geometrical optics ray approximation.

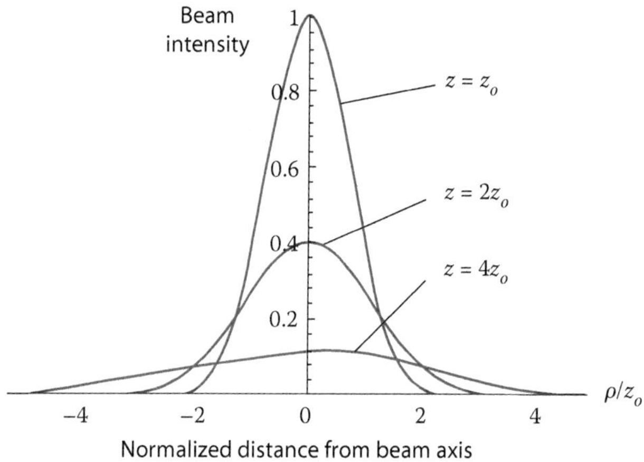

Fig. 9.4 Beam intensity versus distance from the beam waist.

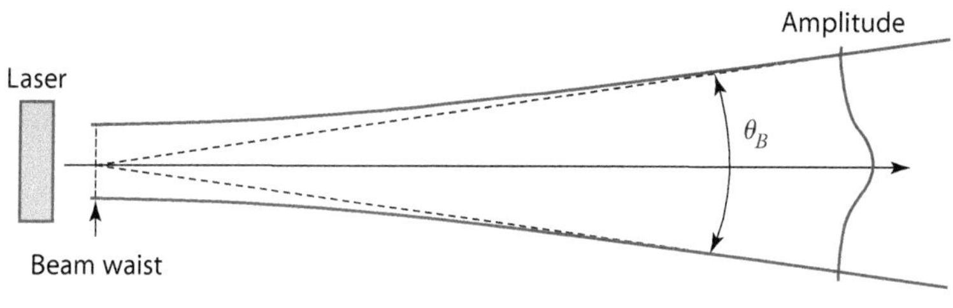

Fig. 9.5 Gaussian beam and divergence.

and they also have less divergence. The beam expansion equation relates the beam diameters and divergences at the input and output of the optics,

$$\frac{D_{\text{out}}}{D_{\text{in}}} = \frac{\Phi_{\text{out}}}{\Phi_{\text{in}}} \tag{9.1}$$

as illustrated in Fig. 9.6.

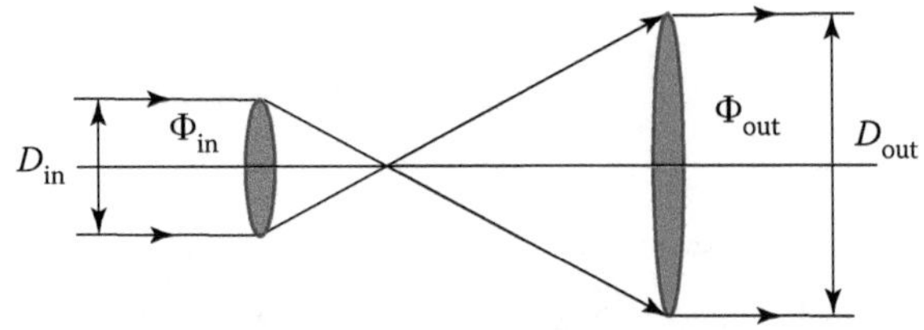

Fig. 9.6 Beam expander.

9.3 Definition of Quantities

9.3.1 Units and Symbols

Light scattering has traditionally been a separate discipline from radar scattering. This has led to separate notation and terminology in the two communities that research these topics. Optical scattering has some terminology in common with *radiometry* (the passive sensing and identification of targets based on their natural microwave emissions). In this chapter, the engineering symbols will be used rather than the physics symbols. The two sets of symbols are compared in Table 9.1. Figure 9.7 defines the basic quantities used in the definitions that follow. Note that these quantities vary from point to point on the surface although the dependence on position vector ($\vec{r}$ or $\vec{r}'$) is not explicitly indicated.

1. *Radiant flux:* The rate of emission of power from a source. The symbols are Φ or P and the unit is watts.
2. *Radiant emittance:* Also called the *excitance*, this is used to characterize extended sources. Most sources of light are large in terms of wavelength and do not emit uniformly in intensity. The radiant emittance is an indicator of the intensity variation across the surface. It is the power

Table 9.1 Summary of Symbols and Terminology (After Ref. [10])

Quantity	Units	Engineering Symbol (LCS)	Physics Symbol (Optics)
Radiant flux	W	Φ	P
Radiant emittance[a]	W/m^2	M	W
Radiant intensity	W/sr	I	J
Radiant flux density	W/m^2	W	—
Irradiance	W/m^2	E	H
Radiance	W/m^{2}sr	L	N

[a]Also called excitance.

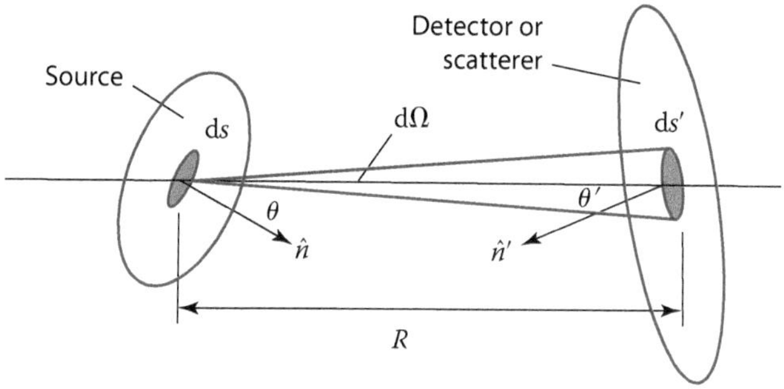

Fig. 9.7 Source and scatterer geometry for LCS definition.

radiated per unit source surface area

$$M = \frac{\mathrm{d}\Phi}{\mathrm{d}s} \tag{9.2}$$

The physics symbol is W and the units are watts per square meter.

3. *Radiant intensity:* The radiation from sources is nonuniform not only in intensity but also in direction. Radiant intensity is the radiant source power per unit solid angle:

$$I = \frac{\mathrm{d}\Phi}{\mathrm{d}\Omega} \tag{9.3}$$

The physics symbol is J and the unit is watts per steradian. This is the *candlepower* of the source.

4. *Irradiance:* This is the power per unit surface area being received by a differential surface $\mathrm{d}s'$:

$$E = \frac{\mathrm{d}\Phi}{\mathrm{d}s'} \tag{9.4}$$

The physics symbol is H and the unit is watts per square meter. This is the *illumination* of the surface.

5. *Radiance:* The intensity per unit area per steradian of a source. This essentially combines the directional and intensity characteristics of an extended source into a single parameter:

$$L = \frac{I}{\mathrm{d}s_n} = \frac{\mathrm{d}^2\Phi}{\cos\theta \mathrm{d}s \mathrm{d}\Omega} \tag{9.5}$$

The unit is watts per square meter steradian, and the physics symbol is N. This is also referred to as the *brightness* of the source.

6. *Radiant flux density:* The Poynting vector W.

Note that, when a target is illuminated, the primed quantities in Fig. 9.7 apply; that is, the target is a receiving surface. Subsequently, the target becomes a source as it scatters the incident light, and the unprimed quantities in Fig. 9.7 are appropriate. Subscripts i and r can be used to indicate *incident* (from a laser) or *reflected* (to a detector).

9.3.2 Frequency Spectrum of a Laser

The wavelength (or frequency) spectrum of a laser is very narrow compared to a microwave source. The bandwidth of a laser is called the *line width.* There will be some dispersion of a waveform after it travels a distance L because the phase velocities at the high and low ends of the spectrum differ.

Referring to Fig. 9.8, let λ_c be the wavelength at the center of the spectrum and $\Delta\lambda$ the line width. After traveling a distance ℓ, the phase shifts at the lower and upper ends of the spectrum are

$$\psi_L = \frac{2\pi\ell}{\lambda_c - \Delta\lambda/2} \quad \text{and} \quad \psi_H = \frac{2\pi\ell}{\lambda_c + \Delta\lambda/2} \tag{9.6}$$

The *coherence length* ℓ_c is defined as the length over which the phase difference $\psi_H - \phi_L$ is 1 rad:

$$\ell_c = \frac{\lambda_c^2}{2\pi\Delta\lambda} \tag{9.7}$$

A *coherence time* can be defined from the coherence length:

$$\tau_c = \frac{\ell_c}{c} \tag{9.8}$$

Coherence length and time define the *spatial* and *temporal* coherences of a source, respectively. Coherence is a critical property for systems that

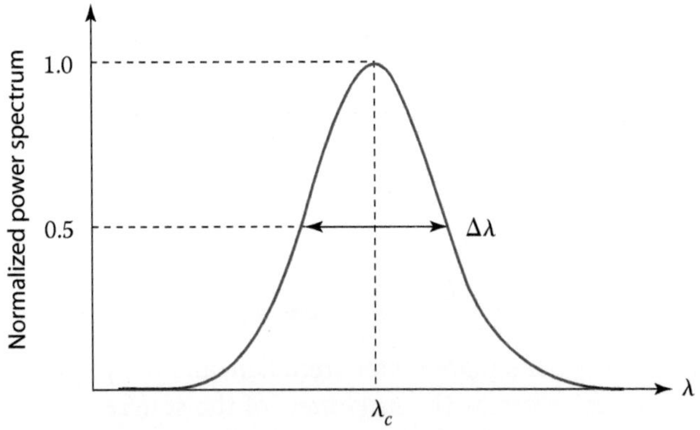

Fig. 9.8 Typical laser spectrum.

Table 9.2 Topical Values of Laser Frequency Spectrum

Source Type	λ_c	$\Delta\lambda$	ℓ_c
He-Ne	633 μm	1	100 m
DFB[a]	840 nm	0.1	1.1 mm
LED	850 nm	10	11 μm

[a]Distributed feedback.

recombine the laser light that has propagated over two or more paths. Typical values are listed in Table 9.2 (after Ref. [6]).

Polarization state is another important characteristic of light, just as it is for microwaves. Depending on the source, light can be polarized with a single state, partially polarized, or unpolarized. Unpolarized refers to an equal combination of all polarizations.

9.4 Laser Radar Equation

9.4.1 System Components

A laser radar consists of the same basic components as a microwave radar. A block diagram is shown in Fig. 9.9. The transmit channel consists of a laser and pulser if range information is to be obtained. The receive channel consists of the optics required to collect, focus, and detect the light scattered from the target [6–9]. The receive optics are frequently based on the classical *Cassegrain* dual reflector system shown in Fig. 9.10. This is the same configuration that is used for telescopes, with a detector substituted for an "eyeball."

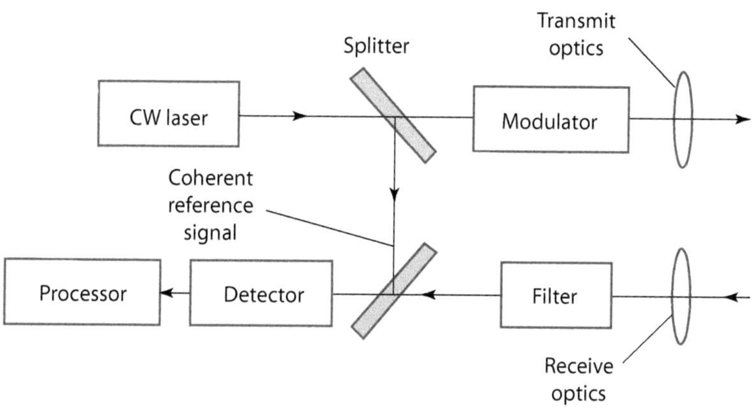

Fig. 9.9 Laser radar system block diagram.

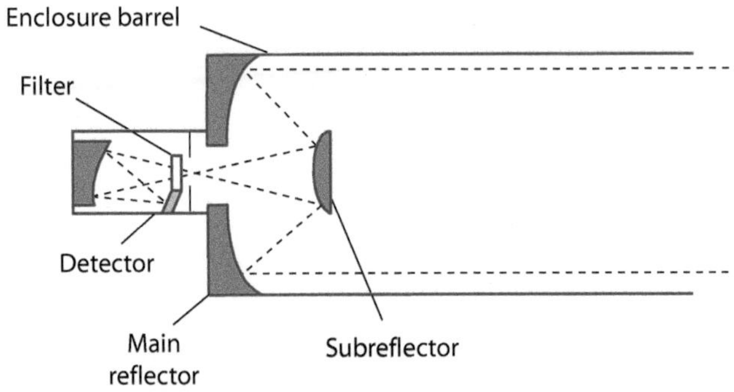

Fig. 9.10 Example of receiving optics and detector.

Laser beams have varying degrees of divergence. A highly collimated beam can have a small spot size at the target and, therefore, illuminate only a portion of its surface. On the other hand, if the beam divergence is great enough and the target range large, the entire target surface can be illuminated just as in the microwave case. The same possibilities exist on receive with regard to the detector *field of view* (FOV). The FOV can contain the entire target or only part of it. These conditions are illustrated in Fig. 9.11 and ultimately affect the form of the laser radar equation.

9.4.2 Derivation for Complete Target Illumination

For the present derivation, assume that the entire target surface is illuminated. The laser beam is circular in cross section with a beam divergence angle far from the waist θ_B, which corresponds to the half-power beamwidth

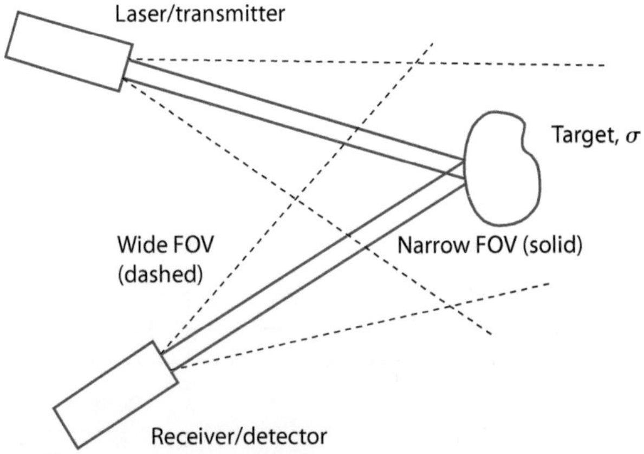

Fig. 9.11 Possible target and radar geometries.

in the microwave case. Thus,

$$\theta_B = \frac{1.02\lambda}{D_t} \approx \frac{\lambda}{D_t} \tag{9.9}$$

D_t is the transmit optics diameter, and it is assumed that the beam intensity is essentially zero outside the angle θ_B.

As illustrated in Fig. 9.12, if the target is at a distance R, the beam diameter in the plane of the target is used to obtain

$$\text{AREA} \approx \pi\left(\frac{R\theta_B}{2}\right)^2 \tag{9.10}$$

and the beam solid angle

$$\Omega_A = \frac{\text{AREA}}{R^2} = \pi\frac{\theta_B^2}{4} \tag{9.11}$$

The gain of the transmit "antenna" is

$$G_t = \frac{4\pi}{\Omega_A} = \frac{16}{\theta_B^2} \tag{9.12}$$

The power received at the radar is

$$P_r = \left(\frac{P_t G_t}{4\pi R^2}\right)\left(\frac{1}{2\pi R^2}\right)\sigma A_r \tag{9.13}$$

The first factor gives the power density at the target and is unchanged from the microwave case. The second factor assumes that the power collected by the target is only scattered into the hemisphere toward the radar. Consequently, 2π rather than 4π occurs in the denominator.

The last factor in Eq. (9.13), A_r, is the receive optics area. It is related to the receive optics diameter D_r by

$$A_r = \pi\left(\frac{D_r}{2}\right)^2 \tag{9.14}$$

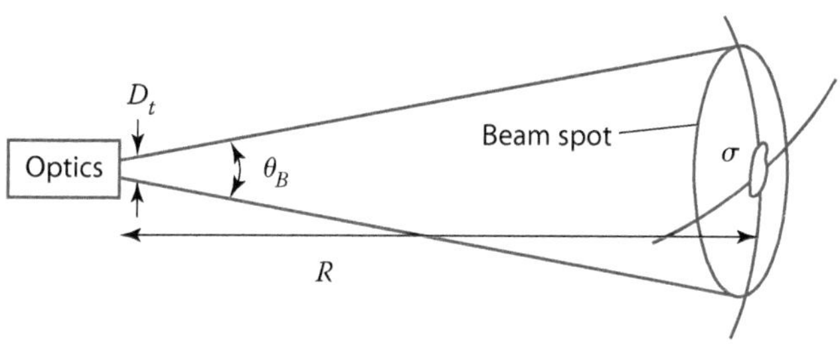

Fig. 9.12 Laser beam and target illumination at range R.

Finally, introducing an optical efficiency factor L_o and assuming that the transmit and receive optics diameters are the same yield

$$P_r = \frac{8P_t \sigma A_r^2 L_o}{\pi^3 R^4 \lambda^2} \tag{9.15}$$

If the beam only partially illuminates the target, the range dependence is $1/R^2$ rather than $1/R^4$ (see Problem 9.5). The optical efficiency accounts for all the system losses and has values in the range of 0.5.

9.4.3 Atmospheric Attenuation

Atmospheric attenuation has not been included in the derivation of the laser radar equation. This loss cannot be neglected and is, in fact, the major factor in determining the radar operating range. Therefore,

$$e^{-2\alpha R} \tag{9.16}$$

must be introduced for a monostatic system, where α is the (power) *extinction coefficient.* Sources of extinction are 1) absorption by gases and aerosols and 2) scattering by aerosols and other particulates. The second source is particularly strong because dust and other particles can be large compared to wavelength.

Generally, the wave will be further attenuated by other processes in addition to extinction. The primary one is turbulence, which causes random fluctuations in the index of refraction resulting in 1) beam spreading, 2) scintillation, and 3) beam wander (refraction/bending). Furthermore, the atmosphere emits its own energy, called *path radiance,* which appears as a noncoherent noise source to the ladar.

The maximum range of a laser radar is usually determined by an iterative process because of the exponential dependence of received power on range R. A "guess" is made for the loss so that $e^{-2\alpha R}$ is known, and then the laser radar equation is solved for R. If the original estimate of loss is inconsistent with the range, the loss is changed and the range recalculated.

9.5 Definition of Laser Cross Section

Laser cross section (LCS) is defined in precisely the same way as RCS [10]:

$$\sigma_{pq}(\theta_i, \phi_i, \theta_r, \phi_r) = \lim_{R\to\infty} 4\pi R^2 \frac{W_{rp}(\theta_r, \phi_r)}{W_{iq}(\theta_i, \phi_i)} \tag{9.17}$$

where the subscripts i and r refer to *incident* and *reflected* and p and q denote the polarizations of the reflected and incident waves, respectively. Given that

$$W_{rp} = \frac{I_{rp}}{R^2} \tag{9.18}$$

the LCS can be written in terms of the radiant intensity

$$\sigma_{pq}(\theta_i, \phi_i, \theta_r, \phi_r) = 4\pi \frac{I_{rp}(\theta_r, \phi_r)}{W_{iq}(\theta_i, \phi_i)} \tag{9.19}$$

For the monostatic case, this simplifies to

$$\sigma_{pq}(\theta, \phi) = 4\pi \frac{I_{rp}(\theta, \phi)}{W_{iq}(\theta, \phi)} \tag{9.20}$$

9.5.1 Parameters

In practice, LCS is a function of many parameters in addition to the target characteristics. These include 1) beam profile, 2) beamwidth, 3) laser temporal and spatial coherence, 4) target surface characteristics, 5) receiver aperture and FOD, and 6) detector averaging. Furthermore, satisfying the far-field criterion [nD^2/λ, where n is generally 2] is difficult at laser frequencies because λ is so small. Thus, the limiting process in the definition of LCS cannot be strictly applied. The net result is that it is difficult to separate the measurement of LCS from the system. It may be tempting to discard the concept of LCS (or, similarly, RCS at millimeter wave frequencies), but LCS does provide useful and repeatable data at distances that can be accommodated by indoor and outdoor facilities.

9.5.2 Polarization States

LCS depends on the polarization of the incident wave as well as the polarization selectivity of the detector. A scattering matrix can be defined as in the case of RCS. The reference directions for polarization are usually one of the following three:

1. Spherical (θ, ϕ)

$$\begin{bmatrix} \sigma_{T\theta} \\ \sigma_{T\phi} \end{bmatrix} = \begin{bmatrix} \sigma_{\theta\theta} + \sigma_{\theta\phi} \\ \sigma_{\phi\theta} + \sigma_{\phi\phi} \end{bmatrix} \tag{9.21}$$

2. Horizontal/vertical (H, V)

$$\begin{bmatrix} \sigma_{TH} \\ \sigma_{TV} \end{bmatrix} = \begin{bmatrix} \sigma_{HH} + \sigma_{HV} \\ \sigma_{VH} + \sigma_{VV} \end{bmatrix} \tag{9.22}$$

3. Rectangular (x, y)

$$\begin{bmatrix} \sigma_{Tx} \\ \sigma_{Ty} \end{bmatrix} = \begin{bmatrix} \sigma_{xx} + \sigma_{xy} \\ \sigma_{yx} + \sigma_{yy} \end{bmatrix} \tag{9.23}$$

The subscript T refers to total, which is obtained by a simple noncoherent sum of the contributions from the two incident polarizations.

9.6 Bidirectional Reflectance Distribution Function

Every point on a large surface does not necessarily scatter uniformly and identically. For instance, the scattering of one patch might be concentrated in a particular direction and that from another part of the surface in a different direction, and these can change with the direction of incidence. Furthermore, the reflectivities could also be different from point to point. The *bidirectional reflectance distribution function* (BRDF) quantifies all the surface scattering characteristics. It is defined as [11]

$$\rho_{pq}(\vec{r}', \theta_i, \phi_i, \theta_r, \phi_r) = \frac{L_{rp}(\theta_r, \phi_r)}{E_{iq}(\theta_i, \phi_i)} \tag{9.24}$$

and has units of steradian^{-1}. It gives the radiance in the direction (θ_r, ϕ_r) as a function of 1) incidence direction (θ_i, ϕ_i); 2) location of reflection point on the target surface $\vec{r}'$; 3) polarization of the incident wave q; and 4) polarization of the detector p.

Consider a differential surface area ds illuminated with a radiant flux density $W_{iq}(\theta_i, \phi_i)$, as shown in Fig. 9.13. The power collected by the differential patch is

$$W_{iq}(\theta_i, \phi_i) \cos\theta_i \mathrm{d}s \tag{9.25}$$

The BRDF of this patch is ρ_{pq} and, therefore,

$$L_{rp}(\theta_r, \phi_r) = \rho_{pq} W_{iq}(\theta_i, \phi_i) \cos\theta_i \mathrm{d}s \tag{9.26}$$

From the definition of LCS,

$$\mathrm{d}\sigma_{pq} = 4\pi \frac{\mathrm{d}I_{rp}}{W_{iq}} = 4\pi \frac{L_{rp} \cos\theta_r \mathrm{d}s}{W_{iq}} = 4\pi\rho_{pq} \cos\theta_i \cos\theta_r \mathrm{d}s \tag{9.27}$$

The explicit angle dependencies have been dropped for notational convenience. This is the fundamental equation for the bistatic diffuse LCS of a surface.

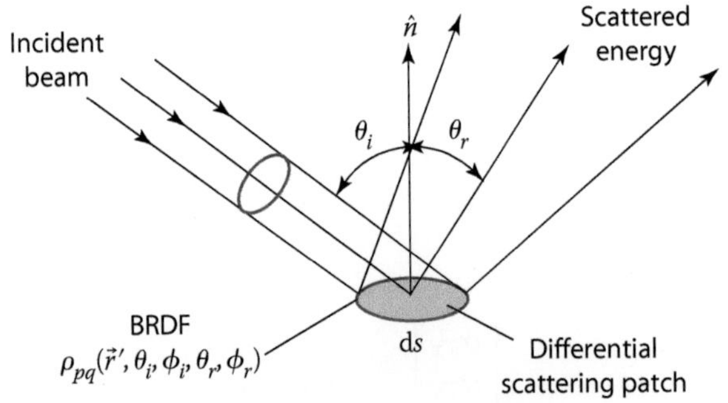

Fig. 9.13 Beam illuminating a differential surface patch.

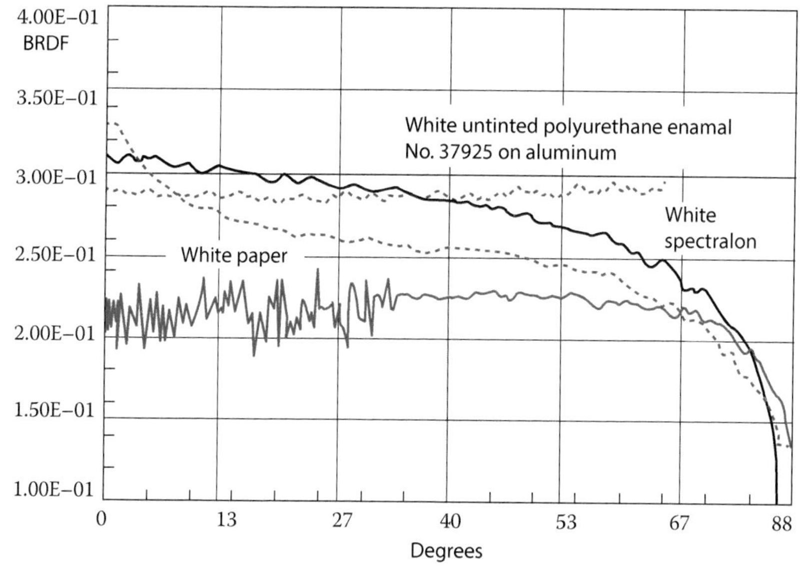

Fig. 9.14 BRDFs of some white surfaces (from Ref. [11], courtesy McGraw-Hill).

Typical BRDFs are shown in Figs. 9.14 and 9.15 for light and dark materials. The difference is several orders of magnitude, and this points to one obvious method of LCS reduction: flat dark paints as opposed to glossy light paints.

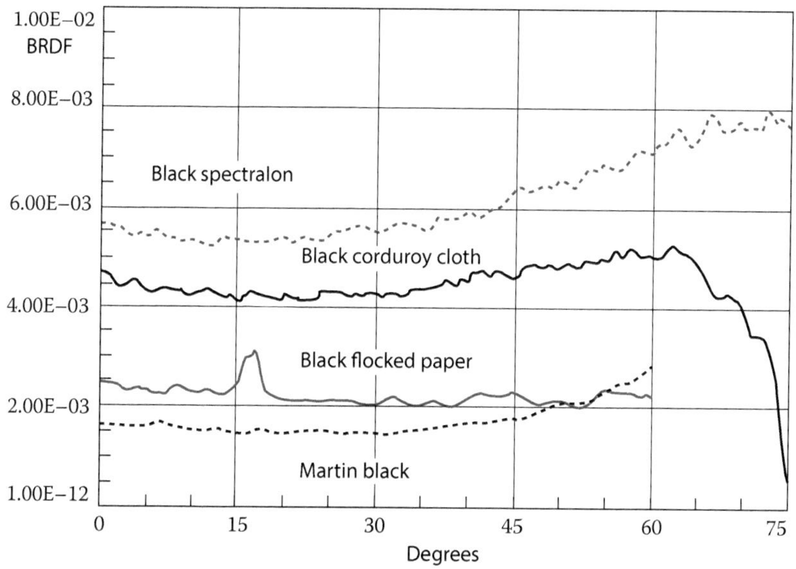

Fig. 9.15 BRDFs of some black surfaces (from Ref. [11], courtesy McGraw-Hill).

According to Eq. (9.24), the BRDF is defined as the differential radiance L divided by the differential irradiance E. The assumptions implied are that the illuminating beam is uniform, that scattering occurs from the surface of the target, and that the differential scattering area is isotropic. When these conditions do not exist, which is typically the situation for a measurement, then the measured quantities are not the radiance and irradiance. Quantities similar to the BRDF have been defined to compensate for these shortcomings. They are the *bidirectional scatter distribution function* (BSDF), the *cosine corrected BSDF* (or simply scatter function), and a related quantity, the *total integrated scatter* (TIS). A thorough discussion of these is given in Ref. [11].

9.7 Diffuse Surfaces

9.7.1 Rayleigh Condition

At optical wavelengths, almost all surfaces can be considered rough. The irregularities are comparable to or greater than a wavelength. The standard criterion for a rough surface is the Rayleigh condition. A surface is considered rough when

$$h \geq \frac{\lambda}{8 \sin \psi} \tag{9.28}$$

where h is the average height of the irregularities and ψ is the grazing angle, as shown in Fig. 9.16. From the previous discussion on imperfections in Chapter 6, increasing the surface roughness is equivalent to introducing path differences relative to a smooth surface. If the surface error is random, the phase error due to path differences will also be random, and the final result is an LCS noise floor. This random component is called *diffuse* scattering, as opposed to specular scattering. For laser scattering, the diffuse component can dominate, and in many instances there may be no specular component present.

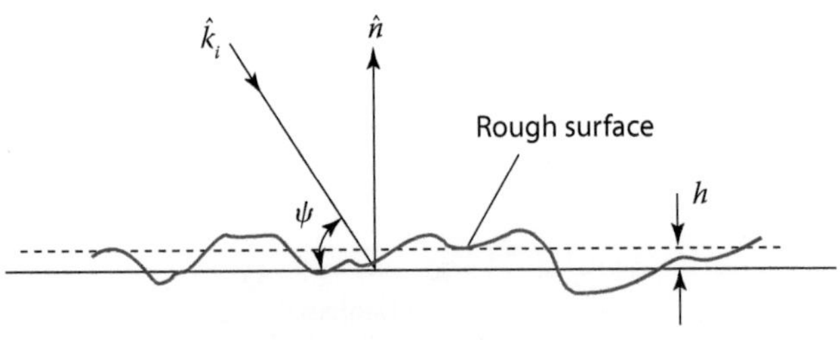

Fig. 9.16 Rayleigh condition for a rough surface.

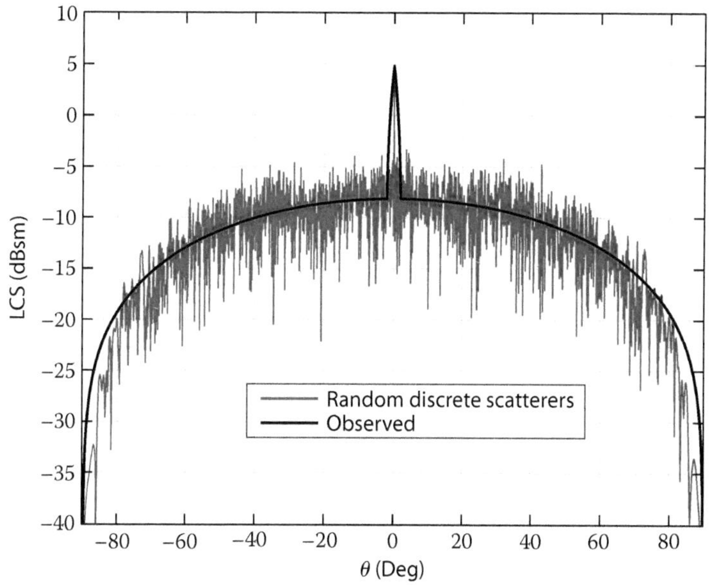

Fig. 9.17 Comparison of scattering from a rough surface with a specular component (top) and one without a specular component (bottom).

The monostatic pattern in Fig. 9.17 is typical of the LCS of a 1 foot square panel with white paint. The curve with rapid variations is generated using a model of closely spaced randomly phased discrete scatterers. For measurements on a continuous diffuse surface the rapid variation of the patterns with angle smooth out and approach the observed curve in the figure. The specular lobe, if it exists, is generally much wider that what occurs for coherent scattering.

Figure 9.18 illustrates three distinct features that can be observed for a bistatic pattern. First is the uniform scattering at most angles that is

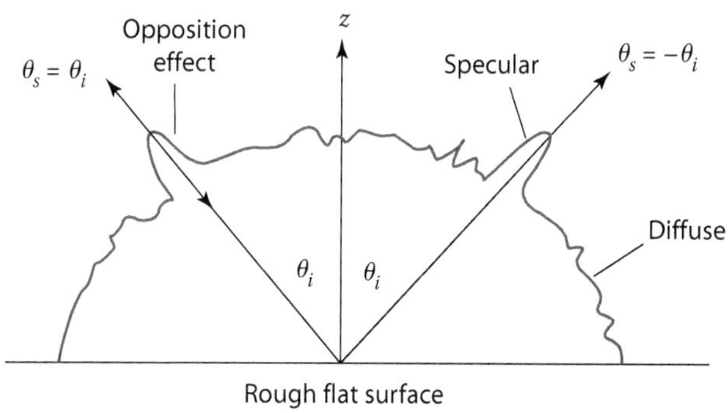

Fig. 9.18 Typical bistatic scattering pattern for a rough flat surface.

characteristic of a diffuse surface. This is true for all angles of incidence θ_i. One exception to this is the specular direction, which satisfies Snell's law, $\theta_s = -\theta_i$. If the surface is smooth enough, a specular lobe occurs, and the height of the lobe increases with surface smoothness. The second angle of enhanced scattering is in the back direction. This is referred to as the *opposition effect* and is due to secondary scattering mechanisms that normally tend to cancel [12,13]. They include localized shadowing and interactions between adjacent surface areas. Because of the small wavelength the reflecting plane is not abrupt; some volume scattering occurs.

An often cited example of the opposition effect occurs when an aircraft is flying slightly above a cloud layer. The opposition effect manifests itself as a halo around the aircraft's shadow cast by the sun on the clouds.

9.7.2 Perfectly Diffuse Surfaces

A *perfectly diffuse surface*, or *Lambertian surface*, is one for which the scattered signal is constant at all viewing angles independent of the angle of incidence. However, when a beam strikes a *finite* diffuse surface, the energy scattered in the direction of the detector will depend on the illuminated surface area and the detector field of view. As illustrated in Fig. 9.19, three cases are possible.

Case 1 Infinite ideal diffuse surface: The measured signal will be the same at any receive angle θ_r, independent of the angle of incidence θ_i.

Case 2 Finite ideal diffuse surface with small detector FOV: Only a portion of the surface is viewed, and no edges are seen. The measured signal will be constant at all receive angles just as in case 1.

Case 3 Finite ideal diffuse surface with large detector FOV: The FOV extends beyond the surface and, therefore, the projected aperture area determines the received signal strength ($\sim\cos\theta_r$).

Similar statements can be made with regard to the incident energy collected by the surface when it is illuminated by a divergent beam (i.e., the beam extends beyond the surface edges). In such cases, a $\cos\theta_i$ dependence occurs.

Diffuse surfaces have a constant BRDF. They scatter equally in all directions independently of the illuminating direction, and all points on the surface scatter identically. Thus,

$$\rho_{pq}(\vec{r}', \theta_i, \phi_i, \theta_r, \phi_r) = \rho_0 \tag{9.29}$$

where ρ_0 is a constant. In this case, it makes sense to define a total *hemispherical reflectance:*

$$\mathcal{R}_d = \int_0^{2\pi}\int_0^{\pi/2} \rho_0 \cos\theta_r \sin\theta_r \, d\theta_r \, d\phi_r \tag{9.30}$$

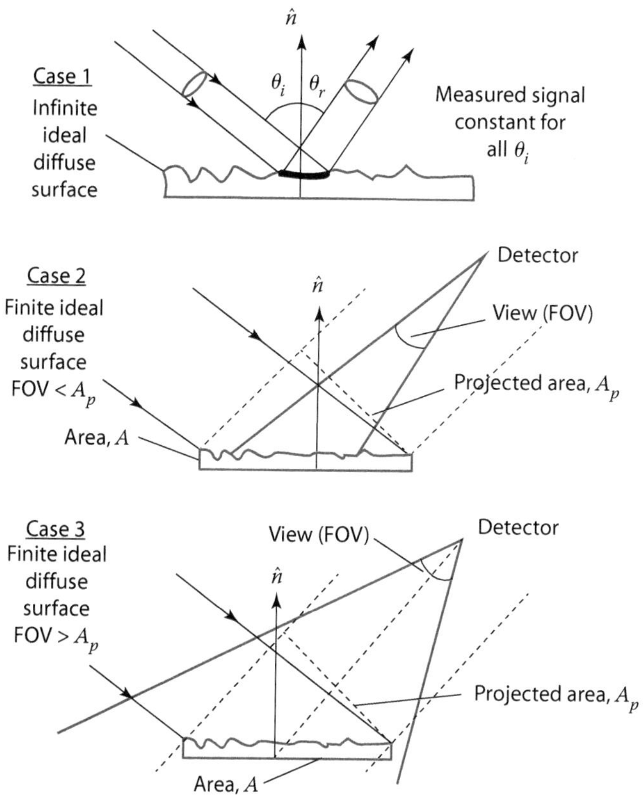

Fig. 9.19 Three cases of illumination and viewing of a diffuse surface.

where the subscript d signifies *diffuse.* For a perfectly diffuse surface, Eq. (9.30) integrates to

$$\mathcal{R}_d = \pi \rho_0 \tag{9.31}$$

The hemispherical reflectance is often the quantity obtained by measurement.

Example 9.1 Laser Cross Section of a Diffuse Sphere

A beam illuminates a sphere of radius a as shown in Fig. 9.20. If the surface is diffuse, the BRDF is a constant, ρ_0. Based on Eq. (9.27), the differential scattered intensity due to an illuminated ring is

$$d\sigma_d = 4\pi\rho_0 \cos^2 \theta ds \tag{9.32}$$

(Continued)

Example 9.1 Laser Cross Section of a Diffuse Sphere (Continued)

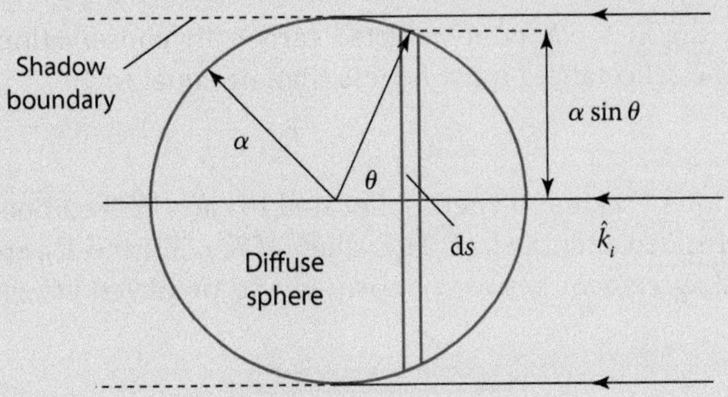

Fig. 9.20 LCS of a diffuse sphere.

where $ds = 2\pi a^2 \sin\theta\, d\theta$. The total scattered intensity is determined by integrating over the illuminated hemisphere:

$$\sigma_d = 4\pi\rho_0 \int_0^{\pi/2} \cos^2\theta \sin\theta d\theta \tag{9.33}$$

yielding

$$\sigma_d = \rho_0 \frac{8\pi^2 a^2}{3} \tag{9.34}$$

In terms of the total hemispherical reflectance, Eq. (9.31)

$$\sigma_d = \mathcal{R}_d \frac{8\pi a^2}{3} \tag{9.35}$$

9.8 Calculation of Laser Cross Section

9.8.1 Components of Laser Cross Section

Empirically, LCS has been found to be composed of three components: 1) specular, σ_s; 2) diffuse, σ_d; and 3) projected area, σ_p. The total LCS is $\sigma = \sigma_s + \sigma_d + \sigma_p$.

The specular component has the potential for extremely large values because of the small wavelength. However, the majority of the incident light is scattered diffusely for most surfaces. Frequently, the specular "flash" cannot even be identified. As Eq. (9.27) shows, the diffuse component varies as $\cos^2\theta$. Measured data for diffuse surfaces generally fall off more slowly than $\cos^2\theta$, approaching $\cos\theta$, which is more characteristic of the projected area component.

Reflectance $\mathcal{R}$, defined in Eq. (9.30), is a measure of reflectivity of a target. Essentially, it is a ratio of all the power scattered by the target to the power incident on the target. A fraction of the reflected light can be associated with each of the LCS components listed earlier. By conservation of energy, the sum of the reflectances must be less than or equal to 1:

$$\mathcal{R} = \mathcal{R}_d + \mathcal{R}_s + \mathcal{R}_p \le 1 \tag{9.36}$$

The portion of scattered energy allocated to each LCS component is frequently determined after the fact. The values of $\mathcal{R}_d$, $\mathcal{R}_s$, and $\mathcal{R}_p$ are chosen to provide good agreement between measured and predicted LCS.

9.8.2 Specular Laser Cross Section

Physical optics (PO) or geometrical optics (GO) can be used to compute the specular component. The expressions for a perfectly reflecting surface must be reduced by $\mathcal{R}_s$:

$$\begin{aligned} \text{GO:}\quad \sigma_s &= \mathcal{R}_s \pi R_1^s R_2^s \\ \text{PO:}\quad \sigma_s &= \mathcal{R}_s 4\pi A^2/\lambda^2 \end{aligned} \tag{9.37}$$

The statistical result for errors derived in Chapter 6 can be used to estimate $\mathcal{R}_s$. For example, if only specular and diffuse scattering occurs for a particular target ($\mathcal{R}_p = 0$), Eq. (6.94) applies. The error-free power pattern corresponds to specular scattering and the error term to diffuse scattering. Thus, from Eq. (6.94)

$$\mathcal{R}_s = e^{-\overline{\Delta^2}} \qquad \text{and} \qquad \mathcal{R}_d = 1 - e^{-\overline{\Delta^2}} \tag{9.38}$$

The specular beamwidths predicted by PO are usually much smaller than those encountered in measurements, especially when the specular lobe intensity is of the same order of magnitude as the diffuse component. Frequently, cosine fits of measured data are used to model the specular component. With the specular peak represented by a cosine shape of power n,

$$d\sigma_s = 2n\mathcal{R}_n \cos^n\theta \, ds \tag{9.39}$$

Assume a circular, flat surface of radius a with no azimuthal (ϕ) scattering dependence,

$$\sigma_s = 2n\mathcal{R}_n 2\pi a^2 \int_0^{\pi/2} \cos^n\theta \sin\theta \, d\theta \tag{9.40}$$

or

$$\sigma_s = 4\pi a^2 \mathcal{R}_n \frac{n}{n+1} \tag{9.41}$$

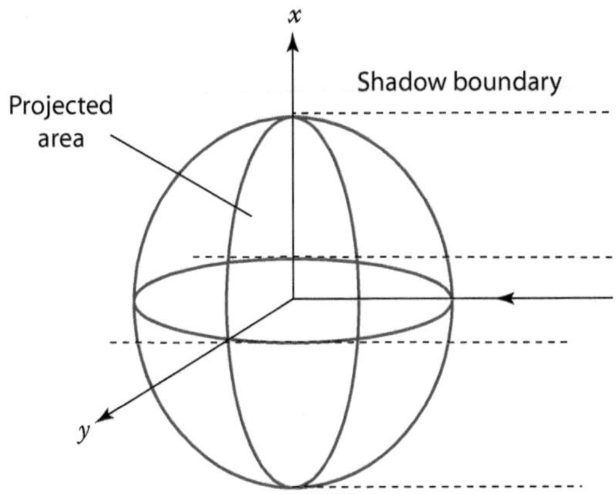

Fig. 9.21 Projected area for a sphere.

Comparing to the PO result gives

$$\mathcal{R}_s = 4\mathcal{R}_n \frac{n}{n+1} \tag{9.42}$$

Note that, when $n = 1$, $\sigma_s \sim \cos\theta$, and Eq. (9.41) reduces to the projected area LCS. When $n = 2$, $\sigma_s \sim \cos^2\theta$, and Eq. (9.41) reduces to the diffuse LCS.

As mentioned earlier, many scatterers do not have the monostatic $\cos^2\theta$ dependence expected of a perfectly diffuse surface. This is usually due to beam profile effects. Adding a projected area component can improve the agreement between predicted and measured data. This term has the form

$$d\sigma_p = 2\mathcal{R}_p \cos\theta ds$$

where θ is the angle from the surface normal and $\mathcal{R}_p$ the projected area hemispherical reflectance. For example, the projected area for a sphere is simply a disk, as illustrated in Fig. 9.21.

Example 9.2 Total Laser Cross Section of a Flat Plate

Consider a square plate of length $L = 6$ in. illuminated by a uniform laser beam with $\lambda = 10.6$ μm. The surface finish is uniform, and the irregularities are random, with an rms value of 0.001 in. In general, there will be three LCS components: specular, diffuse, and projected area. The mean power pattern consists of the error-free pattern (which is the specular), reduced by

(Continued)

Example 9.2 Total Laser Cross Section of a Flat Plate *(Continued)*

the factor $e^{-\overline{\Delta^2}}$, and an error term that increases as $1 - e^{-\overline{\Delta^2}}$ Note that $\overline{\Delta^2}$ is a very large number:

$$\overline{\Delta^2} = 4k^2\overline{\delta^2} = 4\left[\frac{2\pi(0.001)(0.0254)}{10.6 \times 10^{-6}}\right]^2 \approx 906$$

Thus, $e^{-\overline{\Delta^2}}$ is essentially zero, and there is no specular LCS, $\mathcal{R}_s = 0$. The remaining two components are computed as follows:

$$\text{Diffuse:}\quad \sigma_d = 4\mathcal{R}_d A \cos^2\theta$$
$$\text{Projected area:}\quad \sigma_p = 2\mathcal{R}_p A \cos\theta$$

Results are shown in Fig. 9.22 for $\mathcal{R}_d = \mathcal{R}_p = 0.5$.

Note that, if the surface were perfectly flat, the specular lobe height would be very large:

$$\sigma_s = \frac{4\pi(0.1520)^4}{(10.6 \times 10^{-6})^2} \approx 78\ \text{dBsm}$$

In practice, nothing near this level is approached. Furthermore, the lobe width is on the order of a microrad, and the probability of intercepting it in the field

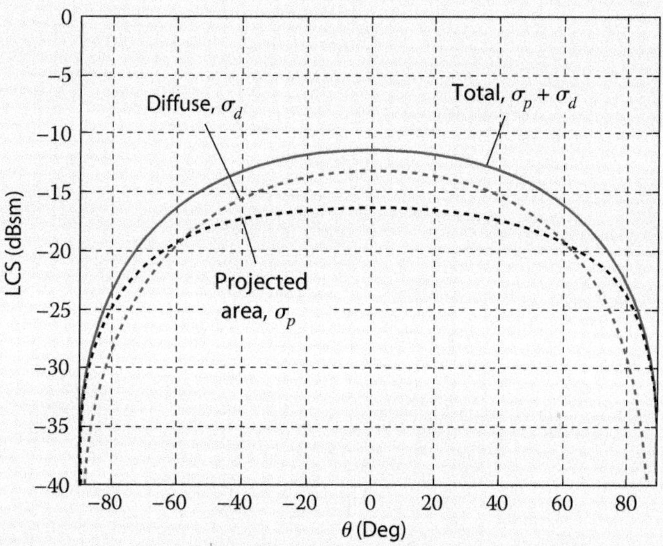

Fig. 9.22 Projected area and diffuse components for a 6-inch plate.

(Continued)

Example 9.2 Total Laser Cross Section of a Flat Plate *(Continued)*

is very small. Truly monostatic laser radars are not possible because of the required separation of the transmit and receive optics. This separation may not allow the receiver to be located in the scattering pattern mainbeam.

Figure 9.22 is featureless compared to the multilobed patterns that occur for RCS. This is a consequence of the nature of the dominant diffuse scattering.

9.9 Multiple Reflections

Multiple diffuse reflections can contribute significantly to the LCS at an observation point. Diffuse reflections do not necessarily have to satisfy Snell's law. As illustrated in Fig. 9.23, a ray incident on the diffuse surface A_1 scatters in all directions. Many of these rays will hit plate A_2 and cause secondary diffuse rays which, in turn, scatter in all directions. Thus, every scatter point on A_1 will completely illuminate A_2.

To use the definition of LCS requires a knowledge of W_r (or I_r) after the second bounce:

$$\sigma = 4\pi \frac{I_r}{W_i} = 4\pi R^2 \frac{W_r}{W_i} \tag{9.43}$$

where

$$R^2 W_r = I_r = \iint_{A_2} L_r \cos\theta_r \, ds \tag{9.44}$$

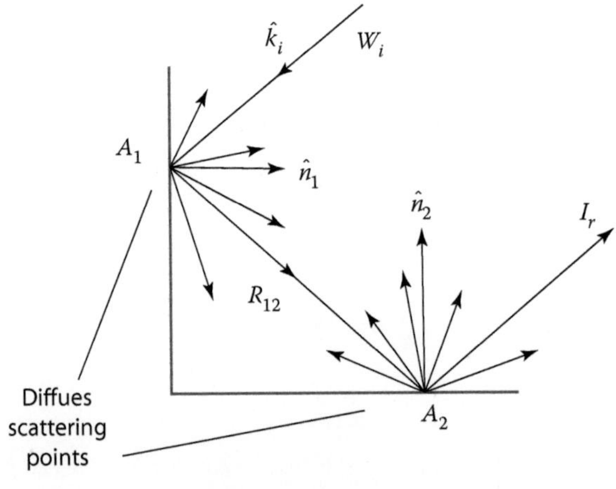

Fig. 9.23 Multiple diffuse reflections.

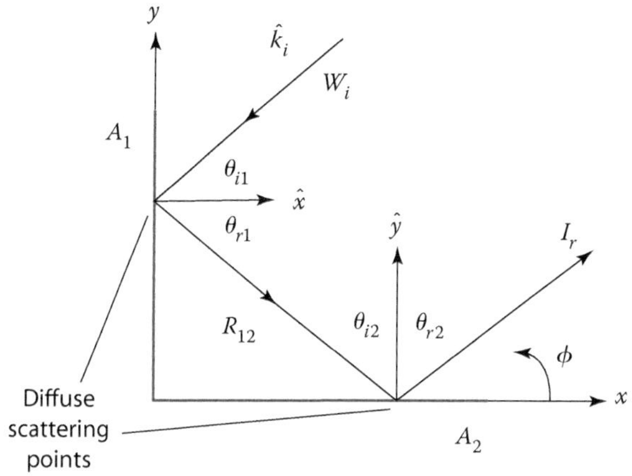

Fig. 9.24 Diffuse scattering by a corner reflector.

and $L_r = \rho_0 W_i \cos\theta_i$. Recall that, for a perfectly diffuse surface, $\rho_0 = \mathcal{R}_d/\pi$, and, for monostatic LCS, $\theta_r = \theta_i$.

The total LCS at an observation point is the sum of the primary scattered field (first reflection) and secondary scattered field (second, third, ..., etc., reflections):

$$\sigma = \sigma_1 + \sigma_2 + \sigma_3 + \cdots \tag{9.45}$$

In the case of a right-angle corner, using the notation in Fig. 9.24 results in

$$\sigma_1 = 4\pi\rho_0 \left\{ \iint_{A_1} \cos^2\theta_1 \,\mathrm{d}s + \iint_{A_2} \cos^2\theta_2 \,\mathrm{d}s \right\} \tag{9.46}$$

where θ_1 and θ_2 are the incident ray angles with respect to the plate normals. The first term is the direct diffuse scattering from plate A_1 and the second from plate A_2. Integrating gives

$$\sigma_1 = 4\mathcal{R}_d(A_1\cos^2\theta_1 + A_2\cos^2\theta_2) \tag{9.47}$$

The second reflection is computed as follows:

$$\sigma_{12} = 4\pi\frac{I_{r2}}{W_{i1}} \tag{9.48}$$

First, the radiant intensity is determined:

$$I_{r2} = \iint_{A_2} L_{r2}\cos\theta_{r2}\,\mathrm{d}s_2 \tag{9.49}$$

or, in differential form,

$$\mathrm{d}I_{r2} = \rho_0 W_{i2}\cos\theta_{i2}\cos\theta_{r2}\,\mathrm{d}s_2 \tag{9.50}$$

This gives the differential radiant intensity due to a small patch on plate 2 illuminated by a single diffuse ray from plate 1. When this formula is expanded to include all the diffuse rays from a differential patch ds_1 on plate 1, Eq. (9.50) becomes

$$d^2 I_{r2} = \rho_0 \cos\theta_{i2} \cos\theta_{r2} ds_2 dW_{i2} \tag{9.51}$$

with

$$dW_{i2} = dW_{r1} = \frac{dI_{r1}}{R_{12}^2} = \frac{L_{r1}\cos\theta_{r1} ds_1}{R_{12}^2} \tag{9.52}$$

and $L_{r1} = \rho_0 \ W_{i1} \ \cos\theta_{i1}$. Combining the preceding relationships gives

$$d^2 I_{r2} = \frac{\rho_0^2 W_{i1}}{R_{12}^2} \cos\theta_{i1} \cos\theta_{r1} ds_1 \cos\theta_{i2} \cos\theta_{r2} ds_2 \tag{9.53}$$

Finally, integrating over both areas gives the total radiant intensity and, subsequently, the LCS:

$$\sigma_{12} = \frac{4R_d^2}{\pi} \cos\theta_{i1} \cos\theta_{r2} \int_{A_1}\int_{A_2} \frac{\cos\theta_{i2}\cos\theta_{r1}}{R_{12}^2} ds_2 ds_1 \tag{9.54}$$

where Eq. (9.31) has been used. This expression must be evaluated numerically. Note that, by reciprocity, $\sigma_{12} = \sigma_{21}$.

Example 9.3 Laser Cross Section of a Diffuse Corner Reflector

As an example of multiple scattering, consider a corner reflector consisting of two 6-in. square plates (Example 9.2). With reference to Fig. 9.24, plate 2 lies in the $x - z$ plane ($\hat{n}_2 = \hat{y}$), and plate 1 lies in the $y - z$ plane ($\hat{n}_1 = \hat{x}$). Points on the two plates are denoted $(x_2, 0, z_2)$ and $(0, y_1, z_1)$, and the angle ϕ is measured from the x axis. The unit vector for two diffuse scattering points, one on each plate, is $\hat{R}_{12}$, where

$$\hat{R}_{12} = \frac{x_2\hat{x} - y_1\hat{y} + (z_2 - z_1)\hat{z}}{\sqrt{x_2^2 + y_1^2 + (z_1 - z_2)^2}} \tag{9.55}$$

The angles that the diffuse ray makes with the two plate normals are easily determined from

$$\cos\theta_{i2} = \hat{y} \cdot (-\hat{R}_{12}) = y_1/R_{12}$$

$$\cos\theta_{r1} = \hat{x} \cdot \hat{R}_{12} = x_2/R_{12}$$

(Continued)

Example 9.3 Laser Cross Section of a Diffuse Corner Reflector *(Continued)*

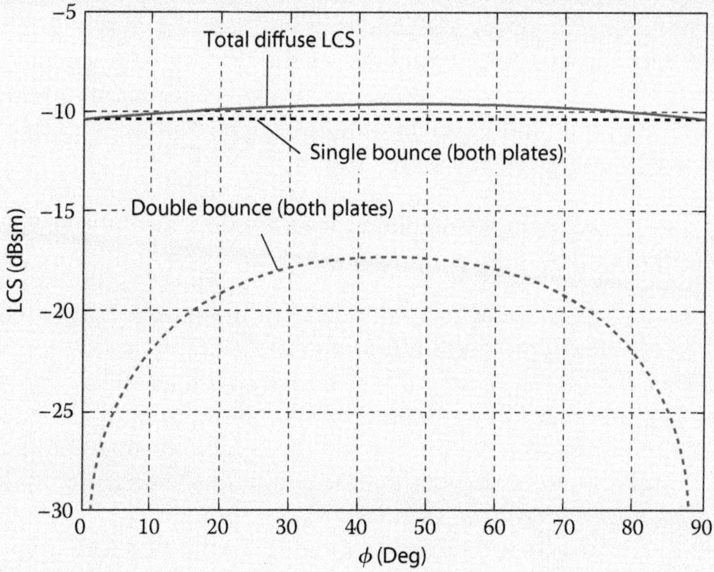

Fig. 9.25 Laser cross section of a diffuse corner reflector.

For the monostatic case ($\hat{k}_i = -\hat{k}_r$)

$$\begin{aligned} \cos(\theta_{i1}) &= \sin\theta\cos\phi \\ \cos(\theta_{r2}) &= \sin\theta\sin\phi \end{aligned} \tag{9.56}$$

The total LCS is the sum of the singly and doubly reflected terms:

$$\sigma = \sigma_1 + \sigma_2 \tag{9.57}$$

Results are presented in Fig. 9.25. Note that the singly reflected term is a constant when both plates are considered. The cosine factor for one plate is compensated for by the cosine factor of the second plate for this particular corner angle. The addition of the doubly reflected field has no effect at the grazing angles but does increase the LCS at other angles. This method can be extended to higher-order reflections as well.

9.10 Laser Cross Section Reduction Methods

In principle, the same four methods used in RCS reduction can be applied to LCS. They are 1) shaping, 2) materials selection and coatings, 3) passive cancellation, and 4) active cancellation. Shaping primarily affects the specular component, which is generally negligible unless the surface is optically polished. Shaping has only a small effect on the diffuse component (which varies as $\cos^2\theta$) and the projected area component (which varies as $\cos\theta$). Active and passive cancellation are effective only when applied to coherent scattering sources and, therefore, are not practical at laser frequencies.

The most effective means of LCS reduction is materials selection and the application of coatings (i.e., controlling the BRDF). One common method for reducing reflection from a dielectric body is the application of quarter-wave films, also known as antireflection coatings. High-reflection coatings can be used in conjunction with shaping.

9.11 Antireflection Films

A thin film can be added to an air/dielectric interface to reduce the intensity of the reflected light. Films are usually applied to optical devices such as glass plates and lenses to improve their transmittance. They can also be applied to laser radar targets to reduce reflectance. Properly designed films cause multiple reflections at the two interfaces (air/film and film/target) to cancel. This is precisely the same approach used in the design of Dallenbach layers and other narrowband radar reduction methods.

Assume that the film thickness is t and its permittivity and permeability are μ_1 and ϵ_1. Similarly, the constitutive parameters of the target material are μ_2 and ϵ_2. A wave is incident from an angle θ, as shown in Fig. 9.26. The objective is to find the total (power) reflection coefficient at the air/film interface R. It will be easier to calculate the power transmission coefficient T first and then use the relationship $R = 1 - T$.

First, define reflection coefficients at each interface for waves incident from the film region. For the air/film interface,

$$\Gamma_1 = -\left(\frac{n_0 - n_1}{n_0 + n_1}\right) \tag{9.58}$$

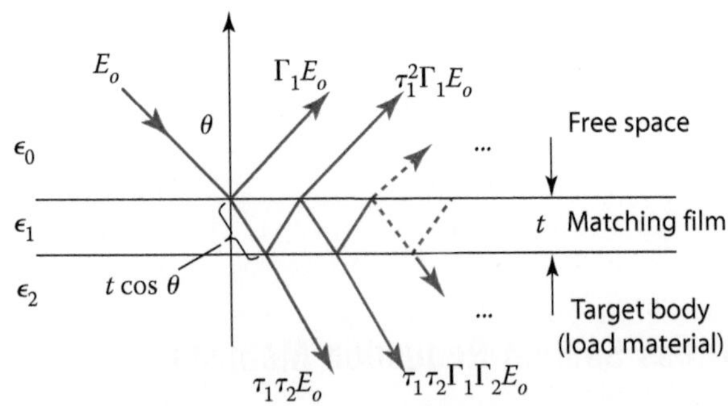

Fig. 9.26 Antireflection film of thickness t.

and, for the film/target interface,

$$\Gamma_2 = \frac{n_1 - n_2}{n_1 + n_2} \tag{9.59}$$

where n_0, n_1, and n_2 are the indices of refraction of the three regions. Define transmission coefficients at these interfaces:

$$\tau_1 = 1 + \Gamma_1 \qquad \text{and} \qquad \tau_2 = 1 + \Gamma_2 \tag{9.60}$$

The total reflected field is the sum of an infinite number of reflected rays. Similarly, the transmitted field consists of an infinite number of rays. After the terms for several transmitted rays have been explicitly written, a pattern emerges:

$$\begin{gathered} \tau_1\tau_2\Gamma_1\Gamma_2 E_0 e^{-j\delta} \\ \tau_1\tau_2\Gamma_1^2\Gamma_2^2 E_0 e^{-j2\delta} \\ \vdots \end{gathered} \tag{9.61}$$

where $\delta = 2\,kn_1t\,\cos\theta$ is the two-way path delay in the film.

The total transmitted field is the sum of all transmitted rays:

$$E_T = \tau_1\tau_2E_0(1 + \Gamma_1\Gamma_2e^{-j\delta} + \Gamma_1^2\Gamma_2^2e^{-j2\delta} + \cdots + \Gamma_1^n\Gamma_2^ne^{-jn\delta} + \cdots) \tag{9.62}$$

This is a geometric series and has a closed-form result:

$$E_T = \frac{\tau_1\tau_2E_0}{1 - \Gamma_1\Gamma_2e^{-j\delta}} \tag{9.63}$$

Now, the power transmission coefficient is

$$T = \frac{|E_T|^2}{|E_0|^2} = \frac{\tau_1^2\tau_2^2}{1 + \Gamma_1^2\Gamma_2^2 - 2\Gamma_1\Gamma_2\cos\delta} \tag{9.64}$$

Finally,

$$R = \frac{\Gamma_1^2 + \Gamma_2^2 - 2\Gamma_1\Gamma_2\cos\delta}{1 + \Gamma_1^2\Gamma_2^2 - 2\Gamma_1\Gamma_2\cos\delta} \tag{9.65}$$

For a normally incident wave, a quarter-wave film thickness causes equal reflections at both interfaces to cancel, yielding a total reflection coefficient of zero. This is due to the fact that the round-trip path difference in the film introduces a 180-deg phase difference between the two reflected waves. Obviously, the reflections at the two interfaces will not be equal in this case (i.e., $\Gamma_1 \neq \Gamma_2$), but we will proceed with the assumption that a quarter-wave thickness provides the optimum cancellation. Thus,

$$n_1t = \frac{\lambda}{4}, \qquad \theta = 0, \qquad \delta = \pi, \qquad \cos\delta = -1 \tag{9.66}$$

will result in a reflection coefficient of

$$R = \frac{(\Gamma_1 + \Gamma_2)^2}{(1 + \Gamma_1 \Gamma_2)^2} = \frac{n_1^2 - n_0 n_2}{n_0 n_2 + n_1^2} \tag{9.67}$$

Forcing the reflection coefficient to zero requires a film index of $n_1 = \sqrt{n_0 n_2}$, which is recognized as the quarter-wave transformer law.

Example 9.4 Performance of Antireflecting Films [14]

Sample calculation: Air ($n_0 = 1$) and glass ($n_2 = 1.5$), which has $R = 0.04$. Using a quarter-wave film, $n_1 = \sqrt{(1)(1.5)} = 1.22$. From Appendix F there are two possibilities that approach this value:

1. Cryolite (Na_3AlF_6) with $n_1 = 1.33$. This would reduce the reflection coefficient to $R = 0.008$.
2. Magnesium fluoride (MgF_2) with $n_1 = 1.384$. In this case, the reflection coefficient is $R = 0.012$.

Figure 9.27 shows the improvement typically achievable with films. Some additional coating materials are listed in Appendix F.

The successful application of films to the reduction of LCS requires a delicate tradeoff between electrical, mechanical, and thermal performance. Mechanical properties of concern are film strength, elasticity, and hardness. Strength refers to the ability to handle stress and strain. Hardness is a measure of the material's resistance to indentation. Airborne platforms with velocities on the order of Mach 4 can have raindrop impact pressures in the

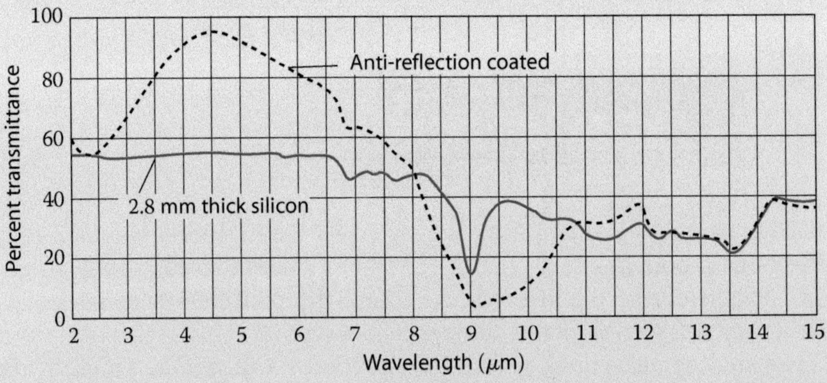

Fig. 9.27 Scattering reduction using an antireflection coating (from Ref. [14], courtesy SPIE).

(Continued)

Example 9.4 Performance of Antireflecting Films [14] *(Continued)*

range of 3 GPa [1 pascal (Pa) = 1 N/m^2]. Materials on supersonic vehicles are also subjected to thermal shock as a result of large thermal gradients across the material and, therefore, thermal conductivity is an issue. Finally, the finishing characteristics and durability of the film must meet the appropriate specifications.

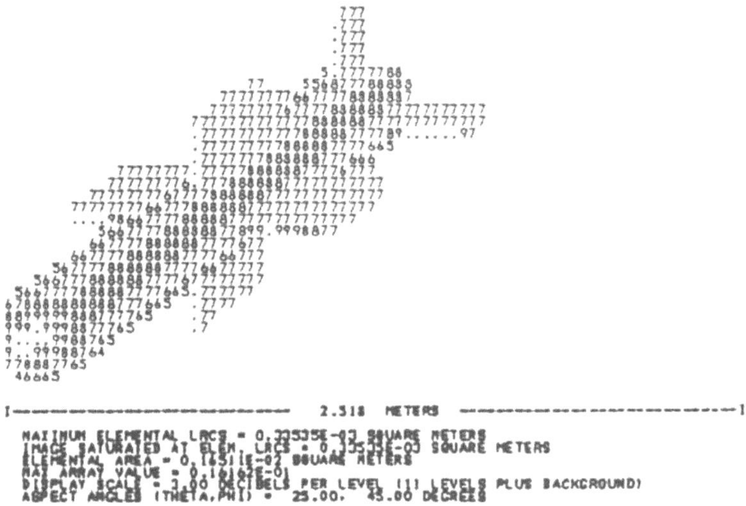

Fig. 9.28 Laser cross section of a missile computed with LSC-2.

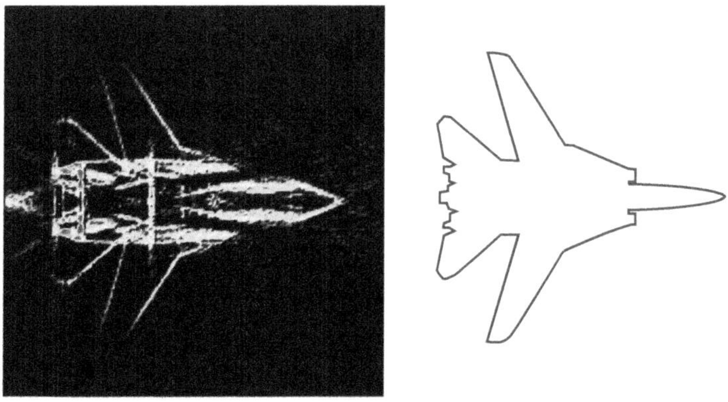

Fig. 9.29 Laser radar image of a F-14 at 0 deg elevation (from Ref. [16]).

9.12 Laser Cross Section Prediction for Complex Targets

Diffuse scattering is the primary mechanism of light scattering by targets with "average" surface characteristics. *Average* refers to maintenance of the target's surface; that is, no special surface preparation or polishing is performed. Relatively accurate LCS predictions for complex targets can be made using the geometrical components method. In many cases, multiple reflections can be neglected and the LCS estimates are still within a couple of decibels of measured data.

The overall LCS level at any particular observation point is determined primarily by the amount of projected target area presented to the ladar. Therefore, it is important to include shadowing. For diffuse scattering, the relative phases of the target components are not important. The total LCS is simply a noncoherent sum of all "subsurface" LCS contributions.

Figure 9.28 shows the LCS of a missile computed using the computer code LSC-2 [15]. The line printer output indicates the intensity of scattered light from the target surface relative to the printed reference. (Dots are off scale.) The output resolution is coarse but serves to identify "hot spots" such as leading edges.

Radar target imaging techniques (Sec. 8.4) can be applied to LIDAR. Figure 9.29 shows an image of a F-14 generated by simulation [16]. The image can highlight LCS hotspots that might need to be treated, such as the engine exhaust ports. A similar image can be generated by measurement.

The restrictions and limitations imposed by the definition of LCS are seldom satisfied. Issues such as coherence, beam profile, volume scattering by the target, and so forth, have a significant effect on LCS and are usually included in simulation codes. As in the case of radar, it is the signal-to-noise ratio (SNR) that ultimately determines system performance, and LCS is just one of many parameters that affect SNR. SNRs can be significantly enhanced by signal-processing techniques. Most LCS computer codes simulate the processing and image formation components of the system as well as LCS computation and atmospheric effects. This provides a more accurate estimate of system performance than is possible from a simple LCS number.

References

[1] Hulme, K. F. et al., "A CO_2 Laser Rangefinder Using Heterodyne Detection and Chirp Pulse Compression," *Optics and Quantum Electronics*, Vol. 13, 1981, p. 35.

[2] Park, D., "Performance Analysis of Optical Synthetic Aperture Radars," *SPIE*, Vol. 999, Laser Radar III, 1988.

[3] Yoder, M. J., and Youmans, D. G., "Laser Radar Wavelength Selection and Trade-Offs," *SPIE*, Vol. 999, Laser Radar III, 1988, p. 72.

[4] Hovanessian, S. A., *Introduction to Sensor Systems*, Artech House, Norwood, MA, 1988.

[5] Saleh, B. E. A., and Teich, M. C., *Fundamentals of Photonics*, Wiley Interscience, New York, 1991.

[6] Buckman, A. B., *Guided-Wave Photonics*, Sanders College Publishing, Philadelphia, 1992.
[7] Ross, M., *Laser Receivers*, Wiley, New York, 1966.
[8] Hudson, R. D., *Infrared Systems Engineering*, Wiley, New York, 1969.
[9] Marcuse, D., *Light Transmission Optics*, VanNostrand Reinhold, New York, 1972.
[10] MacFarland, A. B., Pushkar, R., Wasky, R., and Zidek, P., *Laser Cross Section Handbook*, Mission Research Corporation, WRDC-TR-89-9010, Vol. I, June 15, 1990.
[11] Stover, J. C., *Optical Scattering Measurement and Analysis*, McGraw-Hill, New York, 1990.
[12] Tsang, L., and Ishimaru, A., "Backscattering Enhancement of Random Discrete Scatterers," *Journal of the Optical Society of America*, Part A, Vol. 1, No. 8, Aug. 1984, pp. 836–839.
[13] Gu, Z.-H., Dummer, R. S., Maradudin, A., and Mcgurn, A., "Experimental Study of the Opposition Effect in the Scattering of Light from a Randomly Rough Metal Surface," *Applied Optics*, Vol. 28, No. 3, Feb. 1989, pp. 537–543.
[14] Harris, D. C., *Infrared Window and Dome Materials*, SPIE Press, Tutorial Text Series, Vol. 10, Bellingham, WA, 1992.
[15] Boswell, J., Ferret, L., Kregel, N., Mateiki, T., and O'Dwyer, H., *LSC-2 User's Manual*, General Research Corp., McLean, VA, May 1987.
[16] Berginc, G. "Scattering Models for 1-D–2-D–3-D Laser Imagery," *Optical Engineering*, Vol. 56, No. 3, Mar. 2017 031207-1-031207-13.
[17] Nussbaum, A., and Phillips, R. A., *Contemporary Optics for Scientists and Engineers*, Prentice-Hall, Englewood Cliffs, NJ, 1976.

Problems

9.1 Consider a point P at a distance r from the center of a disk of radius a, as shown in Fig. P9.1 (from Ref. [17]).

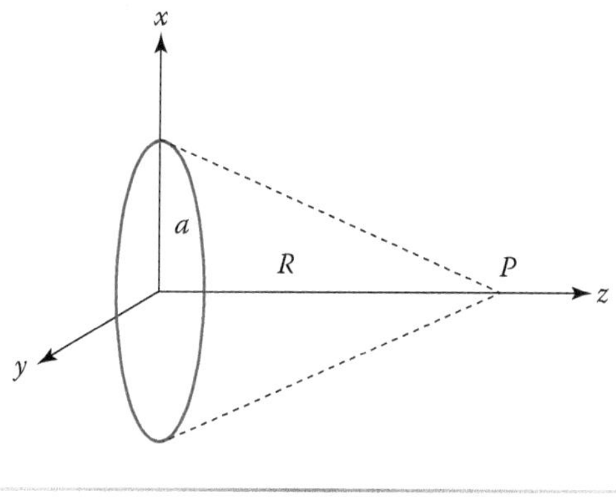

Fig. P9.1

a) Show that the solid angle subtended by the disk is

$$\Omega = 2\pi a\left[\frac{1}{r} - \frac{1}{\sqrt{r^2 + a^2}}\right]$$

b) If $P \to \infty$, show that $\Omega \to 0$.

c) If $P \to 0$, show that $\Omega \to 2\pi$.

d) If $a/r \ll 1$, show that $\Omega = \pi a^2/r^2$.

9.2 A disk of radius a scatters energy toward a surface of area A', a distance d away, as shown in Fig. P9.2.

a) Show that P_r on A' is given by

$$P_r = \pi L A' \frac{a^2}{\sqrt{d^2 + a^2}}$$

b) Show that $E = \Omega L$.

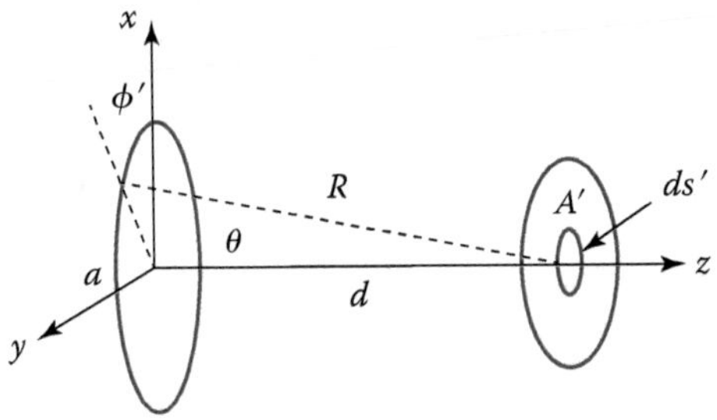

Fig. P9.2

9.3 For a perfectly diffuse surface, there is no absorption and, hence, all incident energy is reflected:

$$L_{rp} = \text{constant} = \rho_{pH} E_H + \rho_{pV} E_V$$

where ρ is the BRDF and $p = V$ or H. The subscript r on L denotes reflected from the surface. Show that

$$\rho_{VH} + \rho_{HH} = 1/\pi$$

$$\rho_{HV} + \rho_{VV} = 1/\pi$$

9.4 Find expressions for the total LCS of a flat circular disk of radius a. Assume that the reflectances for the three LCS components are $\mathcal{R}_s$, $\mathcal{R}_d$, and $\mathcal{R}_p$.

9.5 A monostatic ladar partially illuminates the target in Fig. 9.12.

a) If the target scatters as a single point, show that the received power is

$$P_r = \frac{P_t A \rho L_0}{2\pi R^2}$$

where ρ is the target reflectivity ($=P_s/P_i$).

b) Show that, for a diffuse target,

$$P_r = \frac{P_t A \rho L_0}{2R^2}$$

9.6 A bistatic radar operates as a *floodlight* system with the following parameters:

Transmit :	range, R_t
	optics area, A_t
	beamwidth, θ_{Bt}
	power, P_t
Receive :	range, R_r
	optics area, A_r
	beamwidth, θ_{Br}

Show that the received power is

$$P_r = \frac{8 P_t A_t A_r \sigma L_0}{\pi^3 \lambda^2 R_t^2 R_r^2}$$

9.7 Show that the three LCS components of the cylinder shown in Fig. P9.7 are those given here [10].

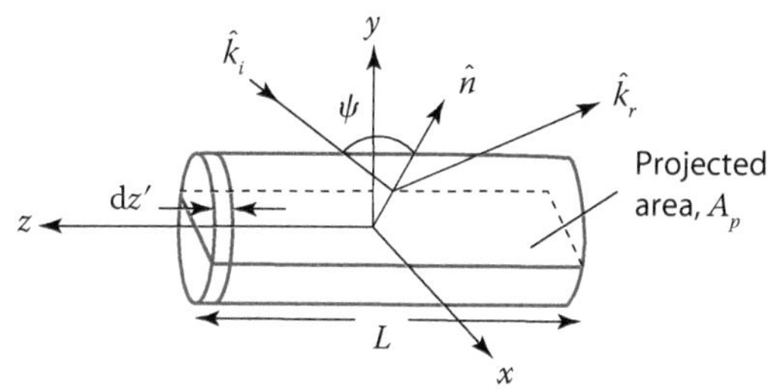

Fig. P9.7

Specular [using the cosine fit of Eq. (9.39)]:

$$\sigma_s = \mathcal{R}_n 2naL \sin^n \theta \begin{cases} \dfrac{\pi n!}{2^n[(n/2)!]^2}, & n \text{ even} \\ \dfrac{2^n\{[(n-1)/2]!\}^2}{n!}, & n \text{ odd} \end{cases}$$

Diffuse: $\sigma_d = \mathcal{R}_d \, 2\pi a \, L \, \sin^2\theta$
Projected area: $\sigma_d = \mathcal{R}_i \, 4a \, L \, \sin\theta$

9.8 A glass surface (index of refraction, $n = \sqrt{\epsilon_r} = 1.5$) is coated with SiO ($n \approx 2.0$). Plot the reflectivity as a function of thickness t for $0 \le t \le \lambda_0$ (λ_0 is the wavelength in free space).

Appendix A Notation, Definitions, and Review of Electromagnetics

A.1 Field Quantities and Constitutive Parameters

The basic electromagnetic field quantities in the meter-kilogram-second (MKS) system are

$\vec{E}$ = electric field intensity, V/m
$\vec{D}$ = electric flux density, or displacement vector, C/m^2
$\vec{H}$ = magnetic field intensity, A/m
$\vec{B}$ = magnetic flux density, Wb/m^2 or T
$\vec{J}$ = volume current density, A/m^2

All of these quantities can be a function of time and space. For instance, in the Cartesian coordinate system, $\vec{E}(x, y, z, t)$.

The *constitutive parameters* relate the intensities to the densities:

$$\vec{D} = \epsilon\vec{E}$$

$$\vec{B} = \mu\vec{H}$$

where $\epsilon = \epsilon_0\epsilon_r$ is the permittivity and $\mu = \mu_0\mu_r$ the permeability of the medium. The quantities with the subscript r are relative values with respect to the values of free space. Another important property of a medium is its conductivity. A general form of Ohm's law is $\vec{J} = \sigma_c\vec{E}$. All the constitutive parameters can be complex quantities.

A.2 Maxwell's Equations

Maxwell's equations, along with the boundary conditions, completely describe the behavior of electromagnetic fields in a region of space. The four equations in differential and integral forms are as follows.

1. Faraday's law:

$$\nabla \times \vec{E} = -(\partial\vec{B}/\partial t)$$
$$\oint_C \vec{E} \cdot \overrightarrow{d\ell} = -\int_S (\partial\vec{B}/\partial t) \cdot \overrightarrow{ds}$$

2. Ampere's circuital law:

$$\nabla \times \vec{H} = (\partial \vec{D}/\partial t) + \vec{J}$$
$$\oint_C \vec{H} \cdot \overrightarrow{d\ell} = - \int_S [(\partial \vec{D}/\partial t) + \vec{J}] \cdot \overrightarrow{ds}$$

3. Gauss's law:

$$\nabla \cdot \vec{D} = \rho_v$$
$$\oint_S \vec{D} \cdot \overrightarrow{ds} = \int_V \rho_v dv$$

4. No isolated magnetic charge:

$$\nabla \cdot \vec{B} = 0$$
$$\oint_S \vec{B} \cdot \overrightarrow{ds} = 0$$

In the preceding formulas, V represents a volume of the medium enclosed by a surface S, and ρ_v is the volume charge density in Coulombs per cubic meter. The differential line, surface, and volume quantities are $d\ell$, ds, and dv. The charge and current densities at a point in space are related by the continuity equation

$$\nabla \cdot \vec{J} = -\frac{\partial \rho_v}{\partial t}$$

Electrostatic and magnetostatic refer to the special case of $(\partial/\partial t) = 0$. An important time variation is the *time-harmonic* or sinusoidal case for which the time dependence is of the form $\cos(\omega t)$, where ω is the angular frequency in radians per second. When one deals with time-harmonic sources, the time dependence is usually *suppressed* (implied), and *phasor* quantities are used. Thus, a phasor is a complex quantity (has both real and imaginary parts) and is independent of time. To obtain the corresponding *instantaneous* (time-dependent) quantity, one must multiply the phasor by $(j\omega t)$ and take the real part of the product:

$$\vec{E}(x, y, z, t) = \mathrm{Re}\{\vec{E}(x, y, z)e^{j\omega t}\}$$

where Re is the real operator. Although the same symbols are used for both the time-varying and phasor quantities ($\vec{E}$ in this case), the time dependence will always be included when referring to a time function to avoid ambiguity.

The entire time dependence for a time-harmonic quantity is in the exponent and, therefore, a derivative with respect to time is equivalent to multiplying the phasor by $j\omega$. Thus, Maxwell's equations in time-harmonic form become

$$\nabla \times \vec{E} = -j\omega\vec{B}$$
$$\nabla \times \vec{H} = j\omega\vec{D} + \vec{J}$$
$$\nabla \cdot \vec{D} = \rho_v$$
$$\nabla \cdot \vec{B} = 0$$

A.3 Current Densities

The current density in Maxwell's equations is a volume current density. It represents the total density at a point in space. In a conducting medium that has *free electrons,* $\vec{J}$ is a conduction current. This current can occur in response to a field or can be *impressed* (forced). The term $j\omega\vec{D}$ in Maxwell's second equation corresponds to a *displacement* current. Therefore, Maxwell's second equation can be written with the three current contributions explicitly listed:

$$\nabla \times \vec{H} = \vec{J}_c + \vec{J}_d + \vec{J}_i$$

As shown in Fig. A.1, the current flowing through an arbitrary surface S is the integral of the volume current density over the area

$$\int_S \vec{J} \cdot \overrightarrow{\mathrm{d}s} = I$$

If the surface under consideration is flat with an area A and the current vector is everywhere normal to it and constant, the current crossing S is $I = A|\vec{J}|$.

The flow of charge can be confined to a *filament* (a line with zero cross-sectional area) or a surface. Because a filament has no cross section, the current flowing in it is simply I. If the current flows on a surface, the density is specified by a surface current density $\vec{J}_s$, which has units of A/m.

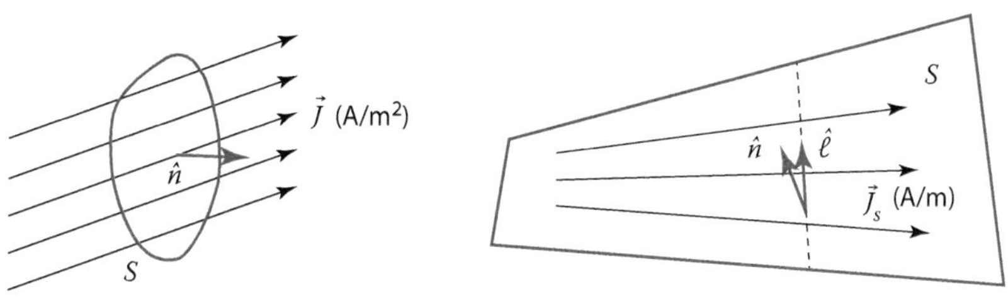

Fig. A.1 Volume and surface current densities.

To obtain the total current crossing a line of length L on a surface, it is necessary to integrate $\vec{J}_s$ along the line

$$\int_L \vec{J}_s \cdot (\overrightarrow{d\ell} \times \hat{n}) = I$$

For the special case of a strip of constant width ω and a constant current flowing parallel to the edges, $I = w|\vec{J}_s|$.

A.4 Boundary Conditions

The field quantities at an interface can be decomposed into parallel and perpendicular components:

$$\vec{E} = \vec{E}_\perp + \vec{E}_{||}$$

where $\perp$ and $||$ are defined relative to the surface. As shown in Fig. A.2, the normal $\hat{n}$ is directed from medium 2 to medium 1, in which case

$$\hat{n} \times [\vec{E}_1 - \vec{E}_2] = 0$$
$$\hat{n} \times [\vec{H}_1 - \vec{H}_2] = \overrightarrow{J_s}$$
$$\hat{n} \cdot [\vec{D}_1 - \vec{D}_2] = \rho_s$$
$$\hat{n} \cdot [\vec{B}_1 - \vec{B}_2] = 0$$

Thus, the tangential components of the electric field must be continuous at a boundary ($\vec{E}_{1\tan} = \vec{E}_{2\tan}$), and $\vec{H}_{1\tan} = \vec{H}_{2\tan}$ if no surface current exists on the boundary. The conditions at an interface between a dielectric and a perfect electric conductor are derived by setting $\vec{E}$ and $\vec{H}$ to zero in the conducting medium.

A.5 Magnetic Current

For convenience, fictitious magnetic charge and current are frequently included in Maxwell's equations. From the boundary conditions, the quantity $\hat{n} \times [\vec{H}_1 - \vec{H}_2]$ is equivalent to a surface current $\overrightarrow{J_s}$. It is possible to choose an arbitrary surface over which the quantity $\hat{n} \times [\vec{E}_1 - \vec{E}_2]$ is not zero. This is mathematically equivalent to a magnetic current on that surface. Allowing for both electric and

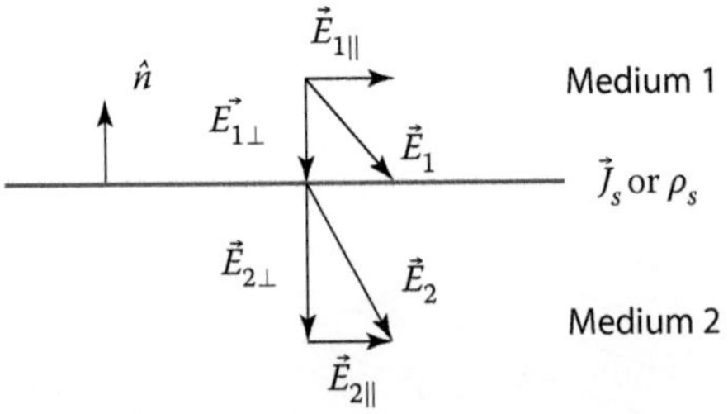

Fig. A.2 Boundary conditions at a planar interface.

magnetic current, Maxwell's equations are rewritten as

$$\nabla \times \vec{E} = -j\omega\vec{B} - \vec{J}_m$$

$$\nabla \times \vec{H} = j\omega\vec{D} + \vec{J}$$

$$\nabla \cdot \vec{D} = \rho_v$$

$$\nabla \cdot \vec{B} = \rho_{mv}$$

and the boundary conditions become

$$\hat{n} \times [\vec{E}_1 - \vec{E}_2] = -\vec{J}_{ms}$$

$$\hat{n} \times [\vec{H}_1 - \vec{H}_2] = \vec{J}_s$$

$$\hat{n} \cdot [\vec{D}_1 - \vec{D}_2] = \rho_s$$

$$\hat{n} \cdot [\vec{B}_1 - \vec{B}_2] = \rho_{ms}$$

The subscript m signifies a magnetic current or charge.

A.6 Types of Media

Media are categorized based on their constitutive parameters μ, ϵ, and σ_c. Some of the important categories are the following.

1. *Linear*: the constitutive parameters are independent of the intensity of the applied field. For a *nonlinear* material, $\epsilon(|\vec{E}|)$, $\mu(|\vec{E}|)$, and $\sigma_c(|\vec{E}|)$.
2. *Homogeneous*: the parameters are independent of location in the material. For an *inhomogeneous* material in Cartesian coordinates, $\epsilon(x, y, z)$, $\mu(x, y, z)$, and, $\sigma_c(x, y, z)$.
3. *Isotropic*: in general, the constitutive parameters can be tensors (matrices or dyads, denoted by boldface). For an *anisotropic* medium, $\mathbf{D} = \boldsymbol{\epsilon}\mathbf{E}$ or

$$\begin{bmatrix} D_x \\ D_y \\ D_z \end{bmatrix} = \begin{bmatrix} \epsilon_{xx} & \epsilon_{xy} & \epsilon_{xz} \\ \epsilon_{yx} & \epsilon_{yy} & \epsilon_{yz} \\ \epsilon_{zx} & \epsilon_{zy} & \epsilon_{zz} \end{bmatrix} \begin{bmatrix} E_x \\ E_y \\ E_z \end{bmatrix}$$

4. *Dispersive*: for most materials, the parameters change with frequency. Thus, for a *dispersive* medium, $\mu(\omega)$, $\boldsymbol{\epsilon}(\omega)$, and $\sigma_c(\omega)$. It is usually a good approximation to consider the parameters *nondispersive* when one is interested only in a narrow band of frequencies. *Simple* media are encountered most often. These are linear, isotropic, and homogeneous.

A.7 Wave Equation and Plane Waves

The *wave equations* for $\vec{E}$ and $\vec{H}$ can be derived from Maxwell's equations in a *source-free region* ($\vec{J}_s$, $\vec{J}_{ms}$, ρ_s, and ρ_{ms} are all zero). For instance, the wave equation (or *Helmholtz equation*) for the electric field is

$$\nabla^2 \vec{E} + k^2 \vec{E} = 0$$

$$k = \omega\sqrt{\mu\epsilon} = \frac{\omega}{u_p} = \frac{2\pi}{\lambda}$$

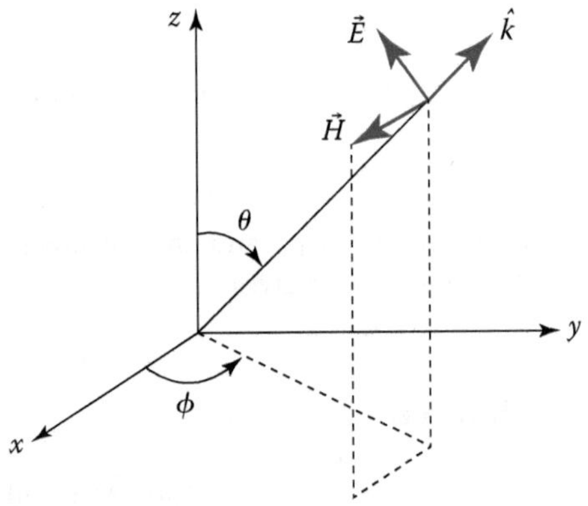

Fig. A.3 Plane wave propagation.

where u_p is the phase velocity in the medium and λ the wavelength ($= c/f$). When the medium is free space, the subscripts 0 will be added. Thus, in free space,

$$k_0 = \omega\sqrt{\mu_0\epsilon_0} = \frac{\omega}{c} = \frac{2\pi}{\lambda_0}$$

The phase velocity in free space is given the usual symbol c and is approximately equal to 3×10^8 m/s.

The vector Helmholtz equation holds for each Cartesian component

$$\nabla^2 E_x + k^2 E_x = 0, \text{ etc.}$$

The most elementary solutions of the wave equation are *plane waves*

$$\vec{E} = \vec{E}_0 e^{\pm j\vec{k}\cdot\vec{r}}$$

As shown in Fig. A.3 $\vec{k}$ is a vector in the direction of propagation, and $\vec{r}$ is a *position vector* from the origin to the point (x, y, z) at which the electric field is evaluated. The vector $\vec{E}_0$ defines the *polarization* (direction of the vector $\vec{E}$ of the wave).

From Maxwell's second equation, $\vec{H}$ is obtained from $\vec{E}$:

$$\vec{H} = \frac{\hat{k} \times \vec{E}_0}{\eta} e^{\pm j\vec{k}\cdot\vec{r}}$$

where $\eta = \sqrt{\mu/\epsilon}$ is the *intrinsic impedance* of the medium. Note that the vectors $\vec{E}$, $\vec{H}$, and $\vec{k}$ are mutually orthogonal for this solution. This type of behavior is characteristic of a *transverse electromagnetic* (TEM) wave.

Example A.1: An x-Polarized Plane Wave Traveling in the +z Direction

For an x-polarized plane wave, $\vec{E}_0 = \hat{x}E_0$. Because the wave is traveling in the z direction, $\vec{k} = k_0\hat{z}$, and the negative sign is used in the exponent. Using $\vec{r} = x\hat{x} + y\hat{y} + z\hat{z}$ gives

$$\vec{k} \cdot \vec{r} = k_0 z$$

Therefore,

$$\vec{E} = \hat{x}E_0 e^{-jk_0 z} \quad \text{and} \quad \vec{H} = \hat{y}\frac{E_0}{377}e^{-jk_0 z}$$

assuming free space ($\eta_0 = 377\Omega$). The instantaneous value of the electric field is

$$\vec{E}(z,t) = \mathrm{Re}\{E_0\hat{x}e^{j(\omega t - k_0 z)}\} = \hat{x}E_0\cos(\omega t - k_0 z)$$

It has been assumed that E_0 is a real quantity.

Planes of constant phase are given by $\omega t - k_0 z =$ constant, which is a plane $z =$ constant at any given t. A plot of the electric field at a constant time t or location z would show that the electric and magnetic field intensities are sinusoidal in both time and space.

A.8 Wave Polarization

Polarization refers to the behavior of the tip of the electric field vector as a function of time at a point in space. As illustrated in Fig. A.4, the tip for a linearly polarized wave oscillates along a line; for a circularly polarized wave, it rotates in a circle. These are special cases of the more general case of elliptical polarization.

A circularly polarized (CP) plane wave is generated by combining two linearly polarized plane waves that are in space and time quadrature. The term *quadrature* implies 90 deg, and therefore *space quadrature* is achieved by rotating the polarization of the second wave 90 deg with respect to the first one. *Phase quadrature* is achieved

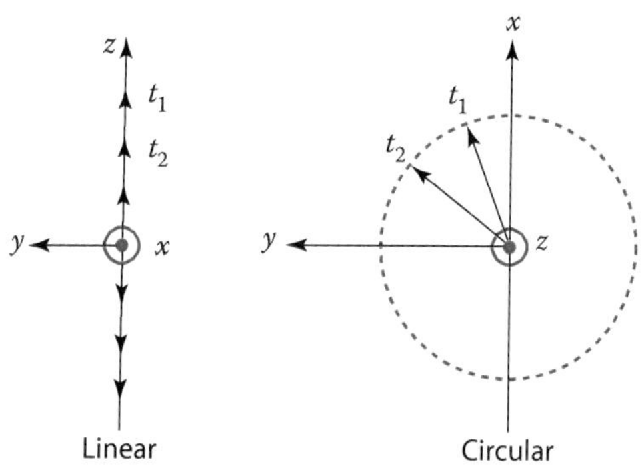

Fig. A.4 Illustration of linear and circular polarizations.

by adding ± 90 deg to the argument of the phasor exponent. An example of the phasor representation of a CP plane wave is

$$\vec{E} = E_0\hat{x}e^{-jkz} + E_0\hat{y}e^{-jkz \pm j\pi/2}$$

Using the fact that $e^{\pm j\pi/2} = \pm j$ gives

$$\vec{E} = (\hat{x} \pm j\hat{y})E_0 e^{-jkz}$$

The negative sign in the exponent implies that the wave is traveling in the positive z direction. The *sense* (direction of rotation) is determined by the sign of j. If $+j$ is chosen, the rotation of $\vec{E}$ is determined by the left-hand rule: with the thumb in the $+z$ direction, the rotation is given by the fingers on the left hand. Conversely, if $-j$ is used, the direction of rotation is given by the right hand. These are called *left-hand circular polarization* (LHCP) and *right-hand circular polarization* (RHCP), respectively.

Circular polarization is desirable when the relative orientation of a transmitting and receiving antenna is changing or unknown. For this reason, CP is usually used on satellite links. However, if an antenna provides perfect circular polarization when it is viewed from a direction normal to the aperture, the polarization at angles off the normal will be elliptical.

A.9 Plane Waves in Lossy Media

The wave number k in the Helmholtz equation can be complex. By convention, the *propagation constant* γ is defined as

$$\gamma = \alpha + j\beta = jk_c = j\omega\sqrt{\mu\epsilon}$$

where the subscript c is used to denote complex k explicitly; α is the *attenuation constant* and β the *phase constant*. Now, the wave equation for the electric field becomes

$$\nabla^2\vec{E} - \gamma^2\vec{E} = 0$$

and its solution is

$$\vec{E} = \vec{E}_0 e^{\pm\vec{\gamma}\cdot\vec{r}}$$

For an x-polarized wave traveling in the z direction,

$$\vec{E} = \hat{x}E_0 e^{-(\alpha+j\beta)z}$$

Note that, if α is not zero, the intensity of the wave diminishes as z increases. This is referred to as *attenuation*.

The constitutive parameters can consist of both real and imaginary parts

$$\epsilon = \epsilon' - j\epsilon'' \quad \text{and} \quad \mu = \mu' - j\mu''$$

where $\epsilon' = \epsilon_r \epsilon_0$ and $\mu' = \mu_r \mu_0$. From Sec. A.3, Maxwell's second equation can be written

$$\nabla \times \vec{H} = \vec{J}_{\text{tot}}$$

where

$$\vec{J}_{tot} = \vec{J}_c + \vec{J}_d + \vec{J}_i = (\sigma_c + \omega\epsilon'')\vec{E} + j\omega\epsilon'\vec{E} + \vec{J}_i$$

The quantity $(\sigma_c + \omega\epsilon) \equiv \sigma_e$ plays the role of an equivalent conductivity. For a good conductor or in the electrostatic case, $\sigma_e \approx \sigma_c$. For dielectrics at high frequencies, $\sigma_e \approx \omega\epsilon''$.

Defining the *effective electric loss tangent* $\tan\delta_\epsilon = \sigma_e/\omega\epsilon'$ allows the total current to be written as

$$\vec{J}_{tot} = \vec{J}_i + j\omega\epsilon'(1 - j\tan\delta_\epsilon)\vec{E}$$

As its name implies, the loss tangent determines the attenuation of the wave as it travels through the medium. A similar formulation applies to Maxwell's first equation and magnetic currents:

$$\vec{J}_{m\text{tot}} = \vec{J}_{mi} + j\omega\mu'(1 - j\tan\delta_\mu)\vec{H}$$

where $\tan\delta_\mu \equiv \mu''/\mu'$ is the *effective magnetic loss tangent.*

When the complex forms of the constitutive parameters are used,

$$\sqrt{\mu_c\epsilon_c} = \sqrt{\mu_0\epsilon_0\mu_r\epsilon_r}[1 - \tan\delta_\epsilon\tan\delta_\mu - j(\tan\delta_\epsilon + \tan\delta_\mu)]^{\frac{1}{2}}$$

Separating the preceding equation into real and imaginary parts leads to expressions for α and β:

$$\alpha = k_0\sqrt{\mu_r\epsilon_r}\left[\frac{1}{2}(1 - \tan\delta_\epsilon\tan\delta_\mu)\left(\sqrt{1 + \tan^2(\delta_\epsilon + \delta_\mu)} - 1\right)\right]^{\frac{1}{2}}$$

$$\beta = k_0\sqrt{\mu_r\epsilon_r}\left[\frac{1}{2}(1 - \tan\delta_\epsilon\tan\delta_\mu)\left(\sqrt{1 + \tan^2(\delta_\epsilon + \delta_\mu)} + 1\right)\right]^{\frac{1}{2}}$$

Similarly, an expression for the intrinsic impedance is found in terms of the relative constants and loss tangents:

$$\eta = \eta_0\sqrt{\frac{\mu_r}{\epsilon_r}}\left[\frac{1 + \tan\delta_\epsilon\tan\delta_\mu + j(\tan\delta_\epsilon - \tan\delta_\mu)}{1 + \tan^2\delta_\epsilon}\right]^{\frac{1}{2}}$$

In polar form, $\eta = |\eta|e^{j\zeta}$, where

$$|\eta| = \eta_0\sqrt{\frac{\mu_r}{\epsilon_r}}\left[\frac{1+\tan^2\delta_\mu}{1+\tan^2\delta_\epsilon}\right]^{\frac{1}{2}}$$

$$\zeta = \frac{\delta_\epsilon + \delta_\mu}{2}$$

For a good conductor, the attenuation constant can be extremely large, and therefore the wave will decay rapidly as it propagates through the material. The *skin depth* δ is the distance through which the amplitude of the wave decreases by a factor of $1/e = 0.368$. From this definition, the skin depth is simply the reciprocal of the attenuation constant:

$$\delta = \frac{1}{\alpha} = \frac{1}{\sqrt{\sigma_c \pi \mu f}}$$

A.10 Group Velocity

In a lossless medium, ω and β are linearly related:

$$\omega = \left[\frac{1}{\sqrt{\mu\epsilon}}\right]\beta$$

For a dispersive medium the relationship is not linear because μ and ϵ are functions of frequency. It is evident that waves of different frequencies will have different phase velocities because $u_p = \omega/\beta$. From Fourier theory, we know that any waveform can be decomposed into spectral (frequency) components. If the waveform is propagated through a dispersive material, each frequency component of the wave will be weighted differently as a result of the frequency-dependent attenuation and phase delay. If the frequency components are recombined at the output of the medium, the wave shape may be substantially different from the one that entered the medium. A common example is the spreading and distortion of a rectangular pulse.

Phase velocity is not a useful description of the velocity of signal propagation. A more meaningful quantity is the *group velocity* defined by

$$u_g = \frac{d\omega}{d\beta}$$

This formula is derived based on a two-term combination of sinusoids of slightly different frequencies. It is possible for the phase velocity to be greater than c (as in the case of waveguide propagation), but the group velocity will be less than c.

A.11 Power Flow and the Poynting Vector

For a plane wave, the direction of power flow is given by the vector $\hat{k}$, which is perpendicular to the plane containing $\vec{E}$ and $\vec{H}$. The power density in watts per square meter is given by the *Poynting vector*:

$$\vec{W} = \vec{E} \times \vec{H}$$

The *Poynting theorem* is a statement of conservation of energy for a volume of space V completely enclosed by a surface S:

$$-\oint_S \vec{W} \cdot \overrightarrow{ds} = \frac{\partial}{dt}\int_V \frac{1}{2}(\epsilon E^2 + \mu H^2)dv + \int_V \sigma_c E^2 dv$$

The term on the left represents power leaving the volume. The first term on the right is the time rate of change of energy stored in the electric and magnetic fields, and the second is the ohmic loss. Finally, the time-averaged power density in a propagating wave is given by

$$\vec{W}_{av} = \frac{1}{2}\text{Re}\{\vec{E} \times \vec{H}^*\}$$

A.12 Reflection and Refraction at an Interface

In this section, reflection and transmission coefficients are derived for an infinite planar boundary between two media. The most basic concept for visualizing the behavior of the RCS of complicated geometries is the reflection and refraction of plane waves at an interface. If a smooth curved surface is electrically large, it can be approximated by a locally planar one, and the scattered field can be represented by a wave that appears to originate from a *reflection point* on the body. As in the case of boundary conditions, parallel and perpendicular field components are defined as shown in Fig. A.5. However, *parallel* and *perpendicular* are now defined with respect to the *plane of incidence* (the plane containing the surface normal $\hat{n}$ and the propagation vector $\hat{k}$). Any arbitrary polarization can be decomposed into parallel and perpendicular components:

$$\vec{E} = \vec{E}_{||} + \vec{E}_{\perp}$$

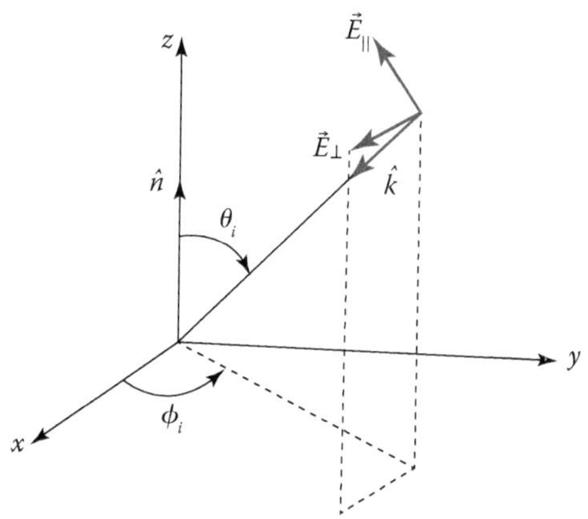

Fig. A.5 Parallel and perpendicular components of a plane wave.

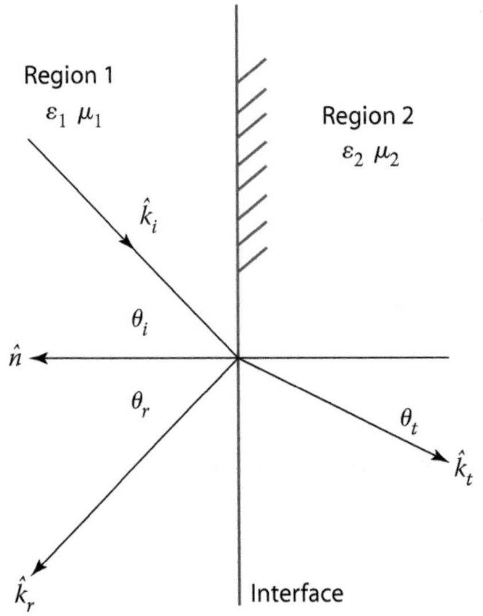

Fig. A.6 Plane wave incident on a plane boundary.

Reflection is considered separately for each polarization type.

A.12.1 Perpendicular Polarization

In the case of perpendicular polarization, the electric field vector is entirely perpendicular to the plane of incidence as shown in Fig. A.6. The wave is incident at an angle θ_i with respect to the surface normal $\hat{n}$ and reflected at an angle θ_r. Region 1 has parameters (μ_1, ϵ_1) and region 2 parameters (μ_2, ϵ_2).

All three waves (incident, reflected, and transmitted) are TEM and expressions can be written for each of the three. (The details can be found in most undergraduate electromagnetics textbooks.)

At the boundary $z = 0$, the tangential components of the electric and magnetic fields must be continuous if it is assumed that no impressed current or charge exists.

Because these quantities are complex, both the real and imaginary parts of each side must be equal. Equivalently, one can equate the magnitude and phase of the polar quantities. Matching phase gives

$$k_1 \sin\theta_i = k_1 \sin\theta_r = k_2 \sin\theta_t$$

The result is *Snell's law*:

$$\sin\theta_i = \sin\theta_r$$

Thus, the incidence angle is equal to the reflection angle. Furthermore, the angle of *refraction* of the transmitted wave depends on the ratio of the constitutive parameters of the two media:

$$k_1 \sin\theta_i = k_2 \sin\theta_t$$

$$\sqrt{\frac{\epsilon_2\mu_2}{\epsilon_1\mu_1}} = \frac{\sin\theta_i}{\sin\theta_t}$$

For dielectrics, $\mu_1 = \mu_2 = \mu_0$, and $\sqrt{\epsilon_2/\epsilon_1}$ is just the ratio of the indices of refraction of the two media.

Matching the amplitudes at the boundary gives the *reflection coefficient* $\Gamma_\perp$. It relates the electric field of the reflected wave to the electric field of the incident wave:

$$\Gamma_\perp = \frac{\eta_2 \cos\theta_i - \eta_1 \cos\theta_t}{\eta_2 \cos\theta_i + \eta_1 \cos\theta_t}$$

Similarly, a *transmission coefficient* can be defined

$$\tau_\perp = \frac{2\eta_2 \cos\theta_i}{\eta_2 \cos\theta_i + \eta_1 \cos\theta_t}$$

Note that $1 + \Gamma_\perp = \tau_\perp$.

A special case of interest is $\Gamma_\perp = 0$, which occurs when

$$\eta_2 \cos\theta_i = \eta_1 \cos\theta_t$$

This particular angle is called *Brewster's angle* and is given by

$$\theta_{B\perp} = \sin^{-1}\left[\frac{1 - (\epsilon_2\mu_1/\epsilon_1\mu_2)}{1 - (\mu_1/\mu_2)^2}\right]^{\frac{1}{2}}$$

A.12.2 Parallel Polarization

The vector components for parallel polarization are shown in Fig. A.6. The reflection and transmission coefficients are

$$\Gamma_\parallel = \frac{\eta_2 \cos\theta_t - \eta_1 \cos\theta_i}{\eta_2 \cos\theta_t + \eta_1 \cos\theta_i}$$

$$\tau_\parallel = \frac{2\eta_2 \cos\theta_i}{\eta_2 \cos\theta_t + \eta_1 \cos\theta_i}$$

with $1 + \Gamma_\parallel = \tau_\parallel \cos\theta_t/\cos\theta_i$.

As in the case of perpendicular polarization, there will be a Brewster's angle at which there is no reflection:

$$\theta_{B\parallel} = \sin^{-1}\left[\frac{1 - (\epsilon_1\mu_2/\epsilon_2\mu_1)}{1 - (\epsilon_1/\epsilon_2)^2}\right]^{\frac{1}{2}}$$

For parallel polarization, the Brewster's angle depends on the ratio of the dielectric constants only if the permeabilities of the two media are the same. This fact is used to design polarization selective surfaces that will reflect parallel polarization but transmit perpendicular polarization.

A.13 Total Reflection

An important condition occurs when $\theta_t = 90$ deg:

$$\frac{\sin(90\text{ deg})}{\sin\theta_i} = \sqrt{\frac{\epsilon_1}{\epsilon_2}}$$

The value of θ_i that satisfies this equation is the *critical angle*,

$$\theta_c = \sin^{-1}\left[\sqrt{\frac{\epsilon_1}{\epsilon_2}}\right]$$

When $\theta_i > \theta_c$, the wave is said to be *totally reflected*. The wave will propagate unattenuated along the interface and be attenuated with distance away from the interface. This is easy to demonstrate using complex angles. For total reflection,

$$\sin\theta_t = \sqrt{\epsilon_1/\epsilon_2}\,\sin\theta_i > 1$$

By trigonometric identity,

$$\cos\theta_t = \sqrt{1 - \sin^2\theta_t}$$

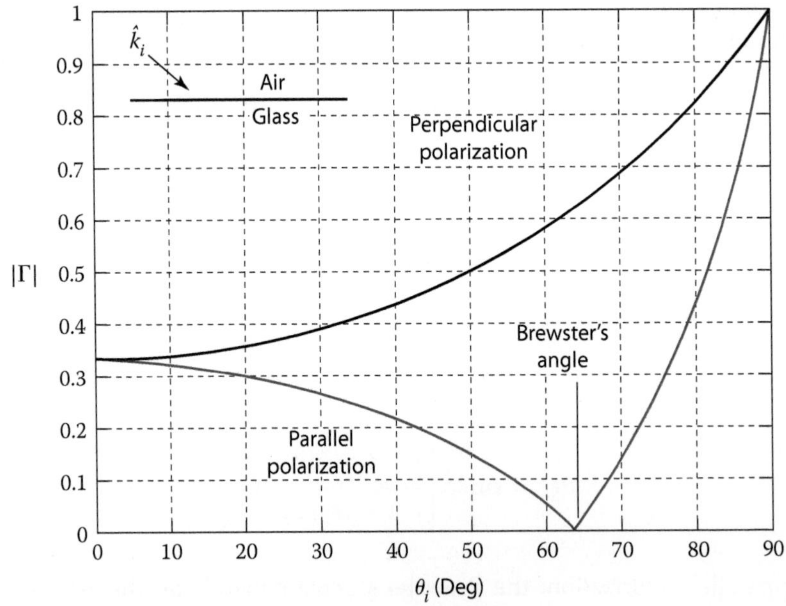

Fig. A.7 Reflection at a glass/air interface (wave incident from the air side).

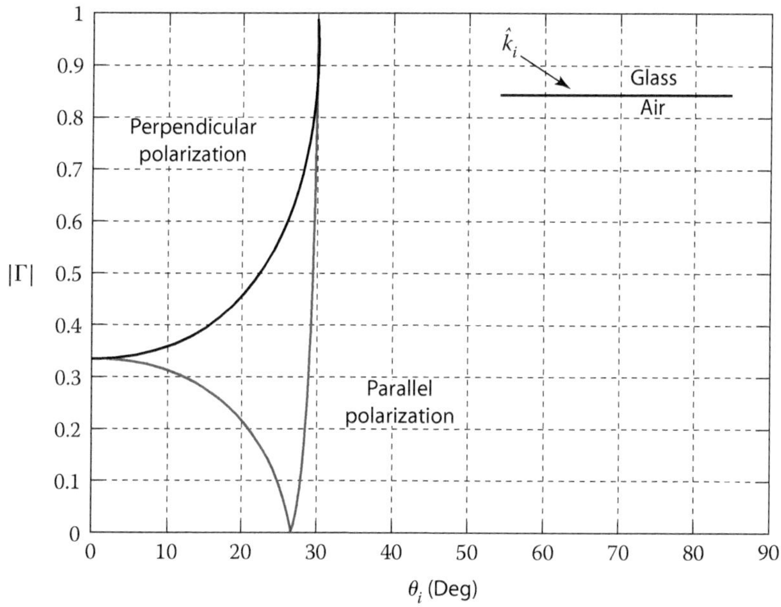

Fig. A.8 Reflection at a glass/air interface (wave incident from the glass side).

When the last two equations and the known form of the transmitted field are used,

$$E_t \sim \exp\left[-jk_2(x \sin\theta_t + z\cos\theta_t)\right] = \exp\ \left[-(\alpha_2 z + j\beta_2 x)\right]$$

where

$$\alpha_2 = k_2\sqrt{\frac{\epsilon_1}{\epsilon_2}\sin^2\theta_i - 1}$$

$$\beta_2 = k_2\sqrt{\frac{\epsilon_1}{\epsilon_2}}\sin\theta_i$$

This type of wave is called a *surface wave* because it is bound to the surface. The rate of attenuation away from the surface depends on the ratio of the dielectric constants and the incidence angle. A wave that attenuates rapidly is *closely bound*, and one that decays slowly is *loosely bound*. Figures A.7 and A.8 show the reflection coefficient behavior as a function of θ_i for an interface between air and glass ($\epsilon_r = 4$). In the first case, the wave is impinging from the free-space side; in the second, from the more dense glass side.

A.14 Standing Waves

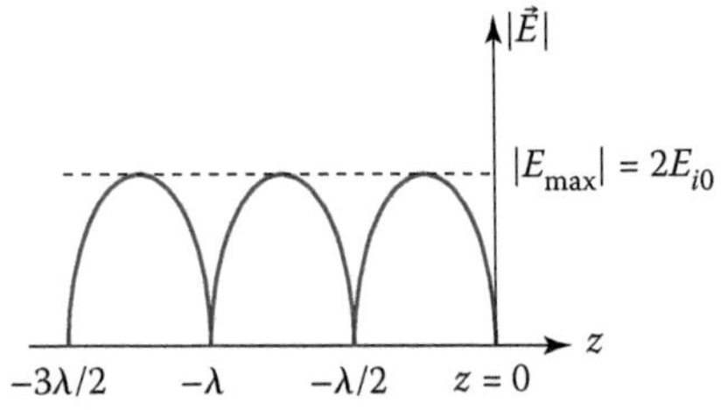

Fig. A.9 Standing wave at an air/conductor interface.

If the difference in the constitutive parameters of the two media forming an interface is large, the reflection coefficient will be large. The total field in region 1 is the sum of the incident and reflected waves. For the case of region 2 being a perfect conductor and $\theta_i = 0$ deg, $\Gamma_\perp = \Gamma_{||} = -1$. Therefore incident and reflected fields in region 1 combine to give

$$\vec{E}_1 = \hat{x}E_{i0}(e^{-jk_1z} - e^{jk_1z}) = \hat{x}E_{i0}2j\sin(kz)$$

where E_{i0} is the amplitude of the incident wave. This is a *standing wave*; the null and peak locations are stationary in space. The nulls are spaced 0.5λ, with the first null located at the boundary as shown in Fig. A.9.

If the material in region 2 is not a perfect conductor, the reflected wave will not have a magnitude equal to the incident wave. Thus, the maximum in region 1 will not be 2, and the minimum will not be 0. The *voltage standing wave ratio*, VSWR (or SWR), is defined as

$$\text{VSWR} = s = \frac{|\vec{E}_{max}|}{|\vec{E}_{min}|} = \frac{1 + |\Gamma|}{1 - |\Gamma|}$$

which is a number between 1 and ∞. As Γ is reduced, the difference between the maximum and minimum electric fields is reduced. In general, it is possible to decompose the total field in region 1 into traveling wave and standing wave components.

Appendix B Coordinate Systems

B.1 Orthogonal Coordinate Systems

The three-dimensional coordinate systems used most often are

Cartesian:	(x, y, z)	$(\hat{x}, \hat{y}, \hat{z})$
Cylindrical:	(ρ, ϕ, z)	$(\hat{\rho}, \hat{\phi}, \hat{z})$
Spherical:	(r, θ, ϕ)	$(\hat{r}, \hat{\theta}, \hat{\phi})$

The quantities and unit vectors are illustrated in Fig. B.1. Some useful relationships that can be derived from the figure are

$$x = \rho\cos\phi = r\sin\theta\cos\phi$$

$$y = \rho\sin\phi = r\sin\theta\sin\phi$$

$$\rho = \sqrt{x^2 + y^2}$$

$$r = \sqrt{x^2 + y^2 + z^2}$$

$$\phi = \tan^{-1}(y/x)$$

$$\theta = \tan^{-1}(\rho/z)$$

These are *orthogonal* coordinate systems because the three *basis vectors* (unit vectors) are mutually orthogonal. For instance,

$$\hat{\theta}\cdot\hat{\theta} = 1, \quad \hat{\theta}\cdot\hat{\phi} = 0, \quad \text{and}$$

$$\hat{\theta}\cdot\hat{r} = 0$$

The Cartesian coordinate system is the only one in which the unit vectors are constant throughout space. In other words, $\hat{x}$, $\hat{y}$, $\hat{z}$ always point in the same direction. This is *not true* for cylindrical and spherical coordinates; the unit vector directions change as a point is moved throughout space.

Consider two vectors in the spherical system. The first one is evaluated at a point P_1 (r_1, θ_1, ϕ_1):

$$\vec{A} = A_r\hat{r}_1 + A_\theta\hat{\theta}_1 + A_\phi\hat{\phi}_1$$

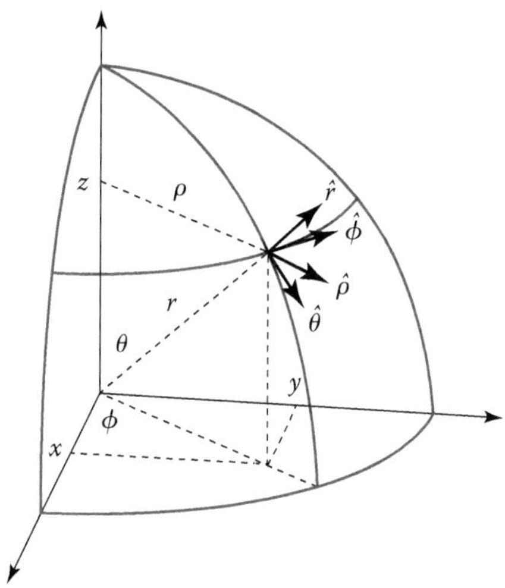

Fig. B.1 Orthogonal coordinate systems unit vectors.

whereas the second one is evaluated at a point P_2 (r_2, θ_2, ϕ_2):

$$\vec{B} = B_r\hat{r}_2 + B_\theta\hat{\theta}_2 + B_\phi\hat{\phi}_2$$

If the sum of these two vectors is to be calculated at the same point in space ($P_2 = P_1$), the spherical unit vectors are equal and the components of the two vectors simply add. However, computing the sum of vector $\vec{A}$ at P_1 and $\vec{B}$ at P_2 requires that $\vec{A}$ and $\vec{B}$ first be transformed to Cartesian coordinates.

B.2 Coordinate Transformations

The transformation of quantities from one coordinate system to another is frequently required. For example, consider a spherical charge that sets up an electric field. Because the charge has spherical symmetry, the field will also have spherical symmetry and, therefore, the electric field can be written concisely in spherical coordinates. However, if the flux passing through a rectangular window is to be calculated, an integration must be performed in rectangular coordinates. Consequently, the electric field must be expressed in rectangular coordinates.

Transform tables that relate the various coordinate system unit vectors (or vector components in general) are derived from the basis vectors. The relationships are summarized in Table B.1. The entries are simply the elements of the *transformation matrices*.

Table B.1 Coordinate Transform Tables

Cartesian (rectangular) and Cylindrical			
	$\hat{x}$	$\hat{y}$	$\hat{z}$
$\hat{\rho}$	$\cos\phi$	$\sin\phi$	0
$\hat{\phi}$	$-\sin\phi$	$\cos\phi$	0
$\hat{z}$	0	0	1
Cartesian (rectangular) and Spherical			
	$\hat{x}$	$\hat{y}$	$\hat{z}$
$\hat{r}$	$\sin\theta\cos\phi$	$\sin\theta\sin\phi$	$\cos\theta$
$\hat{\theta}$	$\cos\theta\cos\phi$	$\cos\theta\sin\phi$	$-\sin\theta$
$\hat{\phi}$	$-\sin\phi$	$\cos\phi$	0
Cylindrical and Spherical			
	$\hat{\rho}$	$\hat{\phi}$	$\hat{z}$
$\hat{r}$	$\sin\theta$	0	$\cos\theta$
$\hat{\theta}$	$\cos\theta$	0	$-\sin\theta$
$\hat{\phi}$	0	1	0

Example B.1: Transformation of a Vector

To transform the vector $\vec{A} = 2x\hat{x} + y\hat{z}$ to cylindrical coordinates, read down the first column of the first transform table:

$$\hat{x} = \hat{\rho}\cos\phi - \hat{\phi}\sin\phi + \hat{z}\cdot 0$$

and use the basic relationships between the coordinate variables

$$x = \rho\cos\phi \quad \text{and} \quad y = \rho\sin\phi$$

in the original expression for $\vec{A}$. Therefore,

$$\vec{A} = 2(\rho\cos\phi)(\hat{\rho}\cos\phi - \hat{\phi}\sin\phi) + (\rho\sin\phi)\hat{z}$$

or

$$\vec{A} = 2\rho\hat{\rho}\cos^2\phi - 2\rho\hat{\phi}\sin\phi\cos\phi + (\rho\sin\phi)\hat{z}$$

Transformations between coordinate systems are conveniently described by matrix equations. For instance, for a vector $\vec{E}$,

$$\begin{bmatrix} E_r \\ E_\theta \\ E_\phi \end{bmatrix} = \begin{bmatrix} \sin\theta\cos\phi & \sin\theta\sin\phi & \cos\theta \\ \cos\theta\cos\phi & \cos\theta\sin\phi & -\sin\theta \\ -\sin\phi & \cos\phi & 0 \end{bmatrix} \begin{bmatrix} E_x \\ E_y \\ E_z \end{bmatrix}$$

With the transformation matrix denoted as $\mathbf{T}$

$$\begin{bmatrix} E_r \\ E_\theta \\ E_\phi \end{bmatrix} = \mathbf{T} \begin{bmatrix} E_x \\ E_y \\ E_z \end{bmatrix}$$

If the spherical components of $\vec{E}$ are known and the Cartesian components are desired, the preceding equation can be inverted:

$$\begin{bmatrix} E_x \\ E_y \\ E_z \end{bmatrix} = \mathbf{T}^{-1} \begin{bmatrix} E_r \\ E_\theta \\ E_\phi \end{bmatrix}$$

Because $\mathbf{T}$ is an orthogonal matrix (because the coordinate systems are orthogonal), the matrix inverse is equal to the transpose. For this reason, the transform tables can be read down as well as across.

B.3 Position Vectors

A position vector, as shown in Fig. B.2, is directed from the origin to a specified point P in space. In Cartesian coordinates, the position vector to a point with coordinates (x, y, z) is

$$\vec{r} = x\hat{x} + y\hat{y} + z\hat{z}$$

This expression for the vector $\vec{r}$ uniquely specifies the point. However, in cylindrical coordinates, the position vector is

$$\vec{r} = \rho\hat{\rho} + z\hat{z}$$

which actually specifies a ring of points a distance z above the $x-y$ plane and a distance ρ from the z axis. Thus, a position vector cannot be specified uniquely in cylindrical coordinates. Similarly, in spherical coordinates, the position vector $\vec{r} = r\hat{r}$ is ambiguous. Consequently, position vectors will always be expressed in terms of the Cartesian variables.

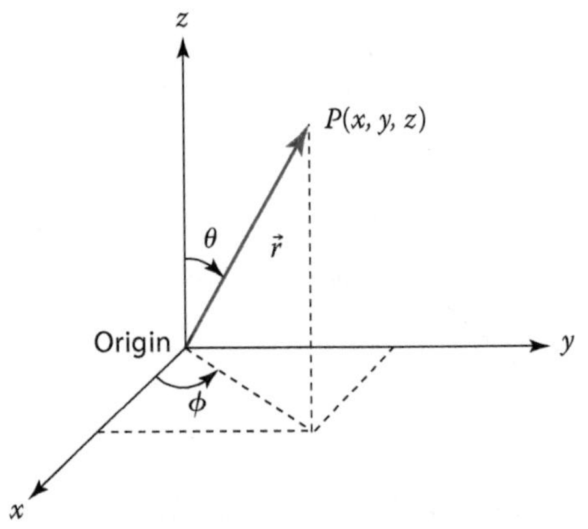

Fig. B.2 Position vector.

B.4 Direction Cosines

Direction cosines are quantities that frequently appear in antenna and scattering calculations that involve spherical coordinates. For a position vector from the origin to the point P, direction cosines are the cosines of the three angles the vector makes with respect to the Cartesian axes. The angles are shown in Fig. B.3 and are defined as follows:

x direction cosine: u or $\cos\alpha$ or $\cos\alpha_x = \sin\theta\cos\phi$
y direction cosine: v or $\cos\beta$ or $\cos\alpha_y = \sin\theta\sin\phi$
z direction cosine: w or $\cos\gamma$ or $\cos\alpha_z = \cos\theta$

In this text, the notation u, v, w are used to represent the x, y, and z direction cosines.

Direction cosine space is shown in Fig. B.4. If a hemisphere of unit radius $r = 1$ is centered at the origin of a Cartesian coordinate system, it defines a unit circle in the $x-y$ plane. The x axis can be relabeled u and the y axis v. If a point on the sphere is given by $(r = 1, \theta, \phi)$, its Cartesian coordinates are

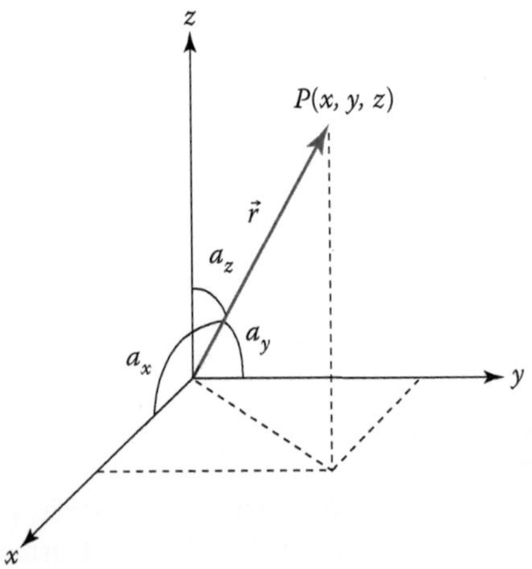

Fig. B.3 Direction cosines.

$$x = r\sin\theta\cos\phi = \sin\theta\cos\phi = u$$
$$y = r\sin\theta\sin\phi = \sin\theta\sin\phi = v$$

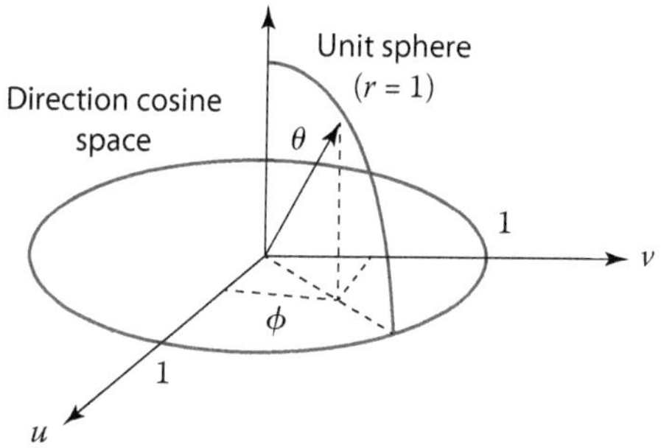

Fig. B.4 Direction cosine space, (u, v).

Therefore, the location of a point in direction cosine space is determined by its direction in space as viewed from the origin. The perimeter is defined by the radius

$$u^2 + v^2 = \sin^2 \theta(\cos^2 \phi + \sin^2 \phi) \\ = 1$$

which corresponds to $\theta = 90$ deg. A point at the center corresponds to $\theta = 0$ deg.

One of the advantages of working in direction cosine space is that antenna beams and, similarly, RCS lobes maintain a fixed shape as the lobes move through space. As illustrated in Fig. B.5, if an antenna beam has a circular cross section in direction cosine space when pointed in the $\theta = 0$ deg direction, it will have the same shape when scanned to $\theta = 45$ deg. Note that the projection onto the hemisphere surface becomes more distorted from a circle as θ is increased. This is the familiar beam broadening that occurs when phased array antennas are scanned to wide angles.

Two other important properties of direction cosines should be noted. First, symmetric points above and below the $x-y$ plane will have the same coordinates in direction cosine space. In other words, the sign of z cannot be determined uniquely by u and v alone. Second, because u, v, and w represent the Cartesian components of a unit vector, *they obey the same transformation rules as a vector.*

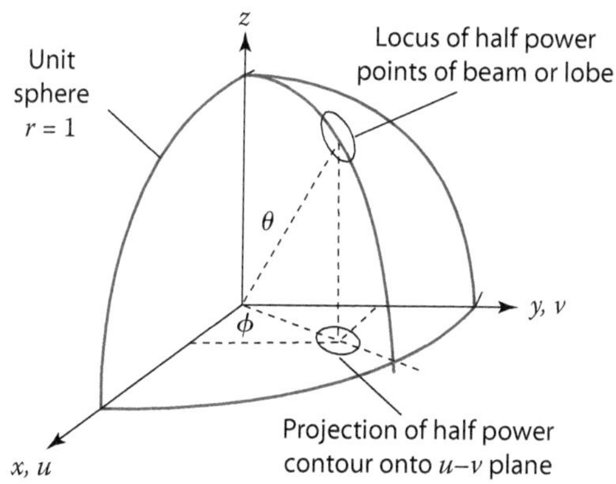

Fig. B.5 Scanning lobe in direction cosine space.

B.5 Azimuth–Elevation Coordinate System

Azimuth and elevation coordinates are commonly used to describe the coverage of radars. The origin of a spherical system is located at the radar as shown in Fig. B.6. The $x-y$ plane corresponds to the surface of the Earth, and the x axis is arbitrarily chosen as North (azimuth reference, 0 deg). Elevation angle is measured from the $x-y$ plane up toward the zenith, which is directly above the radar.

There is no standard notation for the azimuth and elevation variables. Frequently, the designation (AZ, EL) or (α, γ) is used. If the x axis is chosen as an azimuth reference, ϕ can be used as the azimuth variable. However, ϕ is increasing in a counterclockwise direction, whereas azimuth is traditionally measured in a clockwise direction as on a compass so that $AZ = 360 - \phi$ deg. In this text, the elevation angle will be denoted as γ, where $EL = \gamma = 90 - \theta$ deg.

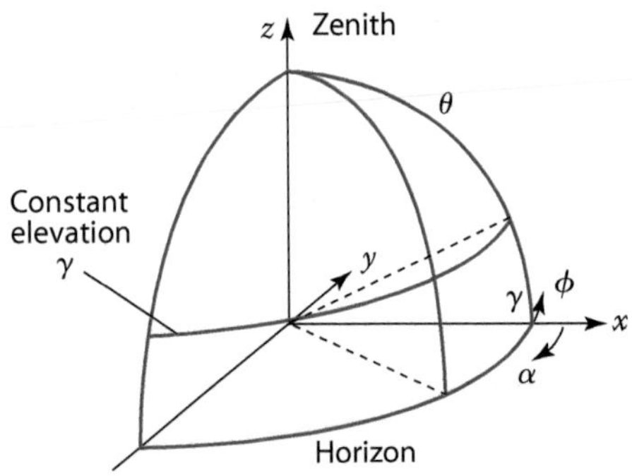

Fig. B.6 Azimuth-elevation coordinates.

Appendix C Review of Antenna Theory

C.1 Antenna Parameters

In this appendix, several important aspects of antenna theory and analysis are presented. Because radar cross section and antenna radiation are closely related, many of the same analysis techniques apply. In a sense, a radar target can be viewed as an antenna structure that is excited by a distant voltage source. Antenna parameters applicable to RCS analysis are directivity (or gain), half-power beamwidth, effective aperture and effective height, and sidelobe level. Antenna performance is generally defined in the *far field* or *far zone.* If the antenna is at the origin, the standard (although arbitrary) far-field condition is $r > (2L^2/\lambda)$, where L is the maximum antenna dimension.

Directive gain D is a measure of how well an antenna concentrates the energy it radiates. Mathematically,

$$D(\theta, \phi) = \frac{U(\theta, \phi)}{P_r/4\pi} = \frac{\vec{W}(\theta, \phi) \cdot \hat{r} r^2}{P_r/4\pi}$$

$U(\theta, \phi)$ is the radiation intensity in the direction (θ, ϕ) in units of watts per steradian (W/sr), and P_r is the total radiated power. The factor 4π in the denominator implies that the reference level for directive gain is an *isotropic source,* that is, one that radiates uniformly in all directions. Directive gain is usually expressed in units of *decibels relative to an isotropic source* (dBi):

$$D, \text{dBi} = 10 \log_{10} D$$

The maximum value of the directive gain over all angles is the *directivity*:

$$D_0 = \frac{[U(\theta, \phi)]_{\max}}{P_r/4\pi}$$

The terms *directivity* and *directive gain* imply a lossless antenna. When losses are included, the term *gain* is used. Although gain is an angle-dependent quantity, it is most often used in reference to the maximum value over all angles, G_0. Thus,

$$G_0 = [G(\theta, \phi)]_{\max} = e[D(\theta, \phi)]_{\max} = eD_0$$

where e is an efficiency factor ($0 \leq e \leq 1$). As a word of caution, the terms *gain* and *directivity* are frequently used interchangeably.

If the antenna is at the origin of a spherical coordinate system, the time-averaged radiation intensity at a distance r is

$$U(\theta, \phi) = \frac{r^2}{2} \mathrm{Re}\{\vec{E} \times \vec{H}^*\} \cdot \hat{r} = r^2 \frac{|\vec{E}(r, \theta, \phi)|^2}{2\eta_o}$$

To obtain the total radiated power in the far field, the Poynting vector is integrated over the surface of a sphere of radius r:

$$P_r = \frac{1}{2} \int_0^{2\pi} \int_0^{\pi} \mathrm{Re}\{\vec{E} \times \vec{H}^*\} \cdot \hat{r} r^2 \sin\theta \mathrm{d}\theta \mathrm{d}\phi$$

Note that, in the far-field, both $\vec{E}$ and $\vec{H}$ have an r^{-1} dependence, so that U and P_r are independent of r.

In the far zone of an antenna, the electric field takes the form

$$\vec{E} = \vec{f}(\theta, \phi) \frac{e^{-jkr}}{r}$$

where the angular dependence is contained entirely in the factor f. For convenience, we now define the *normalized field pattern factor* $\vec{f}_n$ such that

$$\vec{f}_n(\theta, \phi) = \frac{\vec{f}(\theta, \phi)}{|\vec{f}_{\max}|}$$

where $|\vec{f}_{\max}|$ is the maximum value of $\vec{f}$ over all angles. Therefore, $|\vec{f}_n|$ is dimensionless, and its maximum value is one. Now, the directivity is given by

$$D_0 = \frac{4\pi}{\Omega_A}$$

Example C.1: Directivity Calculation

An approximate expression for the radiation field from a horn antenna with its aperture in the $x-y$ plane is

$$\vec{E} = \frac{E_o}{r} e^{-jkr} [a_e(\theta)\hat{\theta} \cos\phi + a_h(\theta)\hat{\phi} \sin\phi] \quad \text{for } \theta < 90 \text{ deg}$$

(Continued)

Example C.1: Directivity Calculation *(Continued)*

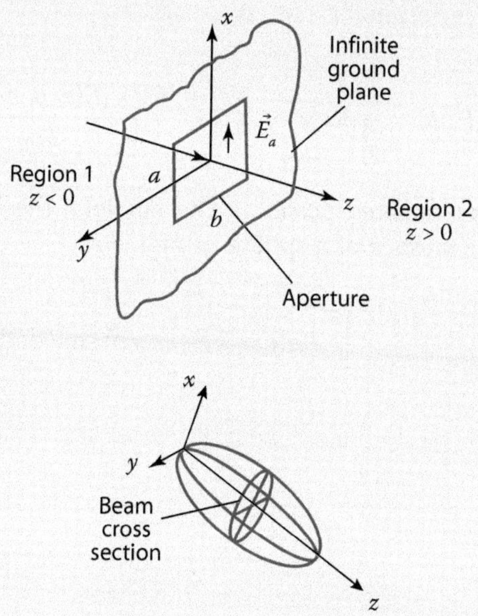

Fig. C.1 A rectangular aperture and three-dimensional polar plot of its radiation pattern.

The functions a_e and a_h are used to specify the pattern shapes in the $\phi = 0$ and $\phi = \pi/2$ planes, as shown in Fig. C.1 For simplicity, assume that both are cosine patterns

$$a_e = a_h = \cos^m(\theta)$$

The field already has the form

$$\vec{E} = \vec{f}(\theta, \phi)\frac{e^{-jkr}}{r}$$

where

$$\vec{f}(\theta, \phi) = E_0 \cos^m(\theta)[\hat{\theta}\cos\phi + \hat{\phi}\sin\phi]$$

To compute the gain, we need the normalized pattern factor. The fact that

$$|\vec{f}| = \sqrt{\vec{f} \cdot \vec{f}^*} = \sqrt{|f_\theta|^2 + |f_\phi|^2}$$

and its maximum value is E_0 give

$$|\vec{f}_n|^2 = \cos^{2m}(\theta)$$

(Continued)

Example C.1: Directivity Calculation *(Continued)*

Using this in the expression for the beam solid angle yields

$$\Omega_A = \int_0^{2\pi}\int_0^{\pi/2} \cos^{2m}\theta \sin\theta d\theta d\phi$$

The $\pi/2$ upper limit on the θ integration is due to the original restriction that $0 \leq \theta \leq \pi/2$. The integral can be evaluated with the help of an integration table,

$$\Omega_A = 2\pi\left[-\frac{\cos^{2m+1}\theta}{2m+1}\right]_0^{\pi/2} = \frac{2\pi}{2m+1}$$

which gives a directivity of

$$D_0 = \frac{4\pi}{\Omega_A} = \frac{4\pi(2m+1)}{2\pi} = 2(2m+1)$$

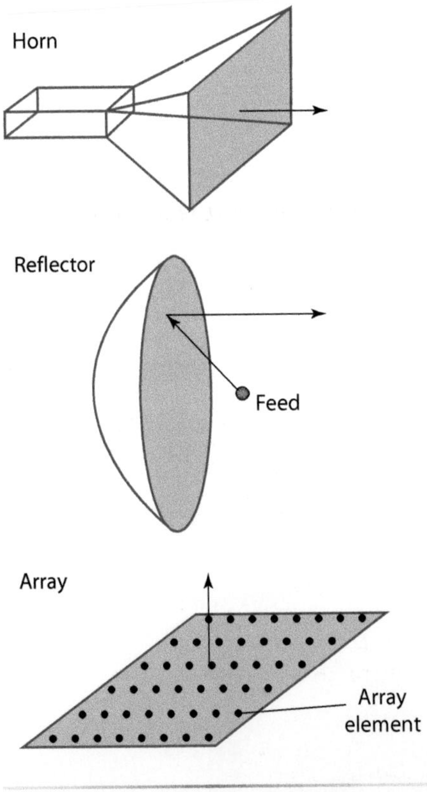

Fig. C.2 Examples of antennas with well-defined apertures.

where Ω_A is the *beam solid angle*:

$$\Omega_A = \int_0^{2\pi}\int_0^{\pi} |\vec{f}_n|^2 \sin\theta d\theta d\phi$$

This is usually the most convenient form for computing the directivity.

Most high-gain antennas have a reasonably well-defined *aperture area*. This is a surface through which the antenna's radiated power flows. Figure C.2 illustrates the concept for horns, arrays, and reflectors. When a plane wave is incident on the antenna, the maximum power that can be collected is related to the effective area of the aperture

$$P_c = W_i A_e$$

where W_i is the power density of the incident wave and A_e the *effective area* of the aperture. For an efficient antenna, A_e approaches the physical area. Again, effective area is an angle-dependent quantity. On account of reciprocity, A_e will have the same shape in space as the directive gain

$$\frac{A_e(\theta, \phi)}{A_{e\ \max}} = \frac{D(\theta, \phi)}{D_0}$$

Effective aperture and directivity are closely related. Effective aperture is more meaningful when the antenna is receiving. The relationship between the two can be derived by examining a transmission link with arbitrary transmit and receive antennas. By considering each antenna in turn to be transmitting and equating the received powers, one finds that *the ratio of the effective aperture to directivity is the same for all antennas*:

$$\frac{D_0}{A_{e\ \max}} = \text{constant}$$

To find the constant, one can calculate D_0 and A_e for any convenient antenna. For a short dipole, $D_0 = 1.5$ and $A_e = 0.119\lambda^2$, giving

$$D_0 = \frac{4\pi A_{e\ \max}}{\lambda^2}$$

Another measure of efficiency that is often used is *effective height*, $\vec{h}_e$. This term is usually used in conjunction with thin, linearly polarized antennas such as wires and slots. It relates the open-circuit voltage at the antenna terminals to the incident electric field intensity across the antenna:

$$V_{\mathrm{oc}} = \vec{h}_e^* \cdot \vec{E}_i$$

Example C.2: Effective Height of a Short, Thin Wire

As shown in Fig. C.3, a short dipole has a constant current I_0 along its length $\Delta\ell$. The incident field is x-polarized:

$$\vec{E}_i = E_{i0}\hat{x}e^{jkz}$$

but the dipole is rotated so that $\delta = 30$ deg.

The magnitude of the effective height is given by the above-mentioned integral:

$$|\vec{h}_e| = \frac{1}{I_0}\int_{\Delta\ell} I_0 \,\mathrm{d}x = \Delta\ell$$

(Continued)

Example C.2: Effective Height of a Short, Thin Wire *(Continued)*

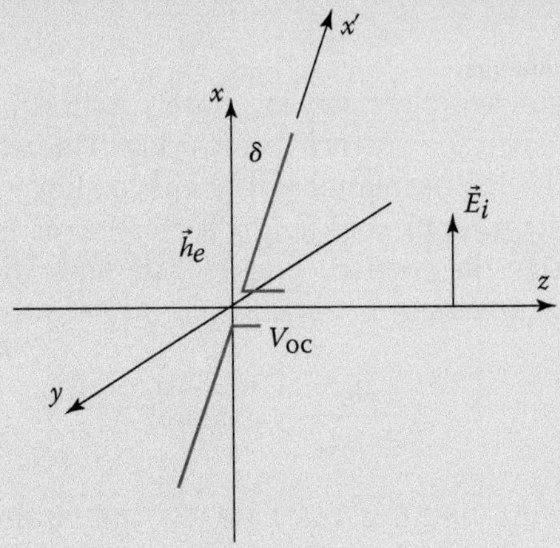

Fig. C.3 Wire antenna with polarization mismatched to the incident field.

and the direction by a unit vector parallel to the dipole:

$$\hat{h}_e = \hat{x}\cos(30\,\mathrm{deg}) + \hat{y}\sin(30\,\mathrm{deg})$$

Combining the last two results yields

$$h_e = \Delta\ell[\hat{x}\cos(30\,\mathrm{deg}) + \hat{y}\sin(30\,\mathrm{deg})]$$

The open-circuit voltage (omitting the phase term) is

$$|V_{\mathrm{oc}}| = E_{i0}\Delta\ell\cos(30\,\mathrm{deg})$$

Mathematically, $\vec{h}_e$ is determined from the current on the antenna:

$$|\vec{h}_e| = \frac{1}{I(0)}\int_{\Delta\ell} I(x)\mathrm{d}x$$

where $I(0)$ is the feed-point current, $I(x)$ the current distribution, and $\Delta\ell$ the antenna's length.

In the preceding example, the open-circuit voltage was reduced by an amount cos(30 deg) as a result of a *polarization mismatch* between the dipole and the incident field. In other words, $\vec{h}_e$ and $\vec{E}_i$ were not parallel.

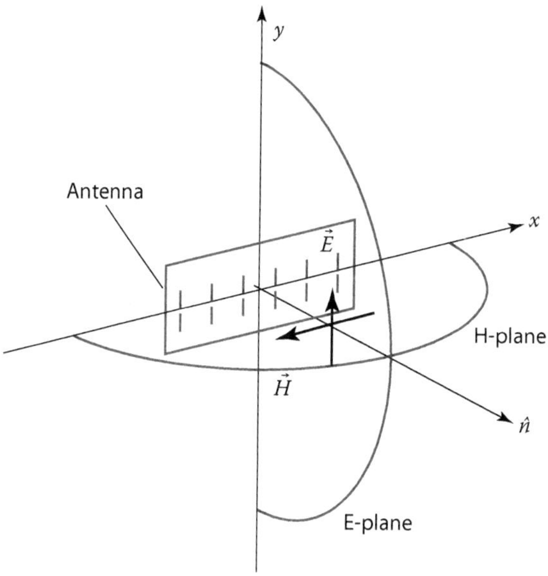

Fig. C.4 Principal planes of a two-dimensional antenna.

In general, a *polarization loss factor* (PLF) can be defined for any antenna:

$$\text{PLF} = \frac{|\vec{h}_e \cdot \vec{E}_i|^2}{|\vec{h}_e|^2 |\vec{E}_i|^2}$$

PLF is an important consideration in designing low-RCS targets with wires and slots. If the incident radar wave is arriving with a known fixed polarization, any wirelike attachments to the target can be oriented so that a large PLF is achieved. For monostatic radars, the PLF will apply on both transmit and receive.

Other important antenna parameters are half-power beamwidth (HPBW) and sidelobe level (SLL). Linearly polarized, planar, two-dimensional antennas have two pattern cuts of particular interest. These are referred to as *principal planes* or *cardinal planes* and are illustrated in Fig. C.4. One is the plane defined by the electric field vector in the aperture $\vec{E}_a$ and the normal to the aperture $\hat{n}$. This is called the *E plane*. The second is the plane orthogonal to the *E* plane, denoted the *H plane*. The *H* plane would be defined by the aperture normal $\hat{n}$ and the magnetic field vector in the aperture plane $\vec{H}_a$ if the field in the aperture were TEM. Other planes are frequently referred to as *diagonal planes* or *intercardial planes*.

A typical principal-plane pattern is shown in Fig. C.5, with various pattern characteristics labeled. One usually chooses the coordinate system so that the aperture normal corresponds to the Cartesian z axis and so that x and y correspond to the principal planes. If the antenna mainbeam is pointed

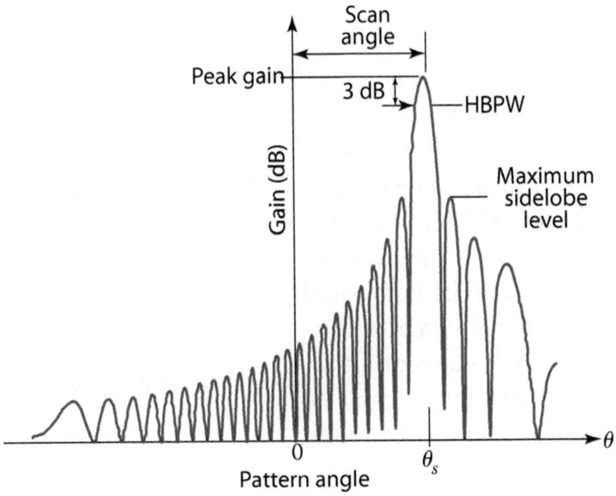

Fig. C.5 Typical principal plane radiation pattern.

in the direction of the surface normal ($\theta = 0$), the *half-power angle* θ_{hp} for a given ϕ plane is

$$|\vec{E}(\theta_{\text{hp}}, \phi)| = 0.707|\vec{E}_{\max}| = 0.707|\vec{E}(0, \phi)|$$

or, in terms of the normalized pattern factor,

$$|\vec{f}_n(\theta_{\text{hp}}, \phi)| = 0.707$$

The HPBW is the total width of the beam at the half-power points. If a symmetrical beam is assumed, HPBW $= 2\theta_{\text{hp}}$.

In the design of radar antennas, sidelobe control is an important issue. If the entire antenna aperture is uniformly illuminated, the first sidelobe level is -13 dB relative to the mainbeam peak. If the aperture illumination is tapered toward the edges (but still symmetric about the center), the sidelobes will be lower, the mainbeam wider, and the gain reduced. Thus, there is a tradeoff between low sidelobes and reduced gain and resolution that must be addressed in the design of the radar.

Target scattering pattern characteristics cannot be controlled as precisely as those for antennas. This is because radar targets are much more complex and because a wide range of excitation conditions is possible. Generally, the RCS designer is content to reduce the RCS below certain threshold levels, independently of the particular pattern characteristics. However, the widths of high-RCS lobes are important because they influence the probability of detection. Because it is rarely possible to change the lobe direction or reduce it below an acceptable threshold, the location of lobes must be considered in mission planning and execution.

C.2 Aperture Theory

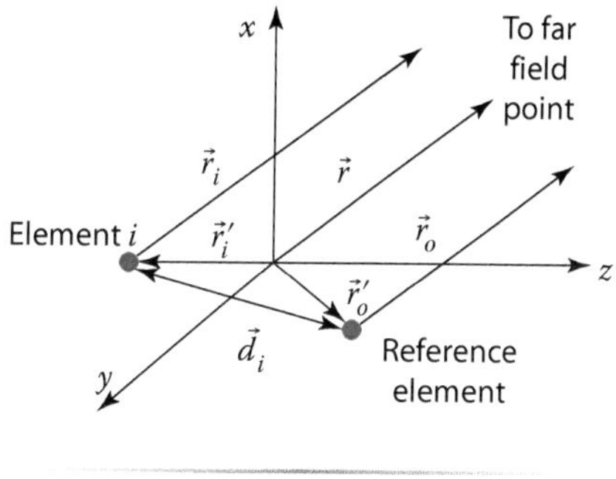

Fig. C.6 Array antenna geometry.

Antennas that contain a continuous radiating surface area are categorized as aperture antennas. Examples are horns, slots, and lenses. If the electric field in the aperture can be determined, the radiated field in the far zone can be computed using a *near-field-to-far-field transformation*. Thus, the details of a particular antenna structure can be ignored as long as the electric field in the aperture is known. Of course, the details of the antenna configuration will determine the aperture field but, for most geometries, a good approximation to $\vec{E}_a$ can be deduced.

Consider a rectangular aperture with dimensions $a \times b$ in an infinite ground plane as shown in Fig. C.6. Assume that the tangential component of the electric field in the aperture $\vec{E}_a$ is known. The transform of the aperture field is

$$\vec{F} = F_x(k_x, k_y)\hat{x} + F_y(k_x, k_y)\hat{y} = \iint_S \vec{E}_a(x, y)e^{j(k_x x + k_y y)}\mathrm{d}x\mathrm{d}y$$

where $k_x = ku$ and $k_y = kv$ (u and v are the x and y direction cosines defined

Example C.3: Radiation from a Rectangular Aperture

A plane wave illuminates the aperture shown in Fig. C.1:

$$\vec{E}_i = \hat{x}E_{i0}e^{-j\vec{k}\cdot\vec{r}}$$

In the aperture (z = 0),

$$\vec{E}_a = \hat{x}E_{i0}$$

The components of $\vec{F}$ are

$$F_x(k_x, k_y) = E_{i0}\iint_S e^{j(k_x x' + k_y y')}\mathrm{d}x'\,\mathrm{d}y'$$

$$F_y(k_x, k_y) = 0$$

The primes are used to designate source coordinates of the aperture field as opposed to observation coordinates. The integrals are easily evaluated to

(Continued)

Example C.3: Radiation from a Rectangular Aperture *(Continued)*

yield the familiar sinc function. For instance, the x' integration reduces to

$$\int_{-a/2}^{a/2} e^{jk_x x'}\,dx' = a\frac{\sin(k_x a/2)}{k_x a/2} = a\,\mathrm{sinc}(k_x a/2)$$

The total far electric field is given by

$$\vec{E}(r,\theta,\phi) = \frac{jke^{-jkr}}{2\pi r} E_{i0} a\, b\, \mathrm{sinc}(k_x a/2)\, \mathrm{sinc}(k_y b/2)[\hat{\theta}\cos\phi - \hat{\phi}\sin\phi\cos\phi]$$

in Appendix B). The far field at large r is

$$\vec{E}(r,\theta,\phi) = \frac{jke^{-jkr}}{2\pi r}[\hat{\theta}(F_x\cos\phi + F_y\sin\phi) + \hat{\phi}\cos\theta(F_y\cos\phi - F_x\sin\phi)]$$

By transforming the components of the electric field in the aperture, one obtains the components of $\vec{F}$, which are used, in turn, to compute $\vec{E}$ in the far field. This is an approximation to the electric field far from the opening. The agreement with the actual field is excellent along the z axis and degrades gracefully as θ increases.

The fact that the far-field pattern is the Fourier transform of the aperture distribution is a powerful tool. A large collection of Fourier transform pairs exists, and it can be drawn upon to design and synthesize antenna patterns and distributions.

Other techniques are available to compute the fields from apertures. Another method is to determine the equivalent currents in the aperture and use them in the radiation integrals (see Chapter 2). For example, the magnetic current at any point in the aperture is

$$\vec{J}_{ms} = -\hat{z} \times \vec{E}_a$$

According to the equivalence principle, one could also use an electric current derived from $\vec{H}_a$ or a combination of electric and magnetic currents. The result will vary slightly between the different approaches. A rigorous result requires solving an integral equation numerically using techniques such as the method of moments (see Chapter 3).

C.3 Array Theory

An *array antenna* is a collection of individual antennas. In most cases, the *elements* of the array are identical. To determine the radiation pattern of an array, the current distribution must be computed on each element and then superposition used to obtain the total field. Calculating the current is a difficult problem because of interactions between the elements. The interaction leads to *mutual coupling,* which has a significant impact on the array's performance. Mutual coupling is not necessarily bad and can actually be used to increase an array's operating bandwidth.

Using the radiation integrals as a starting point, the far electric field radiated by an electric surface current $\vec{J}_s$ is

$$\vec{E}(r, \theta, \phi) = \frac{-jk\eta_0 e^{-jkr}}{4\pi r}\iint_S [\vec{J}_s - (\vec{J}_s \cdot \hat{r})\hat{r}]e^{jk\vec{r}'\cdot\hat{r}}ds$$

The vectors $\hat{r}$ and $\hat{r}'$ are directed toward the observation point (r, θ, ϕ) and aperture point (x', y', z') respectively. The second term in the integrand is present because the component of current in the direction of the observer does not radiate.

The preceding equation applies to any current distribution. For an array, the current source is a sum of currents that exist over discrete surfaces. If there are N identical elements, the expression for the electric field becomes a sum over N integrals. Because the elements are identical, the current distribution will be approximately the same, except for a complex scale factor c_i, determined by how the element is fed. Thus, all the integrations will be the same, and the electric field becomes

$$\vec{E}(r, \theta, \phi) = \frac{-jk\eta_0 e^{-jkr}}{4\pi r}\sum_{i=1}^{N} c_i[\vec{J}_0 - (\vec{J}_0 \cdot \hat{r})\hat{r}] \ \exp[jk(\vec{r}_0' \cdot \hat{r} + \vec{d}_i \cdot \hat{r})]\, ds'$$

The subscript 0 refers to a reference element, and $\vec{d}_i$ is the vector from the reference element to the corresponding point on element i, as shown in Fig. C.6. Regrouping terms gives

$$\vec{E}(r, \theta, \phi) = \left[\frac{-jk\eta_0 e^{-jkr}}{4\pi r}\iint_{S_0}[\vec{J}_0 - (\vec{J}_0 \cdot \hat{r})\hat{r}]e^{jk(\vec{r}_0' \cdot \hat{r})}ds'\right]\left[\sum_{i=1}^{N} c_i e^{jk\vec{d}_i \cdot \hat{r}}\right]$$

Example C.4: Array Factor for an Equally Spaced Array of *N* Elements

For an array of equally spaced elements along the z axis, as shown in Fig. C.7

$$\vec{d}_i = \hat{z}(i-1)d$$
$$\hat{r} = \hat{x}u + \hat{y}v + \hat{z}w$$

so that

$$\vec{d}_i \cdot \hat{r} = (i-1)dw$$

(Continued)

Example C.4: Array Factor for an Equally Spaced Array of *N* Elements *(Continued)*

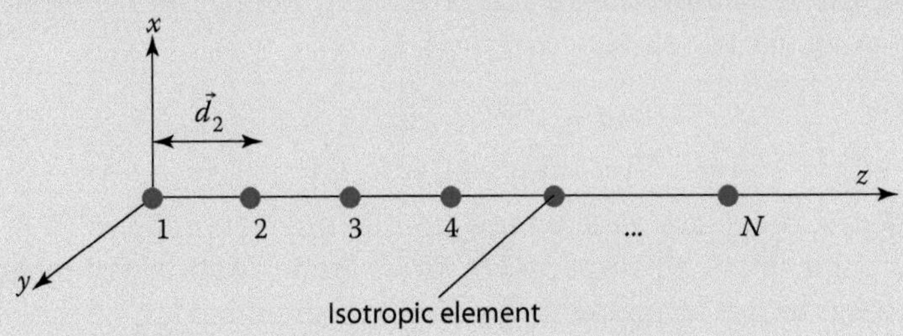

Fig. C.7 Linear array of equally spaced isotropic elements.

where w is the z direction cosine. The AF becomes

$$\mathrm{AF} = \sum_{i=1}^{N} c_i e^{jk(i-1)dw}$$

If the elements are *equally excited,* all the c_i are equal. For convenience, let $c_i = 1$, and change the index ($m = i - 1$):

$$\mathrm{AF} = \sum_{m=0}^{N-1} [e^{jk\,d\,w}]^m$$

The preceding sum is a geometric series that converges to

$$\mathrm{AF} = \frac{1-(e^{j\psi})^N}{1-e^{j\psi}} = \exp\left[j(N-1)\psi/2\right]\frac{\sin(N\psi/2)}{\sin(\psi/2)}$$

where $\psi = k\,d\,w$. This is the *uniform array factor* for an element spacing of d.

The first factor is the radiation pattern of a single element; this is called the *element factor* (EF). The second term in brackets depends only on the array geometry and excitation. This is the *array factor* (AF). The fact that the total field is the product of these two terms is *the principle of pattern multiplication.* Array design is concerned primarily with manipulating the array factor to achieve the required performance. This can be done independently of the type of radiating element as long as the element pattern is "well behaved."

The leading exponential factor arises from setting the origin at one end of the array. If the origin is placed exactly in the center of the array, the

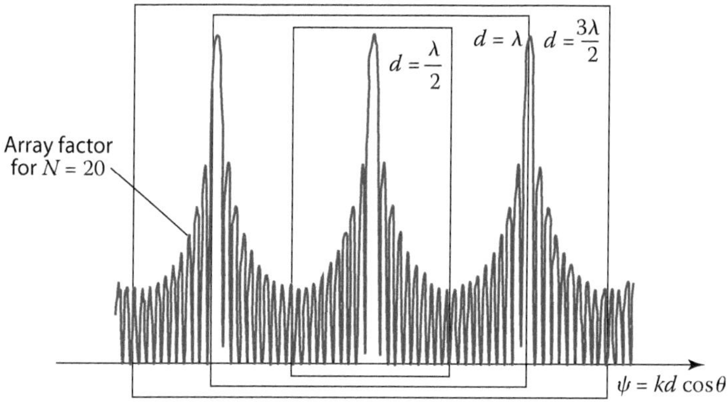

Fig. C.8 Visible regions for a 20-element array for three spacing values.

exponential term vanishes. Because the phase of the electric field is usually not of interest, this leading exponential is simply dropped.

Note that the uniform array factor has a peak value of N at $w = 0$ or $\theta = 90$ deg. If the element spacing or direction cosine is small, the normalized uniform array factor approaches sinc($u/2$). However, unlike the continuous aperture case described by the sinc function, the array factor can have multiple peaks. These are *grating lobes* and are generally undesirable. Their number and locations are determined by $\psi/2 = n\pi$, yielding

$$w_n = 2n\pi \quad \text{or} \quad \cos\theta_n = \lambda/nd \quad \text{for}\, n = \pm 1, \pm 2, \ldots$$

Although the argument of the array factor can take on values between $\pm\infty$, the argument is restricted by the allowable values of θ. Because $0 \le \theta \le \pi$, one finds that $-1 \le w \le 1$. The range of the argument kdw that corresponds to $0 \le \theta \le \pi$ is the *visible region*. Figure C.8 plots the

Example C.5: Array of Subarrays

Consider an eight-element array of isotropic sources. Computing the AF directly for $N = 8$ and $d = \lambda/2$, gives

$$\text{AF}_1 = \frac{\sin[8\pi\cos\theta/2]}{\sin[\pi\cos\theta/2]} \quad \text{and} \quad \text{EF}_1 = 1$$

This array can also be viewed as a four-element array of two-element subarrays. The subarray spacing is $d = \lambda$, and the array factor is

$$\text{AF}_2 = \frac{\sin[4(2\pi\cos\theta)/2]}{\sin[2\pi\cos\theta/2]}$$

(Continued)

Example C.5: Array of Subarrays *(Continued)*

Each element of this second array is a subarray consisting of two isotropic elements spaced $d = \lambda/2$. Thus, the *subarray factor* (SF) is

$$SF_2 = \frac{\sin[2\pi\cos\theta/2]}{\sin[\pi\cos\theta/2]} \quad \text{and} \quad EF_2 = 1$$

It is easy to verify that $(AF_1)(EF_1) = (AF_2)(SF_2)(EF_2)$.

visible region for three values of d for a 10-element array. It is evident that eliminating grating lobes requires a small spacing in terms of wavelength.

Many modern phased arrays are based on small groups of elements called *subarrays*. A subarray generally contains a handful of elements and includes the accompanying power-dividing networks and phase-shifting devices. This approach is more cost effective than fabricating individual elements and then assembling them by hand. It also makes the array easier to maintain because of the modular design. The principle of pattern multiplication can be used to analyze arrays of subarrays as illustrated in the following example.

Two-dimensional arrays are a simple extension of the one-dimensional cases in Examples C.4. and C.5.

Appendix D Review of Transmission Lines

D.1 Waves on Transmission Lines

Transmission lines are structures designed to guide electromagnetic waves efficiently. Potential sources of energy loss on transmission lines include attenuation, reflection, and radiation. Commonly used structures include coaxial, two-wire, and microstrip configurations. The geometry of the line, its dimensions, and the materials of which it is composed determine its characteristic impedance (Z_0), phase velocity (u_p), and attenuation constant (α). These are the three quantities that primarily determine the line's electrical performance.

In practice, the behavior of electromagnetic waves guided by a wide range of objects can be modeled with a transmission-line representation. For instance, at certain angles, a flat conducting plate will capture energy from a plane wave and guide the wave along its surface. This is a *traveling wave*, and its behavior can be modeled by an equivalent transmission line with the appropriate length, impedance, attenuation constant, and phase velocity. This allows use of the standard equations for impedance transformations and matching to design networks to control reflections and radiation.

Figure D.1 shows a *uniform* finite-length transmission line with a characteristic impedance Z_0 and load Z_L. Uniform implies that the geometry and materials are independent of length. Using the coordinate system in the figure, a wave launched at the generator side travels down the line in the $+z$ direction

$$V_i = V^+ e^{-j\beta z}$$

where β is the phase constant of the line (lossless line assumed: $\alpha = 0$). If the *characteristic impedance* of the line is Z_0, there will be a reflection at the load unless the line is *matched* ($Z_L = Z_0$). Let the reflected wave be

$$V_r = V^- e^{+j\beta z}$$

Thus, at any point on the line, the total voltage is

$$V(z) = V_i + V_r = V^+ e^{-j\beta z} + V^- e^{+j\beta z}$$

Similarly, for the current,

$$I(z) = I_i + I_r = I^+ e^{-j\beta z} + I^- e^{+j\beta z}$$

with $V^+/I^+ = -V^-/I^-$.

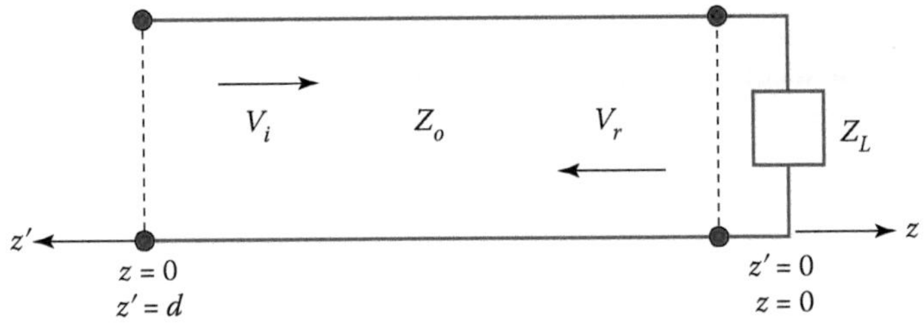

Fig. D.1 Two-conductor transmission line.

As in the case of plane waves, a reflection coefficient can be defined:

$$\Gamma_L = \frac{V_r}{V_i}\Big|_{z=d} = \frac{V^-}{V^+}$$

At the load terminals, the condition $V(d) = Z_L\ I(d)$ holds. Combining this fact with the preceding equations gives

$$\Gamma_L = \frac{Z_L - Z_0}{Z_L + Z_0}$$

There are three special cases of interest:

1. Matched line: $Z_L = Z_0 \rightarrow \ \Gamma_L = 0$
2. Shorted line: $Z_L = 0 \rightarrow \Gamma_L = -1$
3. Open line: $Z_L = \infty \rightarrow \Gamma_L = 1$

The *voltage standing wave ratio* (VSWR) is defined as the ratio of the maximum voltage to minimum voltage on the line:

$$\text{VSWR} = s = \frac{V_{\text{max}}}{V_{\text{min}}} = \frac{|V^+| + |V^-|}{|V^+| - |V^-|} = \frac{1 + |\Gamma_L|}{1 - |\Gamma_L|}$$

The impedance at any point on the line, viewed in the direction of the load, is

$$Z(z') = Z_0 \frac{Z_L + jZ_0 \tan(\beta z')}{Z_0 + jZ_L \tan(\beta z')}$$

where $z' = d - z$, the distance from the load.

Frequently, it is more convenient to work with *admittance Y* than impedance. Admittance is the reciprocal of the impedance and is composed of *conductance* **G** and *susceptance* ***B***:

$$Y_L = G_L + jB_L = \frac{1}{Z_L} = \frac{1}{R_L + jX_L}$$

When there are multiple waves traveling on the line it may be possible to adjust the amplitude and phase of one to cancel another. This is referred to as tuning or *cancellation* and is frequently used to reduce RCS. In practice, a second scattering source is introduced with the proper reflection characteristics, so that its scattering cancels that of another scatterer. The major disadvantage of this technique is that the cancellation depends on the argument βd. Therefore, when the frequency changes, Γ will also change.

D.2 Transmission Line Equivalent of Plane Wave Reflection

Consider a plane wave incident on an interface. In Appendix A, the reflection coefficients for parallel and perpendicular polarizations were presented:

$$\Gamma_{\perp} = \frac{\eta_2 \cos\theta_i - \eta_1 \cos\theta_t}{\eta_2 \cos\theta_i + \eta_1 \cos\theta_t} \tag{D.1}$$

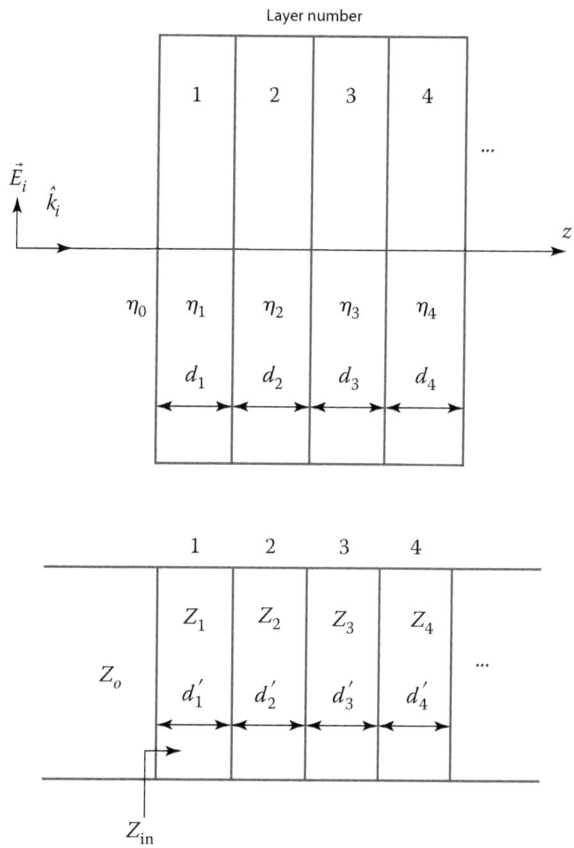

Fig. D.2 Layered media and its approximate transmission line equivalent.

$$\Gamma_{\|} = \frac{\eta_2 \cos\theta_t - \eta_1 \cos\theta_i}{\eta_2 \cos\theta_t + \eta_1 \cos\theta_i} \tag{D.2}$$

Note that these have the same form as the equation for the reflection coefficient on a transmission line if equivalent impedances are defined:

Perpendicular polarization: $Z_0 = \eta_1/\cos\theta_i$ and $Z_L = \eta_2/\cos\theta_t$
Parallel polarization: $Z_0 = \eta_1 \cos\theta_i$ and $Z_L = \eta_2 \cos\theta_t$

The representation of plane wave reflection and refraction described here is very useful when one is dealing with multiple layers of material. Figure D.2 illustrates how the layers of material of intrinsic impedance η_i can be represented as sections of transmission line with characteristic impedance Z_i. The impedance of section i can be transformed back to the boundary with section $i-1$ to obtain a load impedance across section $i-1$. The method is similar to that used in Example D.1. Eventually, a reflection coefficient Γ_{in} is obtained at the first interface.

D.3 Impedance Transformers

The impedance mismatch between a transmission line and load is frequently encountered in practice. To reduce the reflection seen at the input, an *impedance-matching* device can be inserted between the load and the line. Ideally, the device should be lossless, physically small, and broadband. Of course, it is rarely possible to achieve all these criteria simultaneously.

Perhaps the most popular device is a *quarter-wave transformer.* This is simply a $\lambda/4$ length of transmission line with (real) characteristic impedance R_0'. Referring to Fig. D.3, we see that making Z_{in} equal to R_0 requires that

$$Z_{\text{in}} = R_0 = R_0' \frac{R_L + jR_0' \tan(\beta d)}{R_0' + jR_L \tan(\beta d)}$$

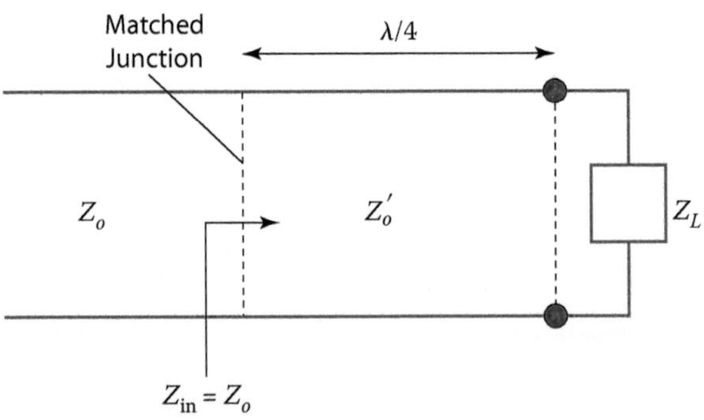

Fig. D.3 Quarter-wave transformer.

Now, using $d = \lambda/4$ and solving for R_0' give

$$R_0' = \sqrt{R_0 R_L}$$

Thus, if the impedance of a quarter-wave section of line is properly chosen and is inserted between the load and the original line, the reflection coefficient can be forced to zero. Electrically, this is equivalent to tuning the mismatches at the two junctions so that their reflected waves cancel.

Appendix E Scattering Matrices

E.1 Scattering Parameters

At microwave frequencies, direct measurement of voltage and current is difficult, if not impossible. Furthermore, as in the case of waveguides, voltages and currents are not unique. The *scattering parameters* defined in Eq. (6.67) are one of several means of characterizing devices based on incident and scattered voltage waves (i.e., electric field). Other methods are *Z parameters, Y parameters*, and *ABCD parameters*. One advantage of scattering parameters is that they are easily measured.

Figure E.1 shows an *N*-port device. At each port, there is an incident and scattered wave, denoted by a_n and b_n, respectively. These are related to the true voltage and current by

$$a_n = \frac{1}{2}\left[\frac{V_n}{\sqrt{R_{0n}}} + I_n\sqrt{R_{0n}}\right]$$

$$b_n = \frac{1}{2}\left[\frac{V_n}{\sqrt{R_{0n}}} - I_n\sqrt{R_{0n}}\right]$$

These are *normalized* waves due to the presence of R_0, a real normalizing constant that is arbitrary but that is usually chosen to be the impedance of the line (or generator) feeding the device. The normalized voltage and current at port n are

$$\tilde{v}_n = \frac{V_n}{\sqrt{R_{0n}}} \quad \text{and} \quad \tilde{i}_n = I_n\sqrt{R_{0n}}$$

or,

$$v_n = a_n + b_n$$

$$i_n = a_n - b_n$$

Traditionally, a and b have been referred to as voltage or current waves, even though their unit is $\sqrt{\text{volt} - \text{amperes}}$. Frequently V_n^+ and V_n^- are used in place of a_n and b_n.

Generalized scattering parameters allow the impedance viewed at each port to be different, for example, as in the case of an impedance transformer. However, most devices are designed to have the same input impedance at all ports, and this is the case we will discuss in detail. In the remainder of this appendix, it will be assumed that the impedance at each port is the same. Furthermore, we will find it convenient to use unit impedances so that

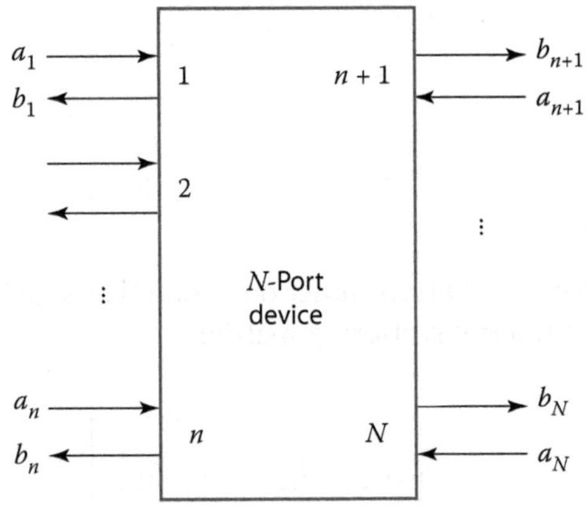

Fig. E.1 *N-port device with incoming and outgoing waves.*

$R_{0n} = R_0 = 1$ and the power incident on each port is

$$P_n = \frac{1}{2}|a_n|^2$$

A scattering parameter S_{nm} relates the outgoing wave at port n to an incoming wave at port m. Thus, if waves are incident on all ports simultaneously, the total outgoing wave at port n is

$$b_n = S_{n1}a_1 + S_{n2}a_2 + \cdots + S_{nN}a_N = \sum_{m=1}^{N} S_{nm}a_m$$

For an N-port device, an $N \times N$ *scattering matrix* relates the vectors $\boldsymbol{a}$ and $\boldsymbol{b}$:

$$\mathbf{b} = \mathbf{Sa}$$

$$\begin{bmatrix} b_1 \\ b_2 \\ \vdots \\ b_N \end{bmatrix} = \begin{bmatrix} S_{11} & S_{12} & \cdots & S_{1N} \\ S_{21} & \ddots & & \vdots \\ \vdots & & \ddots & \vdots \\ S_{N1} & \cdots & \cdots & S_{NN} \end{bmatrix} \begin{bmatrix} a_1 \\ a_2 \\ \vdots \\ a_N \end{bmatrix}$$

E.2 Properties of the Scattering Matrix

The electrical characteristics of a device influence the structure of its scattering matrix and, consequently, lead to relationships between scattering parameters. There are two important properties for lossless reciprocal devices, as follows.

1. *Reciprocal device*: A *reciprocal* junction is one that satisfies $S_{mn} = S_{nm}$, resulting in a symmetric scattering matrix:

$$S = \begin{bmatrix} S_{11} & S_{12} & \cdots & S_{1N} \\ S_{12} & \ddots & & \vdots \\ \vdots & & \ddots & \vdots \\ S_{1N} & \cdots & \cdots & S_{NN} \end{bmatrix}$$

2. *Lossless junction*: For a lossless junction, the power leaving must equal the power entering. If there are N ports,

$$\sum_{n=1}^{N} |b_n|^2 = \sum_{n=1}^{N} |a_n|^2$$

which leads to the requirement

$$\sum_{n=1}^{N} |S_{nm}|^2 = \sum_{n=1}^{N} S_{nm} S_{nm}^* = 1 \quad \text{for} \quad m = 1, 2, \ldots N$$

One can also show that the product of any column of the scattering matrix with the complex conjugate of any other column is zero:

$$\sum_{n=1}^{N} S_{nm} S_{nl}^* = 0 \quad \text{for} \quad m \neq l$$

A *unitary matrix* is one that satisfies the last two properties.

Example E.1: Scattering Matrix of a Lossless Reciprocal Two-Port

For a two-port device ($N = 2$), the scattering matrix has four elements

$$S = \begin{bmatrix} S_{11} & S_{12} \\ S_{21} & S_{22} \end{bmatrix}$$

If a unit voltage excites port 1, with port 2 terminated with a perfect load, S_{11} represents the input reflection coefficient Γ. The signal transmitted from port 1 to port 2 is given by the transmission coefficient τ. For a lossless device, conservation of energy requires

$$|\Gamma|^2 + |\tau|^2 = 1$$

By reciprocity,

$$|S_{22}| = |S_{11}| = |\Gamma| \quad \text{and} \quad |S_{12}| = |S_{21}| = |\tau|$$

To describe the scattering matrix completely, the phases of the scattering parameters must be specified. Let the phases be defined by θ_1, θ_2, and ϕ_1 as follows:

$$S_{11} = |\Gamma|e^{j\theta_1}, \quad S_{22} = |\Gamma|e^{j\theta_2}, \quad \text{and} \quad S_{21} = |\tau|e^{j\phi_1}$$

Using the unitary condition between columns 1 and 2 yields

$$|\Gamma||\tau|\left[e^{j(\theta_1 - \phi_1)} + e^{j(\phi_1 - \theta_2)}\right] = 0$$

or

$$\phi_1 = \frac{\theta_1 + \theta_2}{2} + \frac{\pi}{2} \mp m\pi$$

where m is an integer. Finally, the scattering matrix for a reciprocal lossless device becomes

$$S = \begin{bmatrix} |\Gamma|e^{j\theta_1} & \sqrt{1 - |\Gamma|^2}e^{j\phi_1} \\ \sqrt{1 - |\Gamma|^2}e^{j\phi_1} & |\Gamma|e^{j\theta_2} \end{bmatrix}$$

The phases of the scattering parameters are important in the computation of the RCS of objects such as antennas. The preceding matrix essentially tells us that it is not possible to control the reflection coefficient phase independently at each port. The reflection coefficient phases are related by the exponent of the transmission coefficient ϕ_1, which is generally set by the radiation requirements of the antenna (i.e., beam location). A consequence of this is that the maximum RCS of a reciprocal antenna is in the same direction as the antenna mainbeam.

Example E.2: Scattering Matrix of a Magic Tee

A magic tee is a four-port waveguide power-splitting device (Fig. E.2) used in antenna beam-forming networks. Port 1 is called the *sum arm* and port 4 the *difference arm*; ports 2 and 3 are the *side arms*. If a signal is injected into port 1, equal in-phase signals emerge from ports 2 and 3; a signal entering port 4 results in equal out-of-phase signals from ports 2 and 3. The tee is called *magic* if the input reflection coefficient of each arm is zero. (This condition is generally not possible for most lossless reciprocal multiport devices because of conditions 1 and 2 of Sec. E.2.)

The scattering matrix of a magic tee will have zeros on the diagonal. Furthermore, there is odd symmetry for difference port excitation and even symmetry for sum port excitation. Finally, all power splits are equal, yielding a matrix of the form

$$S = \frac{1}{\sqrt{2}}\begin{bmatrix} 0 & 1 & 1 & 0 \\ 1 & 0 & 0 & 1 \\ 1 & 0 & 0 & -1 \\ 0 & 1 & -1 & 0 \end{bmatrix}$$

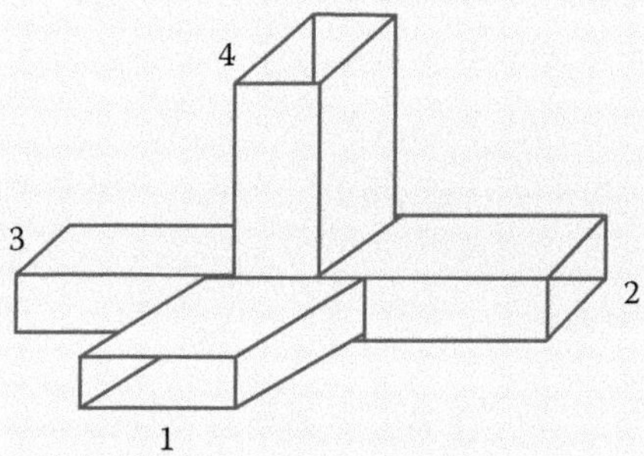

Fig. E.2 Waveguide magic tee.

E.3 Network Analysis

A major disadvantage of scattering parameters is that the scattering matrix of cascaded devices (connected in series) is not the product of the individual device scattering matrices. In general, if the network has a total of N_t ports, N_t equations must be solved simultaneously to obtain the wave vectors.

Example E.3: Cascaded Two-Port Devices

Two two-port devices are connected as shown in Fig. E.3. The scattering matrices are denoted $S1$ and $S2$, with elements $S1_{mn}$ and $S2_{mn}$. The incident and scattered waves are indexed from 1 to 4 as indicated in the figure. The signals a_1 and a_4 provide the excitation; they are known quantities. We want to find the signals everywhere in the circuit.

At each device port, an equation can be written for the outgoing wave:

$$\begin{aligned}
\text{Port 1: } b_1 &= a_1 S1_{11} + a_2 S1_{12} \\
\text{Port 2: } b_2 &= a_1 S1_{21} + a_2 S1_{22} \\
\text{Port 3: } b_3 &= a_3 S2_{11} + a_4 S2_{12} \\
\text{Port 4: } b_4 &= a_3 S2_{21} + a_4 S2_{22}
\end{aligned}$$

Another two equations are obtained from the junction between the two devices: $a_3 = b_2$ and $a_2 = b_3$.

Substituting the continuity relations into the first four equations eliminates a_2 and a_3. Rewriting so that the known excitation quantities are on the right yields:

$$\begin{aligned}
\text{Port 1: } b_1 - b_3 S1_{12} &= -a_1 S1_{11} \\
\text{Port 2: } b_2 - b_3 S1_{22} &= -a_1 S1_{21} \\
\text{Port 3: } b_3 - b_2 S2_{11} &= -a_4 S2_{12} \\
\text{Port 4: } b_4 - b_2 S2_{21} &= -a_4 S2_{22}
\end{aligned}$$

The four equations can be rearranged and cast in matrix form:

$$\begin{bmatrix} b_1 \\ b_2 \\ b_3 \\ b_4 \end{bmatrix} = \begin{bmatrix} -1/S1_{11} & 0 & -S1_{12}/S1_{11} & 0 \\ 0 & -1/S1_{21} & S1_{21}/S1_{21} & 0 \\ 0 & S2_{11}/S2_{12} & -1/S2_{12} & 0 \\ 0 & S2_{21}/S2_{22} & 0 & -1/S2_{22} \end{bmatrix} \begin{bmatrix} a_1 \\ a_1 \\ a_4 \\ a_4 \end{bmatrix}$$

Solving this matrix equation yields the four elements of **b**, from which all the network's signals can be determined via the first set of equations in Sec. E.1.

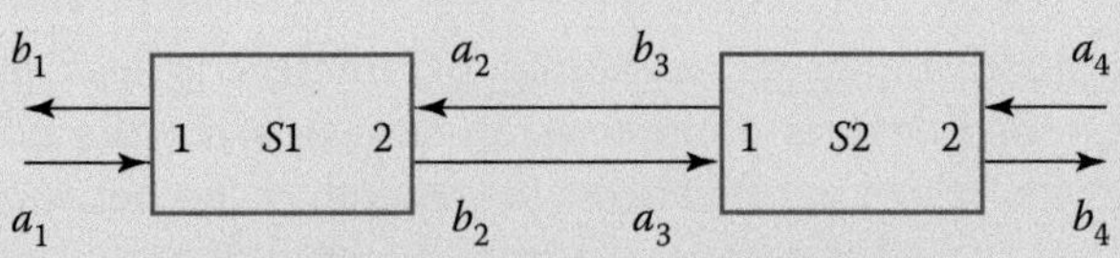

Fig. E.3 Cascaded two-port devices for Example E.3.

Appendix F Properties of Composite Materials

This appendix summarizes the electrical properties of a wide range of engineering materials. Data are compiled from various sources (vendor's literature, materials handbooks, etc.) [1–7]. *All properties are at room temperature and 1 MHz unless otherwise noted.*

F.1 Metal Matrix Composites

Metal matrix composites (MMCs) such as those listed in Table F.1 are usually used in applications in which high strength and light weight are required. Examples include compressor blades and pistons. For applications such as these, the electrical properties are unimportant. However, more recent applications have included external components like missile fins. Although these materials are generally considered to be good conductors, the conductivity can vary widely with the percent of fiber in the composite. The fiber content can be as high as 60%, in which case the conductivity may not be in the "good conductor" range. The resistivities of some good conductors are listed in Table F.2.

F.2 Nonmetallics and Nonmetallic Composites

Tables F.3 and F.4 show some properties of some common nonmetallic materials and nonmetallic composites.

Table F.1 Some MMC Types

Material	Matrix Alloy
Fiber-reinforced alumina	Al
Continuous aluminum oxide (Al_2O_3) fiber	Al, Mn
Continuous boron fiber	B
Continuous graphite fiber	Al, Mn, Cu
Continuous silicon carbide (SiC) fiber	Ti, Cu, Mn, Al
Continuous tungsten fiber	Ni, Co, Fe, Zr, Nb, W, Cr
Discontinuous ceramic fiber	Pb, Ti, Mn, Cu, Al
Discontinuous silicon fiber	Al, Cu

Table F.2 Resistivity of Metals and MMC Alloys

Material	Resistivity, $10^8\ \Omega m$
Aluminum and its alloys	4–15
Brass	3–28
Beryllium and its alloys	3–4
Chromium	13
Cobalt and its alloys	6–99
Columbium and its alloys	15
High copper alloys	8
Copper nickels	15–38
Copper-nickel-zincs	17–31
Gold	2
Iridium	5
Irons	40–110
Magnesium and its alloys	15
Molybdenum and its alloys	19
Nickel and its alloys	135
Platinum	10
Silver	1.5
Stainless steels	98
Tantalum and its alloys	12–20
Tin and its alloys	11–15
Titanium and its alloys	199
Tungsten	5.5
Zirconium and its alloys	74
Zinc	6

Table F.3 Dielectric Constants of Some Common Nonmetallics

Material/Trade Name	ε_r
Alumina ceramic	8–10
Zircon	8–10
Mica	5.4–8.7
Silicon nitride	9.4
Silicone plastics	3.4–4.3
Glass/fused silica	3.8
Acrylics	2.5–2.9
Polystyrene	2.5–2.7
Polyethylene cellular foam	1.05–1.85
Polyesters (general-purpose reinforced moldings)	2.8–4.4
Rigid plastic foams	1.1–2.0
Nylons (general purpose)	3.5–3.8
Epoxies	2.7–5.2

Table F.4 Dielectric Constant and Loss Tangents of Some Nonmetallic Composites

Material/Trade Name	ε_r	tan δ_ε
E-glass	5.9–6.4	0.002–0.005
S-glass	5.0–5.4	0.002
Udel P-1700 (Union Carbide)	3.0	0.003
Valox 420	3.25–4.0	0.02
Kinel 5504	4.75	0.0007
Vespel resin	3.5	0.0035
S-690 Fiberite (glass reinforced)	4.5	0.015
Glass fabric–reinforced epoxy resin	4.3–5.2	0.01–0.02
Glass-reinforced epoxy molding compound	4.0	0.01
Glass-reinforced Teflon (woven)	2.4–2.6	0.0015–0.002
Glass-reinforced Teflon (microfilm)	2.3–2.4	0.0004–0.0008
Polyolefin	2.3	0.0003
Cross-linked polystyrene (unreinforced)	2.5	0.00025–0.00066
Glass-reinforced polystyrene	2.6	0.0004–0.0002
Polyphenelene oxide (PPO)	2.55	0.0016
Ceramic-filled resins	1.7–25	0.0005–0.005
Teflon quartz (woven)	2.47	0.0006
Kevlar 49[a]	3.25	0.03
Quartz fabric–reinforced resin[a]	2.5	0.05
Alumina-aluminal[b]	3.7	0.0045
Boron nitride–boron nitride[b]	3.0	0.002

[a]At *X* band, approximately 10 GHz.
[b]At 300 K.

References

[1] Schwartz, M., *Composite Materials Handbook*, 2nd ed., McGraw-Hill, New York, 1992.

[2] Hoskin, B. C., and Baker, A. A., *Composite Materials for Aircraft Structures*, AIAA Education Series, AIAA, Washington, DC, 1986.

[3] "Composites," *Engineered Materials Handbook*, Vol. 1, ASM International, Metals Park, OH, 1987.

[4] Goosey, M. T., *Plastics for Electronics*, Elsevier, New York, 1985.

[5] Mascia, L., *Thermoplastics*, Elsevier Applied Science, New York, 1989.

[6] Scala, E., *Composite Materials for Combined Functions*, Hayden, New York, 1973.

[7] *Materials Selector*, published yearly by *Materials Engineering*, Penton/IPC Reinhold, 1978.

Index

Note: Page numbers followed by "**t**" indicate table, and "*f*" indicate figure.

N

O

P

S

Supporting Materials

To download supplemental material files, please go to AIAA's electronic library, Aerospace Research Central (ARC), at arc.aiaa.org. Use the menu bar at the top to navigate to Books > AIAA Education Series; then, navigate to the desired book's landing page by clicking on its title. On the landing page, click the link beneath "Supplemental Materials," enter the password **RALCSE3E**, and follow the directions provided.

A complete listing of titles in the AIAA Education Series is available from AIAA's electronic library, Aerospace Research Central (ARC), at arc.aiaa.org. Visit ARC frequently to stay abreast of product changes, corrections, special offers, and new publications.

AIAA is committed to devoting resources to the education of both practicing and future aerospace professionals. In 1996, the AIAA Foundation was founded. Its programs enhance scientific literacy and advance the arts and sciences of aerospace. For more information, please visit www.aiaafoundation.org.